AGRIBUSINESS MANAGEMENT

Steven P. Erickson
(Deceased)

Jay T. Akridge
Purdue University

Freddie L. Barnard
Purdue University

W. David Downey
Purdue University

With contributions from Research Development Editor
Kathleen M. Erickson

Boston   Burr Ridge, IL   Dubuque, IA   Madison, WI   New York
San Francisco   St. Louis   Bangkok   Bogotá   Caracas   Kuala Lumpur
Lisbon   London   Madrid   Mexico City   Milan   Montreal   New Delhi
Santiago   Seoul   Singapore   Sydney   Taipei   Toronto

McGraw-Hill Higher Education

*A Division of The **McGraw-Hill** Companies*

AGRIBUSINESS MANAGEMENT, THIRD EDITION

Published by McGraw-Hill, a business unit of The McGraw-Hill Companies, Inc., 1221 Avenue of the Americas, New York, NY 10020. Copyright © 2002, 1987, 1981 by The McGraw-Hill Companies, Inc. All rights reserved. No part of this publication may be reproduced or distributed in any form or by any means, or stored in a database or retrieval system, without the prior written consent of The McGraw-Hill Companies, Inc., including, but not limited to, in any network or other electronic storage or transmission, or broadcast for distance learning.

Some ancillaries, including electronic and print components, may not be available to customers outside the United States.

This book is printed on acid-free paper.

1 2 3 4 5 6 7 8 9 0 QPF/QPF 0 9 8 7 6 5 4 3 2 1

ISBN 0–07–019637–0

Publisher: *Margaret J. Kemp*
Sponsoring editor: *Edward E. Bartell*
Freelance developmental editor: *Kassi Radomski*
Marketing manager: *Heather K. Wagner*
Project manager: *Christine Walker*
Production supervisor: *Sherry L. Kane*
Coordinator of freelance design: *Rick D. Noel*
Supplement producer: *Sandra M. Schnee*
Executive producer: *Linda Meehan Avenarius*
Compositor: *Shepherd, Inc.*
Typeface: *10/12 Times Roman*
Printer: *Quebecor World Fairfield, PA*

Library of Congress Cataloging-in-Publication Data

Erickson, Steven P.
    Agribusiness management / Steven P. Erickson, Jay T. Akridge, Freddie L. Barnard. — 3rd ed.
        p.   cm.
    2nd ed. published: Downey, W. David (Walter David), 1939– Agribusiness management. © 1987.
    Includes bibliographical references and index.
    ISBN 0–07–019637–0
    1. Agricultural industries—Management. I. Akridge, Jay T. II. Barnard, Freddie L.
    III. Downey, W. David (Walter David), 1939– Agribusiness management. IV. Title.

HD9000.5 .D63   2002
630'.08—dc21
                                 2001041033
                                 CIP

www.mhhe.com

ABOUT THE AUTHORS

DR. STEVEN P. ERICKSON

Steve Erickson was Assistant Head for Teaching and Professor in the Department of Agricultural Economics at Purdue University. After seeing his contributions to the manuscript through to completion, Steve passed away before this book was published. He was a brilliant, award-winning educator. Some of Steve's many teaching awards included Outstanding Undergraduate Teacher in the Purdue School of Agriculture, the American Agricultural Economics Association's Distinguished Undergraduate Teaching Award, and the Charles B. Murphy Award for Outstanding Undergraduate Teaching (Purdue's highest teaching honor). Steve worked extensively with managers through the programs of the Center for Food and Agricultural Business at Purdue. His work in the financial management of agribusiness and retail food merchandising firms was widely recognized. Steve was a leader in developing computer simulations for business and classroom training. Through his writing, the business simulations he developed, his teaching, and his executive education activities, Steve shaped a generation of food and agribusiness managers. This book is dedicated to Steve's memory, and to his family: his wife Ginger, and daughters Abbie and Cassie. One of Steve's students described him best: "Clark Kent in the hallway, Superman in the classroom."

DR. JAY T. AKRIDGE

Jay Akridge is Director of the Center for Food and Agricultural Business and the Executive MBA in Food and Agricultural Business (EMBA), and Professor in the Department of Agricultural Economics at Purdue University. His primary responsibilities include leading the Center and EMBA programs, working with food and agribusiness managers in the Center's professional development programs, conducting research on agribusiness management and marketing issues, and teaching graduate and undergraduate courses in strategic management for food and agribusiness firms. In 1996, he received the Charles B. Murphy Award for Outstanding Undergraduate Teaching (Purdue's highest teaching award). Jay has worked with agribusiness managers in the areas of strategy, finance, and marketing in several countries including Lithuania, Hungary, Poland, Australia, New Zealand, and Argentina. A native Kentuckian, he continues to be involved in the third-generation retail farm supply business owned and operated by his family in Fredonia, Kentucky.

Dr. Freddie L. Barnard

Freddie Barnard is Professor in the Department of Agricultural Economics at Purdue University. He works in the area of agricultural finance and focuses on two major groups: agricultural lenders and producers. In addition, Freddie teaches undergraduate courses in agribusiness management and accounting. He has been, and continues to be, involved in educational projects across the country, to educate lenders on how to improve the quality of decision making. Examples include his work with the industry-wide Farm Financial Standards Council to provide uniformity in financial reporting and analysis, and videotape instructional materials for agricultural lenders and producers. Freddie is originally from Beaver Dam, Kentucky. He and his wife, Shirley, currently live in Lafayette, Indiana with their three children, Jared, Mathew, and Veronica.

Dr. W. David Downey

Dave Downey is Executive Director of the Center for Food and Agricultural Business at Purdue University and Professor in the Department of Agricultural Economics. He has long been a favorite among participants of Center programs and undergraduate students at Purdue. As a professor of agricultural sales and marketing at Purdue, Dave teaches courses in agriselling and agrimarketing strategy to undergraduate students from several disciplines. He has been instrumental in the development of a curriculum in which students can earn a four-year B.S. degree in agrisales and marketing—the first such program in the United States. Dave has received many recognitions for professional excellence, including four major teaching awards from Purdue University and two national awards from the American Agricultural Economics Association.

CONTENTS

PART **II**

PART | III

MARKETING MANAGEMENT FOR AGRIBUSINESS   177

PART IV

FINANCIAL MANAGEMENT FOR AGRIBUSINESS   307

PART V

PART VI

HUMAN RESOURCE MANAGEMENT FOR AGRIBUSINESS   515

PREFACE

INTRODUCTION

Rapidly evolving, maturing, growing, complex. Global, dynamically networked, interrelated, competitive. This handful of descriptors cannot do justice to today's food and fiber production and marketing system. This is an incredibly exciting time to be involved in the food and agribusiness industries. International markets are an inseparable reality for agribusiness managers. Information technology and the Internet make entirely new models for conducting business both possible and practical. Biotechnological developments raise challenging business and policy issues. The new millennium food and fiber end-consumer demands additional attention all along the food production and marketing system. Relationships between players in this system have continued to evolve and as a result, the food production and marketing system is far more complex and interrelated today than it was a mere decade ago.

This rapidly changing, international, high-technology, consumer-focused world is the one in which today's food and agribusiness managers operate. This edition of *Agribusiness Management* was written to help prepare students and managers for a successful career in this new world of food and fiber production and marketing.

WHAT'S NEW

The basic objective of this text has not changed through three editions: to provide students and managers a fundamental understanding of the key concepts needed to successfully manage businesses, adding value to farm products and/or providing inputs to production agriculture. While there are many concepts in this book that will apply to the farm or production agriculture business, the text is focused on the food and input supply sectors of the food production and marketing system.

This edition of *Agribusiness Management* uses four specific approaches to help readers develop and enhance their capabilities as agribusiness managers. First, this edition of the book offers a contemporary focus that reflects the issues that agribusiness managers face both today and likely will face tomorrow. Specifically, food sector firms and larger agribusiness firms receive much more attention in this edition, reflecting their increasing importance as employers of food and agribusiness program graduates. Second, the book presents conceptual material in a pragmatic way with illustrations and examples that will help the reader understand how a specific concept works in practice. Third, the book has a decision-making emphasis, providing contemporary tools that readers will find useful when making decisions in the contemporary business environment. Finally, *Agribusiness Management* offers a pertinent set of discussion questions and case studies that will allow the reader to apply the material covered in real-world situations.

More specifically, the opening section of the text has been completely re-organized to help students better understand the food and agribusiness marketplace, and what agribusiness managers actually do. New in the second section is a chapter on international agribusiness management, an area no contemporary book on agribusiness can ignore. We now start our look at the four functional areas of management—marketing, finance, production/operations, and human resources—with marketing. Ultimately, all business activity revolves around the customer, and the text reflects this customer-oriented philosophy. The marketing and operations/logistics sections were completely re-written to reflect the current thinking in these areas. The finance section also received a complete overhaul, and a new chapter on investment decisions was added. Finally, the human resource section was re-written, again to reflect what we now know about managing people.

Preparing for a new food and agribusiness market requires application of concepts and tools to current situations. This edition of *Agribusiness Management* ends every chapter with a case that is either new or re-written for this text. These cases cover situations ranging from a small cherry cooperative to a Hungarian food company, from a rapidly growing chemical firm, to a mid-size food processing organization. We feel you will find this mix of cases to be a distinguishing feature of the book.

The bottom-line on this third edition of *Agribusiness Management:* this book is contemporary; solid on the fundamentals; and practical and applicable, to provide student and adult learners with an essential understanding of what it takes to be a successful agribusiness manager in today's rapidly evolving, highly unpredictable marketplace.

THE AUDIENCE

Agribusiness Management was written for students. There are tremendous career opportunities in the food and agribusiness industries. In this book, you will be exposed first to the breadth of these opportunities, from research and development manager for a biotech company, to a logistics manager for a major food retail organization. You need to understand the marketplace and some of the unique institutional features of the food production and marketing system before embarking on a career as an agribusiness manager. Preparing for a career in agribusiness management also requires that you understand the fundamentals of management—the basic tasks of planning, organizing, directing, and controlling and the basic functions of marketing, finance, operations and logistics, and human resource management. In this book, you will find all of these topics covered in a straightforward way. We hope the many food and agribusiness examples and case studies bring these concepts to life for you. This is a book you will continue to use as a reference as your managerial career unfolds.

Agribusiness Management was written for managers and soon to be managers who are already in the workforce. We have had the opportunity over the past 15 years to work with thousands of managers through the activities of the Center for Food and Agricultural Business at Purdue University. Most of the case studies and examples in this book come from these industry relationships and interactions. As the business environment changes, and as people assume new responsibilities, we see a need to retool in areas that an individual has not recently been applying in their job. For example, the production manager in a food processing firm who has been asked to serve on a task force focused on helping the firm become more customer oriented may need a "refresher" in marketing management. These individuals (and corporate learning and development directors and training managers) will find this book useful in sharpening their skill set.

Agribusiness Management was written for instructors. Over time, we have found that every instructor has his or her own take on what an agribusiness management course should look like. Some are introductory courses, others have more of a capstone orientation. Some of these courses are part of an entire curriculum in agribusiness management. In other cases, a program may offer only a course or two in the agribusiness area. The organization of this book is structured in a way that instructors will find convenient when developing their course, wherever that course fits in the program's overall curriculum.

An instructor could easily use the material in this edition over a single semester, or over a two-semester or three-quarter course, covering each topic in more detail. Some instructors will find that moving through the book from start to finish sequentially as part of a one semester course makes the most sense. Others may drop chapters on specialized topics like cooperatives, selling, or international business because these topics are covered in other courses. For an advanced course, the book has plenty of rigor. Supplemented with outside case studies, this is an excellent text for a capstone-type course where the material here would be covered more quickly, and more time spent on some of the more advanced ideas. The new and updated cases and the discussion questions will be of value to all instructors as they serve the needs of students hungry for applications and illustrations of the concepts and tools covered in the book.

OUTLINE OF BOOK

In Part I, *Agribusiness Management: Scope, Functions, and Tasks,* we focus on the food and agribusiness industries and the role of the agribusiness manager. In addition, a set of economic concepts of fundamental importance to agribusiness managers is covered. Part I exposes readers to the tremendous variety of firms that comprise the food production and marketing system. The core focus of this book is the four functional areas of management—marketing, finance, operations/logistics, and human resources. Readers will better understand the role that managers of food and agribusiness firms play as they execute the four tasks of management—planning, organizing, directing, and controlling.

Part II, *Agribusiness Management: Organization and Context,* leads readers toward better understanding the context or environment in which agribusiness managers operate by taking them inside the different forms of business organization. A special emphasis is given to cooperatives in this section given their prominence in the food and agribusiness markets. This section also provides a glimpse into the issues an agribusiness manager must face when doing business outside of the United States. The challenges of serving international markets, of sourcing raw materials from international locations, and from competing with international firms have all emerged as key issues in the past decade. Looking to the future, this area promises to be even more important.

We begin our look at the four functional areas of management in Part III with marketing management. This part covers the fundamental concepts and tools an agribusiness manager uses in identifying its target market, and taking its product-service-information offering to the market. The marketing mix— product, price, promotion, and place—is covered in some detail. And, one of the most important elements of the promotion strategy, personal selling, is given special focus. Given that so many individuals begin their managerial career in professional sales, these two chapters will be especially important to many readers.

Financial management is the focus of Part IV. Starting with basic financial statements, and moving through financial ratios, financing the agribusiness, and on to tools for making operating and capital investment decisions, this section addresses the fundamental elements of finance that any agribusiness manager should understand. A series of integrated examples and clear explanations of key terms will help you better understand the language and concepts of finance. More importantly, you will better understand how to use financial information when making managerial decisions.

Part V looks at operations management. This section takes you into areas such as production planning, total quality management, and supply chain management. Production and operations management in agribusiness firms has undergone a profound change over the past two decades. These two chapters will provide a fundamental understanding of the key elements in this important area.

Finally, in Part VI, *Human Resource Management for Agribusiness,* we look at key issues involved in managing a firm's people resources. First, we explore issues around organizational structure and leadership. Then, we turn our attention to the personnel functions of hiring, training, evaluation, and compensation of employees. While held for the final section, the issues surrounding the human resource area are likely the most pressing of all to an agribusiness firm facing a rapidly changing operating environment.

ACKNOWLEDGMENTS

Any large project involves an equally large number of individuals, and this book is no exception. Our most important acknowledgment is to Research Development Editor Kathleen Erickson (no relation to Steve) who served a variety of roles ranging from conducting background research to discussing content and focus as this manuscript moved from idea to reality. Other individuals we would like to acknowledge for their important contributions include Frank Dooley, Anne Duffy, Joan Fulton, Brent Gloy, Allan Gray, Jan Layden, Jess and Carmi Lyon, Betty Ottinger, Lee Schrader, Patsy Schmidt, Joyce Sichts, LeeAnn Williams, and Vida Ziutaike. We would like to acknowledge the helpful review comments of John L. Adrian, *Auburn University;* Dovi Alipoe, *Alcorn State University;* James R. Boyle, *Nebraska College of Technical Agriculture;* John C. Foltz, *University of Idaho;* Conrad Lyford, *Oklahoma State University;* David G. Moorman, *Sam Houston State University;* Dixie Watts Reeves, *Virginia Polytechnic Institute and State University;* Forrest E. Stegelin, *University of Georgia;* and Dawn Thilmany, *Colorado State University.* The thoughtful comments of these reviewers have helped us sharpen the focus of this book. The McGraw-Hill team including Kassi Radomski, freelance editor; Christine Walker, project manager; and Ed Bartell, sponsoring editor were both patient and helpful. Finally, we need to acknowledge the role that the students and managers we have had the chance to work with for the past 15 years have had in shaping the ideas and presentation of this text.

SUMMARY

This is an exciting time for the food and agribusiness markets, and it is an exciting time to be preparing for or retooling for a career in these industries. We have tried to capture some of the excitement in *Agribusiness Management.* We hope that you find this book readable and interesting; challenging and pragmatic; and most of all, helpful as you better prepare yourself or your students for successful careers in the food production and marketing system.

PART I

AGRIBUSINESS MANAGEMENT: SCOPE, FUNCTIONS, TASKS

Productive, diverse, efficient: this section explores today's food, production agriculture, and input supply sectors.

THE BUSINESS OF AGRIBUSINESS

OBJECTIVES

- Describe management's role in agribusiness
- Provide an overview of the functional responsibilities of management
- Describe the unique characteristics of the food and agribusiness industries
- Describe the size, scope, and importance of the food production and marketing system
- Understand the farm-food marketing bill and what it means to producers and consumers
- Provide an overview of the food sector, the production agriculture sector, and the input supply sector
- Outline trends in home and away-from-home food consumption, and trends among the types of firms that serve these markets
- Explore the production agriculture sector, and some of the key changes occurring on U.S. farms
- Outline the major inputs used by the production agriculture sector and key trends in input use
- Understand the types of firms involved in producing and distributing inputs to production agriculture

INTRODUCTION

It is exciting and diverse. It is changing quickly. It relies on the weather, uses an incredible array of technology, is tied in every way to our natural resources, and embraces the world. If you eat, you are involved in it as a consumer of its final products. If you farm, you are involved in it as a producer of the raw materials that ultimately make their way to the end consumer. *It* is the extremely efficient, very complex, global, food and fiber production and marketing system.

This system is vast and it is fascinating: the next time you walk through your local supermarket, think about the number and type of diverse activities involved in growing, harvesting, transporting, processing, and distributing food throughout the 50 states in the United States, and, more broadly, our world. The process by which a 260-pound hog moves from Carroll County, Indiana to a suburban supermarket in Los Angeles (now in the form of a hot dog in a prepackaged children's meal) is very complex, yet it occurs every day in the food production and marketing system.

This food production and marketing system is made up of thousands of businesses, ranging from the small cow-calf producer in western Kentucky to some of the largest corporations in the world. And, it is **management** that drives and directs the firms, farms, and food companies that come together in the food production and marketing system. A retail supermarket, a major corn processor, the local farm supply store and a family farmer each have a person or a group of people responsible for getting things done. These are the **managers.** Their titles range from chief executive officer to president to foreman to son or daughter or spouse. However, wherever they are found within an organization, managers are responsible for assuring successful completion of the functions, tasks, and activities that will determine an organization's success.

This book, *Agribusiness Management,* will focus on the management of food, fiber, and agribusiness firms and we will take a careful look at food and agricultural business management. Our definition of food and agricultural businesses is quite broad; therefore, when we use the term *agribusiness management* in the text, remember that we are talking about the management of any firm involved in the food and fiber production and marketing system. In this book, our discussion of agribusiness management should provide information, concepts, processes, ideas, and experiences that will contribute to your effectiveness in performing the functions and tasks of agribusiness management.

This chapter will first introduce you to the key functions of agribusiness management. Then, we will explore some the characteristics that make the food and agribusiness markets unique places to practice the art and science of management. The ever-changing food and agribusiness industries are then discussed. We will look at firms that (1) move final products through the food and fiber system to the ultimate consumer, (2) transform raw agricultural products into the final products desired by consumers, (3) produce raw food and fiber products, and (4) supply inputs to the farm or production sector.

THE KEY FUNCTIONS OF MANAGEMENT IN AGRIBUSINESS

As you can imagine, the responsibilities of managers in agribusiness are highly varied and can range from ordering inputs for the year ahead, to hiring and firing individuals, to making the decision to sell a multibillion-dollar international subsidiary. A chief executive officer, for instance, is responsible for the overall activities of a large, diversified food or agribusiness firm. In such firms, teams of managers are likely responsible for specialized areas within the firm. On a smaller farm business, one individual may assume roles ranging from chief executive officer, to manager, to laborer, managing multiple projects at different levels simultaneously.

To better understand the form and process by which managers perform the tasks that are required to create and sustain a viable business, the practice of management can be broken down into four key functions:

- Marketing management
- Financial management
- Operations/logistics management
- Human resource management

Ultimately, no matter how large or small the firm, managers have responsibilities in each of these areas. These four functions of management are explored in some detail in this book. However, it is important to have a basic understanding of each area as we begin our look at agribusiness management.

Marketing Management

Marketing, in a broad sense, is focused on the process by which products flow through the U.S. food system from producer to consumer. It involves the physical and economic activities performed in moving products from the initial producer through intermediaries to the final consumer. **Marketing management** involves understanding customer needs and effectively positioning and selling products and services in the marketplace. In agribusiness, marketing management is a key function within each of the sectors of agribusiness: the food sector, the production agriculture sector, and the input supply sector. Marketing management represents an integration of several different activities: selling, advertising, marketing research, new-product development, customer service, physical distribution, pricing—all focused on customer needs, wants, and ultimately, the quest for customer satisfaction.

It is this function of management that is most closely focused on the end user, or the consumer/customer of the product or service produced. It is often argued that without satisfied customers effectively reached through marketing and sales, no business could successfully operate. So marketing management plays a fundamentally important role in most food and agribusiness firms. Marketing management is focused on careful and planned execution of how, why, where, when, and who sells a product or service and to whom it is sold. Decisions here include what products to produce, what services to offer, what information to provide, what price to charge, how to promote the product, and how to distribute the product.

This management function is closely tied to the customer's decision processes, and buyers differ widely in the food production and marketing system—from teenagers for a food manufacturing firm, to a soybean processor for a farmer, to a large integrated swine business for an animal health firm. The ways in which agribusiness buyers (all of the buyers just mentioned and many more) make purchase decisions continues to evolve and change.

Financial Management

Profit is the driver for agribusinesses as they work to generate the greatest possible returns from their resources. Successful achievement of this objective means making good decisions, and it means carefully managing the financial resources of the firm. **Financial management** is involved in these areas and includes generating the data needed to make good decisions, using the tools of finance to analyze alternatives, and managing the assets, liabilities, and owner's investment in the firm.

Financial information allows managers to understand the current "health" of the firm as well as to determine what actions the business might take to improve or grow. **Balance sheets** and **income statements** can provide a wealth of information useful in making decisions. Financial analysis provides agribusiness managers with insights by which to better base decisions. The tools of finance such as budgeting, ratio analysis, financial forecasting, and break-even analysis can be used by agribusiness managers in developing long-range plans and in making short-run operating decisions.

Another way in which the financial agribusiness scene continues to change is in the sourcing of funds. Agribusiness firms are increasingly accessing larger amounts of funds or money from national and international capital and financial markets. And, to be competitive in those markets, they must generate rates of return comparable to other industries. In the past, small agribusiness companies may have been allowed by local lenders to exhibit only modest financial performance. Today, the national and international financial markets expect performance in agriculture comparable to that in other industries, if they are going to provide the agribusiness sector the funding needed for expansion, growth, consolidation, and technological advance.

The sheer amounts of funds needed to finance future operations of a company will continue to increase dramatically. So will the need for managers who understand the tools and techniques used to source and manage those funds. For most agribusinesses, financial management will be a critical component of agribusiness management.

Operations/Logistics Management

New technologies and concepts are rapidly hitting the workplace, changing the way agribusinesses do what they do. The push for quality, the drive for lower costs, changes in the supply chain, and general pressures to be more efficient in meeting consumer demands are swiftly altering the production and distribution activities of agribusiness. **Operations/logistics management** focuses on these areas and provides the tools managers need to meet these challenges. As a result, operations/logistics management has come to the forefront as a key management function for the agribusiness manager.

Operations management focuses on the direction and the control of the processes used to produce the goods and services that we buy and use each day. It involves several interrelated, interacting systems. Operations management involves the strategic use and movement of resources. For instance, a snack food factory begins its process with corn from a food-grade corn producer and ends it with tortilla chips, corn chips, crackers, etc. Managers must think through issues of scheduling, controlling, storing, and shipping as the corn moves from the producer's truck to the supermarket.

Logistics management involves the set of activities around storing and transporting goods and services from manufacturer to buyer. Shipping and inventory costs are huge costs of doing business for many food and agribusiness firms. The logistics management function is focused on new ways to lower these costs, by finding better ways to ship and store product. With the advent of the Internet, the tools of supply chain management, and improved shipping technologies, this has been a dynamic area for food and agribusiness firms.

Successful agribusinesses are those that consistently produce faster, better, more. The management of logistics in food and agricultural supply chains will become increasingly focused on building such time-based advantage. Quicker response to consumer needs, faster delivery times, shorter product development cycles, and more rapid recovery after service problems are all components of time-based advantage in logistics management. At the same time, there is an incredible push for quality, safety, and integrity in food system production processes. Effective operations/logistics management will continue to be crucial in the successful execution of any strategic plan for agribusiness firms.

Human Resources Management

In the end, management is about people. Without the ability to manage the human element—the resources each business has in its employees—businesses do not succeed. By combining efficient management of the marketing/finance/operations/logistics functions of the business, with the thoughtful management of the human side of the business, managers are on the road to successful implementation of their strategy.

Agribusiness managers who can manage people well can significantly impact both productivity and financial success. **Human resources management** encompasses managing two areas—the mechanics of personnel administration, and the finer points of motivating people to offer and contribute their maximum potential. Decisions here include how to organize the firm, where to find people, how to hire them, how to compensate them, and how to evaluate people.

Today's lean agribusiness firms continue to demand more performance from their managers, salespeople, and service and support personnel. For instance, in addition to superb selling skills, salespeople will be expected to have intimate knowledge of technology and a fundamental understanding of the general management problems of their producer customers. Service personnel must be able to maintain increasingly complex equipment. Technical support staff will need to be experts at assimilating and using the massive amount of production data that a large dairy farm or crop farm using site-specific management practices will generate.

These types of demands will require agribusinesses to hire individuals with more initial skills as well as with the ability to grow into different jobs throughout the course of their careers. Agribusinesses will need to be flexible while providing continuing education and development of key skills—skills like general business skills, negotiation skills, problem-solving skills, technical skills, information management skills, and communication skills. Recognition of raw ability, and then development and fine-tuning these skills and abilities will be the human resource challenge. Managed well, that challenge will profitably produce for the company. And, this is the role of human resource management in the food and agribusiness firm.

UNIQUE DIMENSIONS OF THE FOOD AND AGRIBUSINESS MARKETS

It may be easy to argue that management theory and principles are the same for any type of business enterprise. The largest businesses in the country such as General Electric, General Motors, and Wal-Mart and the smallest one-person agribusiness are guided by many of the same general principles. And, in many cases, good management is good management, regardless of the type of firm, or the market in which it is operating.

However, some of the key differences between large and small businesses, between agribusinesses and other types of firms, rest in the specific business environment facing the organization. While there are similarities, the markets facing General Electric's wide range of businesses differ substantially. The automotive industry is different from the retail industry. And, the food production and marketing system has its own unique characteristics. These differences cause the practice of management to differ depending on the business environment facing the firm. Our job is to better understand what is similar about a manager with Microsoft, and a manager with Syngenta or Kellogg's, and how the functions and tasks of a food and agribusiness manager differ from managers in other markets.

As a professional, the manager might be compared to a physician. The knowledge and principles of medicine are the same, but patients differ in such vital details as age, gender, build, and general health. The physician's skill is to apply general medical principles to the specific individual to create the optimal outcome for the patient given the unique set of circumstances at hand. The manager, utilizing specific tools of marketing, financial, operations/logistics, and human resource management, must attempt to solve the problem at hand and create the best outcome for the firm—long-term profitability.

There are a number of ways that the food and agribusiness markets differ from other markets. These differences are important, and influence the business situation that food and agribusiness managers must practice in. Eight key points follow and these points each differentiate food and agribusiness from other markets in some way. While one can find examples of other industries where each point is important (seasonality is important to toy companies, for example), combined, these factors form the distinguishing features of the food and agribusiness marketplace.

Food as a product. Food is vital to the survival and health of every individual. Food is one of the most fundamental needs of humans, and provides the foundation for economic development—nations first worry about feeding their people before turning their attention to higher order needs. For these reasons, food is considered a critical component of national security. And, as a result, the food system attracts attention from governments in ways other industries do not.

Biological nature of production agriculture. Both crops and livestock are biological organisms—living things. The biological nature of crops and livestock makes them particularly susceptible to forces beyond human control. The variances and extremes of weather, pests, disease, and weeds exemplify factors that greatly impact production. These factors affecting crop and livestock production require careful management. But, in many cases, little can be done to affect them outright. The gestation cycle of a sow, or the climate requirements of wine grapes provide examples.

Seasonal nature of business. Partly as a result of the biological nature of food production, firms in the food and agribusiness markets can face highly seasonal business situations. Sometimes this seasonality is supply driven—massive amounts of corn and soybeans are harvested in the fall. Sometimes this seasonality is demand driven—the market for ice cream has a series of seasonal peaks and valleys, as do the markets for turkey and cranberries. Such ebbs and flows in supply and demand create special problems for food and agribusiness managers.

Uncertainty of the weather. Food and agribusiness firms must deal with the vagaries of nature. Drought, flood, insects, and disease are a constant threat for most agribusinesses. All market participants, from the banker to the crop production chemical manufacturer are concerned with the weather. A late spring can create massive logistics problems for firm-supplying inputs to the crop sector. Bad weather around a key holiday period can ruin a food retailer's well-planned promotional event.

Types of firms. There is tremendous variety across the types of businesses in the food and agribusiness sectors. From basic producers to transportation firms, brokers, wholesalers, processors, manufacturers, storage firms, mining firms, financial institutions, retailers, food chains, and restaurants—the list is almost endless. Following a loaf of bread from the time it is seed wheat prepared for shipment to the farmer until its placement on the retail grocer's shelf involves many, many different types of business enterprises. The variety in size and type of agribusinesses, from giants like ConAgra to the small, yet successful family farm shapes the business environment in which food and agribusiness managers operate.

Variety of market conditions. The wide range of firm types and the risk characteristics of the food and agribusiness markets have led to an equally wide range of market conditions. Cotton farmers find themselves in an almost textbook case of the perfect market where individual sellers have almost no influence over price. At the same time, Coca-Cola and PepsiCo have a literal duopoly in the soft drink market. Some markets are global, others local. Some markets are characterized by near equal power between buyer and seller, while others may be dramatically out of balance in one direction or the other.

Rural ties. Many agribusiness firms have important rural links and a number are located in small towns and rural areas. And, these firms are likely the backbone of the rural economy. As such, food and agribusiness has a very important rural development role.

Government involvement. Due to almost all of the previous factors, the government has a fundamental role in food and agribusiness. Some government programs influence commodity prices and farm income. Others are intended to protect the health of the consumer through safe food and better nutrition information. Still other policies regulate the use of crop protection chemicals, and affect how livestock producers handle animal waste. Tariffs and quotas impact international trade. School lunch programs and food stamps help shape food demand. The government, through policies and regulations, has a pervasive impact on the job of the food and agribusiness manager.

Each of these special features of the food production and marketing system affects the environment where agribusiness managers practice their craft. Agribusiness is unique and, thus, requires unique abilities and skills of those involved with this sector of the U.S. economy.

THE FOOD PRODUCTION AND MARKETING SYSTEM

The highly efficient and effective **food production and marketing system** in the United States is a result of outstanding climate and geography; abundant and specialized production and logistics capabilities; intense use of mechanical, chemical, and biological technologies; and the creative and productive individuals who lead and manage the firms that make up the food and agribusiness industries. This U.S. food production and marketing system produces enormous supplies of food and fiber products. These products not only feed and clothe U.S. consumers, but are also exported to the international marketplace to fulfill needs of consumers around the world.

The food production and marketing system encompasses all the economic activities that support farm production and the conversion of raw farm products to consumable goods. This broad definition includes a farm machinery manufacturer, a fertilizer mine, a baby food factory, the crate builder that supplies an apple farmer with crates for shipment, rail and trucking firms, wholesale and retail traders, distributors of food, restaurateurs, and many, many others.

The U.S. food production and marketing system is a staggeringly large, interconnected world, employing roughly 25 million people and producing an output of more than $2 trillion. It generated more than $1.5 trillion worth of value-added products and services in 1999 (Lipton, et al.). Table 1-1 shows that the output of this system represents more than 15 percent of the total gross domestic product (GDP) of the U.S. economy.

Compare productivity in the U.S. food system to other countries worldwide (Table 1-2). Although Japan has a higher per capita gross national income (GNI), the United States has a significantly higher total GNI. Both countries, along with Australia, have a low proportion of their population involved in production agriculture. Note also that farm size varies dramatically, even among developed countries. Geography or limited farmland, climate, crop or livestock focus, or simply the area needed to maintain a viable production unit helps explain this variation.

In contrast, the other countries outlined in Table 1-2 show the characteristics of underdeveloped countries: low GNI per capita and a high proportion of the population involved in production agriculture. In China, 50 percent of the population is involved in farming while in India the figure is 67 percent. As these countries develop, there will be a tremendous demand for new farming technology. As a result, many U.S. firms are attempting to establish joint ventures with firms in these countries to aid in this development process, and capture future markets in the process.

The overall efficiency of the U.S. food and fiber sector is also illustrated in Table 1-2 by the proportion of personal consumption expenditures allocated to food consumed at home. For the average U.S. consumer only about 8 percent of their total personal consumption expenditures are for food consumed at home. In Japan this figure is about 18 percent, while food accounts for over one-half of an Indian consumer's personal consumption expenditures. The efficiency of the U.S. food production and marketing system is really quite remarkable. Consider this: with less than 7 percent of the world's land and 5 percent of the world's population, the U.S. food system produces more than 12 percent of the world's agricultural commodities, including 15 percent of the world's livestock and 11 percent of the world's crops. The United States produces 50 percent of the world's soybeans, 40 percent of the corn, and 25 percent of the world's grain sorghum.

A primary requirement for being a successful agribusiness manager is a solid understanding of this food production and marketing system. Regardless of what specific part of the food system you work in, it is important to understand what happens to food and fiber products both before they reach your firm, and after they leave your firm and head to the consumer.

TABLE **1-1** | Contribution of the Food and Agricultural Industries (FAI) to the U.S. Economy, 1999

	Value added to GDP	Share of FAI contribution to GDP	Share of GDP	Number of workers	Share of FAI employment	Share of total U.S. employment
	Billion dollars	%	%	Thousands	%	%
Farming	**69.8**	**4.6**	**0.8**	**1,714**	**7.1**	**1.2**
Total inputs:	**468.4**	**30.8**	**5.0**	**4,720**	**19.5**	**3.4**
Mining	15.6	1.0	0.2	62	0.3	—
Forestry, Fishing, and ag services	13.2	0.9	0.1	409	1.7	0.3
Manufacturing	160.0	10.5	1.7	1,192	4.9	0.9
Services	279.6	18.3	3.0	3,058	12.6	2.2
Total manufacturing and distribution:	**984.0**	**64.7**	**10.6**	**17,835**	**73.4**	**12.8**
Food processing	177.7	11.7	1.9	1,296	5.3	0.9
Textiles	45.3	3.0	0.5	993	4.1	0.7
Leather	0.3	—	—	4	—	—
Tobacco	34.5	2.3	0.4	29	0.1	—
Transportation	49.6	3.3	0.5	596	2.5	0.4
Wholesaling and retailing	460.3	30.2	4.9	8,306	34.2	6.1
Food service	215.5	14.2	2.3	6,606	27.2	4.7
Total food and agricultural industries	**1521.5**	**100.0**	**16.3**	**24,265**	**100.0**	**17.4**

Source: Edmondson.

Income spent on food by a country's people is heavily influenced by the agricultural production technologies in use.

Photo courtesy of H. David Thurston, Smokin' Doc Photo Collection, Cornell University, Ithaca, NY.

T A B L E **1-2** | Indicators of U.S. Agriculture Sector Efficiency

Country	GNI per Capita[a]	% Population in Agriculture[b]	Average Farm Size[c]	% Personal Consumption Expenditures Spent on Food Consumed at Home[d]
	1999 U.S. $	%	Acres	%
United States	$31,910	2.6	438.0	8.4
Australia	20,950	5	9852.0	19.3
China	780	50	1.0	N/A
India	440	67	4.2	51.8
Japan	32,030	5	3.4	17.6
South Korea	8,490	12	3.2	29.1

Source: [a]The World Bank Country Data.
　　　[b]The World Factbook 2000, CIA (figures for 1998).
　　　[c]Huang, USDA (figures for 1995).
　　　[d]Putnam and Allshouse, USDA (figures for 1994).

The Farm-Food Marketing Bill—A Perspective on the System

An important part of understanding agribusiness comes from understanding just how and what consumers spend on food. At the end of the twentieth century, American consumers were spending more than $618 billion on food for home and away-from-home consumption—up 37 percent from the $450 billion spent in 1990. A better understanding of just what that spending is all about comes from looking at the farm-food marketing bill (Elitzak).

What a Dollar Spent on Food Paid for in 1999

Farm value   Marketing bill

Consumer expenditures, 1999

Figure 1-1
1999 Farm-Food Marketing Bill.
Source: Elitzak, USDA.

The **farm-food marketing bill** is a breakdown of the consumer's food dollar into what the farmer gets for their raw products and what the food industry gets for "marketing" those raw farm products. Marketing here includes what it costs to transport, process, manufacture, distribute, advertise, prepare, etc. the products to make them ready for the consumer. In 1999, for example, for every dollar spent on food, 80 cents was spent on marketing the product, 20 cents went to the farmer (Figure 1-1).

The farm share of the farm-food marketing bill has declined from 41 percent in 1950 to 20 percent in 1999. This trend reflects a number of factors: continuing increases in farm productivity which kept prices for farm commodities relatively low; a huge increase in consumer demand for convenient, highly processed food products; an equally huge increase in food consumed away from home; and increases in prices for many components of the marketing bill including labor, transportation, and energy. As we see more dual income families in the workforce, as we see consumers with higher incomes, as we see consumers who demand more convenience, we see more demand for food in a convenient, ready-to-eat form. And, this demand fuels the increase in the food marketing bill.

As Figure 1-2 shows, marketing costs rose by about 45 percent during the 1990s. Labor accounted for more than 60 percent of the marketing bill increase during this time. In addition, the marketing bill continues to rise because of the trends toward convenience described above. Figure 1-3 shows how the nonfarm components of the marketing bill break down. The cost of labor represents the largest part of the total food marketing bill, accounting for nearly half of all marketing costs. Labor costs were more than $240 billion in 1999, some 56 percent

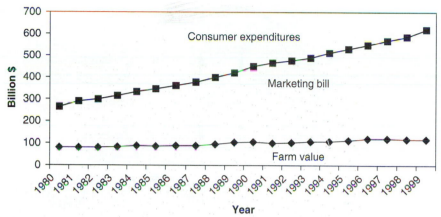

Figure 1-2
Consumer Expenditures for Food, 1980–1999.
Source: Elitzak, USDA.

more than ten years earlier. During the 1990s, employment in the food industry rose 14.5 percent to 7.9 million people. Eating and drinking places showed the greatest growth in employment by booming 21 percent over the decade. What's ahead? The nonfarm share of the marketing bill is likely to continue to rise, as food firms work to meet the needs of a society on the move.

Figure 1-3
Components of 1999 Food Marketing Bill.
Source: Elitzak, USDA.

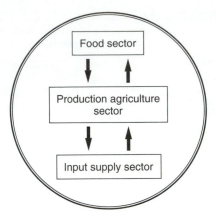

Figure 1-4
The Food Production and Marketing System.

THE THREE PRIMARY SECTORS OF THE FOOD SYSTEM

The U.S. food production and marketing system, for purposes of discussion in this text, is divided into three sectors: the food (and fiber) sector, the production agriculture sector, and the input supply sector (Figure 1-4).

We start with the **food sector.** This is the sector in which food processing, marketing, and distribution occurs (Figure 1-4). Here we have firms such as Kellogg's, Hormel, Kroger, and McDonald's, as well as thousands of other firms, large and small. This group marches to the beat of the consumer, and as the tastes and preferences of consumers change, these firms adapt to serve these changing needs.

Next in line is the **production agriculture sector.** Here purchased inputs, natural resources, and managerial talent come together to produce crop and livestock products. Agribusinesses in this sector vary in size, number, and focus—from the smallest individual operation selling strawberries to neighbors to the 100,000+ head cattle feedlot; from the rice farmer in Louisiana to the canola grower in North Dakota; from the pork producer in North Carolina to the megadairy in Arizona.

The food production and marketing system ends (or begins) with the many, varied activities that take place in the **input supply sector.** This sector is responsible for providing the thousands of different inputs—both products and services—to production agriculture. Firms here include DuPont, Syngenta, John Deere, DeLaval, your local cooperative, as well as hundreds of other firms that manufacture and distribute the inputs that farmers and ranchers need in their businesses. Let's take a closer look at each of these sectors.

THE FOOD SECTOR

At some point, it all comes down to getting raw farm commodities processed, packaged, distributed, and sold to the consumer. A very wide variety of processing and marketing firms are responsible for adding value or utility to commodities as they leave the farm gate. We will start our look at the food sector with food retailers and away from home food firms. Then, we will look at wholesaling firms, food manufacturers and processors, and assembly and transportation firms. We will end this section with a focus on the linkages across firms in the food sector, and between the food sector and the other sectors in the food production and marketing system.

Food Retailing

Food retailing represents one of the largest industries in the United States. Food store sales amounted to $458 billion in 1999 having climbed from just under $200 billion in 1980. Over this same time period, the number of retail food stores fell from about 169,000 to 128,000, or 24 percent. Food stores come in a wide assortment of sizes. Supermarkets, superettes, small grocery stores, convenience stores, and specialized food stores accounted for 82 percent of all food sold in retail stores. Food service outlets, which include restaurants, fast food outlets, cafeterias, and institutions, accounted for 84 percent of prepared food and meals sold in 1999 (Kaufman (b)).

There are several important types, or formats, of retail food stores. One important format is the **supermarket.** This format is defined as any retail food store generating more than $2 million in sales per year. A **chain store** is a set of seven or more stores operating under one management. The number of chain stores has not changed significantly over the past 15 years, however, chain stores have sold an increasingly higher proportion of retail grocery products. Supermarket sales accounted for 70 percent of all food store sales in 1999. Superettes and small grocery stores together accounted for 15 percent of total sales. Convenience stores and specialty retailers (like meat markets) account for the remaining 15 percent of total retail food sales.

Over the past 10 to 15 years a number of new retailing strategies have become popular. No longer is food moved just through the typical supermarket prototype. Some of these formats include:

Supermarket: the conventional or benchmark prototype; carries about 10,000 items; 10,000 to 25,000 sq. ft.

Superstore: increased size and greater variety of products, especially nonfood items; carries around 15,000 items; 30,000 to 40,000 sq. ft. Strategy: one-stop shopping, additional service and selection.

Warehouse store: smaller than the supermarket, carries limited number of items; typically stocks fast-moving, nonperishable products; carries 6,000 to 8,000 items; 10,000 sq. ft.; sometimes called a "box" store. Strategy: low price.

Super warehouse store: more area than warehouse store, additional variety, perishable items, meat, and produce; carries 10,000 to 12,000 items; 15,000+ sq. ft. Strategy: low price with variety.

There are a variety of retail store formats competing for the business of U.S. food consumers.

Photo courtesy of USDA.

Combo store: a supermarket with a pharmacy and more nonfood items, especially health and beauty care products; carries 25,000 items; 40,000+ sq. ft. Strategy: higher service, one-stop shopping.

Hypermarket: huge selection of general merchandise/department store items with a supermarket; carries 50,000+ items, largest of all stores; at least 100,000 sq. ft., many close to 200,000 sq. ft. Strategy: low price and true one-stop shopping.

The shift from supermarket to hypermarket has been an evolutionary process that started in the late 1960s and early 1970s. Such stores are increasingly common in urban areas across the United States.

There has been a great wave of consolidation and structural change in food retailing the last 5 to 10 years. In the 5-year period between 1996 and 2001, more than 3,500 supermarkets were purchased in transactions representing over $67 billion. Widespread consolidation has resulted in fewer, but larger food retailing firms. Table 1-3 identifies the ten largest chain stores in terms of total annual sales. In 1999, Kroger was the nation's largest food retailer with total sales of $45.3 billion. Albertson's, Inc./American Stores, Inc. was in second place with sales of $28.9 billion. In 1999, the twenty largest food retailers captured 51 percent of the nation's food store sales. Supermarket chains are constructing larger stores in a number of different formats. Meanwhile, some nontraditional food retailers have entered the market. Wal-Mart, Sam's Club, and Target, for instance, have expanded their offerings to reach a broader customer base. Another nontraditional food retailer entering the market during the first decade of the new century includes e-commerce providers who have begun offering their online grocery services in some major metropolitan areas.

Food retailing remains an extremely competitive industry with little margin for error. Consolidation trends in the retail food industry in the late nineties continued into the new millennium. In general, retailers merge to help ensure their long-term success in a competitive market. While some argue that fewer and larger retail food outlets will mean a more general selection offered by retailers and higher prices, others insist that larger retailers can offer a broader assortment of more competitively priced products to customers, passing on the savings to them.

TABLE **1-3** | Sales of the 10 Largest U.S. Food Retailers in 1999

Rank/Retailer	Number of supermarkets owned	U.S. grocery store sales (billion dollars)
1. The Kroger Company/Fred Meyer	2,288	45.3
2. Albertson's, Inc./American Stores, Inc.	1,690	28.9
3. Safeway Stores, Inc.	1,659	25.5
4. Ahold, USA	1,063	20.3
5. Wal-Mart Supercenters	721	15.7
6. Winn-Dixie Stores	1,188	13.9
7. Publix Supermarkets	635	13.1
8. Delhaize America (Food Lion, Hannaford Bros.)	1,365	10.9
9. Meijer, Inc.	130	9.5
10. Great Atlantic & Pacific Tea Co.	570	8.0

Source: Kaufman (a).

Food Service

The nation's 844,000 restaurants were expected to generate $399 billion in sales in 2001. The **food service** industry saw steady growth during the 1980s and 1990s as busy people, employed mothers, and a more affluent, mobile society chose the luxury of eating meals away from home. USDA reports that from 1990 to 1999, food spending away from home grew 57.9 percent; spending for food at home grew 32 percent. In 1999, U.S. families spent only about 11 percent of their disposable personal income on food—6.6 percent of that total was spent on food eaten at home, 4.4 percent was spent on food eaten away from home. The proportion of the food dollar spent eating out grew from 44 percent in 1990 to 47.5 percent in 1999 (National Restaurant Association). The National Restaurant Association predicts that 53 percent of the food dollar will be spent on food away from home by 2010. The food service industry, which employs 6.6 million people, is comprised of three major types of firms—traditional restaurants, fast food/quick service restaurants, and institutional food service firms.

Traditional Restaurants   Sales at **traditional restaurants,** also called full-service restaurants, were expected to hit $143 billion in 2001 (National Restaurant Association). Despite a slowdown in the general economy, demographics show that affluent baby boomers are the most frequent customers at full-service restaurants. Today's full-service restaurant may be reflecting a cultural change among Americans. Restaurant industry managers note that eating out at full-service restaurants is an important social function for today's customers. Whereas 25 years ago socializing and entertaining were done at one's home, today's consumer meets family, friends, and coworkers at the local restaurant for leisure, conversation, and convenience. There is growth in the disparity among restaurant types. Substantial growth is expected for casual dining restaurants with per person checks in the range of $15 to $20. These full-service restaurants may cater to families or those looking for a more relaxed dining experience. On the other end of that spectrum, some full-service restaurants, now comfortable with their market, are developing "unique dining experiences" through the use of ethnic cuisine, and unique décor, for example.

Fast Food/Quick Service   The words "fast" and "quick" not only describe the service provided by these restaurants, they also describe the rate at which this industry changes. During the 1990s, quick service food firms such as McDonald's, Taco Bell, and Wendy's more than doubled their sales. Today the **fast food/quick service** segment is the fastest growing part of the food service market. Sales at fast food/quick service restaurants were forecast at $112 billion in 2001, a 4.4 percent increase over the previous year (National Restaurant Association). Now established as part of American culture, fast food restaurants have stepped into the next dimension of customer service by experimenting with the offerings and specials on their menus, the additional services or "perks" they offer returning customers, and faster, better means of providing customers with both fast and nutritious foods. Today's fast food/quick service restaurants often have central wholesale warehouses and buying offices, and many have expanded internationally.

Institutional Food Service   A very wide variety of firms are involved in **institutional food sales** including college dormitories; high school, middle school, and elementary school cafeterias; and preschool and day care food services (Price). This category also includes food offered at government offices, corporate eating establishments, airlines, hospitals, etc. Institutional food services account for an important portion of the food people consume daily. Some types of institutional food service firms expanded rapidly during the 1990s (recreation and entertainment facilities, and retail hosts such as gas stations and bookstores), while sales through other firms in this category were stable or declined (hospitals, vending machines, and the military). Trends and changes in the institutional food service industry will continue to reflect consumer demands for convenience and nutrition. For instance, 30 years ago, it would have been unusual to see tacos, a food now considered common, on the menu of a major university's dorm cafeteria.

Food Wholesaling

One significant industry in the food sector is food wholesaling—representing a 589 billion-dollar business in the year 2000. Today there are more than 2,000 general line wholesalers involved in the U.S. food and fiber sector. There are several ways to categorize the work done by **wholesalers,** however three basic categories capture most firms in this industry. **Merchant wholesalers** represent the largest percentage of food wholesale sales accounting for 56 percent of the total. Merchant wholesalers primarily buy groceries and grocery products from processors or manufacturers, and then resell those to retailers, institutions, or other businesses.

Manufacturers' sales offices and sales branches are wholesale outfits typically run by big grocery manufacturers or processors to market their own products. **Wholesale agents and brokers** are wholesale operators who buy and/or sell as representatives of others for a commission. Wholesale brokers and agents typically do not physically handle the products, nor do they actually take title to the goods.

Most wholesale operations focus on sales to retailers, other wholesalers, industrial users, and, in some cases, the final consumer. A wholesaler may buy directly from the producer and sell to another wholesaler or food processor. More typically, however, the wholesaler buys from the food processor or manufacturer and sells to a retailer. The makeup of the wholesale trade sector involves a large group of varied organizations—some quite small and some very large. Retail chain stores own their warehousing facilities, but typically break this out into a separate operating unit with the objective of generating a profit.

Wholesalers perform a variety of functions for their retail customers. It is important to note that wholesalers are facilitators and that they may take market risk if they take title to the goods they handle. These firms are responsible for handling products and for geographic distribution of products in appropriate quantities. Often wholesalers will finance inventory purchases for the retailer. In this case, the retailer does not have to borrow money from the bank, but uses the wholesaler as a source of operating funds. Many wholesalers offer services like automatic ordering, customer traffic surveys, and suggested shelf-stocking arrangement, among others. The idea is for the wholesaler to forge a profitable working relationship with their retail clients.

Food Processing and Manufacturing

In 1998, **food processing and manufacturing** firms accounted for 12 percent of the value of shipments from all U.S. manufacturing plants (McDonald). The Census of Manufacturers from the U.S. Bureau of the Census counted more than 26,000 food processing plants across the nation at the end of the twentieth century (U.S. Bureau of the Census).

The food processing and manufacturing industry includes a broad range of firms that take raw agricultural commodities and turn those commodities into ingredients for further processing, or into final consumer products. Meatpackers, bakers, millers, wet corn processors, cereal companies, brewers, and snack firms would all be examples of food processors and manufacturers. These firms can be very complex and the markets they serve highly varied. For example, soybean processors in central Illinois break down soybeans into two major components: soybean oil and soybean meal. There are literally hundreds of products that utilize soybean oil as an ingredient, ranging from margarine to cosmetics. The market for soybean meal is primarily for use as a high protein livestock feed supplement. Processors of soybeans are involved in a very complex market. They acquire one input—soybeans—and produce a very wide range of products, each with its own unique market conditions.

Processors of meat products account for about 19 percent of the total value added by food processors and manufacturers (Figure 1-5). Processors of bakery products, dairy products, and fruits and vegetables all account for more

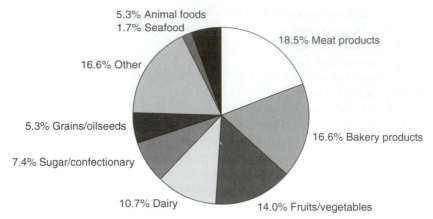

Figure 1-5

Composition of Total Value Added by Food Processors.
Source: McDonald, USDA.

than 10 percent of total processing activity. While some 16,000 firms are engaged in food processing activities, most own but a single processing plant (McDonald). On the other end of the continuum, the 100 largest processing firms account for 75 percent of the total value added in the processing industries. About 30 years ago, the top 100 accounted for about 50 percent of the total value added, so the concentration of value added among the largest processing firms has increased substantially over time.

The sharp increase in concentration among food processing industry firms in the last 30 years has led to a few large companies marketing a wide variety of branded and commodity products. Major commodity processing industries such as grain milling and meatpacking, dominated by giants such as IBP, Smithfield Foods, Cargill, Archer Daniels Midland (ADM), and ConAgra Foods, are examples. Among other large and well-known food processing and manufacturing firms at the start of the twenty-first century are firms such as Philip Morris, Coca-Cola, Frito-Lay, and Suiza Foods.

Transportation and Storage Firms

Our final group of food sector firms includes those that acquire or assemble commodities from agricultural producers, and store and transport these products for food manufacturing and processing firms. **Transportation and storage firms** actually facilitate the marketing and processing phases of the food sector. Firms ranging in size from the grain handling giant Cargill, to local cooperatives that handle grain, are involved in the collection, storage, and transportation of agricultural commodities.

A good example is provided by the way that grain moves from farm to buyer in the Midwest. Grain typically moves from producers to local elevators or to subterminal elevators located near transportation centers. These firms make money on grain movement, not on speculating that the grain they hold will go up in price. The relatively small local grain elevators may sell to a larger subterminal or terminal elevator, or directly to a processor. Terminal elevators are involved in assembling grain from smaller country points to amass a quantity of grain that will fill unit trains or barges. Unit trains could be transported to the southeast and fed to poultry. Or a group of barges might be transported down the Mississippi River and the delivered grain shipped to the international market or sold to a local processor for immediate use.

Managing transportation and storage firms brings a unique set of challenges. A grain buyer for a terminal elevator located close to rail lines and a navigable river must be knowledgeable of rail rates, the availability of rail cars, and barge rates. Often these rates can change significantly overnight. Risk management is paramount, as commodity prices change quickly. Margins for these firms are typically razor thin, hence cost management is a priority.

Linkages in the Food Sector

Many of the organizations involved in the food sector have successfully integrated forward or backward in the food system. Goals for such a strategy include increased operating efficiency and reduced market risk. For example, many firms are involved in both processing and marketing activities and have at least partially integrated back to the production sector by entering partnering or contracting arrangements with producers. Smithfield Foods is a good example of a firm with such a position in the market. The link helps reduce market risk by reducing material quality and supply problems.

This arrangement is also very common in the poultry, fruit, and vegetable production sectors. In the Midwest, for example, the Redenbacher Popcorn Company has producer agreements that guarantee supply before the crop is planted in the spring. Redenbacher contracts with producers by guaranteeing a specific price for popcorn grown on a specific number of acres. The firm provides seed and purchases all popcorn grown on the acreage under contract. Producers deliver production to a local processing facility.

Food retailers such as Kroger own dairies. Brewers like Anheuser-Busch own can manufacturing operations. Firms like ConAgra have a presence at almost every level of the food system. As mentioned earlier, such linkages are common and make the lines between industries very blurry, and the resulting firms quite complex.

THE PRODUCTION AGRICULTURE SECTOR

At the hub of our food production and marketing system is the production agriculture sector. **Production agriculture** includes the farms and ranches that produce the crop and livestock products that provide inputs to the food and fiber sector. And, these farms and ranches are the customers of the firms that make up the input supply sector. As mentioned earlier, a relatively few individuals are responsible for a very staggering quantity of output in the U.S. production agriculture sector. Today's U.S. farmer produces enough food and fiber in a year to feed and clothe 129 people—32 of which live outside the United States. More than 40 percent of the corn grown in the world is produced in the United States. One Nebraska beef producer supplies the beef needs of 600 U.S. consumers.

Every industry in the food system is impacted in some way by production agriculture. And, like the food and input supply sectors, the production agriculture sector has been undergoing profound change in response to a variety of market forces. In this section, we will explore the dynamic production agriculture sector of our food production and marketing system.

Farm Demographics

So what is a farm? The United States Department of Agriculture (USDA) defines a **farm** as "any establishment from which $1,000 or more of agricultural products were sold or would normally be sold during the year" (*Farms and Land in Farms*). This definition includes many part-time farmers with limited acreage and very modest production. The USDA definition of a farm includes the small hobby farmer that sells one horse or 10 lambs per year as well as the commercial operator that farms 8,000 acres and produces 700,000 head of hogs annually. The definition includes those who farm part-time and those who have farmed full-time for generations. USDA reports that nearly 90 percent of farms have some household income from off-farm sources. However, operators of larger farms are much more likely to depend solely on the farm for income (Banker and Hoppe (a)).

The twentieth century was a period of huge change for production agriculture in this country. Changes in farm numbers and farm size are reflective of this change. Historically, individual families or extended families have owned and operated the nation's farms. The family provided the land, labor, and other capital necessary to run the business. As market prices fluctuated, farm families adapted by doing without or diversifying in some way. As mechanization or finances allowed, more land was acquired—ideally to send a child to college, or to provide a living for additional family members coming into the business.

As time moved on, farm expansion required additional inputs such as seed, fertilizer, chemicals, credit, farm machinery, etc. be purchased. Technology-fueled expansion made it possible for farmers to operate and manage ever larger farm businesses, and farm size grew. However, when fluctuations in price or crop losses caused lean years, farmers still had obligations to pay suppliers. Financing became a necessary and critical component of the family farm business. As we will see later in this book, such debt financing also carries a risk. And, farm expansion financed by debt and secured through inflated land values of the 1970s created real problems for some producers. Some of these farms were not able to survive the farm crisis of the 1980s and foreclosures on farms contributed to an overall decline in farm numbers.

As a result of these trends and others, total land in farms in the United States continues to decline slowly (Table 1-4). In 2000, a total of 943 million

TABLE **1-4** | Number of Farms, Land in Farms, Average Size of Farms 1993–2000

Year	Farms	Land in farms (1,000 acres)	Average size (acres)
1993	2,201,590	968,845	440
1994	2,197,690	965,935	440
1995	2,196,400	962,515	438
1996	2,190,500	958,675	438
1997	2,190,510	956,010	436
1998	2,191,360	953,500	435
1999	2,192,070	947,440	432
2000	2,172,080	942,990	434

Source: *Farms and Land in Farms.*

acres was farmed, down from its peak of 1.2 billion acres in 1954. The 2.17 million farms and ranches in the United States in 2000 was down 0.9 percent from the year before. This represented the largest decline in farm numbers in a one-year period since 1991 when 29,000 operations were lost. Farm numbers have dropped almost 10 percent since 1985 (*Farms and Land in Farms*). Texas reports the most farms in the United States with 226,000; Missouri is second with some 109,000 farms.

About 2 million farmers and ranchers make up the U.S. production agriculture sector. However, larger and more specialized operations have continued to evolve since the 1930s. As a result, a much smaller group of producers accounts for the majority of agricultural production. The 1997 Census of Agriculture showed that about 87 percent of all agricultural production came from 18 percent of U.S. farms, and some 55 percent of the total came from only 4 percent of the farms—about 80,000 farmers (Figure 1-6).

Despite all of the changes, today families or family corporations still own nearly 90 percent of the nation's farms (Figure 1-7). However, it is important to note that many of these family-run farming operations are very large and sophisticated businesses. Family-owned corporate farms accounted for about 23 percent of total farm sales in 1997. At the same time, nonfamily-owned corporate farms, particularly in the livestock and poultry industries, are increasing in number. However, such operations accounted for only about 6 percent of total farm sales in 1997 (Banker and Hoppe (b)).

The question of how ownership will be structured in the future is an interesting one. Today's average farmer in the eastern and western Corn Belt is 54 years old. The 1997 Census of Agriculture shows that that average age has been increasing for several years. According to a survey by Rockwood Research, 40 percent of U.S. farmers planned to retire by 2007 and half of those farmers planned to quit farming by 2002 (Hill). The next decade will be one of great change in terms of who farms the land.

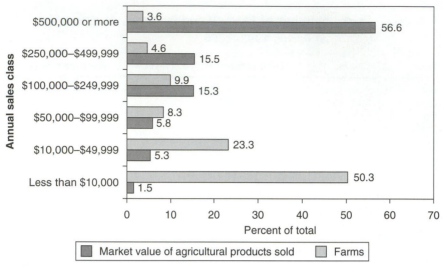

Figure 1-6
Distribution of Farms by Sales Class: 1997.
Source: 1997 Census of Agriculture.

Farm Income

Net farm income in the United States is determined by prices of farm products, production yields, and farm production expenses. Farm prices can be highly volatile, moving upward quickly in response to low processor tomato yields, or plunging if the market doesn't need all of the pork produced by U.S. producers. Note that the commodity nature of most agricultural markets means that individual farmers have little control over the price of their production. They can take steps to manage their price risk, but cannot do much about the actual price they receive. Production expenses can also be volatile. Fuel costs can soar when general market supplies are short. The price of nitrogen fertilizer, which is made from natural gas, can increase very quickly if natural gas is in short supply.

Given the importance of the production agriculture sector, it has been the policy of the U.S. government since the 1930s to support the income of farmers when market conditions are not favorable. There are a wide variety of such programs that assist farmers, and these vary by commodity. Some crop and livestock products, such as peanuts, wheat, and dairy products, have substantial government involvement. Other crop and livestock products, such as soybeans, pork, and lettuce, have much lower levels of government involvement. When combined across all crops, total payments from the government have been an important component of net farm income in the United States (Table 1-5). Net farm income during the first part of the new millennium maintained levels near the average of the 1990s. However, this was due in large part to sizable direct government payments to farmers that offset generally low commodity prices.

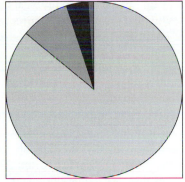

- 86% Individuals
- 9% Partnerships
- 4% Family-held corporations
- 1% Non-family corporations, institutions, other

Figure 1-7
Farm Ownership in 1997.*
Source: 1997 Census of Agriculture.
*Percent of farms.

TABLE **1-5** | **Net Income from Farming: 1970–2000**

Year	Net income (millions $)	Government payments (millions $)	Payments as % of farm income
1970	$14,365	$ 3,717	25.8%
1975	25,546	807	3.1
1980	16,141	1,285	8.0
1985	28,648	7,704	26.9
1990	44,620	9,298	20.8
1995	36,893	7,253	19.6
1996	54,913	7,339	13.3
1997	48,601	7,495	15.4
1998	44,623	12,209	27.4
1999	43,398	20,593	47.5
2000	45,400	23,300	51.3

Source: Strickland.

Future Issues

The production agriculture sector of the food production and marketing system is evolving very quickly. Consolidation in the sector has been rapid to date, and larger farms account for an increasing proportion of total production—especially in the livestock sector. International competition, relatively low commodity prices, volatile input prices, environmental and food safety regulations, and rapid deployment of new technology have all placed substantial pressure on the owner/manager of a production agriculture business (*FoodSystem 21*). Agricultural producers today have responded to these pressures with an increased emphasis on financial and risk management; a focus on control of assets like land and machinery, as opposed to ownership; and the general application of many of the management ideas we will address in this book. For many years people talked about farming as "a way of life." While the "way of life" aspect remains important to many producers, successful commercial agricultural producers in today's business environment are astute, sophisticated business managers running large and complex organizations.

THE INPUT SUPPLY SECTOR

The **input supply sector** of the food production and marketing system is comprised of the firms that manufacture and distribute the myriad of inputs that fuel the production agriculture sector. The agricultural input industry umbrella is a broad one and includes a wide range of firms that provide products and services to agricultural producers. Animal nutrition, seed, machinery and equipment, fertilizer, crop protection, and credit and banking firms are among the enterprises that fall under the general category of input suppliers.

A description of just what these industries look like today only vaguely resembles their appearance just a few years ago. During the last three decades, these industries, like the food and production agriculture sectors, have experienced vast and dynamic change. Taking a quick look at the history book helps define today's agricultural input business. A rapidly growing agriculture fueled a prosperous input supply sector in the 1970s. The massive adjustments of the production sector in the 1980s brought an equally huge reorganization of the input supply industries. In the 1990s, continued consolidation in the livestock sector reshaped the input supply industries serving animal agriculture. The 1990s also brought consolidation to the agronomic industries with mergers and acquisitions leading to fewer, larger players at both the manufacturing and the distribution levels (*FoodSystem 21*).

The productivity of the agricultural production sector continues to increase due, in no small part, to the products and services provided by the input supply sector. Improved varieties of seed may be attributed to advances in biotechnology among other more conventional developments, which allow improved and lower-cost production methods. Animal nutrition, farm machinery and equipment, agricultural pesticides and herbicides, and the many facilitating services offered to producers help increase the productivity of the production agriculture sector. For general purpose of this text, we have categorized the farm input supply sector into three areas:

- Manufacturing
- Distribution
- Services

We will explore each of these three areas in the remainder of this section.

Manufacturing

Input Manufacturers include many company names you may well recognize. John Deere, Bayer, Pfizer Animal Health, and Monsanto are just a handful of examples of large organizations that spend millions of dollars annually in re-

T A B L E **1-6** | **Farm Production Expenses**

Item	1997	1998	1999	2000
Farm origin inputs	**46.8**	**44.8**	**45.5**	**47.1**
Purchased feed	26.3	25.0	24.5	24.7
Livestock purchased	13.8	12.5	13.8	15.2
Seed	6.7	7.2	7.2	7.2
Manufactured inputs	**29.2**	**28.2**	**27.3**	**30.2**
Fertilizer and lime	10.9	10.6	9.9	10.4
Pesticides	9.0	9.0	8.6	8.6
Petroleum and fuel	6.2	5.6	5.8	8.1
Electricity	3.0	2.9	3.0	3.0
Total interest charges	**13.1**	**13.4**	**13.6**	**14.2**
Other operating expenses	**61.4**	**63.0**	**65.9**	**67.7**
Repair and maintenance	10.4	10.4	10.5	10.6
Contract and hired labor	18.6	19.3	20.1	20.7
Machine hire and custom work	4.9	5.4	5.3	5.5
Marketing, storage, and transportation	7.1	6.9	7.3	7.6
Miscellaneous operating expenses	20.3	21.1	22.8	23.2
Overhead expenses	**39.2**	**39.3**	**39.8**	**40.5**
Depreciation	19.6	19.7	19.9	19.8
Property taxes	6.8	6.9	7.1	7.1
Rent	12.8	12.7	12.9	13.5
Total production expenses	**189.8**	**188.6**	**192.1**	**199.7**

Source: McElroy, et al.

search and development to bring improved products to the producer. Virtually everything it takes to run the farm or ranch must be purchased from an input supplier of some sort. Today farmers purchase over 75 percent of all inputs used for production. In 2000, farm production expenditures were $199.7 billion. Table 1-6 shows the breakdown of just how these farm dollars are spent. On average, a U.S. farmer spends about $84,000 dollars per year to run the farm business, and these expenses for farmers represent the sales of the input supply sector. Input manufacturers are responsible for the research, development, production, and manufacturing that makes these products possible.

Inputs for the livestock sector account for the largest proportion of total farm expenditures. Livestock and poultry products include feed, feed supplements, health products, etc., which are required in the production of meat, milk, and fiber products such as wool. Providing producers with products such as these requires access to and use of high technology, often large manufacturing plants and a sizable research and development budget to support efforts of this scale. In addition to products to feed animals and keep them healthy, other input firms manufacture products for housing livestock, storing and managing livestock waste, and moving livestock from farm to market.

Trucks, tractors, combines are all examples of manufactured products supplied to farmers through the input supply sector. Deere & Company (John Deere) is one of the oldest firms in the United States and is a Fortune 100 company. Deere invests more than $1.5 million daily in research and development, focused on bringing new and better products to the world's farmers.

At first glance, the seed industry may not seem like it is an input manufacturing industry. However, seed firms annually invest millions of dollars in developing and producing or "manufacturing" hundreds of new hybrids and varieties of seed. Crop protection chemical companies such as Syngenta and DuPont produce crop protectants that reduce weed and insect pressure, or protect the crop from disease. Recent developments in biotechnology have led to many products that bring these two industries together. Some crops have been modified to resist certain pests. Other crops have been modified to resist a particular herbicide, making weed control much easier. These crops have not been without controversy, and broader market acceptance of these **genetically modified organisms (GMOs)** is evolving slowly outside the United States.

Distribution

Farm production inputs are moved from the manufacturer to the farm through a very wide variety of sales, marketing, and distribution channels. Using the technology developed by manufacturers also requires accurate and timely technical information as well as timely access to the products. This is the job of **input distributors**—individuals, companies, outlets of national organizations, cooperatives, e-businesses, etc.—responsible for getting the products from the manufacturer to the farm, and providing a set of services that insure productive use of the inputs.

These distribution firms represent a very wide range of organizations. Sometimes, a major national manufacturer owns the distribution firm. Local

Advancements in technology by input suppliers contribute to productivity gains in the production agriculture sector.
Photo courtesy of Deere & Company.

agronomy outlets owned by Royster-Clark would be an example. Memphis-based Helena Chemical Company is an international company producing fertilizers, spray adjuvants, and specialty products. Helena reaches farmers through its retail and wholesale locations throughout the United States. In other cases, the distributor is a franchise of a manufacturer. The local dealer network of New Holland provides a good example. In still other cases, the distribution network is independent of the manufacturing sector. And, more recently, e-business firms have emerged in the distribution business.

Local agricultural cooperatives are heavily involved in the distribution of inputs. In Chapter 5, we investigate the organization and operation of cooperatives at length. A **cooperative** is a member-owned, democratically controlled business from which the benefits are received in proportion to use. Local farmers who are cooperative members buy and sell products through the cooperative and thus, the cooperative is a distributor of agricultural products and/or services.

Stores or chains that offer everything from pet food to tractor parts operate as a retail business from which producers make purchases. Tractor Supply provides a good example. Given the very wide range of inputs that farmers use in their businesses, and given the equally wide range of farmers involved in production agriculture, it comes as little surprise that the distribution system involves such a wide range of organizations.

Services and Financing

The **agricultural services and financing** area includes farm management services, veterinary care, consulting businesses, and farm lending just to name a few of the types of businesses involved in this area. Larger farms and fewer farmers, combined with increases in absentee land ownership, have greatly contributed to an increase in the number of farm management firms and services offered in the last decade. Frequently, farm management services are hired to oversee land rental, crop production, financial and tax management for the absentee landlord.

Financing the production agriculture sector is big business. In 1999 the farm production sector controlled about $1,116 billion of total assets—all those resources of value involved in farming. At the same time, total debt amounted to $176.4 billion. This debt includes debt on farmland and machinery, equipment, and buildings, as well as production loans. Countless banks offer agricultural loans and related services to farmers.

The Farm Credit System serves as a lender to the production agriculture sector providing production and farm real estate loans. In 1999, the Farm Credit System had loaned a total of $70 billion with roughly half being in long-term real estate loans and about 25 percent going to agricultural cooperatives. The Farm Credit System has provided approximately 25 percent of the total debt capital needed by the production agriculture sector for farmland and operating funds in recent years. This number is down from a decade ago as commercial banks and nontraditional lenders such as the captive finance companies of other input organizations have become much more aggressive in seeking agricultural loans.

LOOKING AHEAD

Managing in this diverse, multifaceted field of agribusiness requires a wide range of skills and talents. In addition to a strong background in management, agribusiness managers need a deep understanding of the biological and institutional factors surrounding the production of food and fiber. In Part I of this book, we will look more closely at the tasks of management and some key economic concepts that are important to managers. In Part II we will turn our attention to the types of organizations we find in the food and agribusiness markets, and explore some elements of the context within which agribusiness managers make decisions.

With this broader perspective in hand, we turn our attention to a deeper look at the four functional areas of management. First, in Part III we explore the area of marketing and sales. Then, in Part IV we will consider the management of the financial resources of the agribusiness. Parts V and VI deal with the important areas of operations/logistics management and human resource management. It is our goal that this book will provide you with a solid foundation for your further study—and practice—of agribusiness management.

SUMMARY

The food production and marketing system is a complex, dynamic, and extremely productive part of the total U.S. economy. The study of agribusiness includes understanding basic management principles and practices. Four key functions of management are marketing, financial, operations/logistics, and human resources management. An understanding of these functional areas is critical to the successful agribusiness manager.

There are a number of factors that combine to create a unique business environment for firms and managers in the food production and marketing system. This system is a highly efficient one. And, over time, consumers have demanded more and more services with their food products. As a result, the proportion of the food dollar that goes to production agriculture has declined, and the marketing bill has expanded.

The definition of agribusiness used in this book includes three important sectors: the food sector, the production agriculture sector, and the input supply sector. In their combined activities, these three areas provide a tremendous variety of food and fiber products to consumers both in the United States and around the world. In this food system, each sector is tied to the next in the task of developing, producing, and delivering food products to the consumer. The food sector includes a wide range of firms including processors, manufacturers, wholesalers, and retailers. Many of these firms are global firms offering brands we are all familiar with. Others are focused on manufacturing ingredients that are inputs for other food companies. The production agriculture sector is undergoing dramatic change as consolidation continues to occur. Finally, the input supply sector provides the product, service, and information inputs required by the production agriculture sector. Tractors, seed, crop protection products, capital, and advice are examples of products/services provided by input supply firms.

DISCUSSION QUESTIONS

1. List and define the four functions of agribusiness management.
2. What are five reasons the agribusiness sector may be considered unique? How or why could agribusiness firms and firms outside the agribusiness sectors make different decisions in similar situations?
3. The food marketing bill continues to grow each year—the proportion of the total farm-food marketing bill going to the food sector continues to increase. Interpret this trend. What are the reasons for this increase? How does this marketing bill affect farmers? Explain.
4. What are three significant trends facing food retailers? How have these impacted the food retailing industry?
5. Why have away-from-home eating establishments become so popular? Do you expect this trend to continue? Why or why not?
6. The size of U.S. farms has continued to increase over time. What are the positive dimensions of this trend? What are the negative dimensions of this trend? How does this trend impact the food sector and the input supply sector?
7. Environmental regulations and public acceptance of biotechnology were mentioned as important issues facing production agriculture, and hence input supply firms. Take one of these two issues, or the broader social issue of your choice, and outline some of the key implications of the issue for firms manufacturing and distributing inputs to farmers.

CASE STUDY— The Food Production and Marketing System

In this chapter, we look at three key sectors of the food production and marketing system—the food sector, the production agriculture sector, and the input supply sector. For this case study, pick a specific food or fiber product. The product could be cotton jeans, watermelon, frozen dinners, snack chips, beef steak, pizza—pick something that interests you. Then, using the library or the Internet, locate the name of one firm for each of the three sectors of agribusiness that is involved some way in the production and distribution of your product. For pizza, this might be Pizza Hut (food sector), a family-owned tomato grower/shipper (production agriculture), and Dow AgroSciences (input supply sector). After you have identified your three companies, answer each of the questions below:

Questions
1. Briefly describe each of your firms. What markets do they serve? How large are the firms? What products do they produce?
2. How are these firms linked to the product you chose to research?

3. What is similar about the firms you chose? What is different about the firms you chose?
4. How are these firms related in the production and distribution of your product? Is this an explicit link (do they share the same parent company, for instance), or are they loosely related through open markets?

REFERENCES AND ADDITIONAL READINGS

1997 Economic Census, Manufacturing. 2000. U.S. Bureau of the Census. Census of Economics.

"2001 Industry Forecast—Quickservice Outlook." National Restaurant Association. www.restaurant.org/research/forecast_quickservice.html

"2001 Industry Forecast—Fullservice Outlook." National Restaurant Association. www.restaurant.org/research/forecast_fullservice.html

Asia and Pacific Rim Situation and Outlook Series. October 1994. U.S. Department of Agriculture. Economic Research Service. WRS-94-6.

Banker, Dave, and Bob Hoppe (a). December 15, 2000. "Briefing Room: Farm Structure Questions and Answers—How Important Is Off-Farm Income to Farmers?" U.S. Department of Agriculture. Economic Research Service. www.ers.usda.gov/briefing/FarmStructure/Questions/off_farm.htm

Banker, Dave, and Bob Hoppe (b). December 15, 2000. "Briefing Room: Farm Structure Questions and Answers—Are Family Farms Disappearing?" U.S. Department of Agriculture. Economic Research Service. www.ers.usda.gov/briefing/FarmStructure/Questions/Closeup.htm

Edmonson, William. December 8, 2000. "Briefing Room: Food Market Structures—The U.S. Food and Fiber System." U.S. Department of Agriculture. Economic Research Service. www.ers.usda.gov/briefing/foodmarketstructures/foodandfiber.htm

Elitzak, Howard. December 14, 2000. "Briefing Room: Food Marketing and Price Spreads—Current Trends." U.S. Department of Agriculture. Economic Research Service. www.ers.usda.gov/briefing/foodpricespreads/trends/

Farms and Land in Farms. February 2001. U.S. Department of Agriculture. National Agricultural Statistics Service, Agricultural Statistics Board.

FoodSystem 21—Gearing Up for the New Millennium. November 1997. Purdue University Cooperative Extension Service.

Hill, Robert. "Back to the 1970s?" *AgriMarketing.* January 1997. p. 28.

Huang, Sophia. *International Agricultural and Trade Reports.* U.S. Department of Agriculture, Economic Research Service, Agriculture Information Bulletin, WRS-97-4, August 1997.

Kaufman, Phil (a). May–August 2000. "Grocery Retailers Demonstrate Urge to Merge." *FoodReview.* U.S. Department of Agriculture. Economic Research Service. pp. 29–34.

Kaufman, Phil (b). December 8, 2000. "Briefing Room: Food Market Structures—Food Retailing and Food Service." U.S. Department of Agriculture. Economic Research Service. www.ers.usda.gov/briefing/foodmarketstructures/foodservice.htm

Kaufman, Phil, Charles R. Handy, Edward W. McLaughlin, and Kristen Park. August 2000. *Understanding the Dynamics of Produce Markets—Consumption and Consolidation Grow.* U.S. Department of Agriculture. Economic Research Service. Agriculture Information Bulletin No. 758.

Lipton, Kathryn L., William Edmondson, and Alden Manchester. July 1998. "The U.S. Food and Fiber System: Contributing to the U.S. and World Economies." U.S. Department of Agriculture. Economic Research Service. Agriculture Information Bulletin No. 742.

McDonald, James. December 8, 2000. "Briefing Room: Food Market Structures—Food Processing." U.S. Department of Agriculture. Economic Research Service. www.ers.usda.gov/briefing/foodmarketstructures/processing.htm

McElroy, Bob, Roger Strickland, and William McBride. March 26, 2001. "Briefing Room: Farm Income and Costs." U.S. Department of Agriculture. Economic Research Service. www.ers.usda.gov/Briefing/FarmIncome/

Price, Charlene. September–December 2000. "Foodservice Sales Reflect the Prosperous, Time-Pressed 1990s." *FoodReview*. U.S. Department of Agriculture. Economic Research Service. Vol. 23, Issue 3, pp. 23–26.

Putnam, Judy. January–April 2000. "Major Trends in the U.S. Food Supply 1909–1999." *FoodReview*. U.S. Department of Agriculture. Food and Rural Economics Division. pp. 8–15.

Putnam, Judith Jones, and Jane E. Allshouse. *Food Consumption, Prices, and Expenditures, 1970–97*. U. S. Department of Agriculture, Economic Research Service, Food and Rural Economics Division. Statistical Bulletin No. 965, September 2000.

Strickland, Roger. September 25, 2000. "Briefing Room: U.S. and State Farm Income Data." U.S. Department of Agriculture. Economic Research Service. www.ers.usda.gov/data/FarmIncome/finfidmu.htm

The World Factbook 2000. Central Intelligence Agency. www.cia.gov/cia/publications/factbook/index.html

The World Bank—Country Data. www.worldbank.org/data/countrydata.html

2

MANAGING THE AGRIBUSINESS

- Define management and explain the role of a manager
- Understand the decision-making environment for agribusiness managers
- Describe the tasks of planning, organizing, directing, and controlling in agribusiness management
- Understand the steps in the planning process
- Define leadership and compare it to management
- Explain the differences among policies, procedures, and practices
- Describe management by exception, and understand how this idea is used by agribusiness managers

INTRODUCTION

Success or failure? Stellar performer or also-ran? For an agribusiness firm, sometimes success or failure is driven by the broader marketplace—a boom in the general economy, a terribly rapid price hike for fuel. Sometimes winning and losing comes down to chance—a lucky break in the market, a competitor's mistake. The broader marketplace and chance are clearly out of the agribusiness firm's direct control. But, while these external factors are important, the agribusiness firm also has some influence on success and failure, on stellar and mediocre performance. The decisions made by the organization—the allocation of investment funds, the people hired, the products introduced, the plants constructed, the deals cut, and a hundred other decisions—all determine if the firm will be able to capitalize on a favorable market and/or how well prepared the firm is for chance. The firm, and the managers of the firm, have some say in whether or not an agribusiness is successful.

While any firm will take a favorable trend or a lucky break, relying on factors outside the firm to determine performance simply leaves too much to chance. So, we will assert that firm performance hinges in large part on how effectively a manager uses the organization's resources. Managers are hired to utilize firm resources in the best possible manner to achieve the objectives of the owners of the firm. They use resources to capitalize on trends in the market and to manage the risks of a downturn. Managers use resources to take advantage of fortunate circumstances, and to minimize the damage from unlucky ones. Managers drive performance in agribusiness firms.

Who are these managers leading today's agribusinesses? They are each unique individuals who vary in age, gender, background, geographic location, and so on. They are each involved in unique management situations—unique because of their industry, commodity, location, and employees. Because people and situations differ so dramatically, it is difficult to come up with a specific recipe for what it takes to be a successful manager. So, while there are certain management skills and principles that can be learned, these skills must be adapted by each individual to fit the unique situation in which they operate.

Given this responsibility, how do agribusiness managers actually accomplish their task? Perhaps some individuals are just "born managers." But managing is not just something most are born to do. Managing is a learned skill, and this book is about helping you move down the path toward being an effective manager. Business education has come of age. And, it is now recognized that there are identifiable reasons why some organizations succeed while others fail. Today's successful business managers are guided by a set of principles that constitute sound management. In this chapter, we will take a closer look at this business of management. We will explore the key tasks of any manager to provide a foundation for our further exploration of agribusiness management. And, we will introduce some real agribusiness managers to help guide the way.

TODAY'S AGRIBUSINESS MANAGERS

Management is both an art and a science. Managers must efficiently combine available human, financial, and physical assets to maximize long-run profits of the operation by profitably satisfying its customers' demands. Management requires individuals be technically knowledgeable about the organization's product and/or function. They must be good and effective communicators. The ability to motivate people is essential. They must be proficient in the technical skills of management such as accounting, finance, forecasting, and so on.

In addition to a strong background in management, agribusiness managers need a strong understanding of the biological and institutional factors surrounding the production of food and fiber. In other words, not only must they excel at the normal concerns of business management, agribusiness managers must also factor in the uncertainty of the weather, the perishable nature of many of agriculture's products, government policies, and the rapidly changing technology employed in agriculture. They must possess the ability to quickly adapt to changes in market conditions that result from changes in these uncertain factors of weather, product perishability, government policies, and technology. Managers must be able to mix each of these skills and perspectives in the right proportion to deliver the greatest long-run net benefit for the organization.

To help us better understand just who managers are, and what they do, four managers from various sectors of the food production and marketing system will join us through the rest of this chapter. Two are from the food sector (one a technical manager, and one a marketing manager); one is from the production agriculture sector, and one is from the input supply sector (Table 2-1). Throughout this chapter we will get glimpses of their experiences as they manage in agribusiness every day. Their insights relate quite directly to several key management topics presented in this chapter. Each of their experiences helps illustrate some the characteristics and situations that make the food and agribusiness markets unique places to practice the art and science of management. We feel that their observations on this business of management will help you better understand some of these key ideas.

T A B L E **2-1** | **Meet the Agribusiness Managers**

Craig Kirkpatrick is General Manager, Formulated Products with Sabroso Company. Sabroso is an international juice company headquartered in Medford, Oregon. Craig manages five individuals in the business development group, but interacts with individuals throughout the company on both global and national efforts. He identifies attractive markets, imagines what possibilities there are in these markets, develops business plans that work and then, through knowing the resources available throughout the company, he organizes these resources so the plan can be put into action. At that point, he hands over the project to a team for implementation and begins developing a new set of ideas.

Sharon Rokosh is a Food Scientist and Sensory Analysis Manager with Slim-Fast Food Company, a division of Unilever. Her primary job involves overseeing the release of approximately 625,000 cases of product to the trade per week, and ensuring that products meet both company and federal regulatory requirements. She recently implemented a sensory analysis program, trained key users of the plan, and now oversees a panel of 40 employees from all levels of the organization that use the program, as well as managing three individuals in the quality assurance department daily. Sharon also works closely with logistics and operations to ensure that production goals are met.

Dave Anderson is Business Manager for Anderson Farms in Pine Village, Indiana. He graduated from Purdue University with an engineering degree and went to work in the automobile industry at which time he earned an MBA. Dave decided that although he enjoyed the automobile manufacturing world, he didn't want to move every few years as was required in the industry at that time. Anderson has been farming about 3,700 acres of corn and soybeans with his dad and brother since 1977. His responsibilities include managing all the bookkeeping activities for the farm, the financial planning, and managing relationships with various suppliers, agencies, and institutions.

G.W. Fuhr is a District Sales Manager with Pioneer Hi-Bred International. G.W. is located in Waverly, Iowa, and manages 21 sales representatives in a six-county area representing approximately $13 million in sales (110,000 units of corn and 130,000 units of soybeans). G.W.'s primary responsibilities involve planning and organizing for his district and his sales representatives in selling and servicing Pioneer brand products. Motivating individuals is an important element of his responsibilities.

Craig Kirkpatrick, General Manager, Formulated Products, Sabroso Company

What do I like most? This is by far the busiest I have ever been *and* this is the most fun I have ever had in my career. It's fun. I always find myself thinking what can happen, what can we do next. What I like most is the challenge of the markets and making better products available—meeting the needs of the consumer and the market.

Defining Management in Agribusiness

Successful managers feel like managers, see themselves as managers, and are both ready and willing to play the managerial role. When successful managers look in the mirror, they see a leader, a person who is willing to accept the responsibility for change and become the catalyst for action. The success-minded manager is comfortable with this managerial role, and accepts responsibility and authority as a challenge rather than as a curse. The famous educator Nicholas Murray Butler once placed managers in three classes: "the few who make things happen, the many who watch things happen, and the majority who have no idea what has happened!"

We define **management** in this text as the *art* and *science* of *successfully* pursuing desired results with the *resources available* to the organization. Several key words in this definition are italicized to stress the elements of successful management.

Art and *science* are the first two key words, and, as mentioned above, management is both an art, and a science. Because management deals largely with people, we must regard management principles as imperfect equations at best. At the same time, there are a number of such management principles and tools that will help us make better decisions in an imperfect world. Everyone cannot become the top manager for a firm, but everyone can use management principles to foster continual growth and progress toward their personal managerial potential. The third key word is *successful*. Whatever else good management is, it must be successful in meeting desired and predetermined goals or results. Managers must know where they are headed in order to achieve such success.

Finally, consider the *resources available*. Each organization possesses or has at its command a variety of resources—financial, human, facilities, equipment, patents, and so on. Successful managers coax the highest potential returns from the resources available. They recognize the difference between what should be and what is. At the same time, they know how to expand the firm's resource base when resource constraints hamper potential. They use what they have to get what they want and need, and deal in the realm of the possible.

A **manager** can be defined as that person who provides the organization with leadership and who acts as a catalyst for change—he or she is responsible

---

Sharon Rokosh, Sensory Analysis Manager, Slim-Fast Foods

What does it take to manage? I don't know for sure . . . but for me, I've always had the strong desire to manage, to grow—to be able to challenge myself. I've always found dealing with people and working with people truly interesting. I like to think that I could be influential in affecting change. I like having people trust me and know they can rely on me. During the course of my career thus far, I've seen both ends of the spectrum. I've seen managers who have the trust and reliance of the people they supervise, and I've seen the opposite of that also. Observing both those cultures, it's clear to me which is most beneficial—to the company, its objectives and the workers.

---

for management of the organization. Good managers are most effective in an environment that permits creative change. Such managers live to make things happen. Success as a manager, then, necessitates the ability to understand and be comfortable with the managerial role, to accept responsibility, and to provide leadership for change.

What does it take to be a good manager? Herman C. Krannert, an industrialist for whom the Krannert Graduate School of Management at Purdue University is named, summed up some of the necessary requirements for managers this way: "Think creatively; act courageously; be yourself; and your services will be sought by your superiors and by your associates."

Distinctive Features of Agribusiness Management

In many ways, management principles and concepts are the same for any business. Both the largest business in the country and the smallest one-person agribusiness are guided by the same general principles. The differences between managing large and small businesses, between agribusinesses and other kinds of businesses, rests in the art of applying fundamental management principles to the specific situation facing the business. In this chapter many general management principles will be applied to the unique context of agribusiness and agribusiness management. Chapter 1 outlined eight points distinguishing some of the unique elements of the agribusiness management environment (Table 2-2).

Each of these special features of the agribusiness world requires the agribusiness manager to use the principles of management in a very special way. Agribusiness is unique, and requires unique abilities and skills of its managers.

THE FOUR KEY TASKS OF AGRIBUSINESS MANAGERS

Management has been discussed and described using as many concepts as there are authors in the field. Some describe management as a division of areas of functional responsibility, such as finance, marketing, production, and personnel. Others view it as coordinating a series of resources, such as money, markets,

TABLE **2-2**   **Unique Elements of Agribusiness**

Food as a product
Biological nature of production agriculture
Seasonal nature of business
Uncertainty of the weather
Types of firms
Variety of market conditions
Rural ties
Government involvement

materials, machinery, methods, and manpower. Still others view management in terms of approaches or processes—that is, industrial engineering, organizational, and behavioral concepts. Others consider management as a series of tasks, the perspective we will take in this book.

The functions of management discussed in Chapter 1 address four key areas of the firm—any food or agribusiness firm must make decisions in the areas of marketing, finance, operations/logistics, and human resources. While these management functions are important to understand, they don't tell the whole story of what management is all about. We use the term *manager* to describe a very wide range of individuals, from chief executive officer to district sales manager. What does a manager who runs a small, independent crop consulting firm have in common with a senior executive in a multibillion-dollar food company? The answer is simple. In each of these agribusinesses, it's what these managers *do* that makes their jobs similar. Agribusiness mangers execute four principle tasks in their work. These four tasks are:

- Planning
- Organizing
- Directing
- Controlling

Each of these tasks is a part of the agribusiness manager's overall role in managing the people and events within his/her power to generate the best possible outcome for the organization. Each of these tasks of agribusiness management deals with a specific aspect of what managers actually do when they manage.

Figure 2-1 illustrates this task-oriented concept of management as a wheel. The four tasks of management are the spokes that connect the manager with the goals, objectives, and results that are desired by the organization. It is only through planning, organizing, directing, and controlling that the firm's (and manager's) goals are achieved. Overall, management can be no stronger than the weakest spoke in this wheel. Now add **motivation** as the torque, or speed, or effectiveness, with which the tasks are accomplished. Motivation provides the mo-

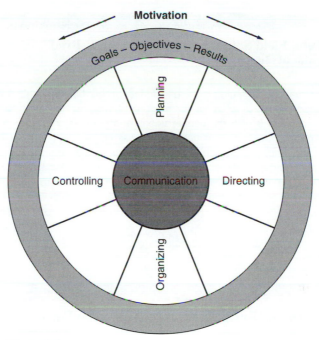

Figure 2-1
The Wheel of Management.

tion by which the wheel either moves forward or reverses, but it is not another task. Strong motivation results in speedy, efficient, successful, forward-moving management, while a lack of motivation can result in a discouraging reversal. The axle on which the entire wheel of management turns is **communication.** Again, this is not another task; however, without effective and timely communication, the wheel of management soon begins to wobble and squeak.

In the sections that follow, each of these four tasks will be discussed. In addition, interviews with our four food and agribusiness managers have been included to help illustrate the importance and nature of these key management tasks. As one will find in the interviews included in the chapter, a manager in one situation may be heavily involved in planning activities and have little or nothing to do with controlling or directing. Another manager's job might be heavily focused on the directing task, while still others are involved in all four tasks. Regardless of which tasks are prevalent in one's job, it is critical for agribusiness managers to understand what goes on in each of the four areas—planning, organizing, directing, and controlling—since someone, somewhere in the organization must undertake these tasks to get a product or service to market. Managers that fully grasp the total picture and where *they* fit in will have greater success at carrying out their tasks effectively.

G.W. Fuhr, Field Sales Manager, Pioneer Hi-Bred International, Inc.

My favorite part of managing is the people interaction. Inherently, I'm more of a people person so that's what I enjoy most. It's funny, but my most challenging task as a manager is also the people side of the business. Anytime you lead a group of people, there will be issues that you need to deal with—"people issues" are what I call them. These might be the nagging personality hurdles, or the quantity of time required by certain individuals in some circumstances, etc. On a day-to-day basis, managing the emotional issues and attitudes is tough even when you have goals, because people's past experiences and capabilities can make it difficult to deal with all those things and still get people to focus on a goal and accomplish it.

Dave Anderson, Business Manager, Anderson Farms

As Business Manager for this operation, my biggest responsibilities are bookkeeping for the farm, the financial planning, and supplying mountains of information to bankers, crop insurance companies, landlords, and the government. I spend a lot of time on spreadsheets and with word processing as well as with e-mail.

Sharon Rokosh, Sensory Analysis Manager, Slim-Fast Foods

My primary job at Slim-Fast focuses on designing and implementing a sensory analysis program, which involves the analysis of all products before they leave the plant as well as evaluation of all the raw materials coming into the facility prior to their use. In terms of management, I am more heavily into directing and controlling tasks. I love the challenges we have at our facility—that includes those challenges presented in terms of dealing with, directing, and controlling these programs and the employees that facilitate them.

PLANNING

Planning can be defined as forward thinking about courses of action based on a full understanding of all factors involved and directed at specific goals and performance objectives. Let's examine this definition in more detail. The first segment focuses on *forward thinking*, which means just that, looking ahead. This is not a

Planning efforts should be forward thinking and directed toward specific goals.
Photo courtesy of USDA.

forecast but an action-oriented statement—thinking about the future. The second element concerns *courses of action,* which implies developing alternatives or methods of accomplishing specific goals and objectives, *based on a full understanding of all factors involved.* Here is where the facts and consequences of the various factors affecting the alternative courses of action are weighed. Finally, there is the most important part of all, as well as the most neglected: *directed at specific goals and objectives,* or, in other words, focused on some end point, some target.

Types of Planning

The planning task represents the preparation of the agribusiness firm for future business conditions. Since the future can be defined to include various time periods that are uncertain, the planning task can take a number of forms. The three types of planning discussed in this chapter are strategic, tactical, and contingency.

Strategic   **Strategic planning** is focused on developing courses of action for the longer term. Long-term may be two or three years for a very small agribusiness, while a major corporate organization may be looking at a 20-year (or longer) time horizon. Strategic plans tackle the broadest elements of an agribusiness firm's strategy: What countries will we operate in? What businesses will we be involved in? What plants will we build? However, it is important to note that the CEO of a major food company will have a strategic plan, but so may a small food processor, and the sales manager for a swine genetics company. It is the long-term time horizon that distinguishes the strategic planning activity.

Any planning activity is focused on achieving some goal, on moving the firm in some direction. Therefore, in many strategic planning situations, one of the first steps is the development of a mission statement or a statement of vision. A **mission statement** helps to spell out this direction for a firm and describes the destination. When developing a mission for a company or even for a department, management must think strategically about the firm's business in a future time period. A mission statement usually describes who the company is, what they do, and where they are headed—in very clear and concise fashion. In many cases, it has three parts:

- Key markets (who we serve)
- Contribution (what we do)
- Distinction (how we do it differently)

TABLE **2-3** | **Examples of Firm Mission Statements**

Pioneer Hi-Bred International, Inc.

Our mission is to provide products and services that increase the efficiency and profitability of the world's farmers.

Our core business is the broad application of the science of genetics.

We will ensure the growth of our core business and develop new opportunities, which enhance the core business.

Ben & Jerry's Homemade, Inc.

Ben & Jerry's is dedicated to the creation and demonstration of a new corporate concept of linked prosperity. Our mission consists of three interrelated parts.

Underlying the mission is the determination to seek new and creative ways of addressing all three parts, while holding a deep respect for individuals inside and outside the company, and for the communities of which they are a part.

Product: To make, distribute, and sell the finest quality all-natural ice cream and related products in a wide variety of innovative flavors made from Vermont dairy products.

Economic: To operate the Company on a sound financial basis of profitable growth, increasing value for our shareholders and creating career opportunities and financial rewards for our employees.

Social: To operate the Company in a way that actively recognizes the central role that business plays in the structure of society by initiating innovative ways to improve the quality of life of a broad community: local, national, and international.

Some examples of firm mission statements are shown in Table 2-3.

A mission statement should be unique to a firm. Generically worded statements that apply to any firm are not useful. Such generic statements do not paint a mental picture of where the firm is destined, and offer no guidance on which activities to pursue and what strategies to follow. In contrast, a good strategic mission statement must be pointed. It must have specific relevance to the company in both its direction and destination. Many agribusiness companies have found that in addition to company-stated missions, teams and working groups supervised by various levels of managers also benefit by developing a "team" mission. Team members can benefit from being involved in developing their own working group's mission. To inspire those involved, this type of mission statement should be expressed in language that attracts attention, creates a vivid image, and provokes emotion and excitement.

Tactical   **Tactical planning** includes short-term plans that are consistent with the strategic plan; tactical plans are a crucial part of implementing the agribusiness firm's strategic plan. Tactical planning can occur at any level in an organization, and is distinguished by the shorter term time horizon. Where strategic planning is focused on what we do in 3 years (or 5 years, or 20 years), tactical planning is focused on what we do tomorrow (or next month, or next year). For example, a chemical company's strategic plan could include a goal of increasing its market share by 20 percent by using a strategy of expanding its geographic

G.W. Fuhr, Field Sales Manager, Pioneer Hi-Bred International, Inc.

As a manager, the biggest thing I do is strategic planning . . . our distribution system is set up with sales agents . . . distributorships. It's my role as a manager to help keep the mission of the company and of the group in front of the people I supervise. My maintaining the big picture helps them understand where we need to go. It is interesting how it works—my keeping the focus on our mission helps them achieve their goals more effectively.

market. The firm's tactical plan may focus on how to increase sales in specific geographic regions that have less competition, and will likely include specific action steps for getting this done.

Contingency   **Contingency planning** is the development of alternative plans for various possible business conditions. It is part of the strategic and tactical planning process for a firm. Hence, the plan that is implemented is contingent on the business conditions that occur. In general, the contingency plan provides guidance when things go differently than expected. Some contingency planning is conducted to prepare for possible crises that may occur. For example, the seed varieties provided by a seed company to a particular area of the country will usually be determined by the expectation of typical weather during the planting season. However, a prolonged wet period during the planting season would cause the seed company to implement a contingency plan based on delayed planting, which may result in providing an additional supply of short-season varieties. On the other hand, a food company may need a contingency plan for the best-case scenario. What do they do if demand for their new, all natural frozen entrée grows at 10 times the initial forecast? How will they handle the extra

G.W. Fuhr, Field Sales Manager, Pioneer Hi-Bred International, Inc.

I would say a big part of my job is helping sales representatives focus on what their job needs to be. Especially when things get busy, really fast and furious in this business during the spring and fall, it's easy for sales reps to focus more on tasks. If I can help them keep in mind the big picture, I can help them deliver more value to the customer. This may mean I help out more with strategic, tactical, and contingency plans. If I can help my sales reps keep these in mind, the bottom line is it helps all of us better position the company and its products positively with customers.

production and distribution? As pointed out here, outcomes that are both bad and good can create the need for a contingency plan.

Levels of Planning

Table 2-4 illustrates the levels at which different types of planning occur. Note that as the level of planning moves from the chief executive to the line worker, several changes occur. At top management levels, plans are strategic and have a tendency toward flexibility, are longer range, are usually written, are more complex, and are broader in nature. At the lower levels, plans are specific in nature, are for immediate action, are usually unwritten, and tend toward simplicity. Although plans tend to be unwritten at the lower levels, almost all managers benefit from having written plans. Upon writing down a plan, the plan becomes more focused and any inconsistencies or gaps are likely to be made visible. Written plans tend to organize and consolidate thoughts, are easier to communicate, and provide a source for further reference.

Depending on the structure of the firm, another reason the nature of planning changes from the top management level to lower levels is the allocation of resources. Top executives develop strategic plans that generally add or subtract resources from the agribusiness, while those plans made at the lower levels are operational and generally relate to using the existing resources in the most efficient manner. Table 2-4 illustrates the need for all these plans to mesh in a consistent manner so that the long-range goals of the organization can be achieved. The challenge is to help those at all levels of the organization understand the long-range goals and how the short-range performance objectives are intermediate steps for accomplishing those goals.

The Planning Process

There are six steps in the planning process. These six steps provide structure to the process and are designed to provide management with as much information as is available when developing the strategic or tactical plan. These steps follow.

1. Gather facts and information that have a bearing on the situation.
2. Analyze what the situation is and what problems are involved.

TABLE 2-4 | Levels and Nature of Planning in the Agribusiness

Strategic level	Tactical level	Day-to-day level
Top management	**Middle management**	**Line employees**
Very flexible	Somewhat flexible	Inflexible
Long-term	Intermediate-term	Immediate
Written analyses	Written reports	Unwritten
Complex, detailed	Less detail, outlined	Simple
Broad	General	Very specific

3. Forecast future developments.
4. Set performance objectives, the benchmarks for achieving strategic goals.
5. Develop alternative courses of action and select those that are most suitable.
6. Develop a means of evaluating progress, and readjust the plan as the process unfolds.

Step 1: Gather Facts   **Gathering facts** and information is the first step of the planning process, even though information gathering is also a constant, recurring part of the process. Its place as a first step is easily justified, since adequate information must be available to formulate or synthesize a problem or opportunity. Fact gathering is subdivided into two parts: gathering sufficient information to identify the need for a plan in the first place, and systematic gathering of specific facts needed to make the plan work once it has been developed. Because of the difficulty of gathering facts, some managers tend to skip or minimize this step of the planning process and resort to the "seat of the pants" or "gut-feel" philosophy, which reduces the likelihood of success. At the same time, a manager should not become so engrossed in fact gathering that inaction results.

Step 2: Analyze the Facts   The groundwork for developing a sound plan is provided during the process of **analyzing facts.** This process answers such questions as "Where are we?" and "How did we get here?" It helps pinpoint existing problems and opportunities, and provides the framework upon which to base successful decisions. An analysis of facts will prevent mistakes and allow for the most efficient use of the organization's resources.

Step 3: Forecast Change   **Forecasting change** is another key element of good planning. It has been said that the ability to determine what the future holds is the highest form of management skill. As managers ascend the organizational ladder, the demands on their abilities in this area steadily increase. The broader, more complex, and more long-range a plan is, the more difficult it is to foresee results accurately. As with all of the steps of planning, forecasting is interrelated with the other steps and it is a logical extension of analysis into a future time setting. Some say, "no one can predict the future in our business." While no one can be expected to predict accurately all future developments, this is hardly a good reason for not attempting to forecast the future at all. Many failures in forecasting result from sloppy, ambiguous, generalized thinking. Forecasting change is not a guessing game; it is part of a disciplined approach to planning.

Step 4: Set Goals/Performance Objectives   The development of **goals** and/or **performance objectives** is the next step in the planning process. **Goals** are the specific quantitative or qualitative aims of the company or business group that provide direction and standards one can use to measure performance. Top management, boards of directors, and/or chief executives often develop these goals which help to bring focus and specificity to the organization's mission (see Table 2-4). The mission statement is the target toward which goals are aimed, and goals are the targets toward which performance objectives are aimed. Well-stated goals should:

1. Provide guides for the performance objectives and results of each unit or person.
2. Allow appraisal of the results contributed by each unit or person.
3. Contribute to successful overall organizational performance.

Performance objectives are then set for specific units and/or individuals. They provide the performance targets at the unit and/or individual levels that are needed to accomplish the broader, longer-range strategic goals. Performance objectives are usually set for shorter time periods than strategic goals and usually are defined by measurable results. Note that all these processes are going on continuously during the planning process. Performance objectives cannot be set in a vacuum, but must be related to the attainable. Therefore, performance objectives must be a consequence of the gathering of, and analysis of, information and facts. They should be clearly aligned with organizational goals.

Some management specialists consider setting goals or performance objectives to be the first step in the planning process. In fact, when we discussed strategic planning, we started the strategic planning process by developing vision and mission statements. In a way, these individuals are correct, but in the approach to planning presented here, accumulating and evaluating information occurs before goal setting and is, in fact, part of the process of setting performance objectives. In practice, many of these steps occur simultaneously—data may be used to develop an initial set of goals, which may lead to a need for more data, which requires refining the goals, and so on.

Step 5: Develop Alternatives   After the performance objectives have been set, agribusiness managers must explore different ways of getting wherever they want to go by **developing alternative** courses of action. Here again, the relationship between performance objectives and results can be seen. The results achieved depend upon the alternative activities selected to meet the objectives. Alternatives must be weighed, evaluated, and tested in the light of the agribusiness' resources.

Imagination is crucial, since new ways and/or new paths may be the key to success. It is important in this step to be creative, yet practical, in generating alternatives. The conditions surrounding each decision must be carefully considered. For example, a firm might believe that equipping its sales representatives with new laptop computers and training them in the use of a new database management package might be the best alternative to improve the productivity of the sales force. However, the firm's budget situation may mean that only one-third of the sales representatives can get new computers. So, the firm may need to look for other (low cost) options to boost sales performance to complement the use of computers and new software.

Step 6: Evaluate Results   Management specialists have found surveillance and monitoring of progress to be a high priority in planning. Carefully reviewing, assessing, or **evaluating results** shows whether the plan is on course and allows both the analysis of new information and the discovery of new opportunities. Evaluation cannot be left to chance. It must be incorporated into the planning process, since a plan is only good so long as the situation remains unchanged. From evaluation one can tell whether results matched performance objectives and where the results fell short or overshot objectives. Evaluation also points out weaknesses in plans and programs so that those portions that are ineffective can be changed. In a fast-changing world, continuous evaluation is essential to planning success.

ORGANIZING

People working effectively; people working toward accomplishing the company's goals—these are objectives of every manager, no matter what industry, field, or organization. There has to be a structure in place to make it possible for people to work effectively toward accomplishing goals. Providing that structure or framework in which to operate is the purpose of the management task of organizing. **Organizing** represents the systematic classification and grouping of human and other resources in a manner that is consistent with the firm's goals. The organizing process is important at each level within a company or firm. And, it is the manager's challenge to design an organizational structure that allows employees to accomplish their own work well, while at the same time accomplishing the goals and objectives of the organization.

The organizing task occurs continuously throughout the life of the firm. Done effectively, this task helps management to establish accountability for the results achieved; it prevents "buck-passing" and confusion as to who is responsible; and it details the nature and degree of authority that is given to each person as the activities of the firm are accomplished. This task is especially important in today's business environment in which many firms are frequently restructuring their operations. This restructuring may be the result of efforts to improve efficiency, reduce costs, or as the result of a merger or acquisition. The manager must develop an effective organizational structure before he or she can implement the strategies needed to achieve the goals developed in the planning task. Organizing involves:

- Setting up the organizational structure
- Determining the jobs to be done
- Defining lines of authority and responsibility
- Establishing relationships within the organization

Until all employees understand their relationships to other employees and to the agribusiness as a whole, cooperation, teamwork, and coordinated action remain impossible to achieve. Just as when the members of a band do not understand their relationship to the whole, discord, rather than harmony, will result. Thus, as part of the organizing task, the agribusiness manager must see to it that each employee has a role that is clearly defined, and that when the employee accomplishes his or her goal, the goals for the organization are furthered.

Craig Kirkpatrick, General Manager, Formulated Products, Sabroso Company

Planning *and* organizing are critical to both what I do now and how I got here. I came to the company as National Sales Manager. It was then we began kicking around ideas about formulated products. I thought, OK, we have these exceptional quality certified products, now what else can we do? What can we develop for the food service, or supermarket area with these valuable products we have? How can we use our abilities as a manufacturer? With whom can we align? This has meant organizing ideas first and foremost. It has involved organizing details and resources through each level of the food marketing system—the input supply sector, production agriculture sector, and the food sector. It has worked. We have become a valuable resource to many other companies and are now sourcing products worldwide. We started Formulated Products, the area I manage, and continue to look for supply chains to plug into.

For example, the overall sales goal for a firm can be broken down to performance objectives for regions or divisions and, ultimately, individual salespeople. The planning process of a seed company is shown in Figure 2-2. All levels of the organization are involved in planning. However, as indicated earlier, the scope of that planning will become more focused as we move from the CEO to the operating level. Likewise, the scope of the goals will become more focused as we move down the organization. In terms of results, the flow is from bottom to top—if Sales Territory 1 achieves the goal of selling 500 units of soybeans at $18/unit, this helps the Soybean Division Sales Department achieve its goal, which helps the Soybean Division. In the end, accomplishment of each person's sales objective contributes directly to the accomplishment of the firm's strategic goal.

Organizational Structure

An **organizational structure** is the framework of a company. It provides the beams, braces, and supports to which the appropriate "building materials" are attached. Different jobs are connected to different parts of the whole framework. It is the formal framework by which jobs are grouped, coordinated, and further defined. Whether they are called groups, departments, or teams, coordinated communication and cooperation between work groups is essential. Consider the Carlson Seed Company, a small seed company in western Illinois. The family-run seed business has a research group headed by Michael, a plant breeder. Michael's research group is deeply involved in developing a particular early season soybean variety.

Paul heads the production group of this small company. While Paul and his people may not necessarily be interested in the specifics of what the research group is coming up with on the genetic level, it is essential they understand what development of this particular early season soybean variety means to their production efforts. Paul realizes the importance of knowing what else is going on in

Figure 2-2
Levels of Planning and Goals for Seed Company.

their business. As a savvy production manager, he realizes that getting this variety ready for sale may mean juggling their current people resources—what time of year to have extra workers available, which projects people work on. It may mean that production redirects mechanical resources, such as changing bagging processes or movement through the dryer system. It may mean devoting more acreage to producing soybean seed, which may displace production of some other product.

Even though the production department may not have anything to do with what group of genes research is working with today, it does benefit them, other departments, and the company in general to understand the organizational structure, and the goals within that structure. A working knowledge of the organizational structure is essential to managers if they are to put plans into action.

An organizational structure exists in all businesses. It might be argued that even a one-person agribusiness has an organizational structure, with one person wearing many organizational hats. Organizing, then, involves formalizing a plan to show the interrelationships of each job and each individual within the organization. Such an organizational structure clarifies who reports to whom. It helps clarify who has responsibilities for specific tasks. An organizational structure also helps clarify who has authority for specific decisions (Figure 2-3).

An **organizational chart** shows the formal organizational structure of a company. It helps capture some important ideas including division of labor, chain of command, bureaucracy, and organizational design.

Division of Labor   The **division of labor** is the manner in which jobs are broken into components and then are assigned to members or groups. The objective is to accomplish more than one group or individual could successfully and efficiently accomplish alone by delegating specific smaller tasks to many.

Chain of Command   **Chain of command** is illustrated in organizational structure by the authority-responsibility relationships or links between managers and those they supervise. This continuum exists throughout the company. The chain of command should be clear so employees know to whom they report and are accountable.

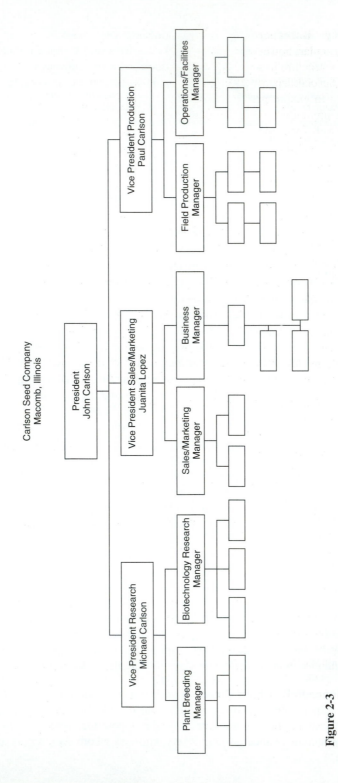

Carlson Seed Company
Macomb, Illinois

President
John Carlson

Vice President Research
Michael Carlson

Vice President Sales/Marketing
Juanita Lopez

Vice President Production
Paul Carlson

Plant Breeding
Manager

Biotechnology Research
Manager

Sales/Marketing
Manager

Business
Manager

Field Production
Manager

Operations/Facilities
Manager

Figure 2-3
Organizational Chart for Carlson Seed Company.

Bureaucracy   **Bureaucracy** is a word that has many negative connotations in today's vernacular; however, bureaucracy was developed as a highly specialized organization structure in which work is divided into specific categories and carried out by special departments. A strict set of guidelines determines the course of activities to ensure predictability and eliminate risk. A bureaucracy is a tightly run, unyielding organizational structure. This organizational structure does work well for some types of businesses. But, given agriculture's unique characteristics and the variability and unpredictability of the weather and other uncontrollable circumstances, many food and agribusinesses do not operate under this form of organizational structure, and go to lengths to take bureaucracy out of their internal processes.

DIRECTING

Directing is guiding the efforts of others toward achieving a common goal. It is accomplished by:

- Selecting, allocating, and training personnel
- Staffing positions
- Assigning duties and responsibilities
- Establishing the results to be achieved
- Creating the desire for success
- Seeing that the job is done and done properly

Directing, then, involves leading, supervising, motivating, delegating, and evaluating those whom you manage. Managers are directing when they see to it that the efforts of each individual are focused on accomplishing the common goals of the organization. Leading is at the very heart of the management process and is founded on a good organizational plan or structure that provides for responsibility, authority, and evaluation.

The task of direction may be described as the task of making the organization take on life, of creating the conditions that make for interest in the job, vigor of action, imaginative thinking, and continuous teamwork. This goal is one that cannot be reached by magic formulas. Its achievement rests in large measure upon the qualities of leadership exhibited by the manager.

The Manager as Leader

In the past, leadership was treated as simply a part of management, but today firms view management and leadership differently. The manager's duties include efforts to perform the various management tasks and functions, whereas, a leader influences the attitudes and behavior of followers and motivates them to do their best work. The increasing importance of leadership in today's business environment requires that we explore it in more detail.

Leadership involves not only providing instructions on how to complete a task and information on the desired results, but it also includes providing incentives to complete that task correctly and in a timely manner. An important aspect

G.W. Fuhr, Field Sales Manager, Pioneer Hi-Bred International, Inc.

Participative leadership sums up my personal style or preference for management type. I really feel that every effort goes more productively if you have buy-in from your group. I try to bring my folks in to the decision-making process and the planning from the beginning. Then when we all see the same vision, we can all work simultaneously toward the same goals, each individual on the team contributing their best characteristics toward that goal. The participative leadership helps me ignite my people.

of leadership is the motivation of employees. Part of motivation is creating a vision for the organization or the group that the members can get excited about. Another method used to motivate employees is to delegate authority by assigning employees more responsibility, which encourages employees to take more pride in their jobs and raise their self-esteem. Of course, such assignments should be combined with monetary incentives that are tied to the organizations' goals and the individuals' performance objectives as determined in the planning task. Leadership is also the process by which the manager attempts to unleash each person's individual potential, once again, as a contribution toward organizational success. Leaders recognize the results of a person's activities count for more than the activities themselves.

Finally, for managers to be effective leaders they need initiative, which is the willingness to take action. Some managers who recognize the need to enact changes are unwilling to take action because making changes takes more effort than living with the status quo.

Other Directing Roles

The manager must also develop and interpret programs, plans, policies, procedures, and practices within the organization. Every department, division, or organization must have a court of last resort. Human nature is such that even those with the best intentions may differ in their interpretations of facts or information. In such cases, it is the manager's task to deliver this interpretation—to become, in effect, the Supreme Court for those who are being supervised.

Part of the manager's directing task involves the encouragement of employees' growth as individuals. An individual who is growing and developing is becoming more productive for the agribusiness as well as for himself or herself. Special provisions for growth in responsibility should be built into the opportunities presented by any job, and job challenge must be maintained for the individual. Promotion policies should be clear, and all concerned should understand them. Each employee should be appraised and evaluated for the unique contributions that the employee can make to the organization.

Sharon Rokosh, Sensory Analysis Manager, Slim-Fast Foods

I really try hard to be connected with my job objectives and my people at the same time. I am genuinely very interested in my work, so I do know a lot of what's going on with people at my work. It's a rapport that I develop with them. As a manager, I feel it is key to have that connection between the projects and programs I oversee as well as the people. I really like to give the people I manage some tools, some more ownership of things. It empowers them to be part of a team; they are an integral part of attaining a goal instead of just being the people who do the work.

In the food industry, we are dealing with some very challenging working conditions. It is an aseptic process which is much more complicated than some processes in the straight manufacturing area. You come to find out that most people want to do their best job. These are factory workers, and they want to do a good job. They are good people and they need the right tools to enable them to do that. I like to be in the role of helping make that happen well. There are not a lot of managers—at least at my level—that seem to put both of those notions together.

The good manager encourages employees to discover hidden talents by stimulating them through varied assignments that offer continually increasing challenges and opportunities.

A manager should not hesitate to provide encouragement for more formalized training in courses, workshops, seminars, educational meetings, and the like. Formal education can benefit both the organization and the employee. A good manager maintains a regular schedule of interaction with employees. Application of this technique will acquaint the manager with the problems and personalities of employees. Such contacts also enable the manager to develop an awareness of each employee's strengths and weaknesses. This, in turn, will help the employee to develop a confidence in "the boss." Both the manager and the employee will be sensitive to issues as they unfold, helping to reduce the chance of unpleasant surprises.

Shaping the Work Climate

It is mainly through the ability to create a work climate for success that a manager can improve employee efforts to the point where employees approach their potential. Without a proper climate, none of the skills and principles of management can flower and bear fruit. Principles used to create the right work climate include:

1. Set a good example.
2. Conscientiously seek participation.
3. Be goals- and results-centered.
4. Give credit and blame as needed: credit in public, blame in private.
5. Be fair, consistent, and honest.
6. Inspire confidence and lend encouragement.

Dave Anderson, Business Manager, Anderson Farms

I really enjoy managing and working with our employees. Having part-time help probably takes more effort than just having maybe two full-time employees, but I really like doing it this way. My management style requires that my employees have a really good attitude and that they be good communicators. It's such an important skill and one I really look for in hiring help. Even if they don't have all the hands-on skills they might need to accomplish a specific job, if they can listen, and understand what I am telling them well enough to ask some good questions, we're in business.

When I say communicators, I don't mean big talkers. I am talking about skillful use of language within our business. This is a dangerous, hazardous business. All you have to do is to look at the Bureau of Labor Statistics to see just how hazardous. If we articulate that we want a certain field tilled, I want someone sharp enough to have the skills to listen to the directions, understand, and carry them out.

Good managers know how to use these principles to create a productive working climate. One key to a productive work climate is the free flow of communication. The agribusiness manager is responsible for designing and implementing the communications process within a given area of responsibility. Free flow of communication means communications must flow not only downward (from management to subordinates), but upward (from subordinates to managers) and laterally (at the same level) to be effective. Too often managers depend almost exclusively on downward communications and then wonder why goals, policies, and procedures are misunderstood.

Successful communications require feedback. Feedback allows the manager to determine whether understanding has indeed occurred. It also allows the good ideas and potential contributions of each employee to be part of the mix of collective wisdom and knowledge found in the organization. The manager must provide the opportunity for this feedback and involvement through a carefully designed communications process involving committees, meetings, memos, and individual contacts.

Policies, Procedures, Practices

As managers complete the directing task, they typically develop guidelines or policies for how tasks should be completed. **Policies** are used to guide the thinking process during planning and decision making. A policy sets the boundaries within which an agribusiness employee can exert individual creativity. Policies make it unnecessary for subordinates to constantly clear plans and decisions with top management. For example, one farm equipment dealer in the Southwest instituted a policy that the general manager must approve all purchases that totaled $500 or more. The purpose of this policy was to protect the business against unexpected large cash outlays.

Sharon Rokosh, Sensory Analysis Manager, Slim-Fast Foods

For new managers, I think it is important to develop a keen sense of observation. Learn to observe, and then observe what you've learned. Observe the company culture as well as those you manage. Then, learn from that. How can you best, most effectively manage in that situation and how can you enhance what you have? Take what you learn and apply it within reason to your own area. You can't be afraid to ask questions, tough questions, of yourself, your workers, and your company.

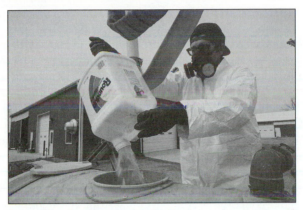

Controlling involves careful monitoring and evaluation of a firm's business and operating activities.
Photo courtesy of USDA.

Policies are best adapted to recurring problems in areas that are vital to achieving the long-range goals of the agribusiness. However, policies are not goals, although they are closely tied to goals. Because they are not goals, policies should never be used to restrict managers as they make decisions about long-range, complex problem situations.

A **procedure** is a step-by-step guide to implement a policy for a specific activity or function. In many cases, there is a definite need to set out just such a precise course of action. A procedure should not, in most cases, be applied to complex tasks of a long-range nature.

When the aforementioned farm equipment dealer sought to implement its new purchasing policy, the procedure involved called for an employee to fill out a requisition form, submit it to the general manager for approval, then send it to the purchasing director. Procedures work best when they are applied to routine and recurring tasks of a relatively simple nature that require control. Both policies and procedures are of tremendous value to the new or newly promoted employee who is learning on the job. They also ensure uniform performance by all employees and prevent unauthorized actions.

Practices represent what is actually done in the agribusiness, and they may conflict with policies and procedures. Managers have to be sure that policies make sense, are relevant, and are enforced, in order for them to become widespread practices. If as a routine practice employees of the farm equipment dealer ignore the policy and procedure regarding the limits on spending, the firm may find itself in a serious cash bind. A course of action that is established on a recurring basis becomes a practice, often by tradition (or habit more than anything else). The status of practices can become as important as that of either policies or procedures, and even more difficult to change, so the agribusiness manager must ensure that practices coincide with policies and procedures.

CONTROLLING

The **controlling** task represents the monitoring and evaluation of activities. To evaluate activities, managers should measure performance and compare it with the standards and expectations they set. In essence, the controlling task assesses whether the goals and performance objectives developed within the planning task are achieved. A detailed discussion of various financial control programs is provided in Chapter 15.

Control in management includes an information system that monitors plans and processes to be sure that they are meeting predetermined goals, and sounds a warning when necessary so that remedial action can be taken. If all people were perfect and their work was without flaw, there would be no need for controls. Everything would come out according to plan. But all people make mistakes; they forget, they fail to take action, they lose their tempers, and they behave, in short, like normal human beings. In addition, even when people do exactly what they are supposed to do, the market, the competition, the weather, the equipment, etc. may not cooperate. In the process, what actually happens may not be what was expected.

Control is complementary to the other three tasks of management. It compensates for the misjudgments, the unexpected, and the impact of change. Proper controls offer the organization the necessary information and time to correct programs and plans that have gone astray. They should also indicate the means of correcting deficiencies. Control requires meaningful information and knowledge, and not that which is outmoded or not germane to meeting organizational goals. Much valuable time can be squandered or wasted by a control program unless a careful check of its real value is made periodically.

When the need for control information is not real, frustration, disrespect, and inaccuracy are likely to result. Employees cannot respect a control program that is not used in a meaningful manner by management. Management can consume valuable time in reviewing useless control data, or it can fail to separate the relevant from the irrelevant, with poor performance as the consequence.

All business activities produce results. These results should be matched carefully to well-conceived objectives that aim toward organizational goals. One

Dave Anderson, Business Manager, Anderson Farms

What I like most about my job is that I can feel a sense of accomplishment in a couple of different ways. I feel that sense of accomplishment in a business sense when I finish a good financial plan that works. I feel that sense in an agricultural sense when I have the opportunity to be outside in the spring and fall and enjoy life around me. What I dislike most about my job is the tedious, misuse of my time by agencies, companies, and others who waste my time by requiring too much paperwork. I find that most annoying. Those things take precious time away from pursuits that could be generating money or other personal reward.

of the most important purposes of control is to evaluate the progress being made toward organizational goals. A good example of this type of control is found in the budget or forecast comparison. Is one over or under the budget? Are forecasts of sales and expenses in line with predictions? The information or control required in all areas must be based on predetermined, written goals. Only then can the agribusiness manager tell whether the reality matches the plans through which success is sought.

Management by Exception

An important management technique called **management by exception** holds the basic premise that managers should not spend time on management areas that are progressing according to plan; rather, they should concentrate on those areas in which everything is not progressing as it should. If sales are in line with the forecast, the manager can devote the bulk of his or her efforts to other areas, such as production, personnel, costs, and expansion. Consider this example of an actual agribusiness company.

The sales forecasts for the Phoenix Fertilizer Company are given in Figure 2-4. Sales are broken down into two component parts, dry fertilizer and liquid fertilizer. Note that the sales forecast is basically the predetermined course of action, or the objective, of the marketing department. These objectives are essential in keeping the sales force focused and in allowing key executives to make remedial decisions related to sales performance, should the agribusiness deviate from its established objectives.

This annual sales forecast for fertilizer was developed by combining all the individual salespeople's estimates of sales for the coming year. Then the marketing manager met with other key individuals, such as the executive vice president, production vice president, new products manager, and purchasing director, to determine whether anticipated input (costs) matched anticipated output (revenue). The committee determined that the control program for sales would be reported on a component basis—that is, by dry and liquid fertilizer sales—and on a monthly basis, with a quarterly review of other specific sales items to be

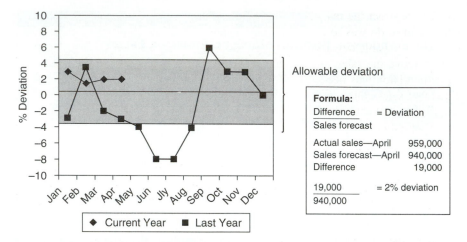

Figure 2-4

Management by Exception Graph for Phoenix Fertilizer Company.

made as a supplement to the overall sales control program. The committee also decided to report the sales in tons because this was the most meaningful measure for all concerned.

The executive vice president and controller agreed that the information needed could be secured from company records, that is, the previous year's sales forecast, past actual sales, and current customer sales invoicing. The office manager would prepare the monthly control information report, and it would be shared with all employees of the company, as well as with the company's banker. The sales manager was given the principal responsibility for accomplishing the objectives of the control program.

Joseph Drake, the sales manager of Phoenix Fertilizer Company, decided that he would use a new approach for the control report: management by exception graphs instead of raw data. He had decided on this reporting technique for the following reasons:

1. It allows one to ignore those management areas where performance is on target, and to concentrate on problem areas instead.
2. It allows one to see at a glance the whole situation, and the comparative data are illustrated clearly as a graph.
3. It is simple and easy to prepare and is easily understood by anyone, even those unskilled in financial management.
4. It allows for a clearer interpretation of the many interrelated factors than raw data can provide.

To construct the management by exception graph, the first thing the committee had to do was to establish an allowable deviation. In terms of the sales forecast, allowable deviation meant the amount by which actual sales could fall or rise above the sales forecast before management had to determine whether remedial action was needed. In the Phoenix Fertilizer Company, the team decided that sales could deviate from the forecast by 4 percent in either direction before remedial action would be needed (Figure 2-4). A deviation of more than 4 percent either below or above the sales forecast would affect such areas as inventory of raw and finished product, production scheduling, capital and cash positions, labor needs, advertising and promotion programs, and pricing policy. Each of these areas would have to be reviewed by the management team if sales deviated from the forecast by more than the allowable range. This, of course, is the major purpose of a control program: to sound the alarm for further study and action.

Once the allowable deviation was established, all that remained was designing the management by exception graph (Figure 2-4). At the bottom is Phoenix's sales forecast for the current year. A graph such as the one in Figure 2-4 could also show several different but related factors, such as total sales, last year's sales, dry sales, and liquid sales, through the use of different symbols or lines. For the sake of simplicity in illustrating the principles involved, only the deviation for total current and last year's sales are shown. And, last year's sales are figured as a deviation from last year's sales forecast, not from the current sales forecast. The shaded area is the allowable deviation from the sales forecast or goals. To the right of the graph is the formula and an example of how the deviations are calculated.

Joseph Drake was aware of the fact that last year several of Phoenix's competitors had suddenly lowered their prices in April, and that because Phoenix had not been alert enough, it had taken them several months to meet this challenge. As a result, many thousands of dollars of sales and consequent profits had been lost. Joseph was determined that this would not happen again.

Phoenix had been forced to lower its prices substantially in order to recover lost customers and was now operating on a narrowing gross margin as cost increased, largely due to higher natural gas prices. Joseph knew that other manufacturers were feeling the same pinch, but each had to fight for market share to meet overhead costs.

To illustrate the value of the management by exception approach assume that September sales were up by 6 percent, or 2 percent over the allowable deviation. The marketing department would have immediately taken a careful survey of competitors' prices. Perhaps some major competitors were trying to adjust prices upward to meet increased cost. If this were true, Phoenix would have several alternatives. Phoenix could increase its price, thereby widening its gross margin and profits but maintaining the same relative market share. Or it could hold prices as competitors increased theirs, to increase the volume of sales and market share, thereby causing increased turnover and increased profits. Plant capacity, growth objectives, available funds, and dealer reaction would all play a part in the final decision. Good financial records and proper analysis of them would also be essential to the final decision.

Many of the tools that the agribusiness manager would need in order to obtain the information and knowledge so necessary to sound decision making will be developed in later chapters, but the final decision will still be somewhat subjective in nature. It will be made by an individual or a team of individuals and will involve the ability to weave the facts together into a complex whole using innate knowledge and experience, using analytic tools where possible, making the best educated decision, and being willing to take the responsibility for the decision.

MAJOR AREAS OF MANAGEMENT RESPONSIBILITIES

Each of the key tasks of management—planning, organizing, directing, and controlling—is used in managing the four major functions of an agribusiness. The management functions are implemented through the use of the various skills, principles, and tools that have become part of the professional agribusiness manager's knowledge and ability. To be successful, the agribusiness manager must be able to execute the four tasks for each of the four basic functions of the agribusiness; that is, marketing and selling, financial management and planning, production and operations, and personnel or human resources. It is important to note that managing the marketing function of an agribusiness involves planning, organizing, directing, and controlling. The same holds for the other three functions. The balance of this book is built around the four basic functions of the agribusiness, and is designed to help students acquire the know-how of the professional agribusiness manager.

SUMMARY

Management is the process of achieving desired results with the resources available. A key to successful management is accepting responsibility for leadership and making business decisions through the skillful application of management principles. Management of the agricultural business is unique because of the biological nature of production, the importance of food to people, the seasonality of food and agricultural markets, and the perishability of agricultural products, among other reasons.

The management process is often divided into four tasks: planning, organizing, directing, and controlling. Planning is determining a course of action to accomplish stated goals; organizing is fitting people and resources together in the most effective way; directing pertains to supervising and motivating people; and the controlling task monitors performance and makes adjustments to stay on purpose. Each of these tasks is a necessary ingredient for accomplishing the established organizational goals.

DISCUSSION QUESTIONS

1. Define management in your own words. What are the differences in the four tasks of management and the four functions of an agribusiness?
2. Pick any food or agribusiness firm. Compare this firm and the market it serves to the list of distinctive features of the food and agribusiness markets. Which of these features seem to be most important for the firm you have chosen?
3. How and why does planning change as one progresses up the organizational ladder?
4. Describe the steps in the planning process. Using these steps, develop a plan for obtaining a summer internship with a food or agribusiness firm.
5. What are the most important components of the agribusiness manager's role as director?
6. What are the advantages of using the management by exception approach to control programs?
7. Assume that Phoenix Fertilizer Company reported actual sales of 1,006 and 1,083 thousand tons for May and June of the current year. Using these data, add two months to the Phoenix management by exception graph. Interpret the deviations for these two months. What actions might be suggested by these results?
8. Reread the comments made by our four agribusiness managers throughout this chapter. What similarities do you see in their jobs? What are the key differences?

CASE STUDY—**Hart Cherry Cooperative**

The Hart Cherry Cooperative was organized two years ago to pit and freeze member-farmers' cherries. The cooperative is experiencing difficulty in keeping grower-members' cherries separate. Most of the cherries are harvested mechanically by shaking them from the trees. They are placed in pallet tanks by the grower and brought into the plant for processing. The cooperative owns all the pallet tanks. When the grower unloads the pallet tanks on a concrete pad, the cherries have to be cooled with running cold water until they are ready for processing. There is considerable variation in quality among the loads of cherries brought in by members. Some loads have large quantities of small twigs and leaves in them, some are rotten or soft, and some have other undesirable qualities.

This is the first year that the cooperative has owned all the pallet tanks. The cooperative board decided that it was best for the cooperative to own them, since growers had been continually taking other growers' pallet tanks whenever their own were unavailable. The policy of the cooperative board and management is that each grower's cherries must be identified so that growers can be paid separately, on the basis of the quantity and quality of their product.

The practice as it had developed this year was for the members to unload their pallets onto the pad. Each member was then supposed to put a name card on each pallet. The problem was that sometimes the growers' new, or inexperienced, or careless truck drivers were failing to put cards on, cards were falling off, or sometimes two or more cards were on the same pallet.

Accurate identification of each grower's cherry crop is a challenge for Hart Cherry Cooperative.

1. Develop a procedure that will help solve the problem by ensuring that each grower's cherries are properly identified.
2. Develop a plan to ensure the procedure you develop will be carried out in practice. Include in your plan how the procedure will be communicated to both employees and grower-members. Include in your plan steps to receive feedback on the procedure from both employees and grower-members.

REFERENCES AND ADDITIONAL READINGS

Krannert, Herman C. 1966. "Krannert on Management." Krannert Graduate School of Industrial Administration. Purdue University. p. 21.

3

ECONOMICS FOR AGRIBUSINESS MANAGERS

- Understand the basic principles of a market-oriented, capitalistic economy
- Explain the reasons profits can exist in a market-oriented, capitalistic economy
- Relate the concept of opportunity cost to economic profit
- Identify three important marginal principles of economics
- Understand how supply and demand interact to determine market price
- Define market equilibrium and what causes it to shift
- Explain the concept of elasticity, its relationship to supply and demand, and how agribusiness managers may use it
- Explain the concepts of form, time, place, and possession utility
- Understand the three functions of a marketing system: exchange, physical, and facilitating
- Understand the factors affecting the operational and pricing efficiency of a marketing system
- Identify sources of market risk, and explain some of the key tools for managing market risk

INTRODUCTION

Agribusiness managers must understand the economics of the world in which they operate. Not only are economic principles useful in predicting business trends, but also those same principles serve as the basis for many management decisions. In Chapters 1 and 2, we outlined the importance of the agribusiness sector and discussed what it means to be a manager. It is also important to understand that agribusiness firms operate within a broader economic environment.

The "economic problem" addressed within that economic environment is that of limited economic resources versus the unlimited wants of society. Providing a decision framework for understanding how scarce resources are allocated is the focus of the social science of economics.

Economics is defined as the study of how scarce resources (land, labor, capital, and management) are combined to meet the needs of people in a society of unlimited wants. These four scarce resources are often referred to as the **factors of production.** When the four factors of production or scarce resources are outlined, economists often refer to "returns to these resources." For example, labor is paid a wage, while management typically receives a salary. Likewise, returns to land are often referred to as rent and returns to capital are represented by interest payments. The way market forces work to determine returns to these factors, and therefore how they are allocated, is at the heart of a market-oriented economy.

This chapter provides a foundation in some of the basic ideas of economics. A professional manager understands and uses the ideas of economics, both to assess the broader marketplace, and to improve the effectiveness of their decision making. In this chapter we will look at a set of economic tools and concepts managers can use for both purposes.

THE U.S. ECONOMIC SYSTEM

A Market-Oriented, Capitalistic Economy

In a free enterprise economy, consumer wants are expressed directly in the marketplace and become the basis for the allocation of scarce resources. Consumers make their wants known by "voting" with their dollars, effectively "bidding" the price up on whatever scarce items they may want. Producers or businesses constantly monitor changing consumer wants as they offer various alternatives, respond to changing prices, and make adjustments in the production process. Producers are keenly aware of the changing availability of limited and costly resources. They continually search for ways to combine resources more and more efficiently in order to attract the consumer's dollar with more desirable goods and services. This market focus is the foundation for the U.S. economic system.

When consumers want more shrimp than is available on the market, restaurant owners and food brokers bid up the price. As menu prices reflect the shortage, fewer and fewer consumers can afford shrimp. But, at the same time, the higher market price encourages Gulf Coast and Pacific Northwest fishers and canners to spend more hours harvesting and processing the now precious delicacy. And if the situation persists over a long enough period of time, fishers may even be encouraged to invest in more shrimp trawlers. Shrimp prices then begin to fall as shrimp availability is brought into line with declining consumer wants; as prices begin to fall, quantity demanded by consumers may gradually increase again.

The economy of the United States is often referred to as a capitalistic system. **Capitalism** is a system in which property is owned and controlled by private citizens. Any profits that can be generated by the use of that private property belong to the owner. This profit motive, along with the corresponding possibility of accumulating wealth, is another key element of the U.S. economic system.

Of course, no economic system is perfect. A completely free market capitalistic system is geared to react only to economic pressures, and so can cause individuals a great deal of pain while it is in the process of making adjustments. An unrestricted free enterprise system does not have an easy answer for the 55-year-old grain elevator owner who loses a lifetime investment when the railroad near the elevator is abandoned; nor does the system immediately respond to public environmental problems. These problems, and others like them, have led U.S. society to place some restrictions on the free enterprise system in order to lessen such hardships. As a country, we artificially support the prices of some farm crops through government farm programs. We impose restrictions on the use of some pesticides in order to protect the health and well-being of citizens. And, government occupational health and safety regulations have made sweeping changes in many agribusinesses. In each case the intervention occurs because society has decided the free market mechanism reacts too slowly or painfully to be acceptable to our society at large. Although some argue that such restrictions have undermined the workings of the free market system, the U.S. market-oriented, capitalistic economic system continues to transmit its feedback on consumer preferences to businesses rather quickly and effectively.

Profits in a Market-Oriented, Capitalistic Economy

In a perfectly competitive free market system, economic profits should not exist. The theory goes that many firms will be attracted into any market that is potentially profitable. Once this happens, production will increase, prices will drop, and profits will eventually fall back to zero. But in the real world profits do exist. And the possibility of earning profit is the motivating force behind most business decisions.

There are several explanations for profit in our market-oriented, capitalistic economy. First, profit is the reward for taking a risk in a business. When a private property owner commits personal resources to a business project, there are no guarantees of a return to this investment. The greater the risk involved, the greater the potential profit for successful ventures if the venture is to attract any investors. Second, profits result from the control of scarce resources. In the U.S. economic system nearly all property is owned and controlled by private citizens. When a citizen owns a resource that others want, the others will bid the price up, which generates a profit for its owner. The greater the demand for a resource, the higher its price will be, and the greater the profit reward to its owner or owners.

Third, profits exist because some people have access to information that is not widespread. Resource owners who have special knowledge, such as secret processes or formulas, can use this information exclusively and can thereby maintain significant advantages over their competition. The concept of patents and copyrights evolved as part of a formal attempt to encourage creativity by ensuring that the creator profits from his or her ideas. Fourth, profits exist simply because some businesses are managed more effectively than others. The managers of such businesses are often creative planners and thinkers whose day-to-day organizations are extremely efficient. The reward for doing the job well is usually profit.

The profit motive is the "spark plug" of a market-oriented, capitalistic economic system. The prospect of earning and keeping a profit serves as the incentive for creativity and efficiency among people. It stimulates risky ventures and drives people to develop ways of cutting costs and improving techniques, always in an effort to satisfy consumers' desires.

ECONOMICS AND RESOURCE ALLOCATION

It is important to understand that the products and services produced by any economic system represent some combination of the four factors of production—land, labor, capital, and management. There are a variety of possible combinations of these factors to produce the same output. That's why one important question in any economic system is how to select the "best" combination of these resources, both for the overall economy and for individual firms. Consequently, the study of economics includes two different areas: macroeconomics and microeconomics. Both areas are briefly outlined below.

Macroeconomics: The Big Picture

Macroeconomics focuses on the "big picture" view of our economic system. If you have taken a course in macroeconomics, you have studied topics such as national income, gross domestic product, and the prime interest rate. The Federal Reserve System, or the Fed, can have a significant impact on the economy by changing monetary policy. Monetary policy focuses on interest rates and the supply of money to the economy. Likewise, Congress can impact our economic system through fiscal policy. Fiscal policy includes government spending and taxing programs.

Agribusinesses are greatly affected by macroeconomic developments because worldwide demand and demand for various food and fiber products is constantly changing. General economic conditions are influenced dramatically by such factors as weather, government policies, and international developments. Macroeconomics is concerned with how the different elements of the total economy interact. An individual firm has relatively little impact on the total economy. However, skill at anticipating and interpreting the macroeconomic environment is critical to the success of any agribusiness manager.

For example, the monetary policy of the Fed may increase interest rates. And, interest rates have an important impact on purchases of tractors, combines, and other farm machinery. Therefore, a farm equipment manufacturer pays close attention to interest rates. Food consumption patterns are impacted by the health of the general economy. In a boom period, some food items, especially more expensive, luxury items, may sell well. However, in a recession, when incomes are lower, food purchasing habits may shift. Managers of food companies follow such developments very closely.

Microeconomics: Economics Within the Firm

Microeconomics is the application of basic economic principles to decisions within the firm. Every agribusiness faces tough questions when it comes to allocating its limited resources. Managers must decide the best way to use physical, human, and financial resources in the production and marketing of goods and services to meet customers' needs and generate a profit. Tools of economic analysis are essential to the manager who must make daily business decisions. In fact, most of the management tools developed in this book are based on fundamental economic concepts.

The successful agribusiness manager must assemble a variety of different types of information, and then use that information effectively, to make the best possible decisions for the short- and long-run financial health of the firm. A few years ago, poor judgment might mean that the firm made a little less profit that year. Today, due to the extremely competitive marketplace, a poor decision may lead to failure of the firm.

BASIC ECONOMIC CONCEPTS

Profit and opportunity cost are terms used by both accountants and economists. Both accountants and economists talk about profits, but each tends to view profits somewhat differently. The accountant looks at profits as the income that remains after all actual, measurable costs are subtracted (see Chapter 12). Economists, however, determine profit by examining alternative uses for resources within the firm.

Opportunity Cost

Opportunity cost is the income given up by not choosing the next best alternative for the use of the resources. It represents the amount that the business forfeits by not choosing an alternative course of action. Since an opportunity cost is never actually incurred, the accountant cannot measure it precisely. But economists argue that from the standpoint of making economic decisions, an accurate determination of the cost involved in choosing one alternative must include the amount that is lost by not selecting the next best alternative.

Economic Profit

Economic profit is defined as accounting profit less opportunity cost. Economic profit, then, forces an examination of alternative uses of resources. Focusing on the opportunity costs helps in analyzing alternative courses of action.

Susan Lambert owns and operates her own landscaping firm. She is 30 years old, has a college degree, and has been quite satisfied with her business. She currently has $400,000 of her own money invested in the business. Susan draws an annual salary of $35,000. Last year her records showed that the business made a profit of $75,000. Her performance in terms of economic profit can be evaluated in the following way.

Starting with her accounting profit of $75,000 and adding her $35,000 salary, we see that Susan realized a total of $110,000 from her business last year. Now Susan must consider the appropriate opportunity costs involved in her decision to operate this business. This requires determining what alternative uses there might be for Susan's **economic resources.** One important resource she contributes to the firm is her own time and talent. She figures that she could sell her business and go to work for someone else for a salary of approximately $30,000 annually, given her experience and contacts. This is an opportunity cost that takes into account Susan's time and management abilities. Then Susan must consider her own investment of $400,000 in the business. What are some alternative uses for this investment? Assume, for example, that she could: (1) put it into a savings account at 5 percent; (2) buy government bonds at 6 percent; or (3) invest in corporate bonds at 8 percent. The opportunity costs would be 8 percent of $400,000 or $32,000 each year, because that is the most profitable alternative use for Susan's money.

Thus, Susan's economic profit would be:

Net income		$ 75,000
Salary withdrawal		35,000
Total accounting profit realized by Susan		$110,000
Less opportunity cost:		
1) Other job	30,000	
2) Best investment alternative,		
$400,000 × 0.08	32,000	
Total opportunity cost		62,000
Economic profit realized by Susan		$ 48,000

This $48,000 economic profit represents the real financial reward that Susan received for risking her own resources in the business. It is sometimes called pure economic profit. As long as this amount is positive, it suggests that the decision to commit the resources to the business is a good one. But what happens when Susan has a bad year and shows an accounting profit of only $5,000, interest rates move to 10 percent, and she is offered a job for $45,000? Economic profit in this situation becomes negative ($5,000 + $35,000 − $40,000 − $45,000 = −$45,000), which indicates that, had it been pursued, the (lost) opportunity to invest her capital in corporate bonds and to take the new job would have yielded higher returns than the landscaping business.

Even though the accounting profit would still be positive, the negative economic profit clearly indicates problems. Should this trend continue for a few years, it would make sense for Susan to sell out, unless, of course, she considers the freedom and psychological rewards from running her own business important enough to offset the negative economic profit.

Economic Profit and Decision Making

Before investing sums of money in specific alternatives, managers must be able to estimate opportunity costs. This estimate helps managers decide whether any given use for their resources of time and money is the very best opportunity available. However, some shortcomings of the economic profit concept should be recognized.

First, there are many "imputed" values in economic profit that in reality are hard to estimate. In the case of Susan Lambert, what interest figure will yield the "correct" opportunity cost for the situation? For some people the 8 percent rate will apply, but other investors have the know-how and time to research other possibilities and invest at rates of 12 percent or higher. The same argument applies to the salary figure specified: this may or may not be realistic, depending on the situation.

Another potentially faulty assumption is made that employment in industry would be currently available for Susan. Furthermore, a change in employment might cause personal or family problems. Weighing adjustments of this kind in dollar terms can be extremely difficult, if not impossible. Finally, different types of investments may be difficult to directly compare with one another in a way that will satisfy the opportunity cost concept. For example, investment in blue-chip or high-quality stocks cannot be directly compared with speculative ventures such as investing in futures and options funds, since the two types of investments involve two completely different classes of risk. Having raised these limitations of the opportunity cost idea, it is also very clear that this concept is a powerful one for managers making resource allocation decisions.

Economic Principles to Maximize Profits

There are three basic microeconomic principles that relate to profit maximization for the individual firm. These principles and the questions they answer are:

Marginal cost equals marginal revenue. This concept answers the question, "How much to produce?" One should continue adding inputs to the point at which the extra cost of producing the last unit of output is just equal to the extra revenue received from that unit of output.

Marginal rate of substitution equals the inverse price ratio. This production principle deals with the question, "What is the least-cost combination of inputs?" A manager should substitute input X for input Y to the point at which the marginal rate of substitution of X for Y is just equal to the inverse price ratio (i.e., the price of Y divided by the price of X).

Equal marginal returns. The final production principle relates to using one variable input of limited quantity over several possible production enterprises. This answers the question, "What combination of products should be produced, or how should a limited input be allocated?" One should spread the variable input among the enterprises until the marginal returns from the last unit of input invested in each production enterprise are equal. In this way, inputs are employed in the enterprises with the highest marginal returns.

Marginal Cost Equals Marginal Revenue

Marginality is an economic concept designed to help managers maximize profits. The concept focuses attention on the changing input-output relationships in a business. Economists have noted that the cost of producing additional units once production has been increased will differ from the costs of producing a unit before the production increase. The additional cost incurred from producing one more unit of output is called **marginal cost (MC).** Usually, the marginal cost falls as more product is produced and sold—up to a point. Then, as the limits of production capacity are tested, it becomes more and more difficult to keep increasing production. Marginal costs may gradually begin to increase as the strain on production capabilities becomes greater.

   Marginal revenue (MR), on the other hand, represents the additional revenue generated by selling one more unit of product. For small producers in a large market, marginal revenue is constant and equal to the market price because it is always possible to sell the additional product for the same price. But for larger producers in a smaller market, producing more may affect the market price adversely. In this case, marginal revenue may fall as more is produced and sold, because buyer needs are being met, and they will only take a larger quantity of product if the price falls.

   A basic principle of economics is that profits will be maximized by increasing production until marginal cost (cost of producing one more unit of output) just equals marginal revenue (revenue generated by selling one more unit of output):

$$MC = MR$$

The idea is that inputs should be added to the production process only until the point at which their costs are just offset by the additional revenue generated by the resulting outputs.

   To illustrate the concept, a few years ago, Robert Sutton, Manager of Riverside Orchards, Inc., was confronted with a problem. Although his apple trees were very healthy and most were in their prime years of production, apple output had been steadily declining. The manager knew that weather had not been a problem. After consultation with extension horticulturists, Robert decided that the trees were probably not being pollinated as well as they might be, so he settled on the plan of adding beehives to the orchard. These two enterprises seemed to complement each other. The bees would help to pollinate the trees and therefore increase apple production, and with a little luck they might also produce some honey to sell with the apples.

   The main problem confronting Robert at this point was to decide on the optimum number of beehives. He decided to experiment on a 1-acre plot of apple trees that was separated from the rest of the orchard. Through experiments he found that an increase in bees tended to correlate well with an increase in apple production. But gradually, as more hives were added, increases in apple production tapered off. As he added hives, he found that past a point, the addition of more hives did not increase apple production at all. Using bees as the only variable input and apples as the only output, Robert plotted the relationship shown in Table 3-1.

   As more hives were added, output of apples increased, up to the level of four hives. Even before knowing anything about costs and revenues it is apparent that the manager would not want to add the fifth hive, since output actually

Determining the optimal use of specific inputs such as beehives is an important decision for agribusiness managers.

T A B L E **3-1** | **Relationship Between Beehives (Input) and Production of Apples (Output) at Riverside Orchards**

Beehives (input)	Bushels of apples (output)
0	200
1	220
2	228
3	234
4	237
5	236

declined at that point. Apparently, adding the fifth hive put so many bees in the orchard that they produced some sort of adverse reaction (perhaps they scared off the pickers!). Even if bees came absolutely free of charge, Robert would be better off holding input to a maximum of four units. Now that the maximum number of hives that could possibly prove useful is known, the important question becomes exactly how many hives should be added to the 1-acre tract of apple trees if Robert is seeking the most profitable level of investment (whereby marginal cost will equal marginal revenue). To discover the solution, study the cost and revenue information presented in Table 3-2.

TABLE 3-2 | Costs and Revenues for Beehive–Apple Relationship

Beehives (input) (1)	Bushels of apples (output) (2)	Total variable cost (3)	Total fixed cost (4)	Total cost (5)	Marginal cost (6)	Total revenue (7)	Marginal Revenue (8)	Profit (9)
0	200	$ 0	$1,000	$1,000	——	$1,200	——	$200
1	220	30	1,000	1,030	$1.50	1,320	$6.00	290
2	228	60	1,000	1,060	3.75	1,368	6.00	308
3	234	90	1,000	1,090	5.00	1,404	6.00	314
4	237	120	1,000	1,120	10.00	1,422	6.00	302
5	236	150	1,000	1,150	——	1,416	6.00	266

Let's look at the data presented in this table more closely. **Total variable cost** (column 3) equals the units of variable input multiplied by the cost of the input. In this example, a hive of bees costs $30. Column 3, then, is simply the number of beehives in column 1 multiplied by the $30 cost per hive. In this example, it is assumed that all other inputs necessary for the production of apples are held constant. Thus the effects on profitability of adding more and more of the single variable input can be examined.

Total fixed cost (column 4) is constant. Fixed costs do not vary with the level of production. These costs must be paid regardless of how many beehives are used. Here fixed costs include all the other expenses of producing apples. The manager estimated these costs to amount to $1,000 for the 1-acre tract. Column 5 is **total cost.** This is the sum of columns 3 and 4. Total cost always equals the sum of total variable cost and total fixed cost.

Marginal cost (column 6) is defined as the extra cost incurred from producing one more unit of output. As the name implies, marginal cost involves a change in cost between two successive levels of output. Marginal cost represents the change in total cost divided by the change in output. For example, the marginal cost of adding the first hive of bees is ($1,030 – $1,000) divided by (220 bushels – 200 bushels) or $1.50/bushel.

In the calculation of variable costs, notice that total variable cost is the only cost that changes over the range of input use. By definition fixed costs do not change with the level of input use or output produced. Thus, marginal cost may also be calculated as the change in total variable cost divided by the change in output. Hence,

$$\frac{\$30-\$0}{220 \text{ bushels} - 200 \text{ bushels}} = \$1.50/\text{bushel}$$

Total revenue (column 7) is equal to the number of units sold multiplied by the selling price. In this case, apples are worth $6 per bushel so the total revenue column is equal to column 2 multiplied by $6.

Marginal revenue (column 8) is the amount of additional revenue generated by selling one more bushel of apples. Marginal revenue equals the change in total revenue divided by the change in output. Again, note the importance of "change" in the analysis. In this example, marginal revenue is constant at $6 because changes in production do not affect price in this particular market.

An important production principle can now be applied by using data from Table 3-2: produce to the point at which marginal cost most nearly equals marginal revenue without exceeding it. One should continue adding input to increase the level of output only until the point at which the extra cost of producing the last unit of output is just equal to the extra revenue that the last unit of output produces.

What level of production should the manager employ in this example? Using the principle of MC = MR, an input level of three beehives should produce the maximum level of profit. At this point the extra cost of producing one more bushel of apples is $5, while the value of that extra bushel is $6. Thus, profit is maximized in this case at three hives and 234 bushels of production.

To check the profit maximizing principle of MC = MR, column 9 shows the amount of profit for each input-output level in our example. **Profit** is defined as total revenue minus total cost. Note that at an input level of 3 hives and 234 bushels of production, profit reaches its maximum of $314. More apples could be produced, but the extra cost of producing the additional output is more than the extra or marginal revenue received.

Marginal Rate of Substitution Equals Inverse Price Ratio

In any business enterprise there are several different combinations of inputs that may produce the same level of output. Thus, a manager who is intent on maximizing profit would be interested in the least-cost combination of inputs that will produce that same level of output. The concept introduced earlier in this section suggested that a manager should substitute inputs to the point at which:

$$MRS = IPR$$

Again, an example is included to illustrate this concept. Curtis Brown feeds cattle and is interested in substituting more hay for corn in the feed ration that he currently uses. Based on past feeding trials, he believes the combinations of hay and corn shown in Table 3-3 will all produce an 1,100-pound steer.

From Table 3-3, the marginal rate of substitution of hay for corn can be calculated as

$$MRS_{\text{hay for corn}} = \frac{\text{change in amount of corn}}{\text{change in amount of hay}}$$

The prices for hay and corn are $0.03/pound and $0.05/pound, so the **inverse price ratio** is

$$IPR = \frac{\text{price of hay}}{\text{price of corn}} = \frac{\$0.03/\text{pound}}{\$0.05/\text{pound}} = 0.60$$

TABLE **3-3** | **Combinations of Inputs to Produce 1,100-Pound Steer***

Hay (pounds)	Corn (pounds)	MRS	Total ration cost
500	1,300		$80.00
600	1,200	1.00	78.00
700	1,125	.75	77.25
800	1,075	.50	77.75
900	1,050	.25	79.50

*Assume that the price of hay is 3 cents/pound and corn is 5 cents/pound.

In this case, the assumed prices for corn and hay result in an IPR of 0.60 and all other variable inputs are assumed to be constant.

To select the least-cost ration, you should continue substituting hay for corn until the marginal rate of substitution is just above or equal to the inverse price ratio. Using the production principle of MRS = IPR, the ration of 700 pounds of hay and 1,125 pounds of corn should be selected. At this point the marginal rate of substitution of 0.75 is as close to the inverse price ratio, 0.60, as possible without being less than the IPR. As a check to the solution, the total ration cost at each input level of corn and hay has been included in Table 3-3. The minimum-cost ration is $77.25, the one selected by our production principle.

Equal Marginal Returns

Another important production decision criterion relates to what should be produced. Often a firm can produce many different products, but inputs are limited or a budget for production expenses must be followed. This final production principle says that the manager should produce among different enterprises to the point where there are equal marginal returns among enterprises, or variable inputs should be used in their highest marginal use to the point at which there are equal returns.

Suppose that Jane Henry is a salesperson for a commercial vegetable seed company. She is preparing for upcoming sales calls. Her calls include existing customers: Taylor Brothers, Johnson Commercial Growers, Schiek and Company, and Bailey Family Farms. Jane has only 30 hours of time to devote to these four accounts and would like to maximize her sales. What should she do, given the estimates of hours and estimated sales in Table 3-4?

Jane allocates her time in blocks of 5 hours and has estimated the increased product sales she will receive with each account if she spends an additional 5 hours of time on that account. Thus, if Jane were to spend all her time on the Bailey Family Farms account, sales for that account would be maximized, but her other accounts would suffer.

TABLE **3-4** | **Sales Call Time and Estimated Sales ($1,000)**

Hours	Taylor Brothers		Johnson Commercial Growers		Schiek and Company		Bailey Family Farms	
	Sales	MR	Sales	MR	Sales	MR	Sales	MR
0	64	—	75	—	79	—	65	—
		6		6		6		10
5	70		81		85		75	
		5		5		4		8
10	75		86		89		83	
		4		4		3		6
15	79		90		92		89	
		3		2		1		4
20	82		92		93		93	
		2		1		0		2
25	84		93		93		95	
		1		0		0		1
30	85		93		93		96	

If Jane utilizes the equal marginal returns principle, she will spend five hours of call time each on the Taylor Brothers, Johnson Commercial Growers, and Schiek and Company accounts and 15 hours on the Bailey Family Farms account. No other combination of call time can give Jane a higher sales volume from her four accounts, as this allocation equates marginal return at $6,000 in sales. You can check this solution by computing Jane's total sales for various allocations of her sales time.

THE ECONOMICS OF MARKETS

Agribusiness managers need to understand the economics of decisionmaking within their organization. But, they also need to understand the economics of the market. In a free enterprise economy, price plays a fundamental role as a coordinating mechanism. Agribusiness managers make operating and investment decisions in response to market signals transmitted through the price of goods and services. Understanding how market prices emerge and what factors impact price levels and changes is the focus of this section. There are two sides represented in any market: (1) supply or how producers or sellers react to different prices and (2) demand or the behavior of consumers or buyers at different price levels. We will discuss both sides separately and then bring these ideas together in our discussion of market equilibrium.

Agribusiness managers must determine how to allocate resources such as sales time and effort across competing enterprises.

Supply: The Seller Side of the Market

Supply, from an economic viewpoint, represents the relationship between two variables: price and quantity supplied. **Supply** may be defined as the quantities that sellers are willing and able to put on the market at different prices. The supply relation can be described in several ways: as a table of prices and quantities called a supply schedule, as a graph (Figure 3-1), or as an algebraic function of prices and quantities (a supply curve) as illustrated below:

$Q_S = \alpha + \beta P$
where
Q_S = quantity supplied,
and α and β = parameters which indicate how variables are related, and
P = price

Supply reflects the direct relationship between price and quantity. The law of supply suggests that as price increases, sellers are willing to put more product on the market. Even though their marginal cost may increase, the higher price increases marginal revenue and thus makes increased production profitable. The supply curve or function shown in Figure 3-1 depicts such a price-quantity relationship. If the price were $20, producers would be willing to supply 270 units. If the price changed to $40, producers would supply 600 units. Of course, a single market price is all that is possible at a given point in time. The supply curve simply shows what sellers would be *willing* to provide at various prices. It is important to note that our definition of supply includes the entire curve exhibited in Figure 3-1, not just a point on this curve.

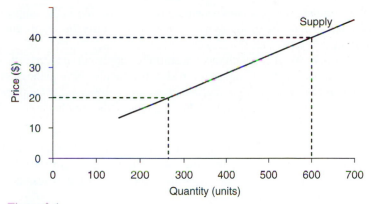

Figure 3-1
Supply Curve.

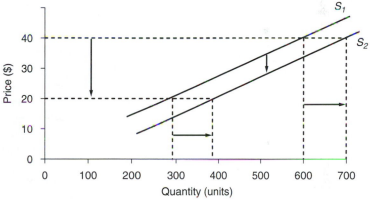

Figure 3-2
Shift in Supply.

The supply curve can shift or change position over a period of time. When supply increases, the entire curve shifts to the right. In this case, the same quantity is offered at a lower price or a greater quantity is offered at the same price. Figure 3-2 shows an increase in supply. The opposite shift in the curve, that is, a shift to the left, would indicate a decrease in supply.

Several different factors can cause a firm's supply curve to shift. Since the amount of product that a company is willing to supply depends heavily on its marginal cost and marginal revenue relationships, anything that changes the basic cost structure of a firm can shift its supply curve. Some of the major factors that cause the supply curve to shift include:

1. *Change in technology* Improved efficiency shifts the supply curve to the right (increase in supply).

2. *Change in the price of inputs* An input price increase shifts the supply curve to the left (a decrease in supply); a price decrease shifts the supply curve to the right.
3. *Weather* Poor weather conditions can shift the supply curve to the left, while very favorable weather conditions can shift the supply curve to the right.
4. *Change in the price of other products that can be produced* If the price of another product the firm could produce increases, the supply curve shifts to the left.

The supply curve is based on the cost curves for the individual producer or seller. Consequently, most factors that shift supply curves relate to direct changes in costs that cause the curve to shift to the right or left. Thus, the supply curve is actually the marginal cost curve for the firm.

Time affects supply, so there are short-run and long-run supply curves for the seller. In the short-run the seller will continue producing as long as variable costs can be covered. In the long-run the seller must cover all costs of production—both fixed and variable. This relationship is shown in Figure 3-3:

Short-run supply = MC above average variable cost (AVC)
Long-run supply = MC above average total cost (ATC)

Finally, two phrases related to supply are often easily confused: a change in supply versus a change in the quantity supplied. A **change in supply** refers to a movement of the entire supply curve. Hence, if the supply of beef was being discussed and corn prices were forecasted to increase, this would result in a shift in the supply of beef to the left. Producers would be willing to put the same quantity of beef on the market only at a higher price. A **change in the quantity supplied** refers to a movement up or down a given supply curve. In this case, the supply curve does not shift. Back to the beef example, the focus here is on what happens to quantity supplied as the price of beef changes, with all other factors held constant.

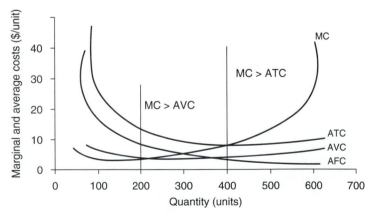

Figure 3-3
Average and Marginal Cost Curves.

Demand: The Buyer Side of the Market

The buyer or consumer side of the market is represented by the demand curve. **Demand,** in economic terms, is the quantity that consumers are willing and able to buy in the market at various prices. As with the concept of supply, we are concerned with a price-quantity relationship. The law of demand states that there is an inverse relationship between price and quantity; that is, as price increases, buyers are willing to purchase less. As was the case with supply, the demand relation can also be described in different ways: as a table of prices and quantities called a demand schedule, as a graph (Figure 3-4), or as an algebraic function of prices and quantities, or a **demand curve,** as shown below:

$Q_D = \alpha - \beta P$
where
Q_D = quantity demanded,
and α and β = parameters which indicate how variables are related, and
P = price

The demand curve that is depicted in Figure 3-4 illustrates a typical series of price-quantity relationships. If the price were $20, buyers would be willing to purchase 500 units. But if the price were $40, buyers would be willing to purchase only 180 units. Of course, only a single price can exist in the market at any time. The demand curve simply shows the amounts that buyers would be willing to purchase at each of the various possible prices. Like supply, our definition of demand includes the entire curve, not a single point on this curve. For a given demand curve, a change in the quantity purchased that is the result of a change in price is called a change in quantity demanded.

A demand curve can also shift or change position over a period of time. Demand increases when buyers are willing and able to buy more at the same price or are willing to pay more for the same quantity. When this occurs, the demand curve shifts to the right. Figure 3-5 shows an increase in demand. Demand decreases

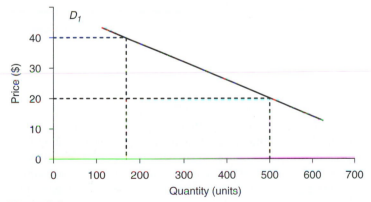

Figure 3-4
Demand Curve.

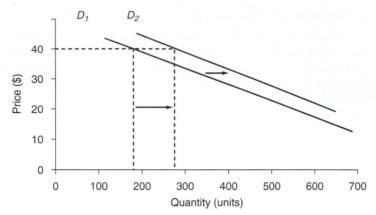

Figure 3-5
Shift in Demand.

when the opposite occurs: buyers purchase less at any given price. When the entire demand curve shifts, a change in demand has occurred.

Several important factors can cause a demand curve to shift. Since shifts in demand are totally dependent on the buyer, the reasons for such shifts are heavily influenced by psychological and emotional factors, and can be quite complex. Some of these factors include the following.

1. *Income* As their incomes increase, consumers can afford to buy more, and this shifts the demand curve to the right if the good is a normal good.
2. *Tastes and preferences* Changes in emotional and psychological wants can shift demand in either direction.
3. *Expectations* When buyers expect the price to fall, they may postpone purchases, thereby causing the demand curve to shift to the left. If buyers expect prices to rise in the future, the demand curve can shift to the right.
4. *Population* Sheer increases in the number of buyers can shift the demand to the right. The opposite is also true.
5. *Price of substitutes or complements* A fall in the price of a close substitute product may shift the demand curve to the left, while a fall in the price of a complementary product may cause demand to shift to the right.

The concept of demand is based on the **law of diminishing marginal utility,** which states that as more and more of a product is consumed, the extra satisfaction of consuming an additional unit actually declines. This translates directly into the negative slope of the demand curve and the inverse relationship between price and the quantity consumed. At higher and higher prices, buyers tend to consume less and less of a given product. At higher prices some goods have substitutes that the buyer would switch to, while at other, even higher, price levels the buyer may give up consumption of that product altogether.

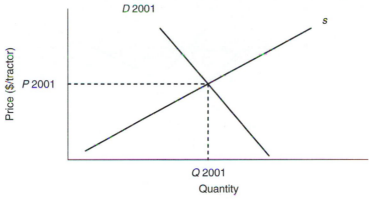

Figure 3-6
Market Equilibrium for 4-Wheel Drive Tractors.

Again, as with supply, it is important to distinguish between a **change in demand** and a **change in the quantity demanded.** A change in demand results from the demand curve shifting to the right or left. A change in one of the factors which shift demand could cause this to occur. A change in the quantity demanded simply refers to a movement along a given demand curve.

Derived Demand

The demand for most agricultural products is a **derived demand.** Derived demand is not based directly on general consumer demand, but rather on the need for a product that indirectly relates to consumer demand. For example, the agricultural producer's demand for fertilizer is derived from the consumer's demand for corn. When export demand for corn increases substantially, the price of corn increases. This also increases the demand for fertilizer, since when the price of corn is higher, producers choose to produce more corn. This is one reason why agribusiness managers and marketing experts are so concerned about general economic trends. Anything that significantly affects consumer demand for agricultural products will have an impact on the demand for farm inputs, through the process of derived demand.

Market Equilibrium and Price Discovery

As producers and consumers meet in the marketplace, an equilibrium of quantity and price is determined. This process is called **price discovery.** There is actually one price and quantity that will "clear the market" at a given point in time. At this point everything that producers are willing to sell will equal everything that consumers want to buy. Figure 3-6 illustrates the market-clearing conditions of equilibrium. Here there is a single price and quantity at which point the quantity supplied and the quantity demanded are equal.

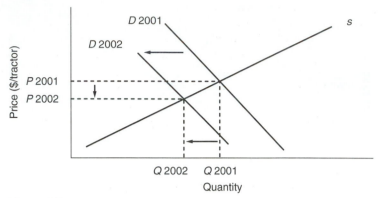

Figure 3-7
Changes in Market Equilibrium for 4-Wheel Drive Tractors.

At market equilibrium the quantity demanded is just equal to the quantity supplied. Often one may hear that supply equals demand at market equilibrium, but this is incorrect. It is important to recall that our definitions of both supply and demand include a series of price-quantity relationships—not a single price or quantity. Since supply and demand can interact at only one price-quantity point it is important to describe this market-clearing condition properly.

In practice, demand and supply are not static, but are in a constant state of fluctuation. At any point in time, "price" represents the market's best guess of the current supply-demand situation. Because conditions are based on buyers' and sellers' judgments of the actual supply and demand, the situation is highly volatile and can change both rapidly and frequently. This is why market prices for a great many agricultural products and supplies change throughout the day. As a quick example, suppose that for four-wheel drive tractors in 2001, the price-quantity relationships for supply and demand are estimated to be those shown in Figure 3-6. This means that for 2001, $Q2001$ tractors were sold at a price of $P2001$. Now suppose that 2002 is forecast to be a bad year for producers, and farm income will be reduced by 15 percent. What would the new supply-demand relationship look like? Figure 3-7 illustrates both the 2001 and 2002 relationships.

The supply and demand curves represent a theoretical relationship between price and quantity at a given point in time. The curves here have been illustrated by straight-line functions. This was done to facilitate explanation of the curves; but most supply and demand curves are actually curvilinear over some range of price and quantity, and may be quite complex. Agricultural economists spend considerable time and effort attempting to estimate demand and supply relationships for specific commodities in order to better understand the dynamics of markets.

Elasticity

Managers are often concerned with predicting the degree of customer reaction to a change in price. **Elasticity** is a good way to measure customer response. Elasticity of demand reflects the percentage change in the quantity demanded when

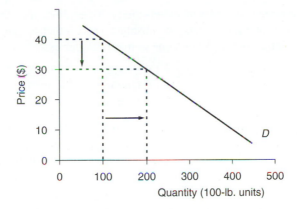

Figure 3-8
Demand for Bluegrass Seed.

the price changes by 1 percent. If the quantity demanded for bluegrass seed increases by 1 percent when its price decreases by 1 percent, the demand elasticity for bluegrass seed is 1.0. The formula for elasticity is:

$$\text{Elasticity} = \frac{\text{\% change in quantity demanded}}{\text{\% change in price}}$$

or

$$\frac{\text{new quantity} - \text{old quantity}}{\text{new quantity} + \text{old quantity}} \times \frac{\text{new price} + \text{old price}}{\text{new price} - \text{old price}}$$

As Figure 3-8 shows, if the price of bluegrass seed decreased from $40 to $30 per 100-pound unit, the quantity demanded would increase from 100 to 200 units.

$$\text{Elasticity} = \frac{200 - 100}{200 + 100} \times \frac{30 + 40}{30 - 40} = \frac{100}{300} \times \frac{70}{-10} = -2.33$$

This means that when the price drops by 1 percent, the quantity demanded increases by 2.33 percent. But if the price fell from $20 to $10 per unit, quantity would increase from 300 to 400 100-pound units. On a percentage basis, demand is less responsive at this level.

$$\text{Elasticity} = \frac{400 - 300}{400 + 300} \times \frac{10 + 20}{10 - 20} = \frac{100}{700} \times \frac{30}{-10} = -0.43$$

These demand elasticities use the formula for arc elasticity of demand. This measures the percentage change over a segment of the demand curve rather than evaluating change at a specific point on the curve. The coefficient for elasticity of demand should always be negative, since change is measured along the demand curve. As price increases, quantity demanded decreases, or vice versa; thus demand elasticities will always be negative.

There are three levels of demand elasticity. To evaluate the elasticity of demand coefficient, one should take the absolute value—that is, ignore the negative sign, and fit the coefficient into one of three categories:

$$| \, e \, | > 1.0 \text{ Elastic}$$
$$| \, e \, | = 1.0 \text{ Unitary}$$
$$| \, e \, | < 1.0 \text{ Inelastic}$$

Thus, if demand elasticity, in absolute terms, is greater than 1.0, demand is **elastic.** This means that a small change in price will result in a relatively large change in the quantity demanded. Types of products for which demand is elastic include products for which there are many close substitutes, luxury goods, and products that make up a large portion of the consumer's budget.

On the other hand, if the absolute value of demand elasticity is less than 1.0, demand is said to be **inelastic.** Here a change in price has a relatively small impact upon the quantity demanded. Goods that fit into this category are typically necessities, products that account for a very small part of the consumer's budget, or products that have no close substitutes. The demand for natural gas to run and heat a large production facility is fairly inelastic—it is essential to the efficient production process and, at least in the short run, it has no substitutes. Thus, this production facility may have no option but to pay higher natural gas bills during the winter to maintain production capacity of the facility. However, over the longer term, the firm may switch from gas to oil, electricity, or some other alternative fuel source. So, the period of time under consideration also affects the elasticity of demand.

Supply elasticity can be calculated in a way that is similar to the calculation of demand elasticity. Supply elasticity is always positive, since changes in price cause the quantity supplied to change in the same direction. It can be less than, equal to, or greater than 1, depending on whether quantity supplied changes proportionately less, equal to, or more than the change in price.

THE MARKETING SYSTEM

With this discussion of some important general economic principles in hand, let's turn our attention to another important idea in economics: the marketing system. As has been hinted markets in the food and agribusiness industries are not independent but highly interrelated. The market for cake mix in the supermarket is related to the markets for flour, chocolate, and sugar, which are related to the markets for wheat and cocoa, which are related to the markets for seed, fertilizer, and chemicals, and so on. We call this interrelated set of market and interfirm relationships such as contracts and alliances a **marketing system.**

Throughout the journey through the marketing system from producer to consumer, a product's value is changed in many ways. The form of the product may change as grain is processed into flour. The flour may need to be stored until needed by a baker. The baked bread may need to be shipped to a supermarket near a consumer's home. The ownership of the product must be transferred from buyer to seller at various points in this journey. The coordination of this marketing system activity is a fundamental role of markets. In this section, we will explore the types of value or utility that are added by firms in the marketing system, and the fundamental functions that any marketing system must perform.

Utility

The market system for most food and fiber products begins with farm supplies moving to the farm, continues with a raw product leaving the farm gate, and culminates with a desired finished product at the consumer level. A hog on a feeding floor in Iowa in January is a far different product from a ham sandwich in Yankee Stadium in July. The changes that transform the live hog into the desired consumer product are referred to as adding **utility** or value to a product. This idea of adding value to a product is significant in our economic system. There are four types of utility: form, time, place, and possession.

Virtually every firm in the marketing system earns its livelihood by contributing utility to products. The farmer adds **form utility** by raising the hog to market weight. But a live hog in Yankee Stadium is of little value to the hungry baseball fan. So meat packers add form utility by breaking down the carcass into various pork cuts and producing canned hams, and the concessionaire adds still more utility by making up sandwiches.

Various other firms in the system facilitate this process by providing other necessary activities to feed our fan at the baseball game. Inventory holders, such as meat wholesalers, add **time utility** to the product by storing the hams, thereby ensuring that sufficient quantities of product are available when needed. Trucking firms and railroads help to add **place utility** by physically moving the hams from Iowa to New York. Many firms in the marketing system facilitate movement of commodities by matching buyers with sellers. Commission agents help the farmer to find buyers for their livestock. Many wholesalers buy from various packing and processing plants. These individuals add **possession utility** to the final product as the ham moves through the system to the concession vendors in Yankee Stadium, and ownership is transferred at various points along the way.

Price and Utility

We will take a closer look at the link between price and utility here. Price differences between two or more locations reflect place utility. As the price of corn in New Orleans increases relative to the price of corn in Chicago, changes in shipping patterns will occur. In this case, more corn will be shipped to New Orleans. If this continues, a point will be reached where there is a glut in New Orleans and a shortage in Chicago, at which time prices will move in the opposite direction, until, theoretically, the market comes to **equilibrium.** As mentioned earlier, in practice the market may never quite reach equilibrium, given the influence of ever-changing market conditions such as new information becoming available to the market. But, markets do trend toward equilibrium. Price differentials of this nature tend to reflect the differing costs of transportation between various points.

Price differentials among different grades of products usually reflect differences in form utility or product quality. Various grades of milk serve as a good example. Some milk may be used for manufacturing purposes, ice cream, cheese, and butter, for example, while a different grade of milk may be used for drinking purposes. The higher grade used in fluid milk for table consumption would command a higher price. Manufacturing-grade milk receives a lower price because the quality standards are less strict, and this difference in price reflects a difference in product form.

Likewise, differences in current prices and futures prices or forward contract prices represent time utility. If the price that an elevator is offering for soybeans delivered in May of next year is much higher than the current fall harvest price, many farmers will store soybeans to sell later. A part of this difference in the future price and the current price represents a premium for adding time utility by storing the soybeans until a future date.

The relationship between market prices and possession utility is usually less clear. However, in the marketing system, any fees, charges, or commissions for connecting buyer to seller must ultimately be reflected in the final consumer price. The purpose of selling directly to consumers is to reduce the charges for possession utility. Any time an intermediary is forced to own and hold inventory, it must be financed. These finance charges for possession utility are in the market price of the product.

MARKETING SYSTEM FUNCTIONS

We've talked about the four types of utility that are added to products as they move from producer to consumer in a marketing system. Now, we turn our focus to three important functions that the marketing system must perform. The way the three functions are accomplished can be very simple, or incredibly complex. In some cases, as with a grower-owned roadside fruit market, all three marketing functions take place in a very short time and in a very direct manner, with only the producer and the final consumer involved. In other cases, such as Texas cotton marketed in the form of shirts at the retail level, the functions are very complex, involve dozens of different firms and hundreds of people, and may require months to complete. But in both cases, the same three marketing functions are required.

There are three major types of marketing functions:

1. The **exchange function** must be performed; that is, the product must be bought and sold at least once in the marketing system.
2. Certain **physical functions** must be performed, such as transporting, storing, and processing the product.
3. Various **facilitating functions** must occur in the marketing system. Somehow there must be at least a minimal amount of market information available; someone must accept the risk of losses that might occur; often the product must be standardized or graded to facilitate the sale of the product; and finally, someone must own the product and provide the financing during the marketing process.

Exchange Functions

Exchange functions involve those activities that are concerned with the transfer of ownership in the marketing system. Certainly, demand and supply analysis (discussed earlier in the chapter) has direct application to this area. Prices are determined in our free enterprise economic system as buyers and sellers meet to exchange commodities. People involved in the process could include brokers and commission agents, whose livelihood comes from matching buyer with seller. These people may or may not take title to the goods they handle; in any event, they are exchanging a service for a fee.

The exchange function receives much attention from economists and managers because price is determined as part of this activity. Much attention is given to the method of price determination. Are there a large number of sellers and buyers; that is, is there competition? Does one group have an economic advantage over the other? What is the competitive nature of markets within this system? Market researchers who study the performance of the marketplace ask these questions and many others. Both parties to any transaction seek the most favorable price possible. Market performance studies focus on the nature of buyer and seller relationships, and measure areas such as product prices and product costs, and levels of firm profitability to determine whether consumers are reaping the full benefits of a competitive economic system.

Buying   At the primary level, buying involves the interaction of the producer or the producer's agent and the processor, the wholesaler, or, less frequently, the consumer. Typically, the buying activity in a marketing system involves a processor purchasing raw material from the primary producer, a wholesaler buying the finished product from the processor-manufacturer, the retailer purchasing from the wholesaler, and finally the consumer buying from the retailer. This final step receives a great deal of attention because it involves the total population; it is also the only phase of the marketing system in which the consumer is directly involved. Thus, the success of the entire marketing process is finally judged by the consumers' buying behavior. The buying function is also important to agribusiness firms. They acquire raw inputs from producers directly and through other marketing agents. Input acquisition or procurement involves a host of questions: "What can we afford to pay?" "How much input do we need?" "Can we store inputs or should we attempt to minimize inventories?" and "How can we protect the firm from adverse market price movements?"

Selling   A product can be purchased in the marketing system only when another market participant is willing to sell, so selling is an integral part of the exchange function. For agricultural producers, deciding when to sell is the primary marketing issue. With some agricultural products, particularly grain, producers have a wide span of time in which to sell—ranging from preharvest forward contracts that promise delivery several months in the future to postharvest decisions to store grain production and sell several months after harvest. With other agricultural products, such as most livestock products and horticultural crops, there is little flexibility in selling time. Once the product is ready for market, there is no possibility of postponing the sale because the product will deteriorate if it is held.

Increasingly, however, there is a trend toward contracting production of agricultural products for a specific price and delivery date, particularly with more perishable products. This system can help create a more orderly marketing process and can be used to manage risk. For the agribusiness firm, selling involves careful consideration of the product to be sold, the promotional programs used to promote it, and price and distribution decisions. Chapters 7 through 11 focus on how agribusiness managers make these important sales and marketing decisions.

Physical Functions

In the overall scheme of producing/developing a product and getting it to its ultimate consumer, there are some physical functions that must occur. Physical functions include the movement and handling of products as they are transported, stored, and processed. Time, place, and form utility are added to the product as it is physically handled.

Transportation   Food and agricultural products are transported from production units through the marketing system to consumers by nearly every conceivable method, because of the wide variation in types of products. Cut flowers and some fresh vegetables are highly perishable and must immediately be trucked or flown to selling points so that they can be consumed within hours or days of harvest. Much agricultural production moves by truck from production units to local or terminal markets. Railroads carry vast quantities of agricultural products, particularly grain, to processors and to port facilities. Often, grain moves in special 100-car trains called **unit trains** in order to reduce marketing costs. Barge traffic on the North American Great Lakes and major rivers is a slower but significantly cheaper means of transportation for much of the grain harvest, as well as other bulky products such as fertilizer.

Storage   Inventory holders in the marketing system handle the storage of products. This function adds time utility to products and is extremely important to the efficient marketing of many commodities. With the adoption of new technologies it is now possible to hold fresh fruit and vegetables for longer periods than just a few days after harvest. A breakthrough in storage technology now allows tomatoes to be stored indefinitely in aseptic holding tanks without any refrigeration. This allows tomato-processing plants to operate at an even rate all year, rather than for just a few weeks during harvest. In fact, this development also allows raw food products to be transported hundreds of miles to centralized facilities that cost less to operate.

Storage is equally important to the agribusinesses that supply producers. Most crop production inputs are bought during a few weeks each year. This presents difficult problems for manufacturers and suppliers, who must produce supplies throughout the year to utilize their manufacturing facilities efficiently, but then compress distribution into a very short time period. Most crop protection chemicals, for example, must be stored in costly temperature-controlled environments. Storage is an expensive marketing function. In 2000, 10 billion bushels of corn were produced in the United States. Certainly, not all that corn could be utilized directly after harvest. Some was exported and some was fed to livestock, but most went into storage on farms or in local elevators so that consumption could be spread over the entire marketing year.

Processing The primary producer adds some form utility to the commodity that moves through the marketing system, changing feeder steers into fat cattle, a single kernel of corn into ears of corn, or an oblong black seed into a big green watermelon. However, this process does not usually include all the form utility that the final consumer demands at the retail level. Cattle must be transformed into packages of steak and hamburger, and kernels of corn must become boxes of corn flakes. This is where the processor/manufacturer plays a vital role in meeting the demands of the consumer. Processors take the raw primary product and turn it into a more desirable form. Processing may involve only a single firm in the marketing system, or it may involve many firms, with each successively adding another type of form utility.

Facilitating Functions

Facilitating functions are those activities that help the market system to operate more smoothly. These functions enable buyers, sellers, transporters, and processors to make their contributions without great concern about risk or financing, and to put together an orderly marketing plan.

Market Information An efficient marketing system requires that all participants be well informed. Buyers need information regarding supply sources, sellers seek information pertinent to prices in different markets, and consumers desire information about quality, price, and sources of product. Inventory holders seek information about current and future prices in an attempt to decide what and how much to store. Wholesalers need information about transportation rates to compare prices of the same commodity in different geographical locations. Market information can be obtained from many different sources. Private firms publish a multitude of market letters dealing with fundamental and technical factors relevant to marketing decisions. The government is responsible for reporting prices, production, disposition, and utilization statistics in different markets periodically to keep all market participants more fully informed of what the future might bring. Market research through private firms and universities is also part of this area of market information.

Risk Bearing Throughout the marketing system, owners of commodities are subject to risks. These risks may be separated into two general categories: physical risks, such as wind, fire, hail, flood, theft, and spoilage; and market risks or the risk that price will change. People usually attempt to limit their vulnerability to physical risks through such means as installing a burglar and fire alarm system in a chemical storage warehouse, or using containers that protect and preserve the quality of grapes in transit. For most physical risks there is a known probability of occurrence of loss or damage, so insurance may be purchased to protect against most physical losses. The purchase of insurance transfers risk to another party—the insurance company. Insurance companies essentially "pool risk" of the insured with the idea that everyone in the pool pays insurance premiums, but not everyone will suffer loss.

Market risks, on the other hand, are not so easily handled in the marketing system. Market risks include the possibility of price declines (or increases), changes in consumer preferences, or changes in the nature of competition. Someone in the marketing system must accept all these risks. Generally, insurance companies will not write a policy guarding against market risks, since it is impossible to actually calculate the probability of loss. Price changes are among the most important market risks to producers and agribusinesses. Because of their importance, market risks and the management of such risks is discussed further in the Appendix to this chapter.

Standardization and Grading   **Grading** agricultural products into specific standardized classes or categories greatly facilitates the buying and selling process and helps the marketing system work more efficiently. The U.S. system of standards and grades is a highly refined system that includes most agricultural commodities. The USDA establishes specific grades for many agricultural products to ensure standardization. A corn processor from Indiana can buy its entire inventory of corn from elevators around the Midwest via telephone, without seeing the actual commodity. The buyer knows within a set of defined limits the quality of corn that will be received because of the system of grades used by industry. For example, an order of No. 2 yellow corn at 15 percent moisture and a low percentage of FM (foreign material) would be standard throughout the industry.

Sellers, in turn, know the quality of their corn and whether it will or will not meet the buyer's specifications. Thus, a large lettuce grower in California can sell a crop of lettuce to an East Coast produce wholesaler via phone or computer. With an effective system of grades and standards, the East Coast wholesaler can specify the quality desired and the California producer can deliver the crop knowing it meets their specific grades and standards. The fact that this buyer and seller never need to actually meet in a physical marketplace to examine the produce adds greatly to the efficiency of the marketing system.

Financing   Someone must own products as they move through the marketing system. Ownership requires the investment of funds into the marketing process for at least a short period of time. The marketing firm that actually buys and takes title to the product provides the **financing.** Of course, some of these funds may be borrowed from a lending institution, so both lenders and investors become involved in financing marketing functions.

Marketing System Efficiency

The term **market efficiency** is often used to describe the performance of the marketing system. It reflects the consensus that the functions of the marketing system should be performed efficiently. New technologies or procedures should be adopted only if they can increase the efficiency with which these functions are performed. Market prices should reflect the fundamental economics of the market, and should send appropriate signals to producers and consumers. **Efficiency** may be defined as increasing the **output-input ratio,** which generally may be accomplished in any one of four ways:

1. Output remains constant while input decreases.
2. Output increases while input remains constant.
3. Output increases more than input increases.
4. Output decreases more slowly than input decreases.

There are two different perspectives on marketing system efficiency. The first is called operational efficiency and focuses on the efficiency of performing the marketing system functions, primarily within the firm. The second dimension, called pricing efficiency, measures how adequately market prices reflect the production and marketing costs and other market conditions throughout the marketing system.

Operational Efficiency

Operational efficiency focuses on the ratio of marketing output over marketing input:

$$(\text{Marketing output})/(\text{Marketing input}) = \text{Operational efficiency}$$

This measure of efficiency is concerned with the physical activities of the marketing system. Output per worker-hour is one often-quoted productivity ratio that measures operational efficiency. Farm supply stores calculate sales per employee-hour to monitor operating efficiency. Fertilizer plants watch throughput per spreader (tons of fertilizer sold per distribution unit) to measure operating efficiency.

In the marketing system, increased operational efficiency is usually synonymous with cost reduction. Typically, a substitution of machinery for labor has been associated with improved efficiency. A seed processor buys new bagging equipment when the equipment is expected to reduce the required amount of labor, thereby improving operational efficiency. Likewise, using 15 dozen plastic totes for distributing eggs to supermarkets reduces delivery and shrinkage or breakage costs. Palletization of farm supplies by wholesalers and manufacturers may also reduce costs significantly. New molded containers for California plums and nectarines may reduce spoilage or damage and allow for larger shipments. Any improvements in quality of output can also enhance operational efficiency. Improvements in these areas and many more allow the functions of the marketing system to be performed at lower cost.

While the focus in this discussion is on efficiency at the individual firm level, it is important to note that we are interested in whether or not the marketing system is organized in such a way as to deliver operating efficiency. Are firms large enough to be efficient? Are firms linked through markets more efficient than vertically integrated firms? Do firms in the market adopt innovative, cost-saving technology in a timely way? Such are the questions to be answered when the operating efficiency of a marketing system is explored.

Pricing Efficiency

Pricing efficiency is concerned with how effectively prices reflect the costs of moving output through the marketing system. The prices that consumers pay for goods delivered by the marketing system should adequately reflect all marketing and production costs. In a perfectly competitive economic environment, prices will adequately reflect all such costs. Firms in the industry are free to enter and exit the marketing-production process as they desire. The demands of consumers are reflected through the marketing system through price signals, resulting in profits and losses for the firms involved.

Pricing efficiency is often measured by comparing the performance of the market to the benchmark of a perfectly competitive market. If a few firms that conspire to maintain high prices dominate the market, the situation would be interpreted as being less than price efficient. Many things can result in pricing inefficiency: consumers who lack sufficient information about alternatives, or firms that dominate a market because of location or excellent personnel, in which case their prices may not reflect cost adequately. Pricing efficiency measures are often less specific than measures of operational efficiency. Norms established in this area tend to evaluate an entire industry and not a specific firm in the industry. The data required to evaluate price efficiency are typically more difficult to acquire. Some governmental agencies, however, do attempt to generate measures of pricing efficiency to examine the relative competitiveness of an industry.

Legislation Affecting Food and Agricultural Marketing Systems

Over the years, there has been considerable legislative and regulatory activity that impacts the food and agricultural marketing systems. Much of this legislation and regulation is developed with a focus on improving the efficiency of the marketing system. Some of this legislation was enacted to inform and/or protect consumers while other legislation impacts the means and modes in which agricultural products are marketed. Some legislation and regulation sets standards that govern trade. Other legislation defines fair and unfair competitive practices. Still other legislation grants exclusive rights to the rewards from new inventions.

There are a myriad of acts, regulations, and agencies that directly affect the marketing systems for agricultural products. And, the amount and complexity of this legislation, particularly legislation that relates to protecting consumers, has increased in recent years. Concerns about healthful foods, cancer-causing food additives, and the impact of pesticides on the environment, among other dangers, are widely reflected in a variety of rules and regulations. Many agribusiness firms not only find their activities carefully monitored and regulated by government agencies, but also must regularly provide extensive reports on various aspects of their business.

It is important to note that development of such laws and regulations almost always involves a cost and benefit trade-off. Trends toward greater consumer protection offer the potential for a safer, healthier food supply. At the same time, they add to the costs of the marketing system. For example, the costs associated with meeting government requirements for bringing a new pesticide on the market have increased manyfold. Ultimately, society, working through the government, must make important choices that balance the costs and benefits of new rules and regulations if the food and agricultural marketing system is to be an efficient and effective one.

SUMMARY

Agribusiness managers must have a clear understanding of economics to be successful. Economics involves the allocation of scarce resources (i.e., land, labor, capital, and management) to meet the needs of society. Macroeconomics focuses on a "big picture" view of our economic system. Microeconomics is the application of basic economic principles to decisions within the firm.

The profit motive is used as an incentive that guides businesses in fulfilling consumer wants as consumers express these wants in the marketplace with their dollars. Several economic concepts provide insights for agribusiness managers. Opportunity cost is the income given up by not choosing the next best alternative for the use of resources. Economic profit is defined as accounting profit less opportunity cost. Economic profit then forces managers to explore alternative uses of resources. The marginal cost equals marginal revenue concept helps managers focus on the most profitable level of production. An agribusiness manager who is intent on maximizing profit would be interested in the least-cost combination of inputs that will produce a specific output. This manager understands the marginal rate of substitution and how this rate is linked to the inverse price ratio. Also, a manager would be interested in what combination of products should be produced, on how a limited input should be allocated.

Supply captures the relationship between price and quantity supplied. Demand curves slope downward from left to right, indicating larger quantities are only purchased at lower prices. Applying supply and demand principles is important in understanding how markets work, and how prices move. Elasticity estimates for demand help agribusiness managers examine the potential impact of price changes.

Products experience numerous changes during their journey through the marketing system to the ultimate consumer. Utility, or value, is added to products in terms of form, time, place, and possession. Several specific market functions must be performed by any marketing system: the exchange function, the physical functions, and the facilitating functions. Market efficiency describes the performance of the marketing system. Both operational efficiency, which measures productivity of performing marketing functions, and pricing efficiency, which measures how adequately market prices reflect the production and marketing costs of the marketing system, are important indicators of system performance.

DISCUSSION QUESTIONS

1. What are some of the strengths of a market-oriented, capitalistic economic system? What are some of the weaknesses? Give at least three examples of ways that U.S. society has chosen to address some of the weaknesses of our economic system.
2. Why should agribusiness managers be interested in macroeconomics? Give a specific example of one macroeconomic development and describe the impact of this development on the food and agribusiness industries.
3. What are your opportunity costs associated with attending college? Be specific.
4. Suppose you have worked through an MC = MR problem and determined the solution. Now, what happens if fixed costs increase by 20 percent? What is the new solution to this problem?
5. Consider the supply of milk. What are at least three different developments that might shift the supply curve for milk. Be specific.
6. Consider the demand for frozen yogurt. What are three different developments that might shift the demand curve for frozen yogurt. Be specific.
7. How might the marketing manager of an agribusiness firm use elasticity of demand coefficients? Be specific.
8. John Ladd owns a "U-pick" strawberry farm. The farm is located next to a large metropolitan area and has experienced steady growth during the past 5 years. He borrows operating funds from the local bank to finance each year's production and repays the loan after the growing season. Each spring John advertises on the local radio and in the local paper that his strawberries are ready to pick. Also, he is selling a small, but an increasing, percentage of his strawberries to local grocery stores on a contractual basis. He keeps the patch clean and provides the containers for customers to use when picking the strawberries. Explain how each of the marketing functions is performed at the Ladd Farm and who performs each function, John or his customers.

CASE STUDY — **Armstrong Agribusiness, Inc.**

A local agribusiness enterprise has employed a consultant to estimate its supply curve for a specific tractor part and to estimate the demand for this part in its local marketplace. The consultant has determined the following:

$$\text{Demand: } Q = 220 - 4P$$
$$\text{Supply: } Q = 40 + 2P$$

Questions

1. Given these two estimates, solve for the market equilibrium price and quantity.
2. Now plot the individual supply and demand curves together for price levels from $5 through $50 in increments of $5 (i.e., $5, $10, $15, etc.). Is the market equilibrium the same on your graph as calculated in question 1?
3. This tractor part is often replaced if producers have extra income. What happens to the supply and demand curves if net farm income is expected to increase by 10 percent?
4. Given the original demand price-quantity relationships, calculate the elasticity of demand when:
   a. Prices increase from $20 to $25.
   b. Prices increase from $35 to $40.
5. Given the original market equilibrium, what would happen to Armstrong Agribusiness if a market price of $50 were set for this part? Explain. Use a graph if you believe this will facilitate your analysis.

APPENDIX: MANAGING MARKET RISK IN AGRIBUSINESS

Managing risks is an inherent part of the agribusiness manager's job. As discussed earlier in this chapter, these risks may be separated into two general categories: physical risks, including natural phenomena such as wind, fire, hail, flood, theft, and spoilage; and market risks or the possibility of price declines, changes in consumer preferences, or changes in the nature of competition. This appendix focuses on the market risks. There are basically four techniques that can aid producers and managers in transferring or reducing their market risks: (1) diversification, (2) vertical integration, (3) forward contracting, and (4) hedging with futures markets.

Diversification

Diversification is the technique of adding to one line of business some other lines of business that pose different risks, so that the likelihood of a loss in one area will be offset by the possibility of gain in another. Since most agricultural products are subject to significant market swings and variable weather conditions, additional products or enterprises that are not likely to experience the same trends can greatly reduce the impact of wide swings in the business. The "don't put all your eggs in one basket" concept has been beneficial to many agribusinesses and producers.

An agribusiness firm that sells fertilizer may contemplate adding feed sales to its product line. Feed sales are typically less seasonal than fertilizer sales, thus generating different levels of market risk. Additional revenues from feed sales can help the firm and may allow the firm to come closer to serving all the needs of their current agricultural customers. Interestingly enough, however, there is a trend away from diversification toward specialization, especially at the production level. This is because the efficiency and cost savings that result from larger more specialized production units are thought to more than offset the disadvantages from lack of diversification. Today's agricultural producers, and many food and agribusiness firms, are increasingly specialized and offset market risks through contracting, hedging, and vertical integration.

Vertical Integration

Vertical integration occurs when an organization takes on other marketing functions in addition to its primary functions, thereby becoming less dependent on other organizations. By controlling its own source of supply or guaranteeing an outlet for products, a farm becomes less vulnerable to market swings. Many large food retailers now label and manufacture their own brands of food. Some large agricultural cooperatives own their own petroleum refineries rather than depend on private oil companies. Other cooperatives not only buy dairy products from producers but process, manufacture, and distribute their own brands to supermarkets.

Thus, vertical integration allows the firm to control more steps or activities through the marketing system which tends to lower market risk. A soybean processing firm may integrate backward in the marketing system when it purchases a group of elevators. Now the firm acquires input or raw materials directly from the producer and adds utility in the processing function. This also allows the firm to guarantee supply sources in short crop years when supplies are tight.

Forward Contracting

Forward contracting is simply the process of making a buyer-seller agreement about a set price for some future delivery date. This agreement completely removes the risk of price fluctuations for both buyer and seller. When producers are sure before the production process even begins what price they will get for their product, only production risks remain. Processors can be assured of a reliable supply at a known price, which allows for more efficient operation of facilities. Of course, the market price will fluctuate and is likely to be either higher or lower than the contract price on the delivery date. Either party may gain or lose on the contract price in relation to the market price, but since the price was forward contracted, the gain or loss is only theoretical, based on a lost opportunity.

This marketing alternative has become more popular with agricultural producers in recent years. If a producer signs a forward contract agreement for the delivery of corn to the local elevator, both price and quantity to be delivered are set. This agreement allows producers to "lock in" a specific profit and not worry about price changes.

However, there are some disadvantages to this arrangement. If a producer forward contracts 10,000 bushels of soybeans in July for harvest delivery, the producer has promised to deliver 10,000 bushels of soybeans to the elevator at a specific price in the fall. If, after this agreement has been signed, the weather turns hot and dry and the producer ends up harvesting only 8,000 bushels of soybeans, what happens? The producer must purchase on the open market an additional 2,000 bushels of soybeans to deliver to the elevator. Hence, given the weather uncertainty, many producers tend to forward contract only 50 to 75 percent of normal production in order to avoid this potential delivery problem. The same logic would apply to almost any forward contracting situation with production risk.

Hedging on the Futures Market

Hedging on the futures markets is another strategy for transferring the risk of price changes from one party to another. Since no one can actually predict the probability of market risks, estimating what may happen to demand and prices becomes mostly a matter of judgment. At best, it is based on someone's studied opinion, and at worst, it is based on a whim or guess about what the market might do. In either case, there are those who are willing to speculate on the direction of future market prices. **Futures markets** provide a mechanism by which trading can be handled swiftly, in standardized units of products that are to be delivered at some specified future date and location.

Let's take a quick look at the futures markets and examine the critical points of this important marketing institution. Corn is traded at the Chicago Board of Trade. This futures market trades in five calendar months through the year: March, May, July, September, and December. The specific contract calls for 5,000 bushels of No. 2 yellow corn at a certain moisture level and foreign material discount. Thus, the contract is standardized in that quantity, quality, delivery time, location, and price are known. There are three delivery points for corn: Chicago, southern Illinois, and northern Ohio.

Thus, if a trader buys one futures contract of December 2002 corn in July 2002, this trader has promised to accept delivery of 5,000 bushels of No. 2 yellow corn in December 2002. Futures markets represent promises. When a trader buys today, nothing happens today. Our trader has simply made a promise to do something in the future, in our case, accept delivery of corn.

There are two market positions taken by traders: buy contracts (or "go long" in futures) or sell contracts (or "go short" in futures). Again, each of these initial positions simply involves a promise to do something in the future. If you believe prices are going to increase, you could purchase or go long futures today (i.e., buy December futures in July). Since you have no desire to physically own this product, if prices head higher, sometime before December you could sell or go short December futures. Now, you have:

Purchased one contract of December corn and then later,

Sold one contract of December corn.

These two trades cancel out and you make money if your selling price is higher than the purchase price, or lose money if the opposite occurs.

As long as futures traders offset each transaction with an equal and opposite transaction, they never have to physically accept or deliver the product. In fact, many speculators might not even recognize the agricultural commodity they are trading. Their interest lies solely in the profits they might make from favorable changes in price while they hold the contract. If they are smart, or lucky, they may do very well. But speculating on futures markets is a highly risky venture, and great losses can occur quickly.

Futures markets are a way to transfer the risk of prices from one party to another so they serve a highly useful purpose. When an agricultural producer or business combines the futures market with the cash market (the market where commodities physically change hands), market risks can be greatly reduced.

There are two types of traders in the futures markets. **Speculators** are traders that do not have nor do they want the physical commodity. Speculators enter the market solely to profit from changes in the price level. Thus, speculators are willing to accept market risk. Alternatively, **hedgers** are traders who have or desire possession of the physical commodity, but do not want to accept price or market risk. Hedgers will enter the futures market by taking a position that is opposite their position in the cash market. Both types of traders are needed to have viable, highly traded futures contracts.

For producers or users who actually intend to deliver or accept delivery of a product, the use of the futures market is quite simple and much the same as forward contracting. If a producer finds the July wheat futures price acceptable in the spring of the year, he or she can sell a contract and guarantee, within some range, that price. A flour miller may also wish to transfer risk by buying a July wheat contract to accept delivery, thereby assuring a known input price for wheat needed in July. Actually, the farmer and miller are not likely to trade with each other directly; speculators may buy and sell the contracts. Since they are simply trading promises, it is not necessary to buy one contract before selling another one. So long as an opposite trade balances out the first transaction before the contract matures, both the producer and the user have locked in a known price and reduced their risk, and speculators have had the opportunity to make a profit or loss.

Agribusinesses that market agricultural products use the futures market to **hedge** their purchases. Most agricultural marketing firms, especially grain elevators, prefer to make a profit from storing, grading, and handling grain, rather than from price fluctuations while they own the product. To do this, they must make immediate and offsetting transactions in both the cash or "spot" market and the futures market.

Futures markets have been established for many agricultural commodities. Trading is always transacted in standardized units or lots so that traders can be confident of what they are trading. Transactions occur in designated markets, such as the Mercantile Exchange and the Board of Trade in Chicago, where hundreds of traders gather daily to buy and sell futures contracts. Their trades are openly cried out as they meet face-to-face in the trading pits. Often the exchanges approach a state of frenzy as traders try to get the attention of other traders to complete a transaction. But what appears to the casual, uninformed observer to be uncontrolled chaos actually is a highly organized free market activity that reflects the broad market expectations of buyers and sellers jointly determining current market prices.

PART II

AGRIBUSINESS MANAGEMENT: ORGANIZATION AND CONTEXT

This section explores the structure and organization of important types of food and agribusiness firms.

THE ORGANIZATION
OF AN AGRIBUSINESS

OBJECTIVES

- Identify some of the important factors involved in selecting the best organizational form for an agribusiness
- Understand proprietorships, partnerships, corporations, limited liability companies, and strategic alliances as forms of business organization
- Summarize the advantages and disadvantages of each of these organizational forms
- Outline and discuss some of the special forms of partnerships and their relative advantages and disadvantages
- Outline and discuss some of the various types of partners
- Understand the role and impact of current individual and corporate tax laws on a firm's organizational structure

INTRODUCTION

An agribusiness may represent a firm with billions of dollars of sales that employs thousands of people, or it may be as small as an individual who is a part-time seed corn salesperson. Agribusinesses may engage in a variety of activities that are related to the production, processing, and marketing of food and fiber products. Though the one-person or one-family agribusiness is not uncommon, most of the actual business volume in agribusiness is now conducted by enterprises that employ hundreds or even thousands of people.

Every agribusiness is owned by someone, and it is the circumstances of ownership that give an organization its specific legal form. There are five basic business forms: the single proprietorship, the partnership, the corporation, the

The right form of business organization is required to cover the varied goals of firm owners, employees, and customers.

limited liability company (LLC), and the cooperative. The form of organization is not necessarily dictated by the size or type of agribusiness: nearly every conceivable size and kind of agribusiness may use any of these five business organization forms. In addition to these five forms, strategic alliances are also being used by agribusinesses as a form of business organization. Strategic alliances can take a variety of forms, and represent an important way that food and agribusiness firms work together today.

The many advantages and disadvantages of each of the five organizational forms must be weighed carefully when attempting to choose the proper one for a specific firm because each form tends to fit some situations better than others. And, even if an agribusiness is organized one way, customers, suppliers, and partners may be organized in another way. So, it is important to understand these different business forms for a number of reasons. This chapter will discuss the important characteristics of each of the first four forms of business organization and strategic alliances, as well as the factors that affect their choice for a specific agribusiness situation. Because of their special prominence in the food and agribusiness markets, we will look more closely at cooperatives in Chapter 5.

FACTORS INFLUENCING CHOICE OF BUSINESS FORM

Each form of business organization has its own individual characteristics. Owners and managers must choose the most appropriate form for their unique circumstances. An agribusiness may want to change its legal form of organization as it grows or as economic and other conditions change. When deciding which form of organization is best, an owner or owners must answer several important questions:

1. What type of business is it, where will it be conducted, and what are the owners' objectives and philosophies for the agribusiness?
2. How much capital is available for the firm's start-up?

3. How much capital is needed to support the agribusiness?
4. How easy is it to secure additional capital for the agribusiness?
5. What tax liabilities will be incurred and what tax options are available?
6. How much personal involvement in the management and control of the agribusiness do the owners desire?
7. How important are the factors of stability, continuity, and transfer of ownership to the firm owners?
8. How desirable is it to keep the affairs of the agribusiness private, and carefully guard any public disclosure?
9. How much risk and liability are the owners willing to assume?
10. How much will this form of organization cost and how easy is this form of agribusiness to organize?

A careful evaluation of these factors will allow the selection of the most appropriate form of business organization in each case.

THE SINGLE PROPRIETORSHIP

The oldest and simplest form of business organization is the **single or individual proprietorship,** an organization owned and controlled by one person or family. In the United States, it is the most popular form of organization. In 1997, there were 17 million nonfarm single proprietorships, which together comprise almost 73 percent of the business entities in the United States. In that year, it was estimated that this relatively large number of small businesses accounted for only about 5 percent of total U.S. business receipts.

Proprietorships tend to be small businesses, although there are notable exceptions. The fortune amassed by the eccentric billionaire Howard Hughes, for example, was largely accumulated from a single proprietorship. When a business reaches a certain size, other forms of business organization usually become more attractive for a variety of reasons. Some very large corporations today are essentially owned and controlled by a single individual, but the advantages of the corporate form at some point resulted in the firm changing from a proprietorship to a corporate form of organization.

Advantages of Proprietorships

The legal requirements necessary to organize as a single proprietorship are minimal. About all that is required is an individual's desire to start a business and the purchase of a license, if one is required for that particular kind of business. If the owner wishes to do business under an assumed name, that is, if the business is to be conducted under a name other than that of the owner, most states require that the assumed name be registered.

Jessica Alverson is a good example. She decided to start a floral shop, which she wanted to call Fragrant Floral because she preferred doing business under that name rather than using her own. Consequently, she registered the new name. Thereafter, those who did business with Fragrant Floral were aware that they were really doing business with Jessica Alverson.

If Fragrant Floral is organized as a proprietorship, Jessica Alverson takes on all risks and rewards for her business.

The **proprietorship** gives the individual owner complete control over the business, subject only to government regulations that are applicable to all businesses of that particular type. The owner exerts complete control over plans, programs, policies, and other management decisions. No one else shares in this control unless the owner specifically delegates a portion of the control to someone else. All profits and losses, all liability to creditors and liability from other business activities are vested in the proprietor. The costs of organizing and dissolution are typically low. The business affairs are completely secret from all outsiders, except for select governmental units such as the Internal Revenue Service.

Whenever **capital** is needed it is supplied by the owner from personal funds or is borrowed against either the owner's business or personal assets. Personal and business assets are not strictly separated as they are in some other business forms; therefore, if the owner as an individual is financially sound, lenders will be more likely to extend funds. The proprietors can sell their businesses to whomever they wish, whenever they wish, and for whatever price they are willing to accept. They can take on as much risk or liability as they wish, but it is important to note that they are personally liable for whatever risk they assume.

A single proprietorship pays no income tax as a separate business entity. All income that the business earns is taxed as personal income even though the IRS (Internal Revenue Service) requires the filing of a separate form to show business income and expenses. Since a proprietor cannot pay him/herself a salary, the amount left over at the end of the year is treated as personal income or salary. The proprietor may choose to keep this money in the business or use part or all of it for personal expenditures.

The proprietorship can conduct business in any of the 50 states without special permission other than whatever licenses are required for that particular kind of business. This is a right guaranteed to individuals by the United States Constitution, which provides that "citizens of each state shall be entitled to the privileges and immunities of citizens of the several states." The person who desires the lowest cost (to organize), simplest, most self-directed, most private, and most flexible form of agribusiness will choose the single proprietorship.

Disadvantages of Proprietorships

Perhaps the most important disadvantage of the proprietorship is the owner's personal liability for all debts and liabilities of the business, which can extend even to the owner's personal estate. In a proprietorship there is no separation between business assets and personal assets. Consequently, this form of business organization is characterized by what is called unlimited liability. The owner's liability does not stop with business assets; it also extends to personal assets. Such assets can be, and often are, used to satisfy financial obligations. So, if Fragrant Floral starts losing money, and the bank demands payment on a loan made to the business, Jessica Alverson is personally liable for the payment of the loan. If the business cannot cover the loan, the bank can typically demand payment from Jessica's personal savings and investments, or other assets she owns.

Another important disadvantage relates to the generally limited amount of capital funds that one person can contribute. Lenders are also somewhat reluctant to lend to an individual owner unless the owner's personal equity can guarantee the loan. Proprietorships often find that they are starved for capital, and this serious disadvantage may do more than stunt growth. Thousands of bankruptcies each year can be traced to a serious shortage of capital when a business is started.

While freedom from business taxes is generally an advantage, it may be a disadvantage. Since business profit in a proprietorship is considered personal income to the owner, a high business profit may throw the owner into a higher tax bracket than would the corporate form of business organization. This is especially disadvantageous if extensive funds are needed for the growth and expansion of the business. Corporate tax rates along with other tax regulations may provide an advantage in such cases.

The concentration of control and profits in one individual may also be a disadvantage. Many highly trained and motivated employees want to participate financially in the business for which they work. They may also be uneasy about the fact that their futures depend on the health and viability of a single person. Thus, proprietorships may experience some difficulty in hiring and keeping good people. Without good, highly motivated employees, the owner may find as the business grows that he or she is "wearing too many hats," with the end result that the business suffers.

Finally, the proprietorship lacks stability and continuity because it depends so heavily on one person. The death or disability of that one person, in effect, ends the business. Proprietorships may be difficult to sell or to pass on to heirs. This is particularly true if they become sizable businesses. Individual shares or parts of the business cannot be parceled out to several individual owners or to heirs in the same way that shares of a corporation can.

For example, Molly's was a regionally noted steakhouse in a small Midwest town. Molly started and managed this business profitably for roughly 30 years. However, Molly was always in charge of everything: food ordering and preparation, tabulating the bill, and collecting money. Food servers had few responsibilities, and turnover among employees was high. Molly refused to bring any family members into the business. But, the food was excellent, and Molly had a very good business, though it remained quite small because she "just couldn't work any more hours." Then, Molly was hospitalized and had a lengthy recovery period. As a result, Molly simply closed the doors and her restaurant was out of business. Some planning and forethought on Molly's part could have averted the closing of this small, but profitable proprietorship.

PARTNERSHIPS

A **partnership** is the association of two or more people as owners of a business. There is no limit to the number of people who may join a partnership. Apart from the fact that a partnership involves more than one person, it is similar to the proprietorship. Partnerships can be based upon written or oral agreement, or on formal contracts between the parties involved. It is strongly urged, however, that if you consider joining a partnership the partnership agreement should be in writing to avoid disagreement and misunderstanding among partners at a later date.

Partnerships can be formed by law whenever two or more people act in such a way that reasonable people would be led to believe they are associated for business purposes. In 1997, there were approximately 1.8 million partnerships in the United States, and they accounted for about 7 percent of total business receipts. Partnerships are the simplest form of business organization by which a number of people can pool their resources and talents for mutual benefit.

There are basically two kinds of partnerships: general partnerships and limited partnerships. A discussion of how each of these partnership arrangements works in agribusiness follows.

General Partnerships

By far the most common form of partnership is what is called a **general partnership.** Continuing with Fragrant Floral, after a couple of years in operation, Jessica Alverson has been very impressed with one of her employees, Erika Lewandowski. Erika has expressed an interest in becoming involved in the business, and has some money from an inheritance she is willing to invest in the business. Jessica and Erika decide to enter a partnership, and rename the business Fragrant Floral and Perfect Gifts. The new name reflects the partner's desire to expand the business into a broader line of gift items. So, with Erika's commitment of capital and her desire to be further involved in the business, a new partnership is formed.

In a general partnership each individual partner, regardless of the percentage of capital contributed, has equal rights and liabilities, unless stated otherwise in a partnership agreement. A general partner has the authority to act as an agent for the partnership, and normally participates in the management and operation of the business. Each general partner is liable for all partnership debts, and may share in profits, in equal proportion with all other partners. If the partnership struggles and has financial problems, all liabilities are shared equally among the partners for as long as sufficient personal resources exist.

However, when one partner's resources are exhausted, remaining parties continue to be liable for the remaining debt. General partners may contract among themselves to delegate certain responsibilities to each other, or to divide business revenues or costs in some special manner (e.g., according to funds invested or job responsibility). Each general partner can bind the partnership to fulfill any business deal made. While the partnership is usually treated as a separate business for the purposes of accounting, it is not legally regarded as an entity in itself, but as a group of individuals or entities. Thus, there is no separate business tax paid by the partnership. Like the proprietorship, partners may not pay themselves a salary. Money left at year's end is divided among the partners and this is their profit or salary from the partnership. The income is taxed at the individual rate.

Limited Partnerships

All partnerships are required by law to have at least one general partner who is responsible for the operation and activities of the business, but it is possible for other partners to be involved in the business on a limited basis. A **limited partnership** permits individuals to contribute money or other ownership capital without incurring the full legal liability of a general partner. A limited partner's liability is generally limited to the amount that the individual has personally invested in the business. The state laws regulating limited partnerships must be strictly adhered to and these acts spell out the limited status of partners: first, the limited partner can contribute capital but not services to the partnership, and second, the limited partner's surname cannot appear in the business's name (unless the partnership had previously been carried on under that name, or unless a general partner has the same surname). Limited partnerships are relatively few in number; therefore, the balance of the discussion of partnerships will apply to general partnerships. However, it is important to note that if a limited partner takes an active role in managing the business, her/his limited liability may cease.

Advantages of Partnerships

Partnerships are just about as easy to start as proprietorships. They require very little expense, although an attorney competent in partnership law should be engaged to draw up the partnership agreement. The partnership may operate under an assumed or fictitious name, providing it is registered in accordance with state laws. A partnership can generally bring together many more resources than a proprietorship because of the increase in the number of people involved. These added resources are not only financial in nature; the business likewise benefits from the variety of unique talents that many different individuals can bring to it. Partners are a team, and because each team member shares in the responsibility and profits, partners are more likely to be motivated than employees of a single proprietorship or corporation. Additional partners can be brought in if more money or talent is needed.

Partners, as individuals, pay taxes only on the income generated from their share of the profits. There is no business tax per se, and this can be a considerable advantage, depending on the income of the partners. Control or management of business decisions and policies is concentrated among the partners. Generally, partners divide the responsibilities of the business; that is, one will head sales, another will head operations, input acquisition, etc. This can be done on either a formal or an informal basis. Partners may sell their interest in the business to others if the remaining partners agree. The business affairs of a partnership are confined to the partnership, and this element of privacy is one of the prime reasons why many people choose to do business as partners. Partnerships share the same privileges of doing business in other states as the proprietorship form of business organization.

Disadvantages of Partnerships

By far the biggest disadvantage of the partnership is the unlimited liability of each general partner. There are many known cases where one partner has generated financial obligations for the partnership, and then because that individual has been personally insolvent, the other partners have had to pay the bills. Even limited partners must be very careful that they do not give any appearance of being active in the management of the business. The law has frequently been enforced on the basis of a person's actions rather than on the basis of the written documents. If a person acts as a general partner would act, then that partner may be forced to accept all the liabilities that such status would incur, even if the formal, written agreement says the individual is a limited partner.

The liability of general partners has created a second disadvantage: partnerships usually have only a limited number of members. Imagine a partnership with 100 members, each able to bind the partnership legally to contracts and to other obligations. Of course, the limited partnership was created to cure this problem, but it has not been very successful. A limited partnership often suffers from the lack of available funds and talented people as compared to a corporation.

Another disadvantage is the lack of continuity and stability of a partnership. When a partner leaves the partnership as a result of withdrawal, death, or incapacity, a new partnership must be formed. The old partner's share must be liquidated, and this can often place a severe burden on the partnership's capital position. Another problem is that which occurs if one of the partners becomes incapacitated by accident, ill health, old age, or for some reason fails to pull a full share of the load. Often the only way to remove such a partner is to liquidate the entire business. When a partner leaves, it is often hard to determine what that individual's share is worth. For this reason, a formula and pay-off method should be incorporated in the original partnership agreement. Then the business may more easily be dissolved and a new one can be formed. If the means for establishing the value of the partner's share and the process for transfer and acceptance of new partners has been firmly established in the written partnership agreement, this transition can be reasonably smooth. While being taxed on income as separate individuals can be an advantage in some situations, it can be a disadvantage in others (just as with the single proprietorship).

Finally, one very important consideration is the need for a carefully drafted, written partnership agreement. Any person entering into a partnership should find the most competent legal assistance possible, an attorney who is familiar with the problems of partnerships, and trust that attorney to prepare an agreement for the agribusiness. When partnerships are formed, the members are all on good terms and cannot imagine feeling any other way about each other, but situations change and people change. Not only will a written contract of partnership provide solutions if problems occur, but also it will serve as a ready reference as important decisions are made in the partnership.

Types of Partners

In forming a partnership, the business entity may include a variety of different partners. Some of the more common types of partners are outlined below with a set of characteristics unique to that particular partner.

General: A **general partner** is one who is active in the management of the partnership, typically has an investment in the business, and is subject to unlimited liability. If nothing is in writing, all partners are assumed to be general partners.

Limited: A **limited partner** has taken steps to limit his/her liability in the partnership. This agreement must be in writing. Also, limited partners may not take an active role in managing the partnership.

Senior: A **senior partner** is typically an individual who helped form the partnership or one who has seniority in the business. Senior partners typically run the business, have an investment in the firm, and tend to receive the major portion of partnership profits.

Junior: A **junior partner** is typically younger and has not been in the business as long as a senior partner. Junior partners tend to receive a smaller portion of the business profits. Junior partners rarely take an active role in managing the business affairs of the partnership.

Secret: A **secret partner** is an individual who desires to take an active role in managing the partnership but is not known to be a partner by the general public. Thus, secret partners do have unlimited liability. An individual may desire to become a secret partner due to involvement in other business ventures. If the partnership fails, the general public will not know of this individual's involvement, thus other businesses controlled by this person would be unaffected by negative public opinion.

Silent: A **silent partner** is known by the public and not active in managing the partnership. An individual may seek to be a silent partner to add name recognition to a firm and to continue having the advantage of limiting liability in the enterprise.

Dormant: A **dormant partner** is not active in managing the partnership and is not known by the general public. Thus, dormant partners retain limited liability. The reason firms may seek dormant partners is to obtain additional investment capital for the business.

Nominal: A **nominal partner** means "in name only." Nominal partners are not active in the business and have no investment. If Jake Smith owned a horse ranch and named it Jake Smith and Sons, assuming they have no investment in the business, Jake's sons would be nominal partners.

THE CORPORATION

A **corporation** is a special legal entity endowed by law with the powers, rights, liabilities, and duties of a person. The corporate form of business organization typically facilitates the accumulation of greater amounts of capital when compared to proprietorships and partnerships. Without the corporate form of organization it is impossible to imagine the creation of today's huge business entities, which employ hundreds of thousands of people and are worth billions of dollars.

In 1997 there were only about 4.7 million corporations in the United States, or about 20 percent of all businesses. However, they generated about 88 percent of the business receipts. Many corporations are multinational giants. The 500 largest corporations, as rated by *Fortune* magazine, generate two-thirds of all industrial sales in the United States. However, most corporations are relatively small, and many are really one-person businesses whose owners have chosen the corporate form of organization as the best for their unique business circumstances. As a matter of fact, in 1997, 82 percent of all United States corporations had annual receipts of less than $1 million.

Nonprofit Corporations

Most corporations are formed for profit-making purposes; however, there are thousands of nonprofit corporations in existence. These **nonprofit corporations** embrace many areas of activity, including those of religious, governmental, labor, and charitable organizations. Federal and state laws specify the numerous forms that these nonprofit corporations may take, along with very specific regulations as to their purpose and operation. A competent attorney can advise whether a nonprofit corporate form of organization is the most appropriate for a particular agribusiness situation. Again, the legal interpretation will be made on the basis of the ways in which the corporation acts, and not on the basis of how it is described in written legal documents.

Examples of nonprofit corporations include some cooperatives, some agricultural trade and research groups, and some farm organizations, such as the National Dairy Herd Improvement Association, Inc. Nonprofit corporations are exempt from certain forms of taxation, and generally they cannot enrich members financially. In many cases, the nonprofit corporation must secure a formal exemption from paying corporate income taxes.

The Nature of the Corporation

The corporation as we know it today is a rather recent innovation compared to proprietorships and partnerships. The early American colonists were very suspicious of this form of organization, and it was not until the 1860s that most states had provided laws allowing for the formation of corporations. A corporation can own property, incur debts, and be sued for damages, among other things. The important distinction to remember is that the owners (stockholders) and managers do not own anything directly. The corporation itself owns the assets of the corporation.

Forming a corporation requires strict adherence to the laws of the state in which the business is being formed. Usually, one or more persons join together to create a corporation. A series of legal documents must be created and examined by the state's designated department for establishing corporations. If the legal formalities are in proper order, and if the proper fee for incorporation has been paid, a charter authorizing the applicants to do business as a corporation is issued. Additionally, a corporation maintains the following legal documents: **articles of incorporation,** which are filed with the state and which set forth the basic purpose of the corporation and the means of financing it; the **bylaws,** which specify such rules of operation as election of directors, duties of officers and directors, voting procedures, and dissolution procedures; and **stock certificates** or shares detailing amounts of the owners' investments.

The laws relating to the formation of proprietorships and partnerships are fairly well established and uniform throughout the nation, but considerable differences in the requirements for forming a corporation exist among the various states. Individual state laws and statutes must be carefully considered by those who wish to form a corporation. Selection of an attorney who is well versed in the corporate law of the specific state of interest is essential to avoid potentially serious organizational problems in the agribusiness corporation.

Stock of the Corporation

When corporations are formed, shares of stock are sold to those who are interested in investing and risking their money in the enterprise. A share of **stock** is a piece of paper, in prescribed legal form, which represents each person's amount of ownership in the corporation. **Common stock** normally carries the privilege of voting for the board of directors that oversees the activities of the corporation. **Preferred stock** differs from common stock in that it is usually nonvoting, and has a preferred position in receiving dividends and in redemption in the case of liquidation. Thus, voting rights are exchanged for lower risk on the investment of capital in the corporation.

Each state has what are commonly called **blue-sky laws,** which regulate the way in which corporation stock may be sold and which protect the rights of investors. Individual state laws must be consulted prior to the sale of stock in a corporation. The most common way of financing corporations is through the sale of stock, but financing through bonds, notes, debentures, and numerous means of borrowing against assets is also practiced (see Chapter 14).

Thus, there are two specific types of stockholders in a for-profit corporation. **Common stockholders** are willing to take risk. They invest in the corporation typically because they believe the value of their stock will increase over time. They are also the true owners of the business. At the annual meeting common stockholders are the ones that vote for the board of directors of the corporation. Each common stockholder has one vote per share of common stock. Some corporations have started issuing nonvoting common stock, but this is not prevalent across most corporations.

Those who desire to control a business may embark on an attempt to acquire the majority of shares of a firm. Also, current stockholders can sign a proxy vote that allows an individual to vote for them at the annual meeting. Many well-publicized attempts to take over corporations are noted as "hostile takeovers." In this case, current corporate management does not desire to lose their control of the firm. Proxy fights may ensue, where the takeover candidate attempts to gain control by enlisting stockholder approval or by simply bidding at a higher level than the current market price for outstanding shares.

In contrast, **preferred stockholders** tend to take less risk on their investment in the corporation. The price of preferred stock typically fluctuates less than common stock in publicly traded corporations. Also, preferred stockholders often invest for the dividends granted by the corporation. Once a firm has established a trend of paying quarterly dividends, management may not wish to break the record. Thus, some firms, even in an economic downturn, may choose to borrow money to pay their quarterly dividends to preferred stockholders.

How the Corporation Functions

The common stockholders in a corporation elect the **directors.** The number of directors may vary according to the bylaws of the organization. It is the responsibility of the **board of directors** to supervise the affairs of the corporation. In a large corporation the thousands of individual stockholders exercise very little actual control. Those they vote for to serve as directors may be unknown to them and are often preselected by a small group of majority stockholders who are allied with top management of the corporation. The board represents the interests of the stockholders, and their major function is to elect officers, hire top management, and evaluate the progress of the business. Also, the board usually has a major role in shaping the vision and mission for the business. In a small corporation there is usually a very close relationship between stockholders and the board of directors; in fact, there could be only one stockholder who, in effect, is in complete control of the corporation.

Advantages of Corporations

The primary advantage of the corporate form of business organization is that the **stockholders** (owners) are not personally liable for the debts of the organization, and in most cases are not responsible for any liability that occurs through the corporation's business activities. The assets of the corporation are all that are at stake in settling most claims. The stockholder can, therefore, only lose the amount he/she has invested in the firm. With the corporate structure it is possible to delegate authority, responsibility, and accountability, and to secure outstanding, highly motivated personnel. Corporations can offer their personnel such benefits as profit sharing and stock-purchase plans, which encourage a high degree of dedication and loyalty to the corporation.

Transfer of ownership is also easier in a corporation than in other business form. Usually, a stockholder can sell shares of stock to anyone for any price that the buyer is willing to pay. An owner can also transfer individual equity to heirs or to others much more easily. However, sale of stock in a small, unknown corporation may not be easy. Finding someone who is willing to risk funds becomes much easier once the stock is traded. As a corporation becomes larger and begins to develop a ready market for its stock, brokerage houses that specialize in the sale of stock and are in constant touch with potential investors may trade the stock. New issues of stock may be purchased by a group of brokerage houses, which, in turn, sell the stock to the general public. Many companies have their stock traded in secondary markets that list daily price movements and facilitate buying and selling of these stocks. Examples of these markets include the New York Stock Exchange, the American Stock Exchange, and NASDAQ.

Stock prices for important markets such as the New York Stock Exchange are widely reported.

Because corporations' ownership rights are traded freely, it is relatively easy for them to raise large amounts of equity capital. The combined investments of hundreds and even thousands of investors made the huge corporate giants of American business possible. However, these investment funds are not automatic just because the corporation decided to sell stock. Buyers invest because of the proven performance or anticipated future performance of the firm.

Finally, the corporation is perpetual in nature. Death, withdrawal, or retirement of its shareholders has little effect on the life of the corporation. This is another advantage that makes investment in a corporation more attractive to those with funds to risk.

Disadvantages of Corporations

The greatest disadvantage of the corporate form of organization is taxation and regulation. The corporation is taxed on funds it earns as profit; then, after it has paid dividends to its stockholders, the stockholders must again pay income tax on the amount that is received as dividends. (This may not always be a disadvantage, as will be discussed later in this chapter.) In addition, there are many states that impose special levies and taxes on corporations, and there are many more laws and regulations controlling the activities of corporations than there are for other organizational forms.

The corporation must withstand a lack of privacy because reports must be made to stockholders and states and because the federal government may require disclosure whenever a stock offering is made to prospective purchasers. A corporation that is chartered to do business in one state may not do business in another state unless it complies with the second state's laws of registration, taxation, and so forth. Finally, individual owners (stockholders) of larger corporations have little, if any, control over management and policies of the corporation. Often their only recourse in the event of dissatisfaction is to sell their stock.

The costs of taxes, records (which must be comprehensive), and operation of the corporate business can be significantly higher than the costs for other forms of organization; for that reason, the corporate form should be evaluated carefully before it is adopted by the agribusiness.

Finally, in smaller corporations, the issue of limited liability is not always so clear. Banks and other lenders may insist on personal guarantees from stockholders for loans before lending a small corporation money. Boards of directors may be sued for a variety of reasons, and the board may well be the owners of the corporation. So, while the corporate form clearly offers more legal protection than either a proprietorship or a partnership, specific circumstances may make such protection relatively thin.

Closely Held Corporations

A special form of corporation has been designed to offset some of the disadvantages of the regular corporation. This form is called the **S-corporation.** Subchapter S of the Internal Revenue Code makes it possible for the owners of a corporation to elect to be taxed as individuals, in the same manner as owners of a partnership or proprietorship. So, the benefits of incorporating as an S-corporation can be significant. The corporation can avoid the double taxation paid by the regular corporation.

Several qualifications must be met: there cannot be more than seventy-five stockholders, the stockholders must be individual persons rather than corporations, and they cannot be nonresident aliens. The election to be taxed as individuals must occur prior to the start of the corporation's fiscal year. Owners, therefore, cannot simply wait to see which method of taxation would be better once the year is over and the returns are prepared. In addition, the S-corporation must be a domestic corporation. It must be a corporation that is organized under state law. The S-corporation can issue only one class of stock if organized after May 27, 1992. Finally, the S-corporation cannot have passive income of more than 25 percent of total receipts in three consecutive tax years. Examples of passive income would be rent, royalties, interest, and dividends.

The rules to qualify for this status essentially guarantee that the corporation is relatively small, and often family-held (number of shareholders), and the corporation is production-oriented. While there are a number of benefits, there can be many pitfalls to this form of organization and agribusiness corporations should consult competent legal and accounting professionals before selecting it.

Fragrant Floral and Perfect Gifts, Inc.

Let's revisit Fragrant Floral and Perfect Gifts. Jessica and Erika have been in business together about a year. And, after attending a management course at a local community college, they are introduced to the corporate form of business organization. They decide they like the idea of limiting their personal liability through the use of the corporation. And, they also like some of the ways that the corporate form of organization eases transitions. What if Erika wanted to move away, but did not want to give up her investment in the business? In addition, they have been talking about expanding the business by establishing another location—so access to investment capital becomes more important. A corporation would help make this possible. After reviewing the advantages and disadvantages, they believe they should become an S-corporation, to help better manage their personal tax situations.

So, working with an attorney, they develop a set of articles of incorporation, and a set of bylaws. They file this paperwork with the state and are granted corporate status. They issue 110 shares of stock, 50 each to Jessica and Erika. In addition, a friend in the landscaping business, Peter Yu, is excited about the possibilities for Fragrant Floral and Perfect Gifts, and he buys 10 of the shares. Jessica, Erika, and Peter are on the board of directors, and Jessica is chairperson of the board. Fragrant Floral and Perfect Gifts, Inc. is open for business!

LIMITED LIABILITY COMPANIES

A **limited liability company (LLC)** is a relatively new type of business organization form that closely resembles a partnership, but provides its members limited liability. Thus, creditors or others who have a claim against an LLC can pursue the assets of the LLC to satisfy debt and other obligations, but they cannot pursue personal or business assets owned by the individual members of the LLC.

Advantages of LLCs

Limited liability companies can include any number of members and ownership is distributed in accordance to the fair market value of the assets contributed. Also, net income generated by an LLC is passed on to its members in proportion to their shares of ownership. The net income is then reported on the members' individual tax returns and taxes are paid by the members and not by the LLC. An LLC is not required to file articles of incorporation as would be true of a corporation. However, it is still a good idea to record contributions and distributions of assets, revenue and expenses, as well as agreements as to how the LLC will operate. These can be included in an article of organization or an operating agreement.

Disadvantages of LLCs

Although the LLC has the limited liability advantage of a corporation, it does not have some of the other advantages associated with the corporate form of organization. The LLC cannot deduct the cost of fringe benefits, such as insurance costs and the use of vehicles by members. Also, it does not automatically continue in the event of the death of a member. Instead, it may be perpetual, end on a set date, or upon an event such as a death of a member, as outlined in the articles of organization or operating agreement.

Limited liability companies are an attractive organizational form for those individuals who desire the simplicity and flexibility provided by a partnership combined with the limited liability offered by a corporation. However, LLCs are a relatively new form of organization, so some questions remain concerning their legal nature. Competent and experienced legal counsel should be sought when considering this form of organization.

STRATEGIC ALLIANCES

Many agribusinesses have formed **strategic alliances** and related cooperative relationships with other firms. Strategic alliances are cooperative agreements between firms that go beyond normal firm-to-firm dealings, but fall short of merger or full partnership and ownership. Such alliances can include joint research efforts, technology sharing agreements, joint use of production facilities, agreements to market each other's products, etc.

Advantages of Strategic Alliances

There are several advantages to firms that form strategic alliances. First, firms can collaborate on technology or the development of new products. The areas of computer hardware and software and biotechnology are examples. Second, firms can improve supply chain efficiency by working together. An alliance between a feed firm and a large integrated swine business would provide an example. Third, firms can gain economies of scale in production and/or marketing. A food firm might enter into an alliance with a broker to distribute its products into a new region of the country. Here, the focus is market expansion and the scale economies this can bring. Fourth, firms, particularly small firms, can fill voids in their technical and manufacturing expertise. A firm in the precision agriculture arena may have an alliance with a major farm equipment manufacturer to collaborate on the development of a new sensor-based system for harvesting melons. Fifth, firms can acquire or improve market access. A firm selling animal health products and an e-business firm might enter a strategic alliance. The e-business firm provides the information technology infrastructure, and access to customers. The animal health firm provides products and an existing customer list. Finally, allies can direct their combined competitive energies into building a competitive advantage and defeating a mutual rival.

Disadvantages of Strategic Alliances

There can also be disadvantages to forming strategic alliances. First, establishing effective coordination between independent companies is both challenging and time consuming. How are responsibilities divided? How will returns be divided? These questions, and a hundred more like them, must be answered in a successful alliance. Second, there may be language and cultural barriers to overcome, as well as attitudes of suspicion and mistrust. This issue includes the need for mutually shared goals. If two alliance partners have different objectives from the alliance, the seeds for long-run problems have been planted.

Third, the relationship may cool at some point in the future and the desired benefits may never be realized, but information may have already been shared. This commonly occurs when there is management turnover among the alliance partners. The situation and personalities that were part of the original deal change over time, and the new management team may look at the world—and the deal—differently. Finally, a firm may become too dependent on another firm's expertise and capabilities and fail to develop its own internal capabilities. Firms must be careful with respect to what they do themselves, and what they depend on from a partner. In the animal health example above, the alliance may be a wonderful move for the animal health firm. Or, it could leave them vulnerable in three years if a competitor purchases their alliance partner and they have no internal e-business capabilities.

A strategic alliance is an attractive organizational form for those firms that want to preserve their independence rather than merge with another firm when trying to either remain competitive or enhance their competitive position. This form of organization enables those firms to collaborate with other firms to enhance their own capabilities, develop new products, and compete more effectively. However, as mentioned above, there are also disadvantages that should be seriously considered before forming a strategic alliance.

TAXATION

As discussed earlier in this chapter, proprietorships, partnerships, and limited liability companies pay taxes on their business profits at the personal rate. Corporations, however, have a separate tax rate for corporate profits. If the corporation distributes dividends to shareholders, then these individuals also pay personal income tax on this amount. Hence, many corporate profits are subject to double taxation—first, the corporate profits are taxed, and then the dividends (paid from after-tax profits) are taxed again.

A proprietorship completes either Schedule C or Schedule F (farms and ranches) to report business income, expenses, and profit or loss. This amount is then carried forward to the individual's Form 1040 where it would be taxed at the individual rate. Likewise, partners and members of a limited liability company complete Schedule K to tabulate their profit distribution from the business and carry this forward to their Form 1040 to pay personal taxes on their share of profits from the business. The partnership and LLC must complete Form 1065 to show all partnership income and profit distributions. Also, S-corporation shareholders report income from the corporation on their Form 1040.

Corporations report sales, expenses, and profits on Form 1120 for ordinary corporations and Form 1120S for S-corporations. Taxes on regular corporations are calculated from the tax rates included in the tax table used for corporations.

The point here is that tax issues do enter in an important way when selecting a form of business organization, especially for smaller businesses. And, when making these choices small business owners and entrepreneurs need competent legal and tax advice.

SUMMARY

Agribusinesses represent nearly every conceivable kind of business organization. A great many are owned and controlled by one person as a proprietorship, or by two or more people as a partnership. These forms of business are the simplest forms and allow their owners complete flexibility, minimize red tape, and incur no corporate profits tax. But owners are personally liable for any debts or lawsuits against their business. Also, the longevity of the business is limited to the life of their owners. The limited liability company (LLC) is a relatively new type of business organization that closely resembles a partnership, but provides its members with limited liability and the opportunity to influence the longevity of the business.

The corporation is an organization created for the purpose of carrying on business. Because it is a legal entity, it can own property in its own right, sue or be sued, and carry on business in its own behalf. Its owners are separate legal entities; thus their liability is limited to the amount of their investment in the firm. Also, the life of a corporation does not depend on how long its owners live. However, corporations generally must pay a special corporate profits tax and must regularly report their activities to federal and state governmental units. The S-corporation is a special type of corporation for small firms; they gain the benefits of a corporation, but are exempt from corporate profits tax.

Strategic alliances are agreements to cooperate that stop short of formal legal combinations. These alliances offer a number of benefits and are commonly used in the food and agribusiness markets. But, there are clearly some limitations and problems to be managed, so firms must explore both the advantages and disadvantages before forming a strategic alliance. What will work in one situation may not work in another situation.

The best form of business organization for a particular agribusiness is an infrequent business decision for both management and owners, but an extremely important one. As an agribusiness firm grows, it must often consider moving toward becoming a regular corporation. And, the timing of that decision is highly important.

DISCUSSION QUESTIONS

1. List and discuss five critical factors in choosing a form of business organization for:
   a. a family farm
   b. a feed firm seeking to expand in the near future
   c. a college fraternity/sorority
   d. a physician who owns farmland
   e. a breakfast cereal manufacturer who wants to collaborate with a juice company on a special promotion emphasizing the importance of a good breakfast
2. Compare and contrast the advantages and disadvantages of a proprietorship and a partnership.
3. You and two college friends want to start a business. You choose a partnership. Carefully outline some of the considerations of setting up this new business. What are some of the specific issues that need to be addressed in your written partnering agreement?
4. Discuss the characteristics of each of the different types of partners.
5. What makes a corporation unique from other business firms? Be specific.
6. When and why would an agribusiness firm choose to change its form of business organization?
7. How would a corporation pursue the purchase of another corporation:
   a. under friendly terms (i.e., both firms want to merge)
   b. under unfriendly terms
   How does your answer differ depending on whether the corporation is publicly traded or closely held?

8. Given the characteristics of an S-corporation, give an example of a firm that would benefit by this classification.
9. Given the characteristics of a limited liability company, give an example of a firm that would benefit by this classification.
10. Would a corporation ever choose to operate as a partnership or proprietorship? Why?

CASE STUDY — Should We Incorporate?

You and two friends, also students, have started a new business enterprise. The business has been organized as a partnership on an equal share basis and produces soil test kits for homeowners. The idea came together during a field trip for a soil science class you were taking. One of your friends was in horticulture, the other in agronomy. Somehow the discussion turned to, How do homeowners do this soil testing stuff? and your idea was born. After some development work, you and your friends have devised a very simple test kit that lets a homeowner do a quick assessment of the fertility requirements for their yard. The future looks bright.

You have the responsibility for the financial side of the business and estimate that the variable cost per test (i.e., the increase in total cost for each test produced) is $18.50 and fixed costs (those costs that are the same regardless of sales) amount to $90,000. Variable costs would include the materials for the test kits, the packaging, and the mailing costs. Fixed costs would include your advertising, rent on warehouse space, etc. (your total cost equals your total variable cost plus your fixed cost). Your average selling price is $25 per test kit. You expect to sell 30,000 test kits this year.

As expected, business has been brisk. Overfertilizing yards has come under fire in the press recently, and homeowners seem eager for your kits, which help them decide precisely how much fertilizer to put on the yard. Then, one day in an agribusiness management class, you hear a lecture on business organization. After listening carefully to the lecture, you decide it might be profitable for you to incorporate. Each of the three current partners would become stockholders in the new corporation, and own one-third of the corporate stock, and work for the corporation.

Not desiring to hire an outside consultant, you are determined to evaluate the incorporation question yourself. You have assembled the following data and relevant facts for the problem. And, you decide the answer should be clear if you address the following questions. Given the assumptions below and your knowledge of the two business forms, answer questions 1 to 11.

Assumptions:
a. The tax rate for individuals is:
   15 percent of personal income 0–$25,000, plus
   28 percent of all personal income in excess of $25,000.
b. The tax rate for corporations is:
   15 percent of profits 0–$50,000, plus
   25 percent of profits $50,001–$75,000, plus
   34 percent of profits in excess of $75,000.
c. The corporation pays a salary of $20,000 to each stockholder for their labor and management input.
d. The average per year "red tape" cost for the corporation is $1,000.

e. All corporate profits are paid out as dividends.
f. Assume all other factors have no impact on your decision.
g. Personal income equals income before individual taxes and personal disposable income equals income after tax.

Questions
1. What is the partnership's net income before withdrawals, profit shares, etc.?
2. What is the corporation's taxable income?
3. The partnership, as a business entity, would pay how much in federal income taxes?
4. The corporation would pay how much in federal income taxes?
5. In the partnership, what is the amount of each individual's profit share (before taxes)?
6. In the corporation, how much would each of the stockholders receive as annual personal income (salary plus dividends)?
7. In the partnership, each individual would have what amount in personal disposable income (after tax)?
8. What is the personal disposable income (after tax) for each of the stockholders in the corporation?
9. Based on these calculations, should the partnership stay a partnership or incorporate?
10. If each individual suddenly decided that the value placed on acquiring limited liability should be $2,000 per year (after all taxes), should the partnership stay a partnership or incorporate?
11. What other issues should you consider as you make this important decision?

5

COOPERATIVES IN AGRIBUSINESS

- Discuss the scope and scale of cooperative involvement in the food and agribusiness industries
- Identify the basic principles that ensure that cooperatives serve the needs of member-patrons
- Develop an appreciation for the heritage that has influenced today's food and agricultural cooperatives
- Explain how local and regional cooperatives are organized to serve members
- Define a New Generation Cooperative and identify how it differs from other cooperatives
- Examine some of the challenges and opportunities that face agricultural cooperatives

INTRODUCTION

You have likely had parent, coach, or supervisor tell you more than once to "cooperate!" When they told you this, they meant "think, help out, get along." In agribusiness the same principle applies—cooperation means working together to solve common problems and/or maximize opportunity. A cooperative is a unique form of business organization that works to meet those goals.

Owned, operated and controlled by members, **cooperatives** are a distinctive form of business enterprise. Cooperatives are committed to helping members improve the prices they receive for the products they produce and/or reduce the prices paid for the inputs necessary to grow those products. Cooperatives are committed to helping members find markets, and/or improve the negotiating position of members. Cooperatives provide economic and/or operational benefits to member-owners, then return the profit to the member-owners based on each member's use of the cooperative. Both the individuals and businesses involved are allowed to reap the benefits of the joint endeavor while maintaining their independence.

Cooperatives are an influential business form in the food and fiber marketing system. And, this chapter explores the role of cooperatives in the food and agribusiness markets, their size and scope, as well as how they work. We'll look at their history, cooperatives today, and what the future may hold for this important type of agribusiness.

THE INFLUENCE OF COOPERATIVES

At the beginning of the new millennium, there were more than 48,000 cooperatives in the United States. They generated more than $150 billion in economic activity. Cooperatives reach the everyday lives of one in three individuals in the United States. There are thousands of cooperatives and hundreds of types of cooperatives. From hardware to books; refrigerators to artwork; livestock feed to ethanol; one can find a cooperative business structure in place. There are cooperatively organized, employee-owned groups, cooperative purchasing groups, and cooperative housing organizations, among many others.

Food and Agricultural Cooperatives

Most food and agricultural cooperatives are organized as business corporations (see Chapter 4). However, their basic purpose—to serve the needs of their patron-members rather than to make a profit for investors—distinguishes cooperatives from other forms of business organization. The cooperative organization gives control to the patron-members rather than to outside investors. Cooperatives are typically owned by those who use them, and cooperatives are legally obligated to benefit their patrons. Because of these two facts, the U.S. political system has traditionally chosen to encourage their growth and development as a logical extension of the independent farm business. Special government policies and regulations have played a role in helping farm cooperatives become a major factor in the input, marketing, finance, and service industries of the food production and marketing system.

Virtually every phase of agriculture is touched by cooperatives, either directly or indirectly. Cooperatives helped market over $25 billion worth of grain and soybeans, over $27 billion worth of dairy products and almost $10 billion of fruits and vegetables in 1998. In total, marketing cooperatives helped their members sell over $76.6 billion worth of farm production in 1998 (National Cooperative Bank). U.S. producer-owned cooperatives account for more than one-tenth of the total value of U.S. agricultural exports—exporting every major commodity from bulk grains and feeds to high-value branded products. Cooperatives are also a major force in the agricultural input industries. In 1998 cooperatives handled over $7.7 billion of fertilizer, more than $4 billion of crop protection chemicals, and more than $8.7 billion in petroleum products.

Finance cooperatives lend almost $150 million per day to help their members buy and operate their own farms and finance their businesses. For example, AgriBank, FCB works with its partner, Farm Credit Services to provide loans and other financial services for farmers and rural homeowners in 11 midwestern and southern states. AgriBank is a wholesale bank that provides funding and services to 30 Farm Credit Services offices. Farm Credit Services, also a cooperative, provides insurance, credit, record keeping and other services to about 200,000 customers. Credit is provided to finance purchases of farm machinery, land, seed, animal feed, and animal housing. Cooperative-provided insurance includes products ranging from crop insurance to credit life insurance.

In addition, a wide variety of service cooperatives aid farmers by irrigating their land, insuring their assets, improving their livestock, and providing electric power and telephone service for rural areas.

In 1999, there was a total membership of nearly 4 million in approximately 4,100 separate food and agricultural cooperatives (National Cooperative Bank). Of these 4,100 cooperatives, approximately 53 percent marketed farm products; 36 percent handled primarily farm inputs, and approximately 11 percent provided services related to marketing or purchasing activities. Obviously, cooperatives are a significant part of the U.S. agribusiness community (Kraenzle, et al.).

Even consumers are directly affected by the wide scope of cooperative activities in agriculture. Although they may not be recognized immediately as such, brand names such as Sunkist (citrus), Land O'Lakes (dairy), Welch's (grape juice), and Ocean Spray (cranberries) are all cooperatives that market their members' produce through these brand names.

Food and agricultural cooperatives have become big business. In 1999, Farmland Industries had sales of $10.7 billion, ranking it 161st on *Fortune* magazine's list of the top 500 U.S. corporations (*Fortune*). Cenex Harvest States Cooperative ranked 267th on *Fortune* magazine's list with revenues of $6.4 billion. Wakefern Food Corporation had revenues exceeding $5.5 billion in 1999. TOPCO Associates ($3.6 billion), Associated Wholesale Grocers ($3.4 billion), and Dairy Farmers of America ($7.6 billion) were all major players in their respective markets in 1999.

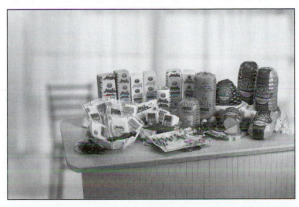

Some cooperatives such as Land O'Lakes have developed important consumer brands.
Photo courtesy of Land O'Lakes, Inc.

Nonagricultural Cooperatives

Cooperatives are by no means limited to agriculture. Mutual insurance companies are cooperatives. More than 1 million cooperative units serve the housing needs of people in the United States. College campus bookstores, which are members of cooperatives, sell more than $1 billion of books and other merchandise each year. Auto parts dealers, hardware store owners, and others are joining together cooperatively to gain the advantage of large-volume purchases. The Associated Press (AP) is a cooperatively organized news agency while the Corporation for Public Broadcasting (a cooperative) provides a wide range of public television programming. Many savings and loan associations and mutual insurance companies are cooperatives. Although these organizations are not agricultural in nature, they are cooperatives and they play an important role in the U.S. economy.

CHARACTERISTICS OF FOOD AND AGRICULTURAL COOPERATIVES

Agricultural cooperatives are organized to serve a basic purpose—to help patron-members increase the profitability of their own businesses. They achieve this goal by providing patron-members with the products and services they need to lower their costs and/or operate more effectively. Marketing patron-members' products or services is another key role of cooperatives. The user-member is their total emphasis. In contrast, generating profit for the owners of the firm is the purpose of the noncooperative business enterprise. This is a very important difference.

Often cooperative and noncooperative businesses operate similarly in the marketplace, but there are many cases where their distinctively different purposes lead to distinctively different operating decisions. For example, in 1974 there was a severe worldwide shortage of fertilizer. At that time, the federal government had imposed price controls to better manage inflation. U.S. and international fertilizer manufacturers had the opportunity to sell their product on the world market at a price that was about three times greater than the controlled U.S. domestic price. Many noncooperative companies quickly moved to reap high profits by selling significant quantities of fertilizer to foreign markets. In the process, they were fulfilling their commitment to their investors to maximize profits.

Cooperatives, however, were not focused on making profits for investors, but on providing their members with all-too-scarce fertilizer. So cooperatives sold their fertilizer production to their members at the controlled domestic prices. In fact, some cooperatives even went so far as to purchase fertilizer from the world market for distribution to members at the lower domestic prices, in order to assure their members of an adequate supply. Thus the orientation of cooperatives toward meeting members' needs triggered a very different set of operating decisions.

Unique Business Features of Agricultural Cooperatives

Cooperatives resemble other forms of business in some regards. They must follow sound business practices and may perform similar functions. Cooperatives have facilities to maintain, employees to hire, advertising to develop, and so forth. There are bylaws, policies, and activities that must be performed to carry out the business at hand. Ultimately, cooperatives must generate a return (member benefits plus direct financial returns) on the investment of their members, which justifies continued membership in the cooperative.

But, in some ways, cooperatives are distinctly different from other businesses. The ownership structure, the way they are controlled, their purpose and how benefits are shared/distributed are unique to cooperatives and how they operate. Three specific features delineate cooperatives from noncooperative businesses:

- Member controlled, member owned
- Operation at cost
- Limited returns on capital

In this section, we will compare and contrast cooperatives with noncooperatives and further define what makes cooperatives a unique business form.

Member Controlled, Member Owned

A fundamental principle undergirding cooperatives is that they must be owned and controlled by the people who conduct business with them. It is imperative that cooperatives maintain their orientation toward servicing those who patronize them. And there is no better way to ensure this than to require that active patron-members who are also owners of the business control the cooperative.

Originally, and even today in most cooperatives, this requirement means one vote for each active patron-member regardless of how much business that member transacts with the cooperative or how much stock the individual member may have accumulated. Democratic control has gained almost universal acceptance as a basic cooperative principle. Some cooperative leaders are extremely strong in their support of this concept.

But recently a small percentage of cooperatives serving a widely diverse membership find that limiting control to one person, one vote creates enough dissatisfaction to threaten the very existence of the cooperative. As a result, some cooperatives determine the degree of individual control in the cooperative on the basis of patronage or investment in the cooperative. The important thing to keep in mind in either case is that cooperative control is always in the hands of the patrons.

Noncooperatives, however, place control in the hands of owner-investors and base the degree of control accorded to each owner-investor exclusively on the amount that the owner has invested in the business. It makes no difference whether the owners patronize the business. In fact, in noncooperatives the owner (or owners) may never see the business or take any role in decision making. This is particularly true in larger corporations.

Members' Board of Directors   Member control of cooperatives is executed through a board of directors, which is selected in open elections from the ranks of active members. The board shoulders the responsibilities of sensing and representing the best interests of all members, setting overall policy, hiring and directing top management, and monitoring the cooperative's performance in achieving its objectives. However, some highly critical decisions involving such issues as mergers or large investments may be taken directly to the membership for a vote.

In a noncooperative business, on the other hand, the owners, whose number of votes is determined individually by the amount of stock they own, elect the directors. Consequently, most board members are stockholders with relatively large ownership interests.

Member Voice Those who are critical of cooperatives say that many of them—even the local ones—have gotten so large that the farmer or member is far removed from having any significant voice in the business. Such critics also argue that the board of directors is sometimes elected on the basis of popularity rather than genuine ability to make policy decisions in a multimillion-dollar business. Critics point to the longevity of some members' terms on the board as further evidence that the total membership is not well represented on a cooperative's board of directors.

While these charges are serious, and one could probably find evidence of situations in which they are justified, it would be unfair and incorrect to assume that they are the general rule. (In addition, investor-oriented corporations are certainly not exempt from these criticisms.) Although there is variation from cooperative to cooperative in the manner in which member ownership–member control is interpreted and carried out, it is unmistakably clear that patron-members maintain their control and ownership of most cooperatives.

Operation at Cost

In most cases, a cooperative's net income is distributed to individual members in proportion to the volume of business that they have done with the cooperative. A cooperative may choose to retain profits rather than pay them out as patronage returns, but when it does, it normally must pay corporate income tax just as any corporation must. (There are some exemptions to this rule, discussed below.) This obligation to return profits to members is a primary factor that separates cooperatives from other forms of business. Noncooperatives are not so obligated and, after paying any tax on profits that may be due, will return profits to the owners of the business in proportion to the owners' investment, or will keep the profits in the business as retained earnings for future growth (see Chapter 4).

Cooperatives may do some business with individuals who are not members. In the case of such nonpatron business, any excess of income over and above the expenses generated specifically by that kind of business does not have to be returned to the customer (although it may be in some cases). Instead, many cooperatives elect to treat this additional income as regular profit, to pay taxes on it just as any business would, and to use the profits to fund growth of the cooperative. (Again, there is a special status, known as Section 521, which allows a cooperative to pass profits on nonmember business back to members without paying corporate profits tax. About 15 percent of farmer-owned cooperatives have Section 521 status.) An agribusiness can maintain its cooperative status as long as not more than half its business is transacted with nonmembers.

Patronage Refunds   **Patronage refunds** allow cooperatives to distribute net returns or margins to members or patrons:

- On an annual basis
- Consistent with standard accounting procedures
- Without regard to how much was earned on individual transactions

Many cooperatives opt to return net earnings to members, however this does not happen on a transaction-by-transaction basis. Instead, cooperatives usually charge market prices for supplies and services furnished to their members and patrons. Operating at normal market prices allows many cooperatives to finance growth by special methods, such as retaining a portion of cash savings and making patronage refunds in the form of stock or other obligations. They also offer competitive prices for products delivered for further processing and marketing. This mode of operation under most circumstances allows them to generate sufficient income to cover costs and meet the continuing needs for operating capital.

When the fiscal year ends, a cooperative then figures its earnings on business conducted on a cooperative basis. If there are any earnings, they are returned to the cooperative's patrons as cash and/or equity allocations on the basis of business volume conducted with the cooperative that year. Distributions of this nature are another example of patronage refunds.

If, for example, cooperatively conducted business earnings generate $3,000,000 for the year, and Vaughn's Produce Farms account for 1 percent of the total business of the cooperative, they receive $30,000 in patronage refund ($3,000,000 × .01 = $30,000).

Financing of Cooperatives   Cooperatives have some unique financing options at their disposal. **Revolving fund financing** is a highly advantageous option that is unique to cooperatives. Based on the principle that cooperative members should be willing to finance the growth of their own organization, this method gives cooperatives the option of issuing patronage refunds in the form of stock or equity allocations, rather than in the form of cash. The idea is that the cooperative is a logical extension of the member's own business, and each member has an *obligation* to contribute directly to its financing in proportion to that member's use of the cooperative's services. The advantage to the cooperative is that since ownership stock represents ownership of valuable assets, it satisfies the cooperative's obligation to return the excess of income over expenses to patron-members, while at the same time the cooperative retains the actual cash earnings for use in financing the business.

Noncooperative businesses argue that this is an unfair advantage, since any excess of income over expenses (profits) that noncooperatives make is taxed first before it is retained for use by the business or distributed as dividends. Since 1962, legislation has required cooperatives to return a minimum of 20 percent of all patronage refunds in cash rather than in the form of ownership stock if they use this form of financing.

Refunds in the form of stock are considered to be income; so cooperative patron-members are required to pay taxes on that stock. Farmers in tax brackets that are higher than 20 percent will find that this type of financing can cause a net outflow of cash, since they must pay out more in taxes on the stock than they actually receive in cash refunds. This situation can create a temporary financial hardship, although in the long run the cooperative stock may increase the farmer's overall net worth. The potential cash-flow problem has caused considerable concern in many cooperative relationships, particularly those where members fall into higher tax brackets. Hence, many profitable cooperatives will pay out 40 to 50 percent of their net earnings in cash.

Theoretically, the cooperative "revolves" the stock periodically, thus allowing older stock to be cashed in. This process involves retiring the stock of an inactive member by giving them cash in exchange for their shares. Sometimes a nominal cash dividend is paid on stock in force. The decision to "call" or cash in older stock, is a decision of the cooperative's board of directors that depends on the cooperative's financial condition and on the availability of cash.

Tax and Finance Regulations   What works for a cooperative in one state may not be legal in another state. Cooperatives have unique tax and finance rules and regulations that apply to the cooperative form of business. In addition to following mandatory federal rules and regulations, states also have laws and codes that must be upheld by the cooperative membership in order to operate. Cooperatives do have some flexibility in designing an equity accumulation program to meet their individual needs. And, understanding the alternatives of cooperatives' tax treatment is significant, particularly regarding patron-based sources of equity, retained margins, and so on.

Limited Returns on Capital

Since the basic purpose of cooperatives is to operate at cost in order to benefit member-patrons directly in their own businesses, it follows that returns on capital invested are limited. **Limiting returns** on member equity to a nominal amount simply ensures that members holding stock in the cooperative are not tempted to view the cooperative as an investment in and of itself, but rather as a service to their own businesses. Typically, returns on capital cannot be greater than the going interest rate or 8 percent (whichever is higher), although this rule does vary state to state. In practice, most cooperatives pay no dividends on their stock; therefore, limiting returns is merely an academic point. But it is important to recognize the limited-returns principle, since it ensures that the purpose of the cooperative—to help members help themselves rather than to become an investment company—is realized.

COOPERATIVES: A BRIEF HISTORY

It is difficult to determine the precise origin of cooperatives. There are traces of cooperative-like organizations in ancient Egypt as early as 3000 B.C. And, there are vestiges of cooperative ideas in Greek, Roman, and Chinese cultures. In the Middle Ages cooperative ideas developed in the form of "guilds" which united workers in an effort to meet common needs. The first farmers' cooperatives are reported to be those of Swiss dairy farmers, who made cheese cooperatively as early as the thirteenth century.

Early Americans also experimented with cooperatives. The first colonists worked jointly on many self-survival projects. Benjamin Franklin organized a mutual insurance company (a cooperative) in 1752. In the early 1800s dairy cooperatives had been organized in the Northeast. By the mid- and late 1800s there may have been as many as 1,000 farm cooperatives—mostly dairy cooperatives—in the United States.

Many people recognize the first formal cooperative of modern times to be the Rochdale Society of Equitable Pioneers in England in 1844. The original twenty-eight members of this early cooperative joined in an effort to purchase supplies for their businesses. Although the Rochdale Society was not the first cooperative in history, their formal principles have served as a model for the development of a great many modern cooperatives.

The Rochdale Principles

The **Rochdale Principles** adopted by the Rochdale Society to guide the operation of their cooperative included the following points:

1. That capital should be of their own providing and bear a fixed rate of interest.
2. That only the purest provisions procurable should be supplied to members.
3. That full weight and measure should be given.
4. That market prices should be charged and no credit given nor asked.
5. That "profits" should be divided pro rata upon the amount of purchases made by each member.
6. That the principle of one member, one vote should govern, and that there should be equality of the sexes in membership.
7. That the management should be in the hands of officers and committees elected periodically.
8. That a definite percentage of profits should be allotted to education.
9. That frequent statements and balance sheets should be presented to members.

While some of these principles have been modified over time, this set of Rochdale principles still provides the basic organizing framework for cooperatives today.

Despite this early activity, it was not until the early 1900s that cooperatives as a way of organizing business in agriculture really began to expand. The growing recognition that farmers could significantly improve their economic condition through cooperatives generated much state and federal legislation that encouraged the growth and development of farm cooperatives. Not only was there a sharp increase in the number of local cooperatives during this period, but many established cooperatives were reorganized and consolidated into larger units as well. Regional cooperatives were formed to support the activities of local cooperatives. The formation of national cooperative organizations, such as the American

Institute of Cooperation, the National Council of Farmer Cooperatives, the Cooperative League, and the American Farm Bureau Federation, supported an increasingly favorable social and political climate for cooperatives.

In the early 1900s many important laws were passed to encourage the growth of cooperatives. The **Capper-Volstead Act of 1922** is among the most significant cooperative legislation, since it ensures the right of farmers to organize and market their products collectively so long as:

1. The association conducts at least half its business with members of the association; and
2. No member of the association has more than one vote, or the association limits dividends on stock to 8 percent.

Balancing Market Power

The early 1900s provided many vivid examples of cooperatives that fulfilled two basic purposes: to set a competitive standard and to act as power balancers in the market. Frequently, rural areas in the United States offered few alternative sources for essential farm supplies. In many cases, farmers had inadequate service, yet were forced to pay high prices. By organizing a cooperative, farmers gained both their own source of supplies and a place to market their products— one that they owned, controlled, and operated to serve their own interests. In the free enterprise system not only did this alternative serve to "balance the power" in the marketplace, but it also encouraged innovations in products and services for farmers. Since the primary objective of the cooperative organization is to serve the needs of its patron-members, cooperatives during this period often "set the standard" for new products and market services.

Cooperatives were particularly effective in these roles during the early years of cooperative development. They are often given the credit for developing open-formula livestock feeds with labels that spelled out nutrient values, and for developing guaranteed-analysis fertilizer. Cooperatives were solely responsible for channeling electric power to much of rural America. Even a hint of the impending formation of a cooperatively owned livestock market or grain elevator was known to cause local prices to move upward. Cooperative lenders who specialized in farm loans were often the only commercial source of funds available for farm expansion projects and annual operating capital.

Over time, the marketplace has changed. And, some opponents of cooperatives argue that noncooperative competition has escalated to the point where the dual roles for cooperatives of product/service innovation and balancing market power are not as important. In fact, some argue that cooperatives themselves have become large, complex organizations in their own right, and can sometimes be slow to respond to member needs, or struggle to serve the broad-based needs of a highly diverse group of members. While such criticisms may or may not be true in a specific cooperative organization, most cooperatives take them seriously. Such challenges are brought on by continued consolidation in the marketplace and the burgeoning size of both cooperatives and their members. These changes force cooperatives to work hard to avoid such problems that could undermine their very purpose.

In the end, the issue of whether cooperatives are needed to serve the roles of market innovator and to balance market power may be academic. The fact is that cooperatives do exist and do play an extremely important role in supplying, marketing, financing, and servicing American farmers and food businesses. And, like any business organization, the long-term viability of cooperatives will hinge on their ability to cost effectively meet the needs of those who buy their products and use their services and those who finance them. In the case of cooperatives, these two sets of needs are captured in one group—the members of the cooperative.

CLASSIFYING COOPERATIVES

The cooperative system in the United States is highly complex and interwoven. It ranges from simple, local, independent cooperatives to the vast interregional cooperatives that link literally thousands of local and regional cooperatives into one complex organization. There are also examples of international cooperatives, which span the globe. The following discussion explores the classification of cooperatives by function and by organizational structure.

Classification by Function

Cooperatives can be classified by the major function of the organization. Three such classifications are noted.

Marketing Cooperatives   **Marketing cooperatives** derive most of their total dollar volume from the sale of members' products. A grain marketing cooperative is an example.

Supply Cooperatives   **Supply cooperatives** earn most of their business volume from the sale of specialized production supplies such as farm inputs and building supplies. In the case of a farm supply or farm input cooperative, these cooperatives may handle a wide variety of supplies ranging from animal feeds, to farm equipment and building materials, to home items such as heating oil, lawn and garden supplies and equipment, and food.

Service Cooperatives   **Service cooperatives** provide specialized business services related to the business of a certain group of individuals. In agriculture, these service cooperatives relate to the business operations of farmers, ranchers, or cooperatives and may include services such as trucking, storage, artificial insemination, and record keeping.

Classification by Organizational Structure

Cooperatives are organized in various ways to best meet member needs. Some cooperatives are organized by the type of commodity they handle. Grain, cotton, ethanol, or pork cooperatives are examples. These cooperatives specialize in marketing a particular product. Some other cooperatives are organized by geographic area. Since some cooperatives serve more than a single local area, perhaps more than a county, state, or even a single multistate region, they must find

Producers of many different agricultural commodities organize cooperatives to benefit their farming operations.

ways to meet the needs of their members within these varying geographic locations. Local cooperative organizations and regional cooperative organizations are two significant types of cooperatives.

Local Cooperatives   The **local cooperative** is usually organized as a corporation; thus, it has the same general structure as any corporate organization. Stockholders (patron-members) select a board of directors who, in turn, select a manager to carry out their policies and operate the cooperative on a day-in, day-out basis. Management then hires appropriate personnel to perform prescribed duties. There has been much consolidation at the local cooperative level over the past two decades. And, what were once modest county level cooperatives may well be $100 million organizations serving very large, multicounty areas. Some call these large, multicounty organizations "super-locals."

Regional Cooperatives   **Regional cooperatives** are "cooperatives of cooperatives." Their primary purpose is to provide manufacturing, processing, and wholesaling services to local cooperatives. They are necessary because local cooperatives are often not large enough to compete effectively in a world of corporate giants. The regional cooperative offers local cooperatives all the advantages of the largest organizations in the market, which in turn lends the locals more power in the marketplace. There are even a few national and international cooperatives. These are typically involved in the basic manufacture of products such

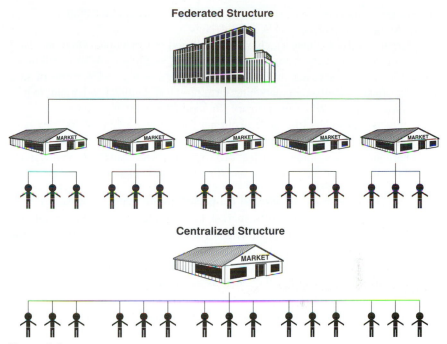

Figure 5-1
Federated and Centralized Cooperative Structures.

as fertilizer. Such national or interregional cooperatives are usually owned by the regional cooperatives.

There are two basic types of regional cooperatives: centralized cooperatives and federated cooperatives (Figure 5-1). Some cooperatives have both centralized and federated elements and are called mixed cooperatives. Each has a different type of organizational structure, which affects the way it operates.

Centralized Cooperatives   **Centralized cooperatives** usually serve a local area or community, county, or several counties. Most usually perform a limited number of initial marketing functions and/or most farm supply sales are at the retail level (Kraenzle, et al.). A few centralized cooperatives, principally regionals, operate over large geographic areas and have members in several states. They often provide more vertically integrated services, such as processing farm products or manufacturing feed and fertilizer.

In a centralized cooperative, the local cooperative outlet is controlled by the regional cooperative rather than by a local board of directors. In this case, local farmers elect regional directors, who control the regional cooperative, which in turn manages and controls the local cooperative outlet. There is no local board of directors per se. Instead, there is usually a local advisory committee that communicates regularly with local management. Good examples of regional cooperatives that are primarily centralized are Agway in the Northeast and the National Grape Co-op Association (Welch's) in Westfield, New York.

Bargaining associations also have a centralized organizational structure. They derive all or most of their business volume from negotiating with distributors, processors, and other buyers and sellers over price, quantity, grade, terms of sale, and other factors involved in marketing farm products. Only a few bargain to purchase farm supplies. While the primary function of a bargaining association is to bring buyers and sellers together to contract for the sale of members' products, many bargaining associations now perform additional functions. Examples include dairy bargaining associations, which at one time only negotiated price. Now, many provide additional services—such as handling milk for spot sales.

Federated Cooperatives   **Federated cooperatives** are made up of two or more member associations organized to market farm products, purchase production supplies, or perform bargaining functions (Kraenzle, et al.). Federated associations often operate at points quite distant from their headquarters. Federated cooperative members are usually local cooperatives, although some are interregional associations with regional cooperative members.

Federated cooperatives are really cooperatives whose members are other cooperatives, usually local cooperatives. Farmers own their local cooperatives. Then each local cooperative, in conjunction with other locals, owns and controls the regional cooperative. The control is from the bottom up, with the regional responding to the voiced needs of the local cooperatives and the local cooperative responding to the needs of the member-patron. Farmland Industries, Kansas City, Missouri is an example of a federated cooperative.

Mixed Cooperatives   A few cooperatives are **mixed cooperatives.** These cooperatives have individual farmer members *and* autonomous cooperative members, which represent a combination of centralized and federated structures. They serve large geographic areas, with members in many states, and provide a variety of integrated services. In some cases, this comes about because a locally owned and controlled cooperative has gotten into financial difficulty and has been taken over by the supporting regional to protect the financial interests of individual farmer members. Southern States Cooperative in Richmond, Virginia is an example of a mixed regional cooperative.

FOOD AND AGRICULTURAL COOPERATIVES TODAY

In 1999, out of the thousands of cooperatives existing in this country, the top 100 cooperatives generated more than $125 billion in revenue, an increase of $5 billion from the year earlier. The National Cooperative Bank, which tracks the top 100 cooperatives annually, notes that during the last 6 years of the 1990s the top 100 cooperatives in the United States increased their revenue base by more than 29 percent.

Food and Agricultural Cooperatives

Food and agricultural cooperatives represent an important group among the nation's top cooperatives. As mentioned earlier, currently there are more than 4,100 food and agricultural cooperatives operating with nearly 4 million members. Food and agricultural cooperatives exist to help producers increase their marketing capability and production efficiency. They are an integral part of the framework of today's food distribution system and provide agricultural producers with credit, supplies, and services.

Notice the names of the organizations shown in Table 5-1, Table 5-2, and Table 5-3 (National Cooperative Bank). Dairy, poultry, crops, and fruit are just a few of the products that cross the tables of U.S. consumers daily which are marketed by cooperative organizations. A point of significance to note in this listing is that of the top 100 cooperatives, 43 are food and/or agricultural cooperatives (Table 5-1), 7 are agricultural finance companies (Table 5-2), and 19 are grocery cooperatives (Table 5-3). Revenues of the top food and agricultural cooperatives tallied $96 billion in 1999, an increase of about $4 billion from the previous year.

Wholesale and retail grocery cooperatives continue to be a major force in the food sector (Table 5-3). After agricultural cooperatives, grocery cooperatives are the second most dominant type of cooperative in the United States. Retailer members are served by their grocery cooperative. These cooperatives help members with marketing strategies, by offering brand name products, and with product identity programs to help them compete with powerhouse supermarket chains and super discount stores.

New Generation Cooperatives

While traditional cooperatives continue to serve member-patrons, there is a new kind of cooperative developing and expanding across the United States. With the organization of the so-called **New Generation Cooperatives (NGC),** a new twist on the concept of cooperatives has emerged. New Generation Cooperatives are defined as a membership organization requiring an up-front equity investment in equity shares possessing both tradable and appreciable properties. Investment in the cooperative is based on a member's anticipated level of patronage and all members adhere to a legally binding uniform marketing agreement (Kraenzle, et al.).

The NGC's work to be market driven in the sense that market demand for the processed end products is used to determine the appropriate scale of the business. That, in turn, limits the size of the membership so that these become "closed" cooperatives; in fact, NGCs require closed membership. Once the stock offering is over, someone new cannot come in without purchasing an existing member's stock. Another unique feature is that membership shares are tradable. These tradable membership shares not only allocate rights and obligations to deliver units of the farms' raw products, but these shares also spread up-front capitalization responsibilities equitably among members. The NGCs, unlike traditional cooperatives, require significant up-front investment and an arrangement

TABLE **5-1** | Largest Agricultural Cooperatives in U.S., 1999

Name	Coop 100 rank	Revenue 1999	Total assets 1999
		Dollars in millions	
1. Farmland Industries	1	10,709	3,258
2. Dairy Farmers of America	2	7,600	7,325
3. Cenex Harvest States	3	6,329	2,788
4. Land O'Lakes, Inc.	4	5,613	2,684
5. Ag Processing, Inc.	13	2,095	789
6. California Dairies, Inc.	15	1,900	364
7. Gold Kist, Inc.	16	1,766	801
8. Agway, Inc.	19	1,484	1,484
9. Southern States Cooperative	21	1,366	682
10. Ocean Spray	23	1,388	894
11. Foremost Farms USA	25	1,301	323
12. PRO-FAC Cooperative Inc.	26	1,239	1,196
13. GROWMARK, Inc.	28	1,156	646
14. WestFarm Foods	29	1,145	293
15. CF Industries, Inc.	31	1,095	1,334
16. North Central AMPI, Inc.	32	1,062	209
17. Prairie Farms Dairy, Inc.	33	1,046	371
18. Dairylea Cooperative, Inc.	36	881	91
19. Sunkist Growers, Inc.	38	862	181
20. American Crystal Sugar Co.	39	844	656
21. Tri-Valley Growers	42	776	558
22. Riceland Foods, Inc.	45	729	329
23. Staple Cotton Cooperative Association, Inc.	48	674	124
24. Plains Cotton Co-op Association	53	633	247
25. National Grape Co-op Association	54	631	351
26. MD & VA Milk Producers Co-op Association, Inc.	55	626	84
27. National Cooperative Refinery Association, Inc.	59	604	486
28. Minnesota Corn Processors, Inc.	60	599	619
29. MFA Incorporated	62	563	297
30. Agri-Mark, Inc.	64	559	174
31. Michigan Milk Producers Association	69	509	99
32. Alto Dairy Cooperative	79	422	66
33. United Dairymen of Arizona	80	412	47
34. Blue Diamond Growers	82	405	174
35. United Suppliers, Inc.	84	403	164
36. Calcot Ltd.	85	395	198
37. Tennessee Farmers Cooperative	88	388	154
38. Citrus World, Inc.	89	381	255
39. Swiss Valley Farms, Inc.	92	371	89
40. Southeast Milk, Inc.	91	371	—
41. Equity Co-op Livestock Sales Association	94	364	42
42. MFA Oil Company	97	329	156
43. Alabama Farmers' Cooperative	100	273	193
Total		**62,248**	**31,275**

Source: National Cooperative Bank.

T A B L E **5-2** | **Largest Agricultural Financial Cooperatives in U.S., 1999**

Name	Coop 100 rank	Revenue 1999	Total assets 1999
		Dollars in millions	
1. Agribank, FCB	17	1,563	21,343
2. CoBank	20	1,466	24,089
3. AgFirst Farm Credit Bank	34	1,000	12,700
4. AgAmerica, FCB	44	738	9,792
5. Western Farm Credit Bank	67	518	7,391
6. Farm Credit Bank of Wichita	75	436	5,497
7. Farm Credit Bank of Texas	77	427	5,402
Total		**6,148**	**86,214**

Source: National Cooperative Bank.

T A B L E **5-3** | **Largest Grocery Cooperatives in U.S., 1999**

Name	Coop 100 Rank	Revenue 1999	Total assets 1999
		Dollars in millions	
1. Wakefern Food Corp.	5	5,501	839
2. TOPCO Associates, Inc.	7	3,600	—
3. Associated Wholesale Grocers	8	3,370	526
4. United Western Grocers	10	2,929	752
5. Roundy's, Inc.	11	2,717	497
6. Associated Grocers, Inc.	30	1,100	270
7. Associated Food Stores	35	952	276
8. Associated Wholesalers, Inc.	37	879	126
9. Certified Grocers Midwest	46	724	119
10. Central Grocers Co-op	47	695	124
11. Affiliated Foods, Inc.	49	671	107
12. Shurfine International, Inc.	50	639	53
13. Affiliated Foods Cooperative, Inc.	57	619	86
14. Affiliated Foods Southwest	61	563	133
15. URM Stores, Inc.	70	490	110
16. Key Food Stores	71	487	—
17. Piggly Wiggly Alabama	72	480	64
18. Associated Grocers	78	424	63
19. Associated Grocers of Florida, Inc.	93	368	61
Total		**27,208**	**4,206**

Source: National Cooperative Bank.

in which members share equitably on a per-unit basis in the revenue stream that has been created. Farmers are required to deliver according to plan—regardless of the open market. The NGCs are not simply trying to keep the prices of inputs and basic commodities fair, they are trying to capture more of the food system revenue stream.

As of 2001, NGCs were formed in all types of commodity markets including soybeans, durum wheat, spring wheat, sunflower seeds, alfalfa, hogs, beef, fish, and edible beans, as well as corn and sugar beets. In Minnesota and the Dakotas alone, the 1990s saw the development of more than 50 NGC projects. Most of them were cooperatively structured and created more than $2 billion in capacity to convert farm commodities into food and nonfood products.

CHALLENGES AND OPPORTUNITIES

Consolidation

Dealing with consolidation has been both a major challenge and focus for agribusinesses during the past 20 years—and cooperatives are no exception. The agricultural prosperity of the 1970s, coupled with a very rapid growth in farm size, caused agricultural cooperatives to grow at unprecedented rates. Many cooperatives doubled and even tripled in size during the 1970s and early 1980s. Like the rest of the agricultural input sector, the mid-1980s and early 1990s brought massive restructuring and consolidation as growth slowed. From 1985 to 1999 the number of agricultural cooperatives in the United States fell from 5,625 to 4,174. Large numbers of local cooperatives merged with other locals, creating "super locals" or "mini-regionals" serving much larger geographic areas. Several regional cooperatives also merged during this period creating even larger organizations.

Delivering value to changing patron-members is key to future success of food and agricultural cooperatives.

Photo courtesy of Growmark®, Inc.

In any business, such consolidation is a challenge. Because of their owner-ship structure and their operating philosophies, consolidation is even more com-plex for cooperative businesses. In some cases such consolidations are driven by the weak financial condition of one of the parties. Such a merger can leave the new joint operation in a marginal position. The attachment of individuals in a geographic region to "their local cooperative" can be very high. This social pres-sure may prevent a merger from a position of strength and also can present major governance problems for the combined operation. For example, such con-solidations often lead to "super local" or "mini-regional" cooperatives with sales of $100 million or more. These organizations often rethink their traditional rela-tionship with their regional cooperative. Managing further consolidation of the cooperative system will remain an important issue for cooperative managers and patron-members in the foreseeable future.

Changing Membership

Another extremely important challenge for cooperatives lies in their rapidly changing membership. Most farmers who choose to become cooperative mem-bers today do so because it is a sound business decision. Few are attracted to co-operatives for the same reasons that their parents or grandparents were. And still fewer seem to be willing to loyally support their cooperative solely on the basis of past performance. Stock patronage refunds that can be redeemed for cash only at retirement offer precious little incentive to a 25-year-old farmer. A young farmer who might have five different fertilizer salespeople calling on them as they vie with each other to make a sale will probably remain unmoved by the ar-gument that a cooperative will protect the farmer's purchasing alternatives. As with any agribusiness, cooperatives must be ready at all times to earn their mem-bers' continued support on the basis of their current performance records.

Along the same lines, size disparity among customers presents major chal-lenges for many cooperative organizations. Large commercial operators ques-tion the one member, one vote principle—when they are buying inputs and mar-keting the outputs from a 5,000-acre operation—why shouldn't they have more say in the governance of the cooperative than a part-time farmer who has a 50-acre operation? In some cases, these larger operations have broken away from the cooperative system and formed their own **buying groups.** The purpose of these groups is to buy inputs as cheaply as possible for their large farmer mem-bers. Losing the volume of such operations can have a major impact on the cost of doing business at a local cooperative. And, like the age issue described above, size disparity will be an important issue in the years ahead.

Capitalization of Cooperatives

Another major challenge facing many cooperatives today is how the patron-members will finance their cooperative. The revolving fund financing concept discussed earlier in this chapter has served cooperatives well by providing a mechanism through which users can finance cooperative growth by retaining earnings. But economic pressures on many cooperatives have been so severe in recent years that some have been unable to rotate or return members' stock—that is, they have not been financially able to pay members on any regular basis for stock they have accumulated through their patronage over the years. It is not uncommon for cooperatives to allow members to cash in their stock only upon retirement or upon death. This inability to rotate stock is a barrier to keeping control of the cooperative in the hands of current users. More seriously perhaps, this signals a limit on the cooperative's ability to generate capital internally. The latter issue has pushed the development of concepts like the New Generation Cooperatives.

Opportunities for the Future

Today, food agricultural cooperatives face the same new and unprecedented challenges that all firms in the food and fiber marketing system face. In addition, the unique organizational structure of cooperatives provides these firms with both interesting opportunities and unique issues to manage. Cooperatives must continue to serve their patron-members profitably. Patron-members expect the same (or better) service and quality from the cooperative that they get from noncooperative businesses, at competitive prices. As the cost-price squeeze continues to put pressure on producers to reduce costs, cooperatives that are unable to meet noncooperative prices for input supplies or to pay existing market prices for commodities they purchase from farmers may lose a significant share of the market.

There are clear signs that cooperatives are up to the challenges discussed in this chapter. The cooperative system of today looks very different from that of 20 years ago. Cooperatives have explored a wide variety of alternative business arrangements to better serve their members. Such openness to new ways of doing business, while never losing sight of their responsibility to the patron-member, will help cooperatives continue to be an important and viable player in the food and fiber production and marketing system.

SUMMARY

Cooperatives are business organizations owned and controlled by those who use them. Their basic purpose is to serve the needs of their members rather than to make a profit for investors. Their growth in agriculture has been encouraged by favorable government policies that are designed to help individual farmers help themselves. Cooperatives are a major force in agribusiness, involved in nearly every stage of the food system, from the manufacturing of fertilizer to the retail food store shelf.

Cooperatives operate at cost, since they are obligated to return any income in excess of expenses to members in proportion to the individual member's purchases or marketing. These savings are partially refunded to members in cash and partly as additional stock. Savings refunded as stock are actually reinvested in the business. This plan, known as revolving fund financing, has been extremely important to the growth of cooperatives.

An outgrowth of some of the challenges facing cooperatives is the New Generation Cooperatives that have developed in the last decade or so. These cooperatives have a defined membership, are member financed but in a nontraditional way, and tend to focus on value-added output markets.

Today, cooperatives remain the largest player in many agricultural markets, yet they face many challenges. Large cooperatives are struggling with problems of size, and larger farmers are placing new demands on their cooperatives. Yet cooperatives have made a myriad of adjustments and continue to play an important role in the food system.

DISCUSSION QUESTIONS

1. List the Rochdale principles. How appropriate do you feel these principles are for today's cooperatives? Justify your answer.
2. Increasingly, cooperatives are being pressured to move away from the one person, one vote concept. Why do you suppose this is so? What impact do you think it might have on cooperatives?
3. List the advantages and disadvantages of cooperative agribusinesses as opposed to noncooperative agribusinesses.
4. Which type of regional cooperative has the greatest control over local cooperative outlets? Is this always an advantage? Why or why not?
5. List some of the challenges facing cooperatives today. Which of these challenges do you believe are the most significant, and why?
6. Define a New Generation Cooperative. How does it differ from a traditional cooperative? Give an example of an NGC operating today.
7. Look up the website for one of the top agricultural cooperatives. Describe its business, its size and scope, and primary business objective.
8. Why would forming a cooperative be of very limited advantage to a farmer if the market the farmer deals in is already highly competitive and products are undifferentiated?
9. Explain why a farmer must make a firm commitment to the cooperative in order to realize gains from coordination of production and marketing.

CASE STUDY — Blueberry Growers' Dilemma

Several blueberry growers in the state have been continually frustrated by their inability to market their blueberries effectively. The growers are scattered over a 300-mile area. Most are small growers with only small patches, but a few of the growers have larger acreages. Some of the smaller patches use "U-pick" type market operations, while the larger growers hire migrant labor and harvest for food processors in the area. All are faced with complicated insect problems and have relied on local chemical-fertilizer businesses to advise them, with varying degrees of success.

Lately some of the growers have been considering developing a cooperative to serve their needs. They have asked you to help them think through some of the problems and opportunities that might result from establishing the cooperative.

1. Explain to the growers what basic cooperative principles they would need to adhere to.
2. What are some of the specific advantages that might be gained from organizing such a cooperative?
3. What problems might you expect the new cooperative to encounter?
4. What additional information would you need before advising the growers whether or not to form a cooperative?

REFERENCES AND ADDITIONAL READINGS

Kraenzle, Charles, Ralph Richardson, Celestine Adams, Katherine DeVille, and Jacqueline Penn. November 1999. *Farmer Cooperative Statistics, 1998*. RBS Service Report 57. Rural Business-Cooperative Service, USDA.

National Cooperative Bank. September 2000. *NCB Co-Op 100*. National Cooperative Bank. Washington, D.C.

"The Fortune 500." April 17, 2000. *Fortune*. www.fortune.com

INTERNATIONAL AGRIBUSINESS

- Understand why agribusinesses may choose to seek out international markets
- Develop an understanding of the importance of international markets to U.S. food and agribusiness firms
- Identify some of the key trends in the export of U.S. food and fiber products
- Describe some of the challenges that agribusinesses face when pursuing international markets
- Understand the elements of a society's culture, and how these elements shape the operating environment for an international firm
- Identify the different methods that food and agribusiness firms might use to enter international markets

INTRODUCTION

Why would an already successful agribusiness consider entering the more risky international marketplace? After all, selling into unfamiliar international markets is full of uncertainties. The agribusiness could be conservative and continue with its strategy of expanding into other U.S. markets. But, despite the unknowns, there are many reasons why a U.S.-based firm would "go global."

In many U.S. food and agribusiness markets, market growth is leveling off and the market is approaching maturity, or a time of slow growth. At the same time, the markets in many other countries are growing very fast, or have the potential to grow fast, offering new business opportunities. U.S. agribusinesses are also finding that their competition in the domestic market is increasingly from foreign-owned and controlled companies. Foreign firms have entered the U.S. market bringing new competition, replacing domestic competitors, and/or have joined forces with domestic agribusinesses. In any case, the competition in U.S. agribusiness markets has become increasingly fierce as each firm—U.S. or foreign—tries to capture a larger piece of the market. Combining the increased competition in the U.S. market from international companies with the opportunities in international markets, the threat is clear: if a U.S. food or agribusiness company doesn't sell it, someone else in the world will!

So, why should an agribusiness manager be interested in better understanding international business? In this chapter, we will explore this question and identify some of the reasons why the international marketplace is of great interest to agribusiness managers. We will also examine some of the reasons doing business internationally may be challenging, and some of the strategies agribusiness firms use to "go global!"

WHY DOES U.S. AGRIBUSINESS NEED INTERNATIONAL MARKETS?

There are many, many reasons for agribusinesses, and the U.S. economy, to be interested in the international business arena. From an economic standpoint, the exporting of U.S. agricultural products boosts employment within the U.S. economy. In 1998, it was estimated that more than 18,000 jobs were created for every $1 billion in agricultural exports. This implies that nearly 800,000 jobs in the agricultural sector were the direct result of farm product exports. Agricultural exports help boost rural economies; approximately one-third of those agricultural export jobs were located in rural areas. Although most of these rural jobs are in the production of agricultural products, others are in more labor-intensive industries such as wholesale and retail trade, services, and food processing and feed processing.

Benefits from Doing Business Internationally

From a managerial standpoint, international markets hold appeal for a number of reasons. Some these reasons are shown in Table 6-1. While each reason in the table may not be part of a specific international opportunity for an agribusiness firm, the list does provide a snapshot of several important reasons agribusinesses find international business attractive.

T A B L E **6-1** | **Benefits from Doing Business Internationally**

- Expand sales
- Take advantage of scale economies
- Capture benefits of a global brand
- Reduce risk by diversifying across markets
- Lower costs of production
- Access lower-cost raw materials through international sourcing
- Broaden access to credit
- Lever experiences from operating in international markets into domestic market

Clearly, one reason to pursue international markets is to expand sales by capitalizing on growth opportunities in other countries. Some newly industrialized countries are experiencing extraordinary growth when compared with established markets such as the United States, Europe, and Japan. Entering a newly industrialized country may offer agribusiness companies a piece of this potential for rapidly growing sales and profits. In addition, product life cycles can often be extended by selling products in foreign countries. The potential for expanding international sales is especially important if the domestic market is experiencing slow, or no, sales growth. A related advantage of serving a larger international market is to achieve scale economies because of larger production volumes.

Name or **brand recognition** is another benefit companies work toward in their globalization efforts. Global recognition of corporate logos and brands increases both the efficiency and effectiveness of a company's advertising efforts. Brands are very expensive to develop, and if a firm can spread those development costs across a broader market, profits can increase. As a result, many companies competitively vie for sponsorship of world-televised sporting events. Reinforcement of brand or logo recognition can be a powerful tool in these cases of sponsorship. For example, in a recent televised World Cup soccer event, sponsor company's logos were shown and/or mentioned in more than 40 countries and in as many as 7 languages. Firms like Coca-Cola, Pepsico, and McDonald's have brands that are recognized in virtually every country in the world.

Agribusinesses may be able to reduce their overall risk exposure by broadening the number of markets from which they purchase inputs and sell products because they are not dependent on a single, local market. Organizations with operations in several countries may stand to be less affected by slow periods in their domestic market when actively engaged in business abroad. Sourcing inputs internationally may also significantly lower costs of production. Such cost advantages allow firms to lower margins, giving them an important price advantage.

Global firms may also have easier access to credit and gain valuable experience from operating in other markets. Firms serving international markets can move technology around the world, locating new markets for the results of their research and development in the process. For example, firms in the hybrid corn seed business regularly draw on genes from seed obtained from such places as Argentina, Italy, and northeast Iowa to develop a superior hybrid that performs well across the same latitudes worldwide. In other cases, firms may identify a new product opportunity in an international market, and be able to bring this idea home to the domestic market.

Opportunities for Smaller Firms

It's not only the multinational conglomerates finding opportunities and success in the international marketplace. Small agribusinesses have found niches in serving needs around the world. Their active pursuit of placing and developing products for international markets has in many cases met with great success. While large companies reap the benefits of deep pockets—economies of scale and greater returns on their investment in research and development—smaller

Food and agribusiness managers source inputs globally as technology and costs of production vary dramatically around the world.
Photo courtesy of H. David Thurston, Smokin' Doc Photo Collection, Cornell University, Ithaca, NY.

agribusiness firms are often more flexible, allowing them to adapt to the changing structure and demands of the international food industry. As with all business in the international marketplace, successful global business endeavors by the small firm require understanding of the unique characteristics and structures of the customer in each given instance (Connor and Schiek).

Although countless factors influence the global marketplace during any business day, a few key factors have helped increase the numbers of opportunities for small agribusiness firms in the international arena. First, emerging markets have entered into the world trade picture at a rate unequaled since post–World War II. Those markets opening up to conduct business with international suppliers, partners, etc., include those in Eastern Europe, other countries of the former USSR, India, Latin America, China, and other Asian countries. Many African countries are moving continually closer to allowing or welcoming business from outside their borders.

The second key factor opening the doors of world markets for smaller businesses is technology. Simply put, today's small food and agribusinesses are often "well wired"—connected via computer, modem, e-mail, telephone, cell phone, fax—making them very competitive with much larger firms for emerging market growth potential. Essentially, the world is truly available to creative, innovative businesses. However, unlike the world market of post–World War II when the multinational companies controlled these markets, the markets today are often open to the best competitors. Companies that will succeed will be flexible enough to adapt to constant change and adjust to an array of challenges. Those companies are often the small, agile companies.

Global marketers promote branded food and fiber products in every corner of the world.

Photo courtesy of H. David Thurston, Smokin' Doc Photo Collection, Cornell University, Ithaca, NY.

A GLOBAL MARKET

To this point in this chapter we have talked about why agribusiness may choose to conduct business in an international location. There is, however, a different viewpoint on international business. Instead of thinking about stepping into an international market, many of today's successful firms approach business from the perspective of conducting business in a global market. This distinction is subtle, but significant. A global perspective is a philosophy, an attitude, and or an approach to conducting business abroad. Global agribusiness management is guided by concepts that view the world as one market. And, firms with a global perspective run their businesses this way.

Let's take a look at one important area, marketing, and explore the differences in these two perspectives. Global marketing is based on identifying and targeting cross-cultural similarities. International marketing management is based on the premise of cross-cultural differences and is guided by the belief that each foreign market requires its own culturally adapted marketing strategy. Today, a global market has evolved on many levels and for many products/services that previously did not exist.

How is it possible to develop a global marketing strategy for some products? Information technology and access to information, goods and services not only in this country, but also across the globe, has changed our world. Truly the business culture of today is more global than in any time in the past. In many ways, our familiarity with the tastes, preferences, attitudes, and cultures of other countries has exploded in the last decade. Just one simple illustration is the fact that salsa sales in the United States have surpassed the sales of catsup. Travel is another factor that has lead to the globalization of today's marketplace. By exposing a greater number of people of various cultures to one another and to other places, an underlying melding of cultures or, at the very least, understanding of cultures occurs.

A more mobile society has changed the landscape of our marketplace at home as well as abroad. Demographics from the state of California offer an example. California, the most populated state in the nation, is also the clear leader in ethnic "mixing" in North America. Consider that in 1970, eight out of ten Californians were non-Latin whites. As of 2000, heavy immigration from Asia and Latin America has changed California's population mix to 49 percent non-Latin whites. California has now joined the states of New Mexico and Hawaii as states where non-Latin whites are not the majority. This type of mobilization of the population is not isolated to the United States alone by any means. It is occurring around the globe. That fact means that people bring their cultures, their traditions, their needs, their connections and networks with them from the lands from which they came.

Each of these factors has provided agribusinesses with more opportunity for marketing global products that would not only be accepted, but also preferred by the new, informed, exposed and traveled, global consumers. In response, global agribusinesses develop a universal strategy for marketing products to the world rather than producing single strategies for individual countries or markets.

INTERNATIONAL AGRIBUSINESS TRENDS

The U.S. food and agribusiness industries continue to gain momentum in international markets even though most food is consumed in the country in which it is produced. Roughly 16 percent of the world's food supply was involved in the international marketplace during the late 1990s and early 2000, compared with only 2 percent 50 years ago. The United States remains the world leader in exports of agricultural commodities and at the same time is an increasingly important player in the import of food and food products.

The Growing Population in the Global Market

The Food and Agriculture Organization (FAO) of the United Nations defines **food security** as the state of affairs where all people at all times have access to safe and nutritious food to maintain a healthy and active life. World population today is roughly 6 billion people. While population is growing more slowly than forecast a few years ago, it is still predicted that it will be a number of years before population growth stabilizes. Projections of future population put the global population between 8 and 12 billion by 2050 with nearly all the growth expected in the developing world. In perspective, during the next 30 years, the increase in the number of human beings is projected to equal the total world population in 1950, or about 2.4 billion people (Figure 6-1).

Given these population growth estimates, there is little doubt that food demand will increase significantly in the years ahead. Many experts estimate the

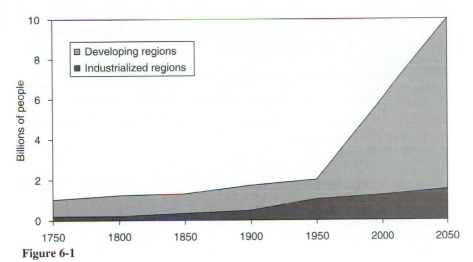

Figure 6-1

World Population Growth.

Source: United Nations Population Division and Population Reference Bureau.

growing population will mean a doubling of food production will be necessary during the next 30 years. Of course, food demand is more than just population—income to buy food is also important. So, the realized increase in demand for food will depend on both population growth and income growth. With respect to income, the world continues to grow more polarized. There is a greater discrepancy among the "haves" and "have-nots." While the number of people in the low-income bracket grows faster than the total world population, the share of income controlled by the upper-income bracket of the population has been rising significantly.

The world is becoming more urbanized as well, with more people moving to cities. The attraction of urban areas is in part due to the jobs available at these business hubs. During this decade, more than half the world's population or an estimated 3.3 billion will be living in urban areas. In 1975, only one-third of the world's population lived in urban areas.

So, not only will the sheer numbers of people on the planet increase, but it is also likely that the numbers of individuals with the ability to pay for products will increase. If current reforms are merely sustained in the countries with food security deficits, it is likely there will be improved food security for an increasing number of people. Translation: opportunities for food and agricultural businesses will continue to grow; and the international marketplace will offer increased possibility.

U.S. Agribusiness Trade and Investment Grows Abroad

The U.S. food industry is growing—and it is growing abroad. Two key factors responsible for that growth are trade and investment. International trade, both imports and exports of food and food products, is increasing faster than domestic

T A B L E **6-2** │ **USDA Baseline Projections—International Trade Summary by Decade**

International trade summary, by decade or indicated period[1]							
Years	Wheat	Rice	Coarse grains	Soybeans	Soybean meal	Soybean oil	Cotton
World trade growth, annual percent[2]							
1960 to 1970[3]	1.1	2.2	4.9	11.4	14.4	11.3	0.8
1970 to 1980	4.7	4.9	8.7	8.2	11.7	12.8	1.2
1980 to 1990	−0.3	0.6	−1.0	−0.4	2.9	0.5	2.5
1990 to 2000	−0.5	7.5	0.5	5.6	4.4	7.4	−1.2
2000 to 2009	2.2	2.2	2.3	0.7	1.7	1.7	1.9
U.S. export growth, annual percent							
1960 to 1970[3]	−0.8	6.3	3.8	12.6	13.0	5.3	−5.4
1970 to 1980	6.4	6.8	12.7	7.2	5.8	5.4	6.1
1980 to 1990	−3.3	−0.5	−0.7	−3.7	−1.8	−5.5	2.3
1990 to 2000	−1.4	2.2	0.9	4.5	3.4	7.9	−1.6
2000 to 2009	3.5	−2.4	2.7	−0.1	0.8	2.2	1.4
U.S. share of world trade, average percent[2]							
1960 to 1970[3]	37.6	19.0	50.0	90.6	65.6	66.6	18.3
1970 to 1980	43.0	22.1	59.4	82.6	43.5	37.5	19.8
1980 to 1990	37.3	20.2	59.4	72.6	23.7	19.3	21.5
1990 to 2000	30.5	14.0	56.8	63.7	18.6	14.5	25.3
2000 to 2009	31.8	9.9	57.4	64.6	19.2	16.0	26.0

[1]Years refer to the first year of the commodity marketing year.
[2]Trade and trade shares include intra–former Soviet Union (FSU) trade for periods starting in 1990 and later; intra–FSU trade for cotton also is included in the 1980 to 1990 and the 1970 to 1980 periods.
[3]Data for soybeans, soybean meal, and soybean oil begin in 1964.

sales. The United States Department of Agriculture reported that U.S. agricultural exports in fiscal year 2001 were $53 billion (*USDA Baseline Projections*). In contrast, agricultural imports into the United States in 2001 were $40 billion, leaving a surplus in U.S. agricultural trade of $13 billion, an 8 percent gain over fiscal 2000.

Table 6-2 shows the dramatic increases in agricultural exports in the 1990s that built on the increases observed in the 1970s. The U.S. agriculture industry is dependent on the export market. Today, roughly 30 cents of every $1 earned on agricultural products comes from exports. Strong export competition continues to influence the projected trends in the export market. Strengthening global economic growth may provide a foundation for gains in trade and U.S. agricultural exports. These factors point to the possibility of rising market prices, increases in farm income, and stability in the financial condition of the U.S. agricultural sector.

TABLE **6-3** | **Characteristics of Developed and Developing Markets for Food**

Developed countries	Developing countries
Saturation	Increasing demand
Food safety	Food security
Buyer's market	Seller's market
Concentration	Fragmentation

Finally, growth prospects for U.S. agricultural exports, particularly in the consumer food products arena, are shifting. As growth in the markets of developed countries such as Canada, Japan, and the European Union slows over time, developing countries and their strong growth economies have become the targets for increased food exports. Interestingly, the competitive market and characteristics of consumer food products between developed countries and developing countries differ (Table 6-3). The successful international marketer must address such differences.

Bulk Commodities, High-Value Products

Agricultural exports are composed of two general categories: value-added products and bulk commodities (Table 6-4). Value-added or **high-value products (HVPs)** are agricultural products that have received additional processing or require special handling or shipping. HVPs include unprocessed foods such as fresh fruit or eggs, as well as processed and semiprocessed grains, oilseeds, animals and products, horticultural products, sugar and tropical products. HVPs have been further categorized into intermediate commodities and consumer products (beverages, meats, and frozen dinners). Intermediate commodities are those that have received some processing (soybean meal), are used as inputs on the farm (animal feeds, seed), or are used by food manufacturers (flour). Bulk commodities, such as wheat, corn, soybeans, rice, and tobacco have historically comprised much of U.S. agricultural exports. However, exports of value-added products increasingly comprise a larger portion of total agricultural exports.

In 1970, wheat, corn, and soybeans comprised 68 percent of the total value of all U.S. agricultural exports. Twenty years later in 1990, the composition of U.S. agricultural exports had changed dramatically and just 53 percent of total export value consisted of these bulk commodities. The other 47 percent of agricultural exports were in value-added products. The rapid growth in HVP exports began after 1985 with the devaluation of the U.S. dollar relative to other currencies. During this time, incomes have increased in countries such as South Korea, Taiwan, Mexico, and Hong Kong and have also spurred demand for HVPs.

While shipping more high-value products every year, the export of bulk commodities continues to be important to U.S. agriculture.

Many of these countries are experiencing changes in lifestyle—convenience and western-flavored foods are increasingly being sought out and preferred by consumers. Import tariffs and other barriers have also declined during this time period providing another reason for the rapid increase in HVP exports. Also, the U.S. government, recognizing the value of advertising, began to fund promotions of agricultural products abroad. This, along with the success of other U.S. firms abroad, has increased the interest of U.S. firms in exporting agricultural products abroad.

As a result of these factors, the early 1990s saw exports of HVPs increasing annually at an average 6 percent rate compared to a declining rate annually of 3 percent for bulk commodities. By 2000, the value of U.S. exports was still hovering at $49 billion, but the value of HVPs had begun to recover. For the first decade of the new millennium, growth in both bulk and HVP exports is expected to climb steadily, exceeding growth of both the 1980s and 1990s and lending strength to total export earnings. HVP export growth overall is projected to average 3.8 percent annually (Table 6-4).

World Trade Organization

Political, social, and economic changes over the last decade alone have dramatically altered the way business is done in the international marketplace as well as the number of entities ready and willing to conduct international business. That growth and excitement has been building significantly over the course of the last five to six decades.

In 1947, an international organization was established to reduce trade barriers through multilateral trade negotiations. That organization was known as **GATT,** the General Agreement on Tariffs and Trade. Beginning in 1988, GATT negotiations intensified with the goal of developing a stronger organization. In

1995, the **World Trade Organization (WTO)** was formed and all GATT agreements of the most recent round of negotiations (the Uruguay Round) were incorporated into the WTO. WTO now has 125 member countries (nearly the whole world except China) and WTO rules apply to over 90 percent of international trade.

The WTO efforts stretch beyond its parent organization's focus on reducing tariffs of manufactured goods. WTO works to eliminate nontariff barriers as well. It can be used to mediate trade disputes, or to challenge environmental, health, and other regulations that may serve legitimate social goals, but may be regarded as impediments to international trade.

A reduction in global trade barriers has led to an increase in trade in the last decade. The most recent round of GATT talks was approved in the United States in 1994. Those tariff reductions, when all is said and done, are expected to result in average tariff reductions of 38 percent for developed economies. This slashes worldwide tariffs from 6.3 percent to 3.9 percent. As a point of reference, tariff rates after World War II were 40 percent. Such changes clearly support the importance of the international marketplace for food and agribusiness firm managers. While some criticize and others champion the WTO, one fact remains: growth in international trade and investment is developing at a faster pace than ever before and faster than would have occurred without the WTO.

CHALLENGES IN INTERNATIONAL MARKETS

The path to successfully doing business in international agribusiness markets certainly has its challenges. Some of the more dominant issues that firms face entering these markets are discussed in this section. These challenges include: cultural differences, exchange rate fluctuations, differences in accounting systems, uncertainty of the political and economic climate, property rights issues, regulations, sanitary and phytosanitary (SPS) rules, trade specifications, and the challenges of management in an international environment.

Cultural Differences

Culture is the set of socially transmitted behaviors, arts, beliefs, speech, and all other products of human work and thought that characterize a particular population. Culture is transmitted from one generation to the next, not inherited automatically at birth. Instead, culture is learned and it helps to define perceptions, beliefs, practices, communication styles, relationships, and family roles. Cultures vary dramatically around the world, and again, such differences must be understood by a firm attempting to operate in an international market.

The concept of culture is very broad and thus some system is needed for describing a particular culture. Anthropologists, for example, consider culture in a total systems approach, and look at a culture based on its systems of education, economics, politics, religion, health, recreation, and kinship. A simpler approach in understanding a culture is by looking at ten general characteristics or

T A B L E **6-4**   **USDA Agricultural Trade Values**

	1998	1999	2000[1]	2001	2002	2003	2004
	Billion dollars						
Agricultural exports:							
Animals and products	11.2	10.1	10.8	9.8	10.5	10.8	11.2
Grains, feeds and products	14.1	14.4	13.4	14.4	15.1	16.7	18.1
Oilseeds and products	11.1	8.7	8.6	8.6	8.8	9.3	10.0
Horticultural products	10.3	10.3	10.5	11.8	12.4	12.9	13.5
Tobacco, unmanufactured	1.4	1.4	1.3	1.2	1.2	1.2	1.2
Cotton and linters	2.5	1.3	1.5	2.2	2.6	2.8	2.6
Other exports	2.9	2.8	2.9	3.2	3.3	3.4	3.6
Total agricultural exports:	53.6	49.0	49.0	51.2	53.9	57.2	60.3
Bulk commodities exports	20.1	17.8	16.8	18.1	19.0	21.0	22.5
High-value product exports	33.6	31.2	32.2	33.1	34.9	36.3	37.8
High-value product share	62.6%	63.7%	65.7%	64.7%	64.7%	63.4%	62.8%
Agricultural imports:							
Animals and products	6.8	7.1	7.2	7.5	7.5	7.6	7.8
Grains, feeds and products	2.9	2.9	2.8	2.9	3.0	3.1	3.3
Oilseeds and products	2.2	2.0	1.9	1.8	1.7	1.6	1.6
Horticultural products	13.9	15.3	15.7	16.4	17.1	17.9	18.7
Tobacco, unmanufactured	0.8	0.7	0.7	0.7	0.7	0.8	0.9
Sugar and related products	1.7	1.6	1.6	1.8	1.8	1.	2.1
Coffee, cocoa, and rubber	6.3	5.2	5.4	5.5	5.5	5.6	5.6
Other imports	2.4	2.5	2.6	2.7	2.8	2.9	3.0
Total agricultural imports	37.0	37.4	38.0	39.3	40.2	41.4	43.0
Net agricultural trade balances	16.6	11.6	11.0	11.9	13.7	15.9	17.3
	Million metric tons						
Agricultural exports (volume)							
Bulk commodity exports	98.5	113.7	109.4	115.5	117.6	121.1	124.5

dimensions. It must be emphasized, however, that this is a simple model, and it is not the only way to analyze a culture. The ten dimensions are the following:

Sense of self and space. Sense of self (or individualism) concerns how much an individual values personal freedom over responsibility to family or national groups. Sense of space is concerned with the physical space we use in our culture. Americans have a sense of space that requires more distance between individuals than that required in Latin America and Arab cultures. Also, the size and location of an executive's office in the United States conveys a great deal about the status of the executive, but it is a poor indicator in some Arab nations.

	2005	2006	2007	2008	2009	1999–2009 Growth rate Percent
			Billion dollars			
Agricultural exports:						
Animals and products	11.7	12.0	12.5	12.8	13.3	2.8
Grains, feeds and products	19.4	20.5	23.3	23.6	25.0	5.7
Oilseeds and products	10.7	11.4	12.1	12.6	12.5	3.7
Horticultural products	14.1	14.7	15.3	15.9	16.5	4.9
Tobacco, unmanufactured	1.2	1.2	1.2	1.2	1.2	−1.6
Cotton and linters	2.6	2.7	2.9	3.0	3.1	8.9
Other exports	3.7	3.9	4.0	4.2	4.3	4.3
Total agricultural exports:	63.5	66.4	71.2	73.3	75.9	4.5
Bulk commodities exports	24.2	25.7	28.9	29.6	30.8	5.6
High-value product exports	39.3	40.7	42.3	43.7	45.1	3.8
High-value product share	62.8%	61.9%	61.3%	59.4%	59.7%	59.5%
Agricultural imports:						
Animals and products	7.9	8.1	8.2	8.3	8.5	1.9
Grains, feeds and products	3.4	3.6	3.7	3.8	4.0	3.0
Oilseeds and products	1.8	2.0	2.3	2.6	2.9	3.7
Horticultural products	19.5	20.3	21.1	21.9	22.8	4.1
Tobacco, unmanufactured	0.9	1.0	1.1	1.1	1.2	4.5
Sugar and related products	2.1	2.0	1.9	1.9	1.9	2.1
Coffee, cocoa, and rubber	5.7	5.7	5.8	5.8	5.9	1.2
Other imports	3.1	3.2	3.3	3.4	3.5	3.3
Total agricultural imports	44.4	45.8	47.4	49.0	50.7	3.1
Net agricultural trade balances	19.1	20.6	23.9	24.3	25.2	8.1
			Million metric tons			
Agricultural exports (volume)						
Bulk commodity exports	127.4	129.8	132.1	134.3	136.6	1.9

[1]The projections were completed in November 1999 based on policy decisions and other information known at that time. For updates of the nearby year forecasts, see USDA's *Outlook for U.S. Agricultural Trade* report. Note: Other exports consist of seeds, sugar and tropical products, and beverages and preparations. Essential oils are included in horticultural products. Bulk commodities include wheat, rice, feed grains, soybeans, cotton, and tobacco. High-value products (HVPs) is calculated as total exports less the bulk commodities. HVPs include semiprocessed and processed grains and oilseeds, animals and products, horticultural products, and sugar and tropical products. Other imports include seeds, beverages except beer and wine, and miscellaneous commodities.

Communication and language. Communications systems, both verbal and nonverbal, distinguish cultures from one another. More importantly, communication is effective when the receiver ends up with the message the sender intended to send. It has been said that as much as 70 percent of a message is nonverbal. Nonverbal messages include gestures, facial expressions, posture, stance, eye contact, and use of color. For example, Far Eastern cultures tend to value politeness over blunt truth.

Dress and appearance. Garments, adornments, and body decorations distinguish cultures. Subcultures also wear distinguishing clothing—managers may dress differently in different cultures.

Food and eating habits. Food is selected, prepared, presented, and eaten differently in cultures. Although Europeans and Americans use the same dining utensils, you can distinguish the two from which hand holds the implement.

Time and time consequences. Different messages may be conveyed by the amount of notice given to or at by a meeting, invitations to future commitments, or the time of departure. To be 30 minutes late for an appointment with a business associate may be considered rude in the United States, but it may be early and unexpectedly reliable in Puerto Rico.

Relationships. The value of relationships in a culture is most easily viewed by the importance of the family unit (and the extended family). Age, gender, status, wealth, power, and wisdom are other aspects that may affect the type and degree of relationships within a culture.

Values and norms. Needs and wants may vary between cultures. Acceptable behavior within a culture reflects the values held by that culture. Underdeveloped cultures may value adequate food, clothing, and shelter. A more developed culture may take these items for granted and may place high value on money, status, and civility.

Beliefs and attitudes. All cultures have some interest in a "higher power" or the supernatural as seen in their various religions. Attitudes toward oneself, others, and the world also distinguish cultures. U.S. business philosophy assumes that people can substantially influence and affect the future. Other cultures may believe that events will occur regardless of what they do.

Mental process and learning. Some cultures emphasize abstract thinking and conceptualization while others prefer learning and memory. However, most cultures do recognize and reward reasoning processes.

Work and work habits. Work and success at work may be viewed differently in different cultures. Work habits, attitudes toward work and authority, and how work is rewarded and measured may all vary within a culture. Some cultures view success at work in terms of the status of the position held within the company, while other cultures view success in terms of the total income earned.

The international agribusiness firm must be sensitive to differences in cultures and make the appropriate adjustments in their business approaches. The experienced agribusiness manager will try to accept the values of the local culture and will seek to work within the accepted behaviors. Adjustments, if necessary, by international agribusiness firms are generally made in three categories:

Product adaptations. Products may undergo modifications and/or the way products are marketed may change.

Individual. A manager overseas for the first time must learn the local language and make adjustments in dealing with people.

Institutional. Hiring practices and organizational structure often must change when working in a new country. For example, an agribusiness must take into account class distinctions, different religions, or different tribes that may require adjustments in hiring and placing people within an organization.

Exchange Rate Fluctuations

Exchange rates determine the relative worth of one country's currency to that of another country. An example of an exchange rate is one U.S. dollar equals 116 Japanese yen. This exchange rate will change, fluctuate, over time. Agricultural trade is very sensitive to changes in the exchange rate relative to some other industries because agricultural goods are more homogeneous (U.S. wheat and Canadian wheat are nearly perfect substitutes). When the value of the dollar rises relative to other competitors' currencies, U.S. agricultural goods become more expensive—and thus less competitive. Therefore, the export performance of U.S. agriculture is closely tied to changes in the value of the dollar relative to other currencies.

Export profits can also be reduced significantly with highly fluctuating exchange rates. For example, a U.S. agribusiness agrees to sell 3,000 tons of corn to a Japanese buyer for $12,500 yen per ton. If the exchange rate increased in terms of foreign currency per U.S. dollar between the time of the contract and the actual time of payment, then the U.S. firm would earn a lower profit because it would receive fewer dollars for the corn.

Agribusinesses can deal with exchange rates in several ways. Sometimes U.S. exporters agree to accept payment in foreign currency because they can manage more effectively the exchange rate risk. This approach can be a bargaining chip as well because foreign firms may be willing to pay a higher price in their home currency. Another method to deal with fluctuating exchange rates between the United States and the importer is to negotiate payment in another country's currency that is relatively stable. Still yet another method is to agree to divide the gains and losses in the exchange market, or to use the futures markets and manage the risk using currency futures.

Accounting Differences

The internationalization of business is also made complex by the different accounting methods that are used in each country. Financial terms are often quite different—"inventory" in the United States is "stocks" in Great Britain; "retained earnings" in the United States is "profit and loss reserve" in Great Britain. Some financial statements do not even exist. The "statement of cash flows," an important accounting statement in the U.S. financial market, does not exist in countries like Germany and Japan.

Financial statement format often varies from one country to another. In the United States, assets are presented in order of their liquidity—the most liquid or current assets are listed first followed by less liquid assets. British financial statements are listed exactly in the opposite order. In fact, assets and liabilities are listed together.

The accounting for noncurrent assets can vary substantially between countries. In the United States, large equipment and capital assets are valued on balance sheets at their historical cost. The value of these assets is depreciated as a means of allocating the costs of these assets to their use over time. However, the original asset value on the balance sheet never changes. That is, an asset's value on the balance sheet rarely reflects the replacement cost or market value of that asset. Countries such as Australia, Canada, and England periodically appraise their assets and make adjustments to reflect their current market value. Depreciation is then estimated using these adjusted asset values.

Research and development (R&D) expenditures are also accounted for differently between countries. In the United States, R&D costs are treated as an expense in the period in which they are incurred. In Brazil, R&D expenditures are capitalized to provide an incentive for companies to invest in developing new products.

As noted, exchange rates between countries fluctuate because of many different factors. These fluctuations may make it difficult to combine the financial data of foreign subsidiaries with domestic subsidiaries. Companies that trade stock or debt securities on exchanges located in other countries (such as the Nikkei in Japan) as well as the United States may have to prepare different financial reports that comply with the particular accounting standards for that country. As one can imagine, these differences in accounting standards between countries have added much complexity to issuing financial statements in a timely, accurate manner.

International Law and Property Rights Issues

A global agribusiness must understand the many legal systems of the countries in which it operates. Once an international firm enters a country, it automatically becomes subject to all of that country's laws. Some governments may monitor daily business activities very closely while others may regulate only the large-scale activities of an agribusiness.

Probably the most important legal issues arising out of operating in the global economy are the interpretation and enforcement of various regional trade agreements. Examples of such agreements include the North American Free Trade Agreement (NAFTA), the World Trade Organization (WTO), and several other large regional trade groups including the European Union (EU), the Association of Southeast Asian Nations (ASEAN), and the Asia-Pacific Economic Cooperation Group (APEC). Each of these represents a restructuring, reorganizing, and refocusing for companies as they respond to the changing competitive mix of the global marketplace.

These alliances, agreements, and trade initiatives will affect the practice of business worldwide. They mean companies will continue to inspect the hows and whys of conducting business efficiently enough and effectively enough to be competitive in the world market. For example, the 1994 North American Free Trade Agreement (NAFTA) spurred U.S. exports to Canada and Mexico. The NAFTA agreement reduced tariffs on agricultural commodities and removed trade barriers.

Most countries protect intellectual property rights such as trademarks, patents, and publications. However, because these patents are issued in a particular country, they are valid and legal only within that country's jurisdiction. Some countries, most notably several in the Far East, do not have—or have not enforced—laws that protect these intellectual property rights.

Private property, and the protection of it, is another important international legal issue. Individual nations, including the United States, reserve the right to take over private property when it is in the best interest of the host country. In the United States, as with other countries, the government may take over private property for a "public purpose" (the United States uses the laws of condemnation or eminent domain to do this). But, the interpretation of what "public purpose" means is often at issue.

Sanitary and Phytosanitary Regulations

Sanitary and phytosanitary (SPS) regulations are intended to protect human, animal, or plant life or health. SPS regulations have been instituted in many countries because consumers in developed countries are increasingly demanding a higher level of food safety. However, in some cases, these SPS regulations have become a tool to use in protecting domestic agribusinesses and producers from competition.

The WTO agreement stresses the desirability of common SPS regulations, and the agreement provides a framework for distinguishing legitimate SPS regulations from those that are not. Three international organizations have been designated as sources of scientific expertise and internationally agreed-upon standards. The organizations will work to promote and address SPS-related regulations in areas such as food additives, pesticide residues, animal health, and plant health.

The benefits of standardized SPS regulations are enormous because firms could develop new products that simultaneously meet the requirements of all markets rather than developing and testing configurations of products that meet different regulations. However, the concerns that brought about many of the SPS regulations must be considered in establishing more general standards.

Management Challenges

Managing employees, clients, or accounts can be a challenge even when all are operating under the same roof. Throw in several thousand miles, a different language, culture, government, and monetary system and one has a management

challenge on hand, to say the least. Astute and deft handling of the numerous political and economic uncertainties in an international market can be among a
manager's most challenging problems. Managing such issues and events may
become heightened or intensified when put in motion in the international arena.

ENTERING INTERNATIONAL MARKETS

As the global marketplace becomes reality to an ever-increasing number of
agribusinesses, an evolutionary process becomes evident. This evolution may
take place by design or by happenstance, immediately or over a course of years.
However, many observers point to three phases of evolution as a firm moves
from a domestic to a global perspective.

Some companies arrive in the international market via detour—some even
by accident. Introduction to the international market for these companies may be
the result of an interaction with an international buyer or perhaps an export company. Often, in these cases, the international market is not treated any differently
than the domestic market. Changes to the product, the marketing, and so forth
are not made to fit the international market—firms simply sell what they have
always sold, but to a new, international customer group.

However, over time, many of those same firms shift to purposeful rather
than incidental marketing of their products abroad and enter the second phase of
the evolution. This means a deeper level of involvement, assigning resources
specifically to developing markets in specific countries, looking for opportunities in production and distribution of the product, etc. This phase of the evolution of the global firm is known as the export marketing phase. Here, the business conducted internationally is viewed as a sideline to the normal domestic
endeavor, but it is recognized as a separate and unique business.

The third phase emerges when a firm truly develops a global perspective,
and manages its business accordingly. Decision making at all levels is done in a
global context. Production and raw material sourcing decisions are optimized
around the world, marketing similarities are exploited globally, and the firm
may focus on building global brands. Managers move across borders regularly
to better understand the nuances of running a global enterprise. For all three
phases of evolution, decisions about market entry must be made—how do food
and agribusiness firms enter other countries?

Modes of Entry

Three general methods are used to enter an international market: exporting, licensing, and foreign production. There are, of course, many variations of these
methods including indirect and direct exporting, franchising, and foreign direct
investment through acquisition, joint venture, or greenfield investment. The
choice of entry method into the global arena depends on factors external and internal to the firm. External forces that influence method of entry are factors such
as the political, economic, and cultural arrangement of the target market. Internal firm factors include the agribusiness's strategic and financial goals, the

firm's financial resources, the type of product that is being produced, the sensitivity of the product or production method to the firm's competitiveness, and the firm's international market experience (Stern, El-Ansary, and Coughlan). Historically, many firms have used a strategy that gradually increases their presence in the global market over time.

Exporting

There are two general means used to export an agribusiness's, products: indirect and direct exporting. Most agribusinesses, especially those with little international marketing experience, initially enter the global market via indirect exporting.

   Indirect exporting uses a trading company or an export management company to handle the logistics of exporting. These trading experts manage the exporting and importing procedures and regulations, and they use their established relationships with buyers and distributors to distribute the product. Advantages offered through working with a trading company include the expertise, knowledge, experience, and connections in the market. These trading companies' networks within the distribution channels can be extremely useful to first-time exporters. Although overall, using indirect exporting may reduce profitability, many firms perceive this to be a low-risk strategy that entails substantially lower investment.

   Direct exporting is where the agribusiness itself handles the details of exporting their products. At this point, the firm conducts research, establishes contacts in the country, and sets up its distribution channels. Firms often open an overseas sales office to manage the operations in that country. Direct exporting involves investments and salaries for the items mentioned previously. In turn, the potential for profits are much higher, and the firm can exert more control over product distribution.

   In general, the advantages of exporting (over the other methods) are lower risk, lower fixed costs (compared to investing in a new plant in a country), and increased speed in reaching the global market. Several U.S. government agencies assist exporters as well. The disadvantages of exporting are primarily managing the trade barriers or protectionism that may exist in a country. Regulations, inspections, tariffs, and quotas are just some of the barriers that may be encountered, as well as less control and long distribution channels. Control over pricing, promotion, distribution, and quality are some of the other problems that may be experienced. As a result, agribusinesses wanting more control and lower costs may decide to look at licensing or direct investment as a means of entering a country's market.

Licensing

Licensing involves contracting with a firm (licensee) in the target market to produce and distribute your firm's (licensor) product. In return, the licensor receives a fee or royalty, one that can be very profitable given that little capital is required by the licensor. Licensing can be especially attractive with agricultural products that are perishable and bulky or where the receiving country has restrictions on imports.

There are also disadvantages of licensing. The license may entail giving secret product formulas or processing technologies to the licensee, a key source of competitive advantage for the firm, and, the licensee may also become a competitor selling products where they are currently being sold. Also, if sales take off, the licensee could decide to establish and sell a competing product, depending on the nature of the agreement that was signed. Finally, the licensor could find its brand image hurt if the licensee has not maintained sufficient control over product quality or promotion.

Direct Investment

Agribusinesses use a variety of investment means to enter and serve a foreign market. These can fall broadly into three areas: greenfield investments, joint ventures, and acquisitions.

Greenfield investments are direct investment by a firm into a particular country. An example would be a new plant expenditure by a multinational firm or foreign subsidiary. Greenfield investments are often the only method of investing in a developing country where no other plants exist. High costs and risks are the drawbacks of these types of investments because unexpected market deterioration may prevent the firm from achieving a return on its investment. Another concern of greenfield investments may be the lack of understanding of the regulations and rules in the existing markets as well as the distribution channels that must be established. In fact, the distribution channel may be "closed" in the sense that other firms may have contractual distribution arrangements that prevent the new firm from selling its products through the existing system.

To counter the difficulties, risk, and costs associated with greenfield investments, an agribusiness may choose to establish a joint venture. A **joint venture** is a form of strategic alliance (discussed in Chapter 4) that involves two or more firms that share resources in research, production, marketing, or financing, as well as, costs and risks. Problems may arise with joint ventures in deciding which firm will have controlling interest; in establishing and maintaining working relationships; and in determining how critical decisions will be made. It also must be noted that joint ventures are not permanent partnerships—they do have an end point. In fact, across all industries, the average life of joint ventures has been estimated to be about 3.5 years.

Purchasing a firm or a controlling interest in the firm, called an **acquisition,** is the third method of direct investment. There are several advantages of an acquisition, but the first step requires finding a suitable company that is available (not likely in less-developed countries) at a reasonable price. Compared to greenfield investments, an acquisition may result in a lower investment to gain control of a production facility. Perhaps importantly, employees, access to distribution channels, market knowledge, experience, and reputation may also be obtained in the purchase. An acquisition can allow a firm to gain country and market-specific knowledge without incurring a long and costly learning process.

However, the acquired firm may contain unrelated or undesirable business lines. And, the firm may have to make significant investments in new equipment, and it may need to change the mindset of current employees to the acquiring

TABLE **6-5** | **Comparing Exporting and Investing in International Markets**

Choose exporting when . . .	Choose investment when . . .
Financial resources are small	Financial resources are large
Little experience in the new market	Experienced in new market
Barriers of trade for product do not exist	High barriers to trade exist
Target market is small	Target market is large
Growth potential is limited	Growth potential is high
Barriers to investment and ownership exist	Barriers to investment or ownership do not exist
Control of market is unimportant	High desire for market control
Globalization is not a high business priority	Firm's goals include an international presence
Foreign market is unstable	Stable political and economic climate
Foreign market is culturally "distant"	Market is culturally "similar"

Sources: Connor and Schiek (1997); Rue and Byars (1980); Krajewski and Ritzman (2000).

firm's style of management. Nevertheless, the number of acquisitions taking place continues to grow. In the food processing area alone, the number of U.S. food processing acquisitions abroad increased from about 5 per year in the early 1980s to 14 per year in the late 1980s (Connor and Schiek). The number of acquisitions of food processing firms in the United States by foreign investors increased from an average of 7 per year to 17 per year during the same time period.

Table 6-5 compares the factors that distinguish the decision to export or invest. It should be emphasized that choosing to export, license, or invest is not an either/or proposition. Firms that decide to invest in production facilities in a country may export products to the country as well. In the case of U.S. firms, many have invested in foreign markets and used their new base market to export to other markets. In contrast, foreign firms investing in the United States have done so to primarily serve the U.S. market.

Start-Up Strategy

In general, most firms follow the same series of steps once they have decided to enter the global market. First, markets are evaluated for their overall profit opportunities for that firm's products. Factors that they consider include market size, rates of growth, and adaptability of that agribusiness's, product to that country's culture, consumer income levels, and the political and economic climate of that country.

If the agribusiness determines that the market is promising, the next step taken is choosing the strategy to enter that market. In general, initial entry is made through exporting the firm's products. If sales of the product start to grow, then the firm follows with some form of licensing or investment into production facilities in the receiving country. This allows for higher profits, more control, and greater flexibility over the distribution and sales of the product.

SUMMARY

International agribusiness is a field that has seen phenomenal growth over the last decade alone. Advantages to conducting business abroad include sales growth, taking advantage of scale economies, risk reduction, lower production costs, access to lower-cost raw materials, access to credit, and experience from operating in other markets. The internationalization of our world, our attitudes, and our abilities to be connected are greatly responsible for the growing trend called globalization. Some companies have done away with their domestic versus international distinctions and instead are competing in global industries with global strategies and approaches.

Exports and international sales have become even more important to the food and agribusiness markets. With population and income increases comes the potential for increased demand for high-value agricultural products. Such trends affect the entire food production and marketing system.

Fueling the internationalization of our business world are the political, economic, and demographic shifts occurring around the world—shifts that in many cases create new business opportunities. The evolutionary steps taken by world leaders and governments to open the doors to international trade include efforts of the North American Free Trade Agreement (NAFTA), and the General Agreement on Tariffs and Trade (GATT), which is now the World Trade Organization (WTO).

Challenges of doing business internationally include exchange rate fluctuations, cultural differences, accounting differences, and regulatory issues. Agribusinesses entering the international market generally do so using one of three general strategies: exporting, licensing, and foreign production. Variations of these methods include indirect and direct exporting, franchising, and foreign direct investment through acquisition, joint venture, or greenfield investment.

DISCUSSION QUESTIONS

1. What are some of the significant trends in today's international agribusiness trade? Why are these significant? What impact do they carry and to whom?
2. Describe the role of the World Trade Organization in international trade today.
3. Pick a country that interests you. Pick a food, farm, or input product you have an interest in that might have a market in the country you chose. Using the ten dimensions of culture outlined in the chapter, describe the culture of the country you have chosen. What are the key features of the culture that would need to be addressed by a manager selling the product you chose into that country?
4. Mark and Angie Lutz farm just outside of town and just west of campus. Mark's primary business is the farm. After being an early adopter of GPS technology on his farming operation, he had an idea. Pushing the idea

further, he developed a valve specifically designed to respond to the GPS signals while applying liquid fertilizer with variable rate technology (VRT). He applied for a patent and was granted one. Angie, a former account executive with an advertising agency, was working in the agricultural communications department at the university. She had helped Mark with strategizing potential marketing plans for their valve. They formed a small business and had a local firm manufacture the valves. A few months later, they introduced the concept to a regional precision farming business and had favorable interest, but no significant orders had been placed. Just enough orders trickled in to keep it interesting during the firm's first 6 months.

One afternoon in December, Angie was hosting a team of Japanese agricultural business executives who were in the United States to learn, among other things, more about precision farming. Part of the agenda involved spouses joining the executives for dinner. Mark and Angie discussed the real-life, practical side of precision farming. The executives were delighted with their company and during the course of the conversation asked specifically how liquid fertilizer could be more effectively applied using variable rate technology (VRT). Mark offered to send plans and information to the executives. The result was that the firm for which the Japanese delegation worked placed an order for Mark and Angie's valves. Almost overnight, they were doing business in the international market.

a. Define and discuss three challenges Mark and Angie may face in their new international business endeavor.

b. What type of benefits might Mark and Angie realize as part of their new international relationship. Why?

c. Using the information presented above, explain how Mark and Angie's business might follow through and expand on this entry into the international market. Use examples based on exporting, licensing, and direct investment. What are the advantages and disadvantages of each approach?

CASE STUDY — **Agrofirma Sidabra**[1]

The managing director of Agrofirma Sidabra walked out of his office and toward the slaughterhouse. He wanted to take one last look around the operation before he headed home for the day. As he walked down the drive, he noticed how calm the facility was at this time of day—one hardly knew that more than 20,000 hogs were housed in the long stretch of low buildings located near the drive.

A thought leaped across the managing director's mind—the calmness of the late afternoon was so far removed from the chaotic business environment he was attempting to guide the Agrofirma Sidabra firm through. Meat consumption in Lithuania was down and export markets were difficult to tap. The steps to privatize the operation had been initiated, but every day brought new challenges in this process. The equipment in his

---

[1]This case study was developed with the assistance of Vida Ziukaite, Lithuanian Agricultural Academy.

operation needed upgrading and he had talked with individuals outside Lithuania about the capital and technology so important to making the firm more efficient. But, inflation was running more than 20 percent a month and potential investors were nervous about all the uncertainty in the country. And, there was the question of marketing. He knew that a new competitive arena was just around the corner where every sale would have to be earned. How should his firm approach this rapidly changing Lithuanian marketplace?

Indeed, he was thankful for the light breeze on this sunny, beautiful day—it helped take his mind off the challenges in front of him. But, he knew his relief would be only temporary since tomorrow he would need to continue his quest for the set of management and marketing decisions that would help guide this business through the transition period and establish it as a profitable enterprise for the longer term.

The Firm

Agrofirma Sidabra is built around a farrow-to-finish pork production enterprise where the hogs are raised in total confinement. The firm also operates a livestock slaughter and processing facility. While both cattle and swine are processed, the focus is on the processing of pork products. The farm has a small greenhouse operation as well, but this is clearly a sideline to pork production and livestock processing. The operation produces about 2,500 tons of pork per year and has had a very successful financial and production record.

The firm was originally organized as a state-owned pork production firm. Prior to 1989, virtually all of the firm's production was sold to state meat enterprises for further processing. In 1989, the operation expanded by building a slaughter facility and adding meat processing capacity to produce sausage and smoked products. This move gave the firm the capacity to slaughter its own production and sell both fresh and processed pork and beef products. The processed products were primarily sausages. The slaughter facility has a capacity of about 10 tons per day. At present, they slaughter all of the swine they produce in their production unit and buy the rest of the needed hogs from area farmers. About 50 percent of the hogs slaughtered are their own and about 50 percent are purchased from other farmers. All cattle slaughtered are purchased from area farmers.

The firm owns about 35 hectares of land, so they purchase feedstuffs as opposed to growing them. They are aggressive buyers of grain at harvest and also must secure storage capacity for the grain. The firm has only enough storage capacity for about two months of their feed requirements at their location. Since grain is in short supply in Lithuania, the government rations its stocks to feed grain users. Agrofirma Sidabra is in a relatively good position in that they have the right to buy up to 50 percent of their feed grain needs from the government.

The management team is currently exploring the purchase of an automatic feeding system from an Austrian firm. This system would help them reduce labor expense and cut their feed costs dramatically. So far, they have not been able to develop a partnership with the firm because of problems with the government and because of a lack of hard currency. Feed quality has been a major concern over time, and problems with feed have hampered efforts to improve their production efficiency, as well as the quality of their products. And, fuel availability is a continuing source of concern. In addition, animal health medications are very expensive, when they can be obtained.

The greenhouses are still in operation, though they have not been profitable over the last couple of years. This is due in large part to high fuel costs. While losses from the greenhouse operations have been substantial, the firm does not want to get completely out of this market yet. Management feels there may be potential here in the long term and would like to be in a position to capitalize on this opportunity when it develops.

A mink growing enterprise was added in 1991. This is a new business for the firm, but looks to have considerable potential. The market for mink pelts is strong and the firm has a ready source of feed in the form of scrap products from the slaughterhouse. In addition, they have individuals already on staff with the expertise to care for the animals.

The firm employs some 260 people. And, these 260 people currently own about 30 percent of the business. The extensive investment in the firm's building and facilities has made it impossible for the current employees to acquire any more of the business to date. And, while employee ownership has its benefits, it greatly complicates labor decisions. The firm probably has excess labor, given current market conditions, and will definitely have too much labor if they are able to acquire the much needed automatic feeding system. The managing director is not sure how to handle this dilemma—how do you fire an owner? And, perhaps more importantly, how do you help an employee understand just what ownership means so that the incentive benefits of employee ownership can be realized? These were difficult questions for the managing director to answer, though he had given them much thought.

The Market

The Lithuanian market for meat and meat products has been a very difficult one over the past year and it appears as if the difficulties are not over yet. With consumer incomes down, demand for inexpensive cuts (hearts, lungs, etc.) has been strong while the demand for cuts like the ham and loin has been weaker in general. The market for expensive processed products has been particularly weak. Some of the firm's sausages retail for as much as 20 *litas* per kilo, while smoked products may retail for more than 10 *litas* per kilo in the Lithuanian meat shops. These prices are beyond the reach of many Lithuanian consumers. However, demand for some of these processed products is very seasonal, with good demand near holidays in most cases.

At this point, the Lithuanian consumer in general is a price shopper. This puts pressure on all types of food firms to be efficient providers of products. Many in the current economic situation view meat as a luxury.

Due to the reduced domestic demand for processed pork products, attempts have been made to reduce retail costs to consumers. In the domestic market it appears that the Lithuanian consumer is willing to accept a bit higher fat content in sausage, for example, if it reduces the cost of the product. However, this fatter product is unacceptable in the export market. Thus, the managing director faces a difficult dilemma: should the firm make a stronger move to reduce the price of its products (by adding fat) to increase sales in Lithuania? How will the export market be addressed? Can two different "grades" of product be produced and marketed?

The market area for Agrofirma Sidabra is defined to a large extent by the transportation costs to get the product to the retail shops. If more efficient means of distribution can be found, the firm's market area could grow.

Agrofirma Sidabra has looked at the export market for some of the reasons outlined above. And, they have exported some fresh pork products to Estonia. However, in general, selling in this export market presents a number of challenges. First, an export license is difficult to obtain. However, this could change abruptly in the future. Export for barter (i.e., for fuel) may be somewhat easier than export for currency. Finally, the firm's products are not of high enough quality at this point for export to many of the Western European or U.S. markets.

The Competition

There are several other pork production complexes in Lithuania, but most do not have further processing capacity. Most of these operations are state-owned enterprises that are just beginning the process of privatization. And, the profitability of these other operations has been questioned. In general, the difficult market for meat has left most of these operations with tremendous overcapacity. Government statistics suggest that the number of hogs on farms fell from 2,705,000 in 1990 to 1,359,000 in 1993—a decline of almost 50 percent over the period. This sharp reduction in the number of animals produced has left many of these large factory-type units with relatively high unit costs. And, the situation has opened the doors for small, flexible operations without the burden of excess capacity to enter the market.

The integrated nature of Agrofirma Sidabra gives it an edge over the competition. In most cases, there are state-owned pork production operations and state-owned meat processing businesses, but these competitors are not integrated. However, there is intense competition for animals—especially at certain times of the year, such as when producers are in the field.

In addition to these traditional competitors, the difficult economic times have created another small, but important group of competitors—the small, rural landowners. Prior to World War II, roughly 70 percent of the Lithuanian population lived in rural areas with the remaining 30 percent living in the urban areas. By the early 1990s, this situation had reversed—30 percent of the population lives in the rural areas and 70 percent of the population lives in the urban areas. Many of the rural residents are farmers and most have a few hogs and cattle. In many cases, they slaughter their own animals and share or sell the meat to their relatives in the urban areas. Such competition cannot be ignored.

The managing director knows that other pork production firms will be looking for ways to survive and has some concerns that they too will integrate and add processing capacity. And, he also worries that smaller operations will enter the market. While he feels he currently has an edge on these operations, he wonders how to position his business to deal with these competitors effectively over the longer term.

Marketing Activities

The firm has pursued few marketing activities to date. This is not an indication of poor management, rather it was a market reality—all of the firm's production was acquired by a state meat enterprise and they did not need to worry about marketing.

Most of the firm's meat is distributed to urban consumers through small shops. There are a large number of such shops the firm works with and the sales per shop are small—about 2 to 4 tons per week on average. In some cases, the firm has agreements with the shops, and in other cases the relationships are much more informal. Sales are primarily made over the telephone. All sales are cash; the firm does not have a credit policy of any type. Increasingly, the firm provides delivery service for its product to the shops, with the shops paying the firm for the cost of the transportation.

The firm has worked hard to establish itself as a reliable supplier of meat to any shop at any time. Much of this reputation was built during a recent period when pork was in especially short supply in Lithuania. The integrated nature of the business allowed the operation to continue to supply its shops when other firms could not. In addition, they were able to pick up new customers when other firms were unable to supply their regular customers. And, the managing director feels they have a reputation with their customers for delivering a high-quality product. Quality here is defined as freshness and leanness. He feels that both of these characteristics of his business are important strengths.

There are, however, some problems to address. Once the meat is in the shops, the firm has little control over the quality of the product. The firm's primary way of maintaining quality is in the choice of the shops to which they sell. Given this circumstance, insuring consistency to the end consumer is very difficult. Their smoked sausages and hams are sold almost exclusively through private shops and restaurants since consumers who shop at such stores or frequent these restaurants typically have money to spend. They have discussed the idea of a branded product, but have not pushed this notion very far. They do little or no advertising—this, in general, is expensive and they don't know how much good it would do. They provide no marketing support to the shops—Agrofirma Sidabra's marketing efforts stop when they transfer the meat to the shop. Any additional meat sales efforts are the responsibility of the shops themselves.

The Future

Management has a number of problems on the table. First, there is the lack of available capital to expand. While a constraint, there is little the managing director can currently do except keep a constant watch for sources of capital and try to make the business as profitable as possible to create capital internally. Second, they would like to form partnerships with other firms to help them open up export market opportunities as well as to obtain supplies like veterinary products that are difficult to acquire in Lithuania. The firm is driven to become more efficient, yet labor costs are difficult to reduce given the ownership interest of the 260 employees. In addition, it has proven difficult to secure the automated feeding equipment needed to reduce feed costs.

Finally, the managing director knows that he/she needs to work on the firm's marketing efforts. The managing director realizes the firm is well positioned to get through the current difficult situation and would like to position the business to really grow once the economy begins to turn around. The question is how to get this done.

Questions

1. What are the strengths and weaknesses of Agrofirma Sidabra?
2. What are the opportunities for this firm in the market? What specific threats for this firm appear on the horizon? How does the Lithuanian market differ from the U.S. market for meat products?
3. What target markets should the firm pursue? Consider both domestic (Lithuanian) and international markets.
4. What suggestions do you have for the firm manager? What steps can he take to pursue the markets you feel are most likely to be profitable?
5. Specifically, which international markets would you pursue and how would you enter these markets?

REFERENCES AND ADDITIONAL READINGS

Connor, John M., and William A. Schiek. 1997. *Food Processing, An Industrial Powerhouse in Transition.* Second Edition. John Wiley and Sons.

Krajewski, Lee J., and Larry P. Ritzman. *Operations Management: Strategy and Analysis.* Reading, MA: Addison-Wesley Publishing Co., 2000. p. 435.

Rue, Leslie W., and Lloyd Byars. *Management Theory and Application.* Homewood, IL: Richard D. Irwin. 1980. p. 550.

Stern, Louis W., Adel I. El-Ansary, and Anne T. Coughlan. 1996. *Marketing Channels.* Fifth Edition. Prentice Hall.

U.S. Department of Agriculture. February 2001. *USDA Baseline Projections to 2010.* Office of Chief Economist. USDA. Staff Report WAOB-2001-1.

MARKETING MANAGEMENT FOR AGRIBUSINESS

Marketing and sales activities of food and agribusiness firms, the focus of this section, must continually shift with the changing direction of customer needs and wants.

7

STRATEGIC MARKET PLANNING

- Outline the marketing concept
- Review the evolution of marketing in the food and agribusiness industries
- Present the market planning framework
- Discuss the components of a SWOT analysis
- Examine the market segmentation concept
- Develop the fundamental idea of positioning

INTRODUCTION

Ask people what they think of when the term "marketing" is mentioned and most will answer "advertising" or "selling." While advertising and selling may be two of the most visible marketing activities, marketing within food and agribusiness firms involves far more than these two methods of marketing communication. In reality, marketing includes a wide spectrum of decisions and activities that center on satisfying customer needs and wants. The full marketing process involves identifying customer needs, developing products and services to meet these needs, establishing promotional programs and pricing policies, and designing a system for distributing products and services to customers. **Marketing management** is concerned with managing this total process.

THE MARKETING CONCEPT

Marketing can be defined as the process of anticipating the needs of targeted customers and finding ways to meet those needs profitably. There are several key ideas in this definition. Marketing is about anticipation. Good marketers are always working to anticipate what their customers' needs will be in the future.

This may mean anticipating the features farmers will be looking for in a new tractor or it might involve anticipating the type of seasonings consumers will want in a new frozen chicken entree. Good marketing involves having the right products and services available when the customer is ready to buy them. It follows that good agribusiness marketers know a lot about their customers.

A second key idea is the notion of a **target market.** Clearly, "one size does not fit all" in the food and agricultural markets. Good marketers understand this and focus their efforts on the unique needs of specific target markets or market segments. They know that the small livestock farmer in Kentucky needs a very different set of products and services than does the large turf seed operation in the Willamette Valley of Oregon. Good marketers understand that a high-income, dual career couple with no children has different needs than a young, middle-income family with two children. While agribusiness firms may pursue more than one target market, their approach to any single market segment involves a set of decisions tailored to the unique needs of the segment.

The final key idea in the definition of marketing is that of profitability. After careful study of a particular target market, the agribusiness marketer will likely generate a long list of products and services that the customer might be interested in. Such a list for the small Kentucky livestock farmer might include feed delivery services, extended credit, a 1-800 phone number for questions, a well-trained, professional salesperson who makes on-farm calls, a staff nutritionist who is always available for questions, a website with links to useful information sources, and so on. The challenge for the agribusiness marketer is deciding which of these things the customer will actually pay for. Agribusiness marketers must provide a set of products and services to their customers at prices that generate an acceptable rate of return for their firm.

The Evolution of Marketing

Traditionally, marketing was viewed as "selling what you have" and some agribusinesses still approach marketing in this way. More effective agribusiness marketers, on the other hand, focus on "having what you can sell"—anticipating customer needs. The starting point for any marketing program must be the identification of customer needs—and satisfying customer needs is the primary focus for any market-driven organization. However, marketing has evolved in agribusiness firms over time and most of today's effective agribusiness marketers didn't start with a focus on customers.

Many agribusinesses that achieved early success usually did so because they had a successful and unique product that satisfied a specific customer need. And, some agribusiness firms continue to operate with a central focus on the product. Here, the idea is to create a product that is so good customers will seek it out. This approach to marketing is known as being **product driven.** The old saying "Build a better mousetrap and the world will beat a path to your door" reflects this marketing philosophy.

In a product-driven organization, product development, research, engineering, and operations are the primary focus. These firms produce a product in high demand and sales are good, so customer needs aren't a primary concern. Given this, what is the problem with this marketing philosophy? Make no mistake—good products and services are a fundamental part of any agribusiness firm's success. But it is not uncommon for a product-driven organization to become so focused on producing its product or service that it becomes insensitive to changes in farmer or consumer needs. The drive for internal operating efficiency may get more attention than new features consumers may want. Sales may slow as competitor products that are similar are introduced to the market. Marketers in these firms then begin a search for solutions—ideas to help generate increased sales.

When sales growth slows, many firms will adopt a new approach to marketing that centers around intensifying the sales effort. Organizations that focus primarily on communicating the benefits of their products are called *sales driven.* This approach may involve taking the firm's core product and introducing a number of variations or extensions to serve new groups of customers. It may involve a search for new geographic markets where the firm has not been before. The sales-driven organization may add more salespeople and ask them to work harder selling the features and benefits of the firm's products and services. Or, it may spend increasing amounts on promoting the firm's products through a variety of advertising activities.

The idea behind the sales approach to marketing is that customers just don't know enough about the product—if the message is delivered effectively, sales growth problems will be solved. Like good products, effective sales and market communications efforts are important to the success of any agribusiness. But, sales-driven agribusinesses fail to ask one important question—do we have what the customer wants to buy? Failure to carefully consider this question leaves a sales-driven organization highly vulnerable to competitors who are more in tune with changing customer needs.

Frustrated with sales efforts that are ineffective, successful firms ultimately turn their total attention to customer needs. Truly understanding what the customer needs to run their farm or ranch more efficiently and profitably, or how consumer food tastes and convenience demands are changing becomes central to everything they do. This focus on customer needs drives all decisions in the organization, from product development efforts, to production location decisions, to asset allocation—decisions are made with a clear vision of how the firm intends to satisfy the customer in mind. This type of organization is called **market driven.**

A market-driven agribusiness is one with a good product and a good sales effort, but also one with a clear understanding of what type of customers it is trying to serve and what these customers want and need from the firm. It is a firm that takes ideas like "the customer is king" and "getting close to the customer" very seriously. Market-driven firms invest in market research to better understand their customers, then use the information generated to guide decision making. These

TABLE **7-1** | **Characteristics of Product/Sales and Market-Driven Organizations**

Product/Sales driven	Market driven
Focus on product and sales volume	Focus on customer needs, customer satisfaction and profits
Make a sale	Solve a problem
I've got a deal you can't refuse	I think I have something that can really help you
Sell what you have and do	Have and do what you can sell

are agribusiness firms that are focused on building a long-term relationship with their targeted customers and firms that are willing to make some short-run sacrifices to do this. Some other key differences between product/sales-driven firms and market-driven firms are summarized in Table 7-1. The remainder of this chapter focuses on the keys to building a market-driven strategy.

COMPONENTS OF A STRATEGIC MARKETING PLAN

The **strategic marketing plan** integrates all business activities and resources logically to meet customers' needs and to generate a profit. It involves five sets of marketing activities and decisions that must fit together in a consistent fashion (Figure 7-1):

1. Conduct a SWOT analysis.
2. Choose a target market.
3. Choose a position.
4. Develop the appropriate marketing mix.
5. Evaluate and refine the marketing plan.

Building a strategic marketing plan involves a careful assessment of the marketplace (opportunities and threats) and an objective evaluation of the firm's strengths and weaknesses. This is called a **SWOT analysis** (**s**trength, **w**eakness, **o**pportunity, **t**hreat). The focus of this analytic phase is to uncover business opportunities and to understand what advantages the firm brings to the market and where the firm is at a disadvantage.

Based on an assessment of market opportunities and an evaluation of the firm, an agribusiness will select a target market (or target markets)—a group of customers who will respond in similar fashion to a given offer. Consistent with the needs of the target market, the firm will define what **position** it wants to take in the market. Here the focus is deciding what the firm wants to be known for among the target customers.

This position is created by a set of key decisions a marketer must make—the marketing mix. The *marketing mix* is often referred to as the four P's of marketing: product decisions, price decisions, promotion decisions, and place

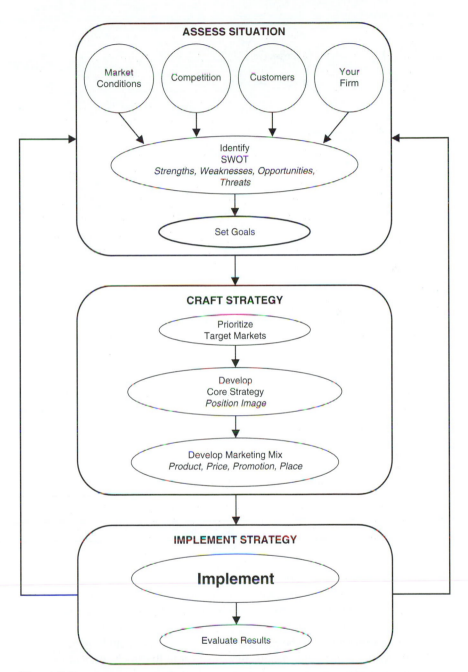

Figure 7-1
The Strategic Market Planning Process.

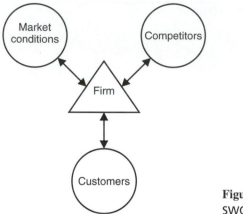

Figure 7-2
SWOT Analysis.

(or distribution) decisions. Finally, since markets are dynamic with customer needs continually changing and competitors always on the move, good marketing always involves an evaluation and refinement stage. Here actual results are checked against forecast results and appropriate adjustments made. The first three steps will be the focus of the remainder of the chapter. The marketing mix will be explored in Chapter 8 and evaluation of the marketing strategy will be discussed in Chapter 9.

CONDUCTING A SWOT ANALYSIS

Assessing the competitive environment requires careful study of (1) general trends in the market, (2) strengths and weaknesses of key competitors, (3) current and anticipated customer needs, and (4) the firm's strengths and weaknesses. The results of such study are summarized as a SWOT analysis. Strengths and weaknesses represent a look at what is going on inside the firm while the opportunities and threats capture what is happening outside the firm in the market (Figure 7-2). This analysis phase of the market planning process is done with a single purpose in mind—to identify the business opportunities the competitive environment presents.

General trends in the market that may be of interest include key technological trends, the general economic situation, important political developments, and weather patterns. Marketers may rely on government statistics, the trade press, industry experts, company research staff, and outside consultants to help them frame the general trends in the market. The results of such study might tell the livestock feeding equipment manufacturer that large livestock farmers have aggressively invested in production technology, that the long-term forecast for interest rates is stable to declining, that government reductions in farm program outlays will lead to more variable grain prices, and that growing conditions in the Corn Belt are likely to be excellent next season. Such an assessment of general trends tells the equipment manufacturer much about how the business environment will greet their next new product introduction.

Evaluating the strengths and weaknesses of competitors is a fundamental part of strategic market planning.

Assessment of competitor strengths and weaknesses is a second key part of understanding the competitive environment. Customers will always evaluate an agribusiness firm's products and services relative to competitive products and services. This may involve a comparison of tangible features like horsepower, yield, and price, and/or it may be a comparison of more intangible features like reputation, product knowledge, and convenience. It is important for agribusiness marketers to understand what they are going up against in the marketplace.

Good marketers know how their products and services compare with competitor products and services in those areas that are important to the target customer. Such insights help marketers understand what they should focus on in their marketing strategy. For example, a hybrid seed corn company might have a particular variety with average yield characteristics, but exceptional drought tolerance when compared against hybrids from competitors. Here, the firm may focus on those regions of the country more vulnerable to drought and emphasize this characteristic of their product, leaving regions with more normal weather to companies with superior yielding, but less drought-tolerant hybrids.

A thorough competitor analysis may also turn up business opportunities—a group of customers the competitor doesn't service well, a product feature that interests customer but is not currently offered, or an entire market segment that has been overlooked and which may offer new business to an opportunistic firm. Competitor intelligence may involve market research, use of secondary data like trade press articles, or informal data collection by the firm's field sales force. Such data is collected and evaluated with a key question in mind, "How does the competitor stack up against our own efforts in this area?"

The final and most important part of assessing the competitive environment is understanding customer needs and wants. Here the focus is a fundamental understanding of what the customer will need from the organization, both today and in the future. Food companies study trends in demographics of consumers (older consumers have different needs than younger consumers) and they study cultural changes affecting food choices—trends toward eating lighter, eating away from home, and consuming less fat have important implications for marketing strategies. For agribusiness firms selling to farmers and ranchers, "knowing the customer" typically means thoroughly understanding the customer's business and the role the firm's product plays on the farmer's or rancher's operation.

Back to the livestock feed equipment example. If the firm is targeting pork production operations with rapid growth potential, it will be critical to understand the types of changes in feeding systems farmers make as they expand their production capacity. By carefully studying pork operations with growth potential, the firm may find that equipment capacities must be reengineered, that there is a big demand for engineering expertise to assist in planning equipment needs, and that financing programs are well received given that capital for expansion can be tight. Armed with this information, the agribusiness marketer can incorporate features that will address these issues into the firm's marketing plan.

Firms can gain insights into what their customers want from them through a wide variety of techniques and activities. Some of these techniques are very informal—a breakfast meeting with six key customers, for example. Some are terribly sophisticated—a virtual shopping trip where consumers shop a "virtual" retail food store on a computer, with every movement and action recorded and analyzed. Many larger agribusinesses employ professional marketing researchers or outside consultants to study their customers, their competition, and trends in the marketplace. Even many smaller firms develop their plans only after they have conducted special studies of their customers. Marketing research can be based on complex statistical techniques, or it can simply result from informal interviews and observations. But, in any case, it should provide objective, analytical information on which to base marketing decisions. Additional insights on the tools used to better understand a firm's market are given in Chapter 9.

A second key part of the market planning process involves understanding what the firm does well (strengths) and areas the firm could improve (weaknesses). This evaluation plays an important role in the market planning process.

First, the marketer may identify a unique strength that can be cultivated into an advantage. Perhaps the firm has the most extensive distribution system in the market and therefore has an advantage in terms of moving product to customers quickly. Second, the firm may identify an area where improvement is needed if a particular customer group is to be served. An example here might be a retail plant food operation that wants to target larger grain producers, yet finds its key sales personnel lacking in agronomic expertise. To be successful in this market the firm will need to invest in people and training to enhance the expertise of its field sales presence. Finally, the firm may identify a weakness of such magnitude that it becomes a constraint on decision making. A small feed mill may find

T A B L E **7-2** | **Assessing Internal Strengths and Weaknesses**

Marketing

How effective are our product development activities?

How do our prices compare with the competition?

How much do we spend on advertising and promotion compared to the competition?

How satisfied are our customers?

How does our sales force compare to the best sales force in our industry?

Finance

How profitable is our organization?

What is our debt position?

Do we use our investment in current assets such as receivables effectively?

How well do we manage cash flow during the seasonal swings of our business?

Do we do an effective job of investing in new technology and replacing old assets?

Operations/Logistics

Do we deliver to customers on time and accurately?

How well do we manage our investment in inventory?

How well do we manage quality through our production process?

What are our transportation costs relative to those of our competitors?

Have we effectively managed the risks in our supply chain—both risks to our employees, and risks to our customers?

Human resources

How does our training program compare to those of the best firms in the market?

Do we have an effective compensation program?

What do our turnover rates look like relative to industry standards?

Do we provide advancement opportunities for employees?

Do we have an effective work environment—do employees enjoy working here?

that it does not have the scale of operations to serve a very large, integrated pork production operation in its local market area—and investment in such capacity may not be a prudent business decision.

Like assessing the marketplace, the focus here is identifying business opportunities or those constraints that stand in the way of business opportunities. Typically, such an evaluation involves a careful review of the firm's entire operation. Some of the key questions to be asked are shown in Table 7-2. A challenge of the marketer is making an objective evaluation. It is easy to be biased when assessing the firm's operation. Securing outside feedback is very important here. Many of the tools described above for competitor and customer analysis also can be used here to get an outside perspective on firm strengths and weaknesses.

Benchmarking is another technique that can be used for this purpose. Benchmarking involves identifying a noncompetitive firm that is known for excellence in a particular area, and then carefully studying that firm to see how they deliver this excellence. The findings of this study are then compared against the agribusiness firm's current practices and key differences are explored further. Such a benchmarking study can provide the firm with extremely valuable information that is helpful in both identifying strengths and weaknesses as well as suggesting strategies for dealing with each.

MARKET SEGMENTATION

Market segmentation involves grouping customers into segments or categories according to some set of characteristics. Here, the agribusiness manager is interested in developing groups of customers that respond in a similar fashion to a given offer. These groups are called market segments, target markets, or market niches. By recognizing the common characteristics, needs, or buying motives of each unique segment in the total market, an agribusiness can design specialized marketing strategies that may appeal to the particular segments it wants to serve. This idea is illustrated in Figure 7-3 and Figure 7-4. Figure 7-3 shows the "market" for a particular farm input as viewed by a firm using a mass market strategy. Here, the firm takes one offering to the entire market. In Figure 7-4, a firm using market segmentation has identified four distinct groups or segments in the market, and the shading indicates that this firm has chosen to focus their marketing resources on two of the four segments.

The number of ways that a market can be segmented is limited only by the agribusiness marketer's imagination. (To prove this statement, just think about the variety of restaurants and the market segments they aim to serve that are present in any midsize city.) A few particular ways to segment a market are discussed here. A **geographic segmentation** may be appropriate for some markets. Food tastes and preferences clearly vary geographically and certain products have much more appeal in some regions than others.

Demographic segmentation is used in many markets. Age, income, size of household, education, number of children, type of employment, and so on can all be important market segmentation variables. Such demographic segments are heavily used in the food industry. The marketing plan developed to promote hot dog consumption among children under the age of 16 might look far different than the marketing activities developed to reach the 22 to 29-year-old consumer.

For the farm market, segmenting customers by **operating characteristics** is typically useful. This may involve such characteristics as type of operation (crop versus livestock, for example), size of operation, production technology used (no-till versus conventional till), and form of ownership (owner/operator versus cash rent versus crop share). Demographic variables like those mentioned above can be used in concert with the characteristics of the farm operation to further define the market segments.

In other instances, **psychographic or behavioral market segments** may be developed. Here, the focus is on more than physical characteristics. The marketer may be interested in wine buyers who value image and prestige in their

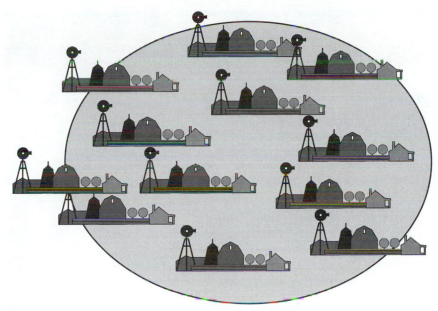

Figure 7-3
Mass Market Strategy.

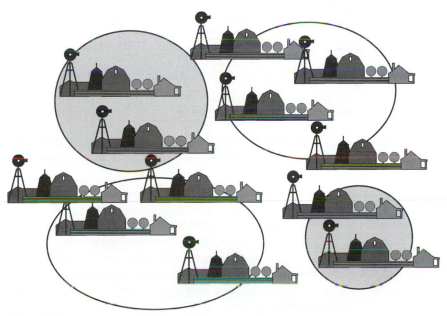

Figure 7-4
Market Segmentation Strategy.

Market segmentation requires careful assessment of the specific needs of different types of customers.

purchase. Or, the farm lender may be interested in those buyers who value a relationship with their lender, as compared to those who don't. Such segmentation approaches are relatively sophisticated, and are heavily used by branded food companies to market products aimed at "the healthy lifestyle" segment (Healthy Choice), the "gourmet" segment (Opus One wine), or the "young and rebellious" segment (Mountain Dew), among many others. Such segmentation approaches require a variety of demographic, behavioral, and psychological data to construct—all tied back to some common set of needs the group is seeking in the product or service.

To illustrate the segmentation concept, consider a feed manufacturer. This firm will recognize dairy, beef, hogs, and poultry as clearly different market segments with different needs. Further, dairy farmers' operations may be classified as small, medium, or large, according to herd size. The needs (and therefore the marketing strategy employed) of a small, part-time farmer who owns his property and milks 50 cows will be far different than a large commercial dairy with 5,000 cows using the latest computerized feeding system. With a thorough understanding of the needs of the target segment in hand, special promotional programs and pricing strategies can be developed. Marketing programs for such targeted market segments are thought to be far more productive than mass marketing efforts aimed at the total market.

A key point here is that there are clearly definable differences between market segments. If such differences do not exist, a mass market strategy works just fine. Typically, designing marketing plans for different segments involves a substantial commitment of firm resources. So, careful thought must go into the market segmentation employed. Well-defined market segments will pass the following test:

1. Measurable. Can the market segment be identified and evaluated? In some cases, a market segmentation can be developed easily using readily available demographic data. In other cases, it may be very difficult to identify how many and which farmers want to buy based on a relationship with the supplier as opposed to those who simply want to buy on price basis.
2. Substantive. Is the market segment large enough to serve? Some segments may be well defined and measurable, but be simply too small to justify the required resource commitment.
3. Actionable. Can the firm effectively serve the segment? In some cases, the firm may identify a large, well-defined segment, but simply not have the people, products, or services needed to effectively serve the segment.

In addition to these questions, agribusiness firms look at several other factors as they make their choice of target market. The level of competition is important. A segment may pass the three-question test posed above, but still not be an appealing segment if every competitor in the marketplace is aggressively pursuing the segment. Growth is another key issue. A segment may be too small to be worth the firm's effort now, but if it has substantial growth potential, it may make sense to stake a claim on the segment early and then grow sales with the segment. Finally, the firm has to be focused on the profitability of serving any given segment—in the end, can the firm deliver value to the segment at a profit?

Every agribusiness firm must decide on its optimal level of emphasis on a given segment, since limited resources must be directed to the area where they will be most productive. Decisions about the segments that represent the most efficient use of resources rest on data about the type of product, competitive behavior, size of the company, and other factors.

Some firms may choose a single area of concentration, using the rationale that concentrating all their efforts on a single class of customers allows them to do a better job and may effectively block out new competition. A farm supply store that caters only to larger commercial farmers would be one example. Another farm supply store might develop a "dual segment" concentration by building a special display area for lawn and garden equipment and seasonal nursery stock, so that it appeals to suburban homeowners. In another area of the facility, they may have a special office area to work with commercial farmers. Of course, there are a host of factors involved in market strategy decisions, including firm size, location, experience, and competition. But once the market segment or segments have been targeted, the agribusiness must concentrate on products and services that will succeed there.

POSITIONING

You know the customers you will target, you understand their needs, and you have a clear picture of the marketplace, your product, and the benefits it offers to your customers. *You* clearly understand the value you are delivering—but how do you communicate that value to your prospects and customers? To answer this question, marketers use a concept called **positioning.** Positioning has been defined as the process of creating the desired image in the customer's mind. Al Ries and Jack Trout in their marketing classic, *Positioning: The Battle for Your Mind* state, "Positioning starts with a product, a piece of merchandise, a service, a company, an institution or even a person. But positioning is not what you do to the product. It is what you do to the mind of the prospect."

Note that the emphasis is not on the product itself but on customer perceptions. The idea is to secure a place in the customer's mind for your product based on some factor or factors that differentiate your product from the competition. For example, you may want a place in the customer's mind that says that you are the yield leader, that you have the most complete line of products, or that you have best product support.

When an agribusiness brings a product or service to the market, it is important that they be very clear on what image or position they are trying to create. The desired position of the product or service becomes the bridge between the needs and wants of the target market and the specific actions the firm will take to satisfy those needs and wants (the marketing mix).

In many cases, the firm summarizes the desired position with a simple statement that captures the essence of what the firm is trying to accomplish. Some examples include:

"Nothing Runs Like a Deere"—John Deere

"Plant the Leader"—Pioneer Hi-Bred International

"The Other White Meat"—The National Pork Producers Council

Deere is communicating reliability and quality. Pioneer is telling producers to go with the best. The National Pork Producers Council is urging consumers to look at their product as another white meat and a substitute for chicken. In every case, these statements of position communicate a specific message to the customer—hopefully about a trait, characteristic, or feature that is important to the customer.

To better understand the idea of position, it is important to understand the idea of competitive advantage. An agribusiness firm's **competitive advantage** is that set of competencies where the firm has a clear and distinct advantage over the competition. Competitive advantage is the firm's edge in the marketplace, it is the reason customers choose to do business with the firm as opposed to buying from another organization.

Harvard Business School Professor Michael Porter suggests there are two ways to build a competitive advantage. First, the firm can attempt to provide customers with unique products and services unavailable from other firms. This approach is called a **differential advantage.** Pursuing a differential advantage means that you focus on being a value-added firm by providing services that enhance your product and capitalize on what makes your firm unique. Agribusiness firms differentiate themselves in a variety of ways—product performance, delivery, product quality, taste, packaging, customer service, technical expertise, image—the list could go on. Any area that customers perceive to be important becomes a potential basis for differentiation. Firms pursuing a differential advantage stress their uniqueness in their positioning.

The other way to claim a competitive advantage is to be a low cost leader. **Cost leadership** involves meeting the market's product offering with an offering of comparable quality and features, but beating the market on price. Hence, cost leaders work hard at running extremely efficient operations. Cost management is a central focus, and every potential efficiency that does not undermine the product/service offering is pursued with a vengeance. Firms positioning themselves as cost leaders will emphasize price in their message, while assuring the market that their quality is acceptable.

While many agribusinesses manufacture and market products with commodity-like characteristics, it does not follow that differentiation is impossible. According to Tom Peters in *Thriving on Chaos:* "Any product or service can become highly value added; that is, there is no such thing as a nondifferentiable commodity. The more the world perceives a product/service to be a mature commodity, the greater the opportunity to differentiate it through the unending accumulation of small advantages." If an agribusiness firm's product/service offering is perceived as identical to another firm's, the sale will go to the firm with the lower price—a commodity transaction. In the end, the challenge is to identify just what things add value for the customer and then to deliver them flawlessly. Such a fundamental understanding of the market combined with superb execution can put distance between an agribusiness firm and its competitors.

Note that the position is generally built on the firm's differential advantage (i.e., things like yield advantage, breadth of inventory, quality of sales force, etc.). The challenge here is to take those things that make the firm unique—the competitive advantage—and turn them into the desired image with the target customers—the position.

How does this work in practice? Let's take a look at how a seed firm might apply some of the ideas to develop its position. Consider the following scenario: A turfgrass seed company deals primarily in proprietary seed lines. The firm's research department has finished work on a new variety of dwarf fescue. The variety has fared very well in public evaluations. This variety has a distinctive dark green color, produces a dense, durable turf, and is hardier than other varieties. The firm has decided to try to build on these features and establish a differentiated product that will command a premium price over commodity fescues. How can the concept of positioning be used to accomplish this objective?

The first questions to ask are what position does the firm own and what position do they want to own? Ries and Trout call this "thinking in reverse" because when answering the questions you don't focus on the product, but on the target market. How do customers see your firm? Going back to the example, let's say that after some serious thought and several long focus group sessions with the potential customers, the firm finds their current position is based on their ability to provide premium varieties of seed which meet the changing needs of customers. This position has been developed through a heavy emphasis on research that has enabled the firm to consistently release high-performing new products.

Here, the current position is consistent with the desired position, given the goals for the new variety. The next question relates to the competition—is this a position the firm occupies by itself? Here, the firm may find that two of their competitor's positions overlap their own. This is important information—the firm will want to make sure their marketing plan strengthens and enhances their position and puts some distance between their firm and the competition.

In the case of the new dwarf fescue, what position should the firm take? For the residential market, two key ideas seem to surface—easy care and environmental awareness—and the position could build on these two benefits. The variety does not grow as fast or as tall as other varieties. The fact that the homeowner spends less time mowing is an important benefit. To complement the easy care benefit, the firm could also position the product as environmentally sensitive. The variety needs less fertilizer and water than other varieties and the slow growth of the product means fewer clippings to dispose. This new product could be positioned as the grass that delivers more free time to an environmentally conscious consumer.

Having established the positioning, let's take a quick look at how the firm might actually communicate these ideas to the market. (This process—developing the marketing mix—will be considered in more detail in Chapter 8.)

You can communicate tangible attributes—those characteristics that can be identified with some precision—of the product by using university test plot data, public evaluations, the firm's test plots, on-farm weigh tests, etc. For the easy to care for and environmentally sensitive turfgrass, data on number of mowings required, the amount of fertilizer and water needed, and the number of bags of clippings removed help communicate, in a tangible fashion, the two ideas in the position. These two notions can't be fully supported with facts and figures however, and an agribusiness marketer must think about communicating the more intangible side of these concepts.

Communicating the intangible takes a little more creativity. Going back to our dwarf fescue example, the firm might pursue a number of activities to demonstrate the intangible benefits and support distributor marketing of the product. The firm could develop point-of-purchase materials for the dealer that showed the homeowner resting over a dwarf fescue lawn in a hammock while a neighbor was busy mowing. Or, a direct mail piece targeted for new homeowners could show piles of clippings and fertilizers that compared the dwarf variety and regular varieties. The firm could arrange to get one of its researchers interviewed on regional television or radio about the environmental features of the new variety. Finally, the seed could be packaged in a bag made of recycled paper. (Note also that the product's name should be consistent with its position.)

After deciding on the market position the firm will seek, they must then develop a marketing mix that will support the position. Claims ring hollow if the firm can't deliver the promised product/service bundle at an acceptable price. And, the market communications effort is fundamental to communicate and support the desired message. Finally, the product/service bundle must be made available in a way that supports the position as well.

In the end, nothing is more critical to communicating value than delivery. There is a significant difference between saying you are doing something for your customers and actually getting the job done. If you ask a manager or a sales representative if they add value to their product, they'll tell you "yes" every time. Adding value is equated with meeting customer needs—and what firm does not think that they are striving to meet their customers' needs? Given the similarity of value-adding activities among seed firms, the bottom line is that delivery of the bundle you communicate may well be the way your firm is able to differentiate itself in the marketplace.

SUMMARY

Marketing is the process of anticipating the needs of targeted customers and finding ways to meet those needs profitably. The strategic market planning process begins with a careful assessment of the business environment and a critical evaluation of the firm's strengths and weaknesses. The results of this assessment are summarized in a SWOT analysis. The focus here is identifying key business opportunities in the market—those areas where the agribusiness should focus its marketing resources.

The market assessment provides data for a fundamental choice the agrimarketer must make—what target market(s) should the firm pursue? A target market is a group of customers and prospects that will respond in similar fashion to a given offer. Agribusiness marketers can segment a market on a wide range of characteristics—demographic, geographic, business, and psychographic. The key for the marketer is to identify a target group with unique needs that the firm can serve profitability.

An important part of the strategic market planning process is determining the position the firm wants to take with its target market. The position serves as a focal point for the firm's marketing mix—it captures the essence of the message the firm is trying to send the market. An agribusiness firm's position is typically drawn from its competitive advantage—the firm's unique edge over rival firms.

The set of decisions agribusiness managers make as part of the strategic market planning process may be among the most fundamental choices facing an agribusiness firm. Decisions about what markets to pursue and what position to take in these markets drive much of the rest of the firm's business activities.

1. Explain the difference between a product- or sales-oriented firm and a market-oriented firm. Why is this difference important to an agribusiness manager?
2. A key part of the SWOT analysis is evaluating the impact of broad market factors such as government policies, international trade developments, new technologies, and so on. For the agribusiness industry of your choice, identify five key trends in the industry marketplace and describe the opportunities and threats each trend represents for agribusiness firms in that industry.
3. How can an agribusiness manager evaluate the strengths and weaknesses of their organization? Why is it important to understand what an agribusiness does well and those areas where performance is not as strong?
4. What is meant by a market segment? U-pick fruit and vegetable farms (farms where customers harvest their own fruit and vegetables) probably appeal more to some market segments than to others. Identify and describe two market segments that you believe might be important to this type of agribusiness. What are the key needs of these two segments that a marketing manager for the U-pick farm would need to focus on?
5. Focusing on one of the target markets you identified for the U-pick farm described in question 4, what position should the firm take in the market? Based on the key needs you identified for one of the market segments, write a positioning statement for the firm. What key themes will you stress in your statement?
6. Two types of competitive advantage are described in the chapter. Identify a food or agribusiness firm pursuing a differential advantage. Also, identify a food or agribusiness firm pursuing a cost advantage. Compare and contrast these two strategies. What is similar about the two firms? What is different?

CASE STUDY — **Richmond Supply and Elevator, Inc.**

Richmond Supply and Elevator (RS&E) is a family-owned feed, fertilizer, and grain company located in central Iowa. The firm is owned and managed by Lance and Paul Taylor, the sons of the original owner, Joe Taylor.

RS&E handles the Rapid Gain line of animal feeds. The firm currently employs eight full-time personnel, in addition to Lance and Paul. In 2000, roughly 15 percent of their total $6.0 million in sales was feed, 60 percent was grain, and 25 percent was fertilizer. Richmond sold about 2,000 tons of feed in 2000. Over time the firm has been profitable, but not exceptionally so. In 1999, the firm reported a loss of $50,000 on sales of $6.0 million while in 2000 the net income after taxes was $65,000 on about the same level of sales.

The Market

Consolidation has been the key word in the central Iowa market and a smaller number of larger farm customers have been the result. RS&E had more than 180 feed accounts in 1989. In 2000, RS&E was serving fewer than 50 farmers (because farm size had increased), but was selling twice as much feed (through some good management decisions and some competitors going out of business). The average customer farms about 750 acres, typically one-half in corn and one-half in soybeans. Hogs are essentially the only livestock in Richmond's market area, and hog feed makes up 90 percent of Richmond's feed tonnage.

Competition

Richmond has a number of other firms handling feed in its market area. RS&E's toughest competition comes from a Lake Feeds dealer, Princeton Elevator, Inc. Lake Feeds entered the market five years ago and has pursued expansion aggressively. They market a quality product and sell aggressively on the farm.

The Lake Feeds dealer has held its price up over the period, taking some of the price pressure off Richmond. Lance feels that Lake Feeds' growth has primarily come by taking business away from companies selling cheap, low-quality feed. In fact, RS&E's tonnage has increased over the past five years, despite the presence of Lake Feeds. Lance believes that Lake Feeds' aggressive quality push has actually helped Richmond increase sales.

The Firm

Lance (38) is the older of the brothers and handles the "inside" jobs—waiting on customers, writing orders, ordering, monitoring the books, and so on. Paul (36) handles the firm's operations—overseeing the fertilizer plant, feed mill, and grain elevator. Paul also handles RS&E's outside selling effort. His cellular phone has greatly improved communications. The firm takes a very low-pressure approach to selling on the farm. Paul's customer visits are viewed as problem solving and public relations—not really sales calls. Paul makes few calls on farmers RS&E doesn't currently sell. Lance and Paul considered hiring an outside salesperson, but feel they just "can't make it pay" at this point. They are also looking into more e-mail communications with their customers but have not made this move yet.

Of the other eight employees, one drives the feed truck, one runs the feed mill, one is a bookkeeper, and the other five are general warehouse employees/drivers. With exception of one warehouse employee who has been with Richmond four years, all employees have been with the firm at least five years. Lance and Paul also hire seasonal labor to help them through the spring and fall peak selling seasons.

The firm owns two bulk feed delivery trucks. Their feed mill was constructed in 1985 and is in excellent condition. Paul figures the mill runs at about 75 percent of capacity. The firm is located about 60 miles from the nearest Rapid Grow feed plant. RS&E sells little pelleted complete feed, choosing instead to push concentrate feeds and keep their feed mill in use, reducing hauling costs in the process.

The brothers are looking into the construction of overhead feed holding bins to reduce truck waiting time. They would also like to get more bulk bins on the farms of their customers to reduce unloading time. Lance feels they "must take steps to increase efficiency over the longer run." The brothers are considering remodeling the office/showroom and acquiring a computer system to handle accounting tasks and keep customer records.

Marketing

Quality and service are key components of RS&E's business philosophy. They focus their sales effort on Rapid Grow's premium products and feel they get a good gross margin on what they sell (16 percent in 2000). The brothers are very conservative about growth—about 5 percent a year is all they want. "We cannot let our current level of customer support slip," says Lance, "and, we are willing to let some opportunities go by to make sure that we can keep our current customers happy."

Customer service is delivered several ways, according to Lance. Quality feed products are the cornerstone and the brothers have been very pleased with the Rapid Grow line over time. Nutritional support is another component of customer service. This is done by drawing on Lance's excellent knowledge of swine nutrition and by making heavy use of a talented local veterinarian who doesn't sell feed or animal health products. In addition, they draw on Rapid Grow's nutritional experts whenever necessary. To date, they have not put up a company website. But, they have been thinking about how the Internet might help them build even better relationships with their farmer customers.

Richmond has worked hard to maintain margins, attempting to increase gross margins on feed over time. For the most part they have been successful at increasing per-ton margins on feed. RS&E is not the cheapest feed company in the market, nor do they intend to be. "We will make money on what we sell," Lance says. They offer a 2 percent discount for cash. In addition, they have a volume discount program that Lance admits is a bit "informal." Richmond rarely advertises and holds farmer meetings only occasionally. "Unless there is a new product to push, I'm just not a great believer in meetings," Lance has said.

Summary

Lance and Paul have built a good business over time. However, their market is changing rapidly and the competition is getting tougher. The brothers wonder just what position they should take in the market. What will they need to do to support this position? How will they communicate the value their products and services provide to their customers? Clearly, Lance and Paul have many tough decisions to make.

QUESTIONS

1. What are the key opportunities and threats facing RS&E in the marketplace?
2. What are the key strengths and weaknesses of RS&E?
3. Based on the limited information in the case, describe the firm's target market. In your opinion, does the firm have a good grasp of who their target market is and what this market wants from the firm?
4. What position should the firm take with their target market? What key themes should they emphasize in their positioning statement?
5. Is RS&E a product-driven firm, a sales-driven firm, or a market-driven firm? Explain your answer.

REFERENCES AND ADDITIONAL READINGS

Peters, Tom. 1987. *Thriving on Chaos.* New York: Harper and Row.

Porter, Michael E. 1980. *Competitive Strategy: Techniques for Analyzing Industries and Competition.* New York: The Free Press.

Porter, Michael E. 1985. *Competitive Advantage: Creating and Sustaining Superior Performance.* New York: The Free Press.

Ries, Al, and Jack Trout. 1982. *Positioning: The Battle for Your Mind.* New York: Warner Books.

CHAPTER 8

THE MARKETING MIX

OBJECTIVES

- Discuss the elements of the marketing mix
- Explore how agribusinesses create value for their customers
- Review product adoption and product life cycles in agribusiness markets
- Identify key features of agribusiness pricing strategies
- Summarize various methods of product promotion and market communications
- Examine channels of distribution in agribusiness

INTRODUCTION

When an agribusiness marketer thoroughly understands the needs of the target market and has identified the appropriate position for their product or organization, the truly creative work begins. At this point, the agribusiness marketer must take a general idea, such as "Nothing Runs Like a Deere," turn that idea into a tangible product/service/information offering, and then communicate the desired position to the target market.

In this chapter, we'll examine how agribusiness marketers must translate the position they want to create for the target market into a product/service/information bundle that serves customer needs and communicates the desired image. The set of decisions involved in accomplishing this task is called the marketing mix.

THE MARKETING MIX

Countless decisions face today's agribusiness marketer—particularly once the target markets are identified, and a company position is chosen. As an example, let's use large commercial vegetable growers in the San Joaquin Valley of California as

199

Figure 8-1
The Marketing Mix.

a potential target market for John Deere. What specific products, services, and information will be provided to these growers? How will this bundle be priced?
How will the value provided by these products and services be communicated to
the target market? What is the best distribution channel to use to effectively service this group of large California growers? This set of critical decisions—the four
P's of marketing: product, price, promotion, and place—is called the **marketing
mix** (Figure 8-1). The ultimate job of the agribusiness marketer is to craft a marketing mix consistent with the firm's desired position that creates good value for
the customer.

Each of these four elements of the marketing mix, the four P's, are explored
individually in this chapter. But, keep in mind that each of these elements must
complement one another in order to create the intended position in the market.
For example, how would the California vegetable growers respond if they were
called on by a John Deere salesperson who did not understand the technical features of the high-quality tractor the growers were interested in? Or, how would
the growers react if the salesperson was very knowledgeable, but the tractor
failed three times during the first week of operation? What would these growers
think about John Deere if the promotional literature they received on the tractor
was sketchy and unprofessional? For an agribusiness marketing program to be
successful, all elements of the marketing mix must communicate a clear and
consistent story—the position—to the target market.

Successful marketing requires the right marketing mix: the correct combination of product, price, place, and promotion strategies.

PRODUCT/SERVICE/INFORMATION DECISIONS

Product/service/information decisions form the heart of the marketing strategy and are among the most fundamental decisions an agribusiness makes. (We will use the term "product" instead of "product/service/information" in this section, but don't forget that service and information is an important part of the firm's strategy in this area.) Decisions that must be made here include the mix of different products and services offered, the extent of each product line, the specific characteristics of each product sold, and the level and type of information provided in the bundle. Products selected for the product mix should complement each other technically, in the distribution channels, and with customers to take full advantage of marketing efficiencies.

The Value Bundle

When developing a product strategy, the agribusiness marketer begins with the needs of the target market clearly in mind. What set of products, services, and information can the firm deliver to meet the needs of the target market? One especially useful concept that can help answer this question and guide development of a product strategy is the idea of the **value bundle.** The value bundle is the set of tangible and intangible benefits customers receive from the products and services an agribusiness provides. The fundamental challenge when developing a

Figure 8-2
The Value Equation.

product/service/information strategy is determining what to include and what not to include in the value bundle for the target market.

What is value? And, more importantly, how do customers define value? **Value** to the customer is defined as the ratio of what they receive (benefits) relative to what they give up (costs) (Figure 8-2). As customers make purchasing decisions, they compare the perceived benefits of the purchase against the perceived costs. If the value equation is tipped in favor of the benefits (i.e., the benefits exceed the costs), the customer has found a good value and will make the purchase, provided another supplier isn't providing more value on the same item.

In the definition of value, note the importance of the word "perceived." Some benefits and costs will be tangible and easily measurable—the horsepower of a tractor or the per-unit cost of fruit juice, for example. But, intangible benefits and costs—the prestige or the peace of mind that comes with owning the "best" farm equipment, or the convenience that comes with a special type of juice package—can be just as important (or more important) in the purchase decision.

There is no easy way to measure benefits like prestige or convenience, but customers will form highly unique opinions—perceptions—about the value of such benefits. Because many benefits are intangible, the agribusiness marketer must not only know how the customer defines value, they must make sure the value their bundle provides is clearly communicated to the customer. In the end, no matter what the agribusiness marketer thinks about their products and services, it is the customer's perception of benefits and costs that will determine whether or not a purchase is made.

To relate the customer's definition of value to the agribusiness firm's product strategy, it is helpful to visualize the firm's product/service/information offering as consisting of four separate parts. This way of looking at the value bundle has been called the **total product concept** by some authors (Manning and Reece). The total product concept looks at the product at four levels: the *generic product,* the *expected product,* the *value-added product,* and the *potential product* (Figure 8-3). Understanding these different levels can help an agribusiness marketer better understand how value is added, and ultimately, can guide development of the product strategy.

The foundation for a firm's product/service/information offering is the **generic product.** This is the standard product with no special services or features attached. For DEKALB Genetics, this is a unit of hybrid seed corn; for Ocean Spray, it might be cranberry juice. There is little or no differentiation at

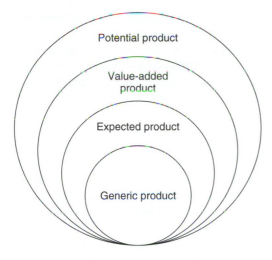

Figure 8-3
The Total Product Concept.
Source: Manning and Reece. *Selling Today:*
An Extension of the Marketing Concept.

the generic product level—at this level customers see all agribusiness firms as alike in the market. A firm stopping here in developing a product strategy will compete solely on price, since the firm is not providing customers with any value above and beyond what any other firm is doing.

Let's use a retailer of fertilizer to illustrate the total product concept. The generic product for this firm might be bulk fertilizer and crop protection products. There are no services attached and the set of products carried by the firm can be purchased from any other firm in the market. The generic product is a true commodity in the mind of the customer.

Most agribusinesses don't stop with the generic product. The **expected product** builds on the generic product by surrounding it with the bare minimum set of features/services that the customer expects when they make a purchase. Like the generic product, the firm doesn't get much credit by providing the expected product—at this level the firm has simply met customer expectations. On the other hand, firms not providing the expected product are discounted heavily in the mind of customers because even the bare minimum is not being provided.

The expected product for the fertilizer retailer might include standard custom application service, standard agronomic recommendations, a 30-day credit policy, some minimum set of equipment available for rent, standard complaint handling service, and a clean and neat facility. Again, the focus is on the minimum set of product-related services the customer expects to receive. While the firm has added value to the generic product at this level, the customer still sees a commodity since only their minimum expectations have been met.

To this point, most of the focus has been on tangible products and service. For the value-added and potential products, value is driven almost totally by customers' perceptions about the more intangible attributes surrounding the products and services. The **value-added product** is the first opportunity for the agribusiness firm to truly exceed the customer's expectations. Here, the firm goes beyond the tangible, physical properties of the product and the minimum services that are typically provided with the product. The focus at the value-added product level is satisfying customer needs in a way that isn't expected and providing services or a level of service that isn't being provided by competitors. Higher levels of customer service, unique information, and the intangible benefits that go along with features such as reputation, trust, integrity, and safety may all be part of the value-added product.

Following our retail fertilizer example, retail customers may expect standard custom application service as part of the expected product. However, custom application service provided by the most highly trained drivers in the market and a guarantee that the driver will be in the field within 15 minutes of the promised time is likely to exceed customer expectations. A comprehensive agronomic plan developed by a trained agronomist and updated annually with a personal visit from the agronomist may send signals that this firm is adding more value than competitors. Or, weekly e-mail messages to key accounts giving updates of changes in growing conditions and pest and disease problems may clearly communicate that this firm goes above and beyond for their customers. It is this level of attention to customer needs that drives the value-added product and starts to more clearly delineate firms in the market.

One last dimension of the product exists—the **potential product.** Here, the focus is on the future: What is the next benefit that customers will seek from the firm? How will the product/service/information bundle be managed to add even more value for customers? Once the value-added product is successful, the firm must actively look for other possibilities in the marketplace because competitors will seek to emulate any success the firm may enjoy.

The potential product for the retail fertilizer firm might include investing in a new information management system to allow the firm to track comprehensive yield, fertility, and weed and pest data by field for customers. Or, it might mean acquiring a multiple-nutrient variable rate applicator guided by a global positioning system (GPS), which will allow the firm to apply precise levels of fertilizer on an as-needed basis as the rig moves across a field. It could be an interactive website that allows producers online access to a databank of key agronomic information. The potential product may mean bundling up these types of services with even more professional advice to deliver a precision agronomic package far superior to any other offering in the market. The focus here is innovation and making sure the firm stays ahead of competitor attempts to deliver the value-added product and the expected product.

Note that the ability to meet and exceed customer expectations is heavily influenced by what competitors offer. Unfortunately, if any of the value-added attributes provided by a firm are also provided by competitors, in the customer's mind these benefits may become expected, taken for granted, and will not be as effective in differentiating the product/service/information offering. In this case, which is probably the typical one, actual delivery of the benefit becomes the deciding factor: if the firm does the same thing as a competitor, but can do it better in a way customers truly perceive to be unique, then there is still an opportunity for differentiation.

Value is a matter of customer perception—it is based on what the customer believes to be true. As managers, it is sometimes easy to lose sight of what customers value. Marketing games of one-upmanship can lead to a "bells and whistles war" with little regard for whether the customer actually wants the new feature provided. Just because a major competitor has added another service does not mean that it is time to add the same service. Failing to carefully consider the new feature from the customer's perspective may add cost to the product for something that is not valued by the customer. Letting customer needs—current and anticipated—guide construction of the value bundle is a fundamentally important marketing concept.

Product Adoption and Diffusion

The manner in which customers adopt a new technology, a product, or a service is another important part of a firm's product strategy. The adoption process is tied closely to a product's life cycle and suggests how new products should be introduced into the market. E. M. Rogers, using hybrid seed corn as the focus, did the classic research into the adoption-diffusion of new products. Rogers suggested that ideas are diffused through the market in systematic stages:

1. *Awareness.* At this stage, people have heard about the product but lack sufficient information to make a purchasing decision.
2. *Interest.* A potential customer becomes interested enough to learn more about the product.
3. *Evaluation.* The customer decides whether or not to try the product.
4. *Trial.* The customer samples the product.
5. *Adoption.* The customer integrates the product into a regular-use pattern.

Some individuals who adopt products more quickly than others tend to be **opinion leaders.** These people are watched carefully and followed by other customers in the market. Some opinion leaders actively attempt to influence others within their sphere. This group of opinion leaders is an extremely important one to agrimarketers. And, many agribusinesses spend considerable time and money identifying opinion leaders and working closely with them to build favorable relationships.

As a new idea or technology is introduced into the market, Rogers found it will be adopted in a systematic way as more and more users adopt the idea. He classified users into five distinct categories according to how quickly they adopt a new idea.

Rogers's research (Figure 8-4) suggests that the total number of individuals willing to try a totally new idea is very small, perhaps about 2.5 percent of a given market. **Innovators** are these venturesome people who like to try new ideas. They are not necessarily opinion leaders; since they try so many new things, their peers may regard them as a bit unconventional.

The second wave of individuals who try a new technology are called early adopters. **Early adopters** (about 13.5 percent of a market) are respected individuals who adopt new ideas quickly but with some caution, usually after observing the experience of the innovators. They are usually key opinion leaders in the community and are therefore highly important to the agribusiness marketer.

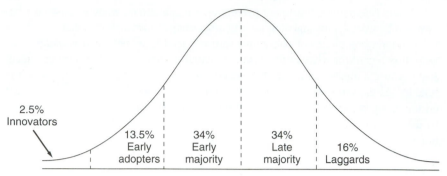

Figure 8-4
Categories of New Technology Adopters.

The **early majority** (about 34 percent of a market) form the third group of adopters. Individuals in the early majority are deliberate people who see themselves as progressive, but not generally as leaders. They form a large and important market for any product or service. Following the early majority in the adoption process are the late majority. The **late majority** (another 34 percent) tend to be skeptical in their view of new ideas, and adopt them only after considerable evidence of performance and/or satisfaction has been shown. This group follows the majority opinion. Finally, **laggards** (16 percent) are tradition-bound individuals who take so long to adopt new ideas that by the time the ideas are adopted, they are no longer new.

This pattern of new technology adoption holds for most new ideas, both within the food and agricultural markets and outside these markets. Agrimarketers who introduce new products can initially focus their total marketing program toward the innovators and early adopters, gradually changing their marketing strategy as the product is accepted by the other adopter categories over time.

Product Life Cycles

A final concept that is important when developing a product strategy is the product life cycle idea. **Product life cycles** relate to the sales and profits of a product or service over a period of time. There are several distinct phases in the life of a product, from its development and initial introduction to its eventual removal from the market (Figure 8-5).

The **development stage** is that period in which the market is analyzed, and both the product and the broader marketing strategy are developed. During this time there is no revenue, but there are significant expenditures for product and market development. For example, this is the period when a new rabbit feed is researched, formulated, and tested, while plans are developed for its introduction into the market.

The **introductory stage** is that period in which the new product first appears on the market. Usually, there are high costs associated with introducing a new product. The new rabbit feed may require a heavy promotional effort and

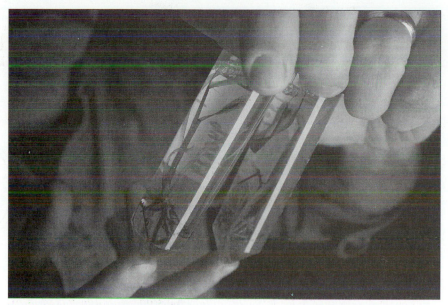

Even in the earliest stages of development, the product life cycle is an important consideration for agribusiness marketers.
Photo courtesy of Pioneer Hi-Bred International, Inc.

special offers to dealers to stock the product, both of which reduce the probability of early profits. At some time during the introduction period, the product should begin to show a profit, depending on the degree of its success. A number of introductory strategies are possible. The firm might choose to make heavy introductory expenditures to reduce the time for consumer acceptance. Or, it might introduce the product without fanfare, simply adding it to the product line at perhaps a low initial price, in the hope of minimizing introductory costs.

The **growth stage** is a period of rapid expansion, during which sales gain momentum and prices tend to hold steady or increase slightly, as firms try to develop customer loyalty. The distribution system is expanded, which makes the product available to a larger market. Profits expand rapidly because fixed costs are spread over a larger sales base. However, at some point during the growth stage, it is common for costs to begin to increase as the firm attempts to reach increasingly difficult new markets. In addition, the visibility of increasing profits often attracts new competitors to the market. As the firm attempts to reach new, less familiar markets and the effects of competition begin to be felt, growth slows, and profits, while still increasing, begin to increase at a decreasing rate.

The **maturity stage** is characterized by slow growth or even some decline of sales as the market becomes saturated. Sales lag because most of the potential customers have been tapped and because competitors have entered the market, leaving only new sales to the late majority and laggards, and replacement sales to established customers. This stage usually lasts longer than the others, so most products in the market at any point in time are in this stage.

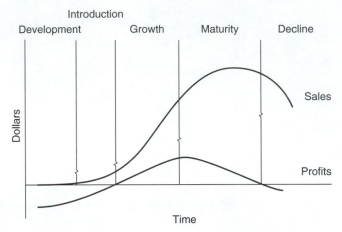

Figure 8-5
The Product Life Cycle.

Much marketing activity is designed to prolong the maturity stage by propping up sales and protecting profits. Competition often becomes intense in the maturity stage as competitors battle for market share, often using price as a weapon. Firms struggle to refine their product by changing design, adding features, making new advertising claims, and developing promotional campaigns and incentives to protect or enhance market share. Strategically, firms hope to differentiate their product sufficiently to push it back into the growth stage. But all this additional marketing activity drives up costs, and profits begin to decline. How rapidly profits decline in the maturity stage depends on a host of factors, but this decline is always of major concern to management.

The **decline stage** finds sales declining more rapidly. Changes in consumer preferences or new substitute products may hasten the death of the product. As the product dies, profits slip to zero or worse, and some firms withdraw from the market. The remaining firms may reduce marketing expenditures, until eventually the product may disappear totally from the market.

Many firms find the decline stage very difficult to manage. Executives who have built their professional careers around the growth of a product are sometimes emotionally involved with the product and are thus reluctant to admit its decline. Marketing managers may legitimately expect sales growth to resume as economic conditions change. In any case, to drop a product from the line and to decide on the best timing for this action is a difficult decision. Yet, attempting to prolong the life of a product may drain the firm financially and preclude the development of new products.

The evolution of a product through its life cycle presents the essence of product decisions for the marketing manager in the agribusiness firm. Although each product has a life cycle, the life cycle can look very different from one product to another. Some products have short life cycles lasting only a year or two, while others may have life cycles that span dozens of years. One of the marketing manager's jobs is to use the marketing tools at their disposal to prolong the profitable life stages.

PRICE DECISIONS

Pricing is a critical marketing decision because it so greatly influences the revenue generated by an agribusiness. Pricing decisions do this in two ways:

1. Price impacts revenue as a component of the revenue equation:

$$(Revenue = Price \times Quantity\ Sold)$$

2. The price level itself greatly affects the quantity sold, through its effect on demand relationships for the product or service.

Complications arise because these two price effects work in opposite directions. Lower prices produce less revenue per unit, but usually generate an increase in quantity sold, while the opposite is true when price is increased. (Recall the concept of demand elasticity discussed in Chapter 3.) Of course, increased sales mean that fixed costs are spread over more units; therefore, per unit costs may be reduced, at least to a point. The net result is that pricing decisions are a real challenge to marketers.

Some price decisions involve highly complex mathematical methods, while others depend on simple rules of thumb or intuitive judgments. The type of product, customer demand, competitive environment, product life-cycle stage, and product mix are some of the factors considered in price determination. A successful pricing strategy is made after giving careful consideration to the value delivered, the cost of the product/service/information bundle, the goals of the pricing strategy, and the pricing strategies of competitors.

The perceived value of the product/service/information bundle becomes a ceiling on the price charged (Figure 8-6). If the firm sets the price higher than the perceived value, the customer's benefit/cost calculation moves in the wrong

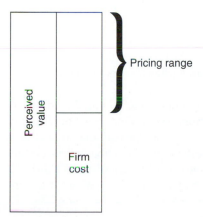

Pricing range

Perceived value

Firm cost

Figure 8-6
Upper and Lower Limits on Pricing Decisions.

direction and the customer won't buy the product. Obviously, firms must under-stand the value they provide to customers when developing a pricing strategy.

The firm's total cost of providing the product/service/information bundle provides a floor on price. At least over the longer run, the firm must cover full costs through the sale of the product if it is to remain in business. However, the firm can never focus entirely on costs when making a pricing decision. An ex-ample from Hungary provides a vivid illustration of pricing economics in action. Some budding entrepreneurs decided there was money to be made by importing rice into Hungary. They found the needed capital, purchased a barge-load of rice, and transported the rice to Budapest. The entrepreneurs knew what they had paid for the rice, they knew what it cost them to ship it to the country, and they knew how much profit they wanted to make on the deal, so they priced their rice accordingly. Three months later, the stubborn entrepreneurs still had not sold a single kilogram of rice.

In retrospect, the problem was obvious: while they were acquiring the rice and shipping it to Budapest, the world price for rice declined substantially. And, given that plenty of rice was available at the lower world price, potential buyers could care less what the entrepreneurs paid for the rice or how much it cost them to ship it. While a bit dramatic, the story does illustrate a key point—customers really don't care about production and marketing costs. Few agrimarketers would approach a pricing decision as naively as the entrepreneurs in the story. But, it is easy to let internal cost considerations dominate pricing decisions, forc-ing the customer's perception of value into the background. An ineffective pric-ing decision is the likely outcome.

The firm's pricing goals are important considerations in any pricing decision—what does the firm want to accomplish with the pricing strategy? Is the goal to maximize market-penetration, getting as much market share as possible, as quickly as possible? Or, is the agribusiness less concerned about market share and more in-tent on skimming profits in the short run, knowing that a premium price and strong margins will attract competitors over time? A variety of other goals are possible, each with a unique impact on the firm's pricing strategy.

Finally, the firm must evaluate the competition as it makes pricing deci-sions. In the end, the agrimarketer wants to know what value competitors add and what pricing goals they have for their bundle of products and services. This is important since the value the competitor provides sets a standard against which the agribusiness will be measured. As mentioned earlier, value is a rela-tive idea and any value bundle provided by an agribusiness is judged relative to that of their competition.

In the end, if a profit cannot be made by pricing the product based on the value it delivers, at least one of three conditions exists. First, costs may simply be too high—the firm needs to become more efficient. Second, the product may not deliver enough value compared to the (efficient) production and marketing cost of the product, and continued production should be questioned. Third, the product may support other products in some way and the firm may want to con-tinue production at a loss.

Pricing Strategies

Armed with a thorough understanding of the value provided, the costs of producing and delivering the product/service/information bundle, the firm's pricing goals, and the competitor's strategies, an agribusiness marketer is ready to make a sound pricing decision. Let's take a look at some ways to put all of this information together by exploring several commonly used pricing strategies.

Cost Pricing   **Cost-based pricing,** or **cost-plus pricing,** is a simple strategy that involves adding a constant margin to the basic cost of the individual product or service. This margin is intended to cover overhead and handling costs, and leave a profit. In retail businesses, such as farm supply or food stores, it is a simple matter to "mark up" merchandise by some percentage:

Cost ($1.00) × 1.30 (30% markup) = $1.30 (Selling Price)

Farm construction firms sometimes operate on the basis of the cost of materials plus a percentage. The markup theoretically represents the cost of handling the product or performing the service; it therefore varies with different product lines and among different industries. In reality, the markup may not reflect costs accurately and markups may be more likely to be based on tradition rather than logic.

A major problem with cost-plus pricing is the difficulty of allocating fixed or overhead costs to a specific product or service. Many accounting systems simply are not adequate to determine how much of the delivery cost should be allocated to delivering each product. And, even if a method for this kind of determination existed, the cost of keeping track of time spent on such related product-specific expenditures is prohibitive in many cases.

Yet, because of its simplicity, the cost-based pricing method is popular, especially for the retailing of large numbers of products. In addition, computerized management information systems allow cost data to be measured and allocated more precisely, which facilitates the use of this strategy.

ROI Pricing   **ROI (return on investment) pricing** is similar to cost-plus pricing. This strategy begins with the cost of the product just as cost-plus pricing does, but instead of focusing solely on fixed or overhead costs, the focus is a target ROI. For example, a firm calculates how much profit will be needed to earn 15 percent ROI. Then, assuming a reasonable sales volume, it estimates costs and calculates the gross margin that must be generated to hit this target.

The major drawback of ROI pricing is that the price calculation assumes a certain level of sales volume, when, in fact, the sales will be at least partly determined by price. In fact, the price chosen may not generate the sales volume assumed, in which case the strategy breaks down. Yet the concept of pricing to achieve an acceptable ROI has considerable appeal.

Competitive Pricing   **Competitive pricing** lies at the other end of the spectrum. While cost-based and ROI pricing methods tend to ignore market conditions, competitive-pricing methods essentially base price on competitors' prices.

This method involves setting price at the "going rate," according to some general market average. An agribusiness employing this strategy may simply follow the price lead of a competitor.

Competitive pricing does not always involve matching the competitors' price; a price may strategically be held above or below that of competitors. A local lawn and garden store might choose to hold its price consistently above that of a large chain discount store. Or a small, independent grain elevator with very low overhead costs may choose to regularly offer 2 cents per bushel more than a large cooperative competitor who dominates the local market.

Because competitive pricing is the norm in commodity-based markets, the strategy is widespread in agribusiness, and is used many times by smaller firms whose markets are dominated by larger firms. This works well as long as the smaller firm has a favorable cost structure relative to other firms or is small enough not to be a major threat to the larger firm.

When one firm's value bundle is quite similar to another firm's in the market, price usually becomes a major factor in the buying decision. When firms struggle to differentiate their products and services, it is difficult to price above the market level. Consequently, most agribusinesses keep a close eye on the market price and vary from it only in subtle ways. Of course, this causes major problems for any agribusiness that does not have an efficient cost structure. Less efficient businesses are forced into financial difficulty because of the necessity of keeping prices competitive. And it is not uncommon for agribusinesses that are in a low-cost position to intentionally exert pressure on their higher-cost competition in order to gain an increase in market share. While there is nothing clandestine or illegal about this, the effects can be devastating for the less efficient agribusiness and beneficial for the customers.

CTO Pricing   CTO (contribution-to-overhead) pricing is a method of encouraging extra sales by selling additional products above and beyond some base sales projection, at a price slightly greater than the additional out-of-pocket costs of handling the product. In other words, CTO pricing, which is also called marginal-cost pricing, ignores the full cost of producing and selling a product, and focuses on the incremental cost of making the sale. This strategy assumes the overhead costs will be covered by normal sales as projected, so that if additional products are sold at any price that is above their variable cost, the additional sale will make a contribution to overhead and profit that would not otherwise exist. (See Chapter 16 for a detailed description of volume-cost relationships.)

The logic of this pricing method is quite compelling when viewed in terms of the marginal or extra sales opportunity. Since many agribusiness sales are made on a negotiated basis, there is ample opportunity for using CTO pricing methods to increase sales. Whenever fixed costs are a major component of total costs, as they are in many agribusiness industries, there is great temptation to utilize this method of increasing sales, making additional contributions to overhead, and increasing total profits.

The big problem with this strategy is limiting CTO pricing to only marginal or extra sales. In reality, there is a great tendency for competitors to react to the lower "spot" price, which causes the average market price to tumble, leaving the market unstable at best or in shambles at worst. The key is holding CTO pricing to marginal sales.

Value-Based Pricing   **Value-based pricing** involves pricing at a level just lower than the estimated perceived value of the product/service bundle. This idea makes sense—if the product is brought to the market at a price just lower than the perceived value, it should receive a favorable response from customers. The challenge is determining just what the perceived value of the bundle is to the customer. **Economic value analysis** is one way to formulate a price that reflects the economic value in a product. Here, the price that should be charged for a product is determined by dividing the product's economic value into two parts: the reference value and the differentiation value.

The **reference value** of a product is the price of a competing product or the closest substitute. This reference value forms a starting point for the price calculation. The **differentiation value** is the perceived value of the new product's unique attributes. To determine the price, positive perceived values of attributes are added to the price of the closest substitute (the reference value). Any negative perceived values of attributes are then subtracted from the reference value to find the total economic value of the product. The marketer may use this value-based price directly in the marketing strategy, or the marketer may use this price as a starting point from which to initiate some of the other pricing strategies described below.

Penetration Pricing   In a **penetration pricing** strategy, a product is offered at a low price in order to gain broad market acceptance quickly. These strategies are used primarily to introduce new products into a market, particularly price-responsive products that must sell in large volume to reduce per-unit costs. Penetration pricing can quickly cut into the sales of established competitors, even in cases where brand loyalty may be a factor. This strategy is also used in situations when a competitive product is expected to follow quickly, with the logic that the firm in the market first may be able to stake an initial claim on customer loyalty. After a new product has gained customer acceptance, the price may be gradually raised to a more profitable level.

Skimming the Market   A **skimming the market** strategy is virtually the opposite of penetration pricing. Skimming involves introducing a product at a high price and making excellent profits on the sales that are made initially. Then, as this relatively limited market becomes saturated, the price is gradually lowered, bringing the price into a range affordable to more customers. The appeal of this strategy is that it affords the opportunity to maximize profits on new products as quickly as possible. Skimming the market works best with products that are new, unique, fairly expensive, hard to duplicate quickly, and sold by firms that are well known and respected in the industry. New, superior varieties of hybrid seed are often introduced this way, with prices falling as the variety gains wide acceptance.

Discount Pricing   **Discount pricing** offers customers a reduction from the published or list price for some specified reason. Volume discounts are common among agribusinesses. The purpose of a volume discount is to encourage larger purchases, which reduce per-unit costs and promote more sales. Volume discounts can be on a per-order basis, but they more commonly accumulate throughout the season or year, and may take the form of a rebate at the end of the season. This kind of discount is frequently associated with customer loyalty programs.

Cash discounts are designed to encourage prompt payment for products and services. There are infinite varieties of cash discount programs. Many use terminology such as "5/10, net 30," which means that the customer will receive a 5 percent discount if the invoice is paid within 10 days, but the full amount is due in 30 days, regardless. Because of the tremendous seasonality in most agribusiness industries, cash discount programs are widely used.

Early-order discounts are often given by manufacturers of agribusiness products as an incentive to order and/or take possession of products in the off-season. Because of storage and shipping problems in peak season, many manufacturers aggressively promote early shipment with very significant price reductions. Crop protection chemical firms make wide use of early-order programs, sometimes offering a host of nonprice incentives to accompany price discounts, such as "one free with each pallet ordered." Sometimes price protection programs are included to protect the early buyer against late-season price reductions by guaranteeing customers that if the price falls later during the heavy-use season, the buyer will receive a rebate, effectively giving them the lower price.

Loss-Leader Pricing   **Loss-leader pricing** involves offering one or more products in a product mix at a specially reduced price for a limited time. The idea is to encourage long-term adoption of that particular product. In a retail store setting, the featured item is also expected to draw customers to the store and increase sales in all other product lines. Featured items are sometimes even sold at cost in order to boost store traffic. Loss-leader pricing is common in consumer retail settings, such as farm-home stores, nurseries, and food stores.

Psychological Pricing   **Psychological pricing** involves establishing prices that are emotionally satisfying because they sound lower than some virtually equivalent price. Odd prices, such as 99 cents, sound a great deal lower to customers than $1. "Two for $1.99," instead of "$1 each," gives the illusion of a special deal, and appeals to the customer's instinctive attraction to a bargain.

Prestige Pricing   **Prestige pricing,** on the other hand, appeals to a high-quality, elite image. Many people have a strong belief that "you get what you pay for" and tend to equate price and quality on an emotional basis. Here, even prices— $50 or $100, for instance—are often used to communicate a prestige image. Prestige pricing is used in agribusiness, especially by any firm that is pursuing a high-income, discriminating consumer. Premium dog food, unique food items, and quality wines would all be products likely to carry a prestige price.

An Example   To illustrate some of the concepts involved in developing a pricing strategy, let's consider an example. The focus is a manufacturer of agricultural chemicals that is bringing a new nitrification product, Ni-Tri, to the market. Nitrification products are used with anhydrous ammonia to enhance yield and improve dry-down in corn. Farmers look for four attributes in nitrification products: (1) increased yield; (2) drier corn; (3) easier harvest; and (4) environmental safety. This manufacturing firm is looking to employ perceived value pricing, and needs to estimate the perceived value of their new product.

Ni-Tri's closest competitor is a product called Enforce, which is currently priced at $36.00 per gallon. This $36.00/gallon becomes the reference value for the new product. Based on an economic assessment of Ni-Tri's benefits relative to those of Enforce, the following information is obtained:

Selling Price of Enforce	$36.00/gallon
Perceived positive value of drier corn Ni-Tri attribute	$ 2.50/gallon
Perceived positive value of easier harvest Ni-Tri attribute	$ 2.50/gallon
Perceived positive value of environmental safety Ni-Tri attribute	Unknown
Perceived negative value of yield Ni-Tri attribute	–$ 3.00/gallon
Total Economic Value	$38.00/gallon

Here, the makers of Ni-Tri could price the product somewhere around $38.00 per gallon—$36.00/gallon reference value and $2.00/gallon differentiation value—to reflect the economic value the producer would receive from using the product. Note that it proved impossible to put an economic value on the environmental benefit in the example above. The product manager for Ni-Tri will need to make a judgment call as to just what this benefit will mean to the target market. Accordingly, the final selling price may be a little higher than $38.00 per gallon.

At this point, the firm's pricing goals haven't been considered and the strategy doesn't really address the competition to any degree. Let's consider some other ways of looking at the problem that incorporate pricing goals and competitors in the decision-making process.

One strategic alternative would be to build market share. To increase market share through pricing, it is important to price at a level where the ratio of perceived value to selling price is higher than that of the closest competitor—a form of penetration pricing. If Ni-Tri has a perceived value that is 5 percent higher than Enforce, then Ni-Tri justifies a higher price than the competitive product. As long as the actual price premium is less than 5 percent, Ni-Tri will be in a position to gain share on Enforce:

$$\frac{\text{Perceived Value Ni-Tri}}{\text{Price Ni-Tri}} > \frac{\text{Perceived Value Enforce}}{\text{Price Enforce}}$$

for example:

$$\frac{\$42/\text{gallon}}{\$37.40/\text{gallon}} > \frac{\$40/\text{gallon}}{\$36/\text{gallon}}$$

Here, Ni-Tri will take market share away from Enforce since customers will find they get more value from every dollar they spend on Ni-Tri relative to Enforce. As the price of Ni-Tri is reduced, Ni-Tri should pick up more market share on Enforce.

The same idea could be applied to a situation where Ni-Tri's makers were more concerned with skimming the market and less concerned about market share. Here they would price the product so that the selling price of Ni-Tri was more than 5 percent higher than that of Enforce. Ni-Tri will show a higher margin on each sale, but customers will find Enforce to be the better value over time, and Ni-Tri will eventually lose market share.

The firm could combine the pricing approaches on page 215 with volume and cash discount programs, and an early-order discount program to complete its pricing strategy. One final dimension that should also be mentioned is the issue of competitive response. The examples have intentionally been oversimplified to illustrate the key concepts involved. But in reality, anticipating and managing the price response of the competition is a crucial part of a successful pricing strategy.

Legal Aspects   Pricing is a delicate issue, from both marketing and legal standpoints. There is a great deal of federal legislation that clearly prohibits any form of collusion among firms in establishing prices or discrimination against particular customers. The Sherman Act (1890), the Federal Trade Commission Act (1914), and the Clayton Act (1914) provide the foundation for U.S. antitrust law and prohibit acts (including pricing) that substantially reduce competition among firms.

The Robinson-Patman Act (1936) strengthened the Clayton Act and strongly forbids pricing acts that discriminate among buyers without a cost justification. Any product of like kind and quality that crosses state lines must be offered to all purchasers at the same price, unless it can be clearly demonstrated that the price differences are based on differences in the cost of serving various customers. Similarly, any discounts or allowances made to one customer must be available to all customers. Local businesses that are not involved in interstate commerce and that are the final seller may sell at any price they choose to any customer, so long as they are not in collusion with other sellers. These laws are enforced by the federal government through the Department of Justice and the Federal Trade Commission. In addition, competitors and/or customers can sue firms violating these rules. The financial penalties for violations of these pricing laws can run into the hundreds of millions of dollars.

PROMOTIONAL DECISIONS

Promotional activities in the agribusiness are designed to do one thing—communicate the value of the firm's products and services. Any marketing strategy incorporates a variety of methods for informing customers of the value of the firm's products and services and convincing them to buy from the firm. This part of the marketing strategy is basically a communications process intended to modify customer behavior and encourage a positive buying decision.

Figure 8-7
The Promotion Strategy Process.

The promotion mix chosen by the agribusiness firm is typically a combination of advertising, personal selling efforts, general public relations activities, and sales support programs. The mix must consider the life-cycle stage of the product, the stage of product adoption in the marketplace, competitors' actions, and available budget. The promotion mix is designed to support and complement the total marketing program (Figure 8-7).

Developing a promotion strategy typically involves the following steps:

1. *Identify the target audience.* In general, this is the target market. But, the target audience for a specific promotional effort could be any number of different groups, including specific groups of prospects, a group of **key influencers** (people who have some influence on the purchase decision, but don't actually make the decision), or a group of former customers the firm wants back. The target audience determines what to say, how and when to say it, and where to say it.

2. *Determine the communications objective.* Here the firm decides the action they want to encourage as a result of the communication. This desired action may be an immediate purchase, it may be to remind the customers of something, or it could be to change a noncustomer's attitude. This step is clearly related to the product adoption model described earlier. The firm would have a different objective for customers in the awareness stage as compared to customers in the adoption stage.

3. *Design the message.* Here the marketer creates a message consistent with the communications objective. Ideally, the message will get the audience's attention and elicit the desired action. This is a truly creative phase of the market communications process.

4. *Select the communications channel.* There are a variety of channels available to the agribusiness marketer and these are discussed in some detail on page 218. In this step of the process, the marketer decides which of these channels to use. In most cases, a variety of channels are employed. For example, a promotional campaign for a new breakfast cereal might involve television advertising, coupons in the print media, personal sales calls to food retailers, and in-store merchandising support such as signs and display racks.

5. *Manage the implementation of the program.* This step involves allocating the promotional budget to the various activities, coordinating the market communications process, and measuring the results of the process. While heavily centered on administrative activities, this is a fundamentally important part of any successful market communications effort.

Virtually any marketing communications strategy, no matter how large or small, will go through these steps either formally or informally. Discussion about key market communications channels follows.

Advertising

Advertising is mass communication with potential customers, usually through public communications media such as television, radio, newspapers, magazines, or the Internet. Some advertising is **institutional** or **generic advertising,** or intended not to promote a particular product but rather to build goodwill for the total company or industry. Most advertising is **product advertising,** designed to promote a specific product, service, or idea.

Advertising performs several important functions. First, it creates awareness about the product, which facilitates personal selling efforts. In some cases, public exposure through communications media lends a degree of credibility to the product. Psychologically, a potential customer comes to feel that a nationally advertised product must be worth considering. An advertisement can motivate a customer to seek out the product, or can at least serve as a reminder of the product's existence. Advertising also performs an educational function, helping customers to learn more about the product and its use. Finally, advertising can reinforce the value of a purchase that has already been made. Research suggests that recent purchasers of a product are among those most likely to read an advertisement for that product.

Much advertising is initiated and sponsored by manufacturers of agribusiness products. Some of this advertising is oriented toward the dealer or distributor to influence ordering and selling decisions. But much manufacturer-sponsored advertising is aimed toward the product's end user; in this case, the farmer or the consumer. The idea is to encourage ultimate demand to "pull the product through the pipeline," hence, the term **pull strategy.**

Geographically, broad-based advertising is usually the sole responsibility of the manufacturer. But local advertising, cooperatively sponsored by the manufacturer and the local dealer or distributor and called **cooperative advertising,** is also very common in agribusiness. The manufacturer, with the name of the local agribusiness inserted and the cost jointly shared, usually prepares local advertisements. The manufacturer may also supply the local agribusiness with flyers, posters, or product brochures, and may even send direct-mail advertisements—with the local business name imprinted on the advertisement—to local customers.

There are a wide variety of communications media ready to carry the agribusiness's message. Television is used heavily by national food manufacturers, but is much less used by manufacturers of farm inputs. Radio is used more commonly by local agribusinesses, since it is less expensive and can target the farm audience by tying into agricultural programs such as market news and farm broadcasts. Many aggressive local radio stations in rural areas employ professional farm broadcasters who do such a good job of covering agricultural news that they attract a strong following for the products they advertise.

Magazines are an important media for both food companies and agricultural input companies. A variety of local, regional, and national magazines provide an important method of communication with customers. An enormous number of specialty magazines appealing to tightly defined groups enable marketers to focus on publications of specific interest to their target market. Likewise, agricultural input firms can purchase advertisements in general farm publications serving all of agriculture or very specific publications serving only particular types of producers. Trade magazines geared to specific types of agribusinesses are also widely used to communicate with dealers and distributors.

Newspapers are an especially important medium for both food retailers and food manufacturers. Inserts and coupons represent an important form of advertising for both groups. Likewise, local agribusiness firms usually find advertising in local newspapers to be a cost-effective way to reach their clientele.

Direct mail is an increasingly important media for agribusiness marketers. Advances in computer technology have enabled agribusiness marketers to build huge databases containing detailed information on customers and prospects. Using these databases, marketers can tightly define targeted direct-mail promotions to very specific groups of individuals. For example, a feed company might desire to promote a new feed program only to Iowa pork producers with at least 1,000 sows, who use all-in/all-out production methods. Using the firm's database, contacting these groups with a letter, a promotional audiotape, or a promotional video is a relatively simple process. Such efforts are very cost effective and well received by the target audience as the information is targeted to their specific needs. Likewise, such databases can drive telemarketing efforts with similar, positive responses.

The Internet has emerged as an important promotional tool for food and agribusiness firms. Company websites provide useful product information, including the location of dealers who carry the product. Food companies have created highly educational and entertaining websites to encourage frequent access to the sites. Password protected areas are used to allow customers to access personal data, or to allow access to a higher level of information that the firm wants to make available only to qualified individuals. Banner ads can be used to promote goods and services on other sites. The ability to collect detailed information on customers allows the agribusiness marketer the ability for personalized e-mail campaigns, and other more sophisticated website strategies. When the information management, communications capabilities, and convenience of the Internet are considered, it is clear that the Internet is part of a firm's product/service/information and place/distribution strategies, in addition to the promotion strategy role.

Determining how much to invest in advertising is a challenge for most agribusinesses, since it is difficult to determine the actual effect of advertising on sales. Large manufacturers of agribusiness products and their advertising agencies spend a great deal of money attempting to measure the effectiveness of different advertisements. As one executive put it, "I know half of my advertising is wasted—and if I could determine which half, we'd cut it out!" This is a big problem for smaller agribusinesses because one person is generally not able to devote exclusive attention to advertising. Most local agribusinesses tend to do little media advertising, tending to focus on other promotion strategies. These firms argue that their best advertising in the local market is "word of mouth," and choose to allocate their promotional dollars to customer meetings and personal contacts, or simply to sell at a lower cost. Advertising budgets for local farm-oriented businesses are generally limited to 1 to 2 percent (or less) of total sales.

Sales Promotion

Sales promotions are programs and special offerings designed to encourage interested customers and prospects into making a positive buying decision. There is an almost infinite variety of such tools, and many are used extensively by agribusinesses to support personal selling and advertising activities. These tools range from the very expensive to the very inexpensive, but are felt by many to strongly influence customer decisions.

Some of these programs are aimed directly at the final consumer or the farmer. Giving prospects and customers hats, ski caps, jackets, belt buckles, pens, note pads, and a myriad of other "freebies," all imprinted with the company logo, has become so common that farmers and ranchers have come to expect it. And more and more often, as a means of getting their undivided attention, larger producers are being "entertained" over lunch or dinner by marketing people. Food shoppers are the focus of a vast array of promotional activities including free samples, coupons, special offers (buy 1, get 1 free), contests, games, incentive programs (e.g., accumulating "purchase points" to redeem for prizes), and rebates. Loyalty programs, which encourage repeat buying by offering various rewards based on accumulated purchases, have become an important form of retail sales promotion.

Such sales promotion programs are an integral part of agribusiness marketing strategies. However, many of these sales promotion programs are easily copied, which can reduce their effectiveness. As a result, some suppliers and retailers feel trapped by such promotions, but because of competitive pressures, are afraid to cut back on them. Thus expensive promotions and incentives should be entered into carefully with a clear objective in mind, and not simply as a quick reaction to a competitor offering.

Other promotional programs include educational sales meetings sponsored by suppliers and retailers for farmers, ranchers, and growers. Evening meetings and day-long seminars, often including a complimentary meal, provide much important technical information to producers, while subtly promoting the use of chemicals, seed, feeds, animal genetics, and the like. Field days, where farmers see products and services demonstrated in the field, are highly popular and effective. Regional trade shows where large numbers of suppliers attract thousands of farmers are also common. Farmer meetings and shows have become an important communication and educational link between suppliers and producers. Similar activities are sometimes sponsored for food consumers. Cooking classes, wine tasting parties, and educational seminars on new foods all can encourage sales and build the firm's image.

The variety of sales promotions used in selling to dealers and distributors is even more elaborate. Sales contests and incentive vacation trips are common. Some especially productive agribusiness managers may "earn" trips to an exotic resort for themselves and their spouses in the slack season, at the expense of their supplier. Dealer and distributor meetings during the winter months are usually well planned, educational, and often elaborate and entertaining. Some suppliers develop educational and technical training programs for their customers' employees. Other incentives may link purchases with a key need of the dealer—buy a certain amount of product and receive a fax machine, a computer, or a cash rebate toward a particular type of chemical application equipment, for example.

Not all agribusinesses become extensively involved in well-developed promotional programs. Some choose instead to concentrate on service and price. However, sales promotions are an important part of the market communications strategy for most food and agricultural input firms.

Personal Selling

In most agribusinesses, the salesperson plays an important part in the market communications process. Promoting a product through personal selling provides the most flexible and highest impact possible, since the salesperson can tailor the communication to meet the individual needs of the customer or prospect. This flexibility is especially important for complex products where usage and benefits may vary dramatically from customer to customer. For many agribusinesses, establishing a long-term relationship with priority customers is the focus of the marketing effort. And, personal selling plays a fundamentally important role in establishing such relationships. The agribusiness salesperson has the responsibility for not only taking the firm's product/service offering to the field, they also have the responsibility for keeping the firm informed about what customers and prospects want and need.

While personal selling can provide an important impact as a market communications tool, it is also an expensive form of communications. Individual sales calls are costly, and agribusinesses have invested considerable resources in technologies such as laptop computers, cellular phones, fax machines, e-mail, and message services to make salespeople as productive as possible. In addition, considerable effort is invested in targeting accounts that will be served through a salesperson, versus those that might be served via a less costly approach such as telemarketing.

Given the importance of this form of market communications, even the smallest local agribusinesses now have an individual with the title of salesperson on their staff. And, those that don't have a "salesperson" still rely heavily on personal contact and interpersonal communications to promote their product. Often the managers of such firms spend much of their time promoting products and services through personal contact. In larger agribusinesses, the personal selling process is far more formalized and highly structured. Management of the sales force and its organization take on great significance. Selecting, training, motivating, compensating, and territory allocation all become critically important decisions in the management of personal selling activities in the larger agribusiness. Because personal selling is so important in agribusiness, this topic is treated in much greater detail in Chapter 10 and Chapter 11.

Public Relations

Public relations is another important form of market communications. Public relations activities are somewhat unique as a form of market communications as they typically influence the target audience in an indirect way. Such influence may be through favorable news stories that create a positive image for the firm. In many cases, such activities involve an outside, objective party in some way to carry the message. As such, public relations activities are likely to be viewed as more objective than other forms of market communications.

Public relations activities can range from very direct—food company expenditures on nutritional education programs for school children—to very indirect—a favorable news story that shows how a local agribusiness helped coordinate the fund-raising effort for an injured farmer. Tools used for public relations by agribusiness firms include press relations, or getting news stories carrying the company's desired message publicized through the television, radio, and print media; company communications, or using company announcements (internal and external), speeches, and press releases to publicize the firm's desired message; and lobbying, or working with legislators to encourage laws and regulations favorable to the firm and its products.

In general, public relations activities help solidify the position or image the firm is trying to create in the market. For the nutritional campaign, the message may be that the food company is concerned about the nutritional needs of children. For the fund-raising story, the message may be that the local firm really cares about the local community. Savvy agribusiness firms do not leave public relations activities to chance—they incorporate such activities into their promotional strategy. This may mean having the news media attend the rollout of a new product, it may mean a column in the local newspaper on agronomic advice written by their field agronomist, or it may mean a well-publicized community service activity in which the firm's employees participate. Such efforts may require little monetary investment, but may pay major dividends for the firm.

PLACE DECISIONS

Every agribusiness must decide how to move products to its customers. Marketing channels are systematic ways of transferring both the physical product and the ownership to the customer as efficiently as possible. In many ways, the channel is a communication system linking the manufacturer or producer to the customer. Consumer demand signals flow through the distribution system to manufacturers, while supply and availability signals in the form of prices and costs flow to consumers. In many cases, independent agribusinesses such as wholesalers or distributors facilitate the transfer of ownership and physical product. In other cases, manufacturers find it more efficient to sell directly to the customer, handling these distribution functions inside the organization.

Physical Distribution

Physical distribution decisions are particularly important to much of agribusiness, since so many farm supplies and food items are bulky and highly seasonal in demand. Literally millions of tons of phosphate fertilizers must be moved from Florida mines, and millions of tons of potash fertilizers must be shipped from western Canada to farms and ranches all over the United States and Canada. Likewise, highly perishable crops like lettuce, tomatoes, and peppers must find their way from California, Arizona, Texas, and Florida to food stores all over the United States. At the same time, millions of bushels of grain must be moved from local storage to the coasts for shipment to overseas markets.

The costs of physical distribution activities are high. And, in season, rail car and truck shortages can cause spot supply shortages, creating bottlenecks for agribusiness products in the distribution system. Consequently, a firm's marketing strategy increasingly incorporates plans for assuring an adequate physical distribution system. Policies encouraging larger storage facilities at local levels, company ownership of rail cars, company-operated truck fleets, unit trains with 100-car shipments of the same commodity, transcontinental barges and pipelines, and partnering relationships with suppliers of logistics services are becoming common as a means of ensuring timeliness and reducing costs.

Even at the local level, agribusinesses are highly concerned about physical distribution problems. Bulky products must be broken down into prescription quantities to meet the needs of individual farmers quickly during intense high-use periods. Bulk feed delivery, on-farm storage, custom application of chemicals and fertilizer, on-farm breeding services, and a myriad of other products and services must be delivered effectively. For food retailers, Internet shopping and home delivery of purchases is increasing in frequency. Convenience is a major issue on the minds of many food shoppers, making store location decisions among the most fundamental a food retailer can make. Much of this physical distribution requires a great deal of highly specialized equipment and labor, as well as major investments in facilities. Indeed, physical distribution at all levels is a high cost activity and is critical to the agribusiness.

Channel Management

Channel management decisions are concerned with who owns and controls the product on its journey to the customer. These decisions have important implications for how the marketing functions will be carried out and who will carry them out. The market channel selected is closely tied to the physical distribution problem. But the question of who owns the product and who performs the various market functions required to transfer a product from the manufacturer to the consumer is much broader.

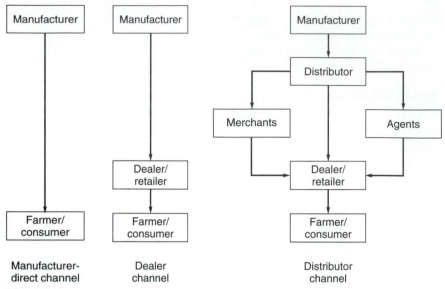

Figure 8-8
Major Agribusiness Distribution Channels.

There are three basic systems used by most agribusinesses (see Figure 8-8). Each system differs in the extent and type of involvement of the middleman. In each system, all marketing functions are performed. That is, products are owned, transported, financed, and stored. But in each system these functions are performed by different parties.

In a **manufacturer-direct distribution system,** the manufacturer sells directly to the farmer or food customer. The original manufacturer or producer, who owns and controls the product until the final user purchases it, performs all marketing functions. This system has the advantage of ensuring that the product will be priced, promoted, and sold in ways that are acceptable to the manufacturer. This method is used widely by smaller local or regional farm input manufacturers, who can develop their own sales force as they grow, and by catalog merchants, U-pick fruit and vegetable operations, and farmers' markets in the food market.

Some firms believe this strategy allows them to be more aggressive in the marketplace, because they can create and control their image with customers. This channel also works better where customers are concentrated. As the business grows and customers are distributed more widely, with increasingly different characteristics and needs, it becomes more difficult to manage the direct-selling system. Marketing to widely different customers often requires at least regionalized marketing policies, which become difficult to administer effectively. However, when organizations give their sales force considerable autonomy, the system can be effective even on a regional or national scale. Likewise, major developments in customer databases which allow manufacturers to "know" their end users have facilitated this channel. In addition, developments in e-commerce have allowed manufacturers to go directly to the end user employing the Internet and the firm's website as their "virtual" sales force. This method of direct marketing will increasingly find application in the food and agribusiness markets.

Note that many companies that sell directly do so by establishing their own wholesale and retail systems. The product must be physically transported and distributed as in any other system, but in the direct system, company-owned warehouses and/or retail stores will be used, which allows manufacturers to ship and store products whenever and wherever they choose. This freedom provides some real economic advantages.

In a **dealer-distribution system,** manufacturers sell their products to dealers, who, in turn, sell products in their own local market. The big advantage is that the local dealer is more in touch with the needs of local customers and can maintain the flexibility to serve both manufacturers' and customers' needs quickly and efficiently. This eliminates the tendency for an executive 1,000 miles away to fill local storage space with a product simply to gain more room in a full warehouse, even though local farmers may not need the product in large quantities that season. In addition, some feel that independent dealers are more highly motivated to serve the local customer than are company employees.

Some manufacturers give their dealers (who may be called distributors, even though they sell directly to farmers or consumers) the right to sell their products. Although the manufacturer cannot control the sales areas of dealers, giving some dealers this right and refusing it to others can strategically limit the number of dealerships. To maintain the right to sell the manufacturer's products, dealers may agree to maintain certain levels of service to customers and promote the products in special ways. There may even be a contractual agreement, sometimes called a franchise, spelling out the responsibilities of both the dealer and the distributor.

Other manufacturers encourage distribution through as many dealers as possible. Products that are frequently purchased and less expensive generally tend to be mass-marketed in order to maximize convenience to the buyer and saturate the market. Animal health products and many hard lines of farm supplies are marketed in this way. Products are sold to the dealer almost as they would be to a farmer-consumer. As farms and ranches grow bigger, many manufacturers are beginning to recognize them as "dealers," especially if the farmer resells at least some product to a neighbor. In effect, the traditional manufacturer-dealer-farmer system seems to be breaking down in some product lines.

In the food area, this channel has taken a different form with the emergence of major retail chains such as Kroger and Albertson's that control much of the retail food market. Such retailers fill an important need by bringing the products of a myriad of manufacturers together under one roof for the convenience of the food consumer. While small independent food stores remain important in some areas, national and regional chains dominate the scene. These chains, by having access to food shoppers, are a very powerful force in the marketing channel. And, many of the larger chains are also integrated into manufacturing their own private label food products that compete with the brands from manufacturers.

The **distributor system** uses both distributors, or wholesalers, and dealers to market products to farmers and consumers. This more complex marketing system usually evolves for economic reasons. Larger organizations can more easily afford to develop their own marketing system. Independent distributors may already have a well-developed network of transportation, salespeople, and customers, and their size allows them to operate quite efficiently. They can often add a new product line much more cheaply than the manufacturer can establish an entire distribution system. This allows manufacturers to concentrate on what they do best—manufacturing product—and allows distributors to do what they do best—support distribution activities.

Merchant-distributors, who actually take title to the manufacturer's product, are called distributors, jobbers, wholesalers, or cooperative buying groups. They physically order, receive, and distribute products. As totally independent businesses, they make their own decisions, relying on manufacturers only for technical assistance and information. Many agricultural chemicals and food products are sold through regional wholesalers whose sales force sells to dealers and food stores, who in turn sell to farmers and food shoppers.

Agent-distributors perform the function of helping products move through the system, but they do not take title to the product and usually do not physically handle the product. Agent-distributors are known as brokers, sales agents, and manufacturer's representatives. They are independent businesses that simply locate customers, negotiate a deal, and receive a commission for the transaction. Agents and brokers often specialize in certain product lines and types of customers. Their primary tool is communication. In markets in which shortages are occurring, brokers are usually able to locate extra product for a special price. They perform the important function of bringing buyers and sellers together.

E-commerce firms have emerged as a new type of dealer/distributor. These firms represent a variety of manufacturers and have no physical "storefront," relying instead on sophisticated websites to assist consumers and producers in making purchase decisions. The customer or producer places the order over the Internet, financing is arranged over the Internet, and the merchandise is shipped to the home or place of business, or is made available for pickup at a convenient location. These new "e-dealers" will continue to search out the products and customers where they provide a more effective and efficient marketing channel than the traditional "bricks and mortar" channel.

The total distribution system can be far more complex than is described here. Many manufacturers may use several different systems simultaneously. For example, Agrium, a large fertilizer manufacturer based in Canada, markets fertilizer through its own retail outlets and through independent fertilizer dealers. Pepsi products are available virtually anywhere people get thirsty—from food and convenience stores, to sporting events, to study lounges, to restaurants and institutional food service outlets, to drink machines on almost every corner. Such distribution systems gradually evolved as a result of changing products, customers, and the competitive environment. And, this evolution can be expected to continue, especially in light of the rapid developments in the e-commerce arena.

SUMMARY

Agribusiness marketers must translate the position they want to create for the target market into a product/service/information bundle that serves customer needs and communicates the desired image. The set of decisions to accomplish this task is called the marketing mix.

Product decisions determine what products, services, and/or information are offered by the agribusiness. The needs of the target market are at the center of any decision in this area. Agribusiness marketers create a value bundle that will satisfy customer needs. This bundle includes both the tangible attributes of the product or service as well as any intangible benefits the firm may offer the customer. As products are introduced they move through a systematic adoption and diffusion process. Each new product has a life cycle, passing through distinct phases from its introduction to its demise. Both the adoption-diffusion process and the product life cycle have important implications for market planning.

Pricing decisions are critical to marketing success. Pricing strategies are based on the perceived value of the firm's products and services, the firm's cost of doing business, the marketing goals of the firm, and competitive actions. A wide range of pricing strategies are available, from simple "rules of thumb," to far more sophisticated approaches that involve carefully measuring the value delivered by the firm to the target market.

Promotional decisions determine how the agribusiness will communicate with the market. These decisions involve determining the proper mix of advertising, sales support, personal selling, and public relations needed to communicate the firm's desired image to the market. Developing a market communications strategy involves identifying the target audience, determining the communications objective, designing the message, selecting the communications channel, and managing the implementation. A coordinated set of activities, which communicate the desired image to the target market, is the goal of the agribusiness marketer in this area.

Finally, place decisions concentrate on the methods and channels of distribution that will optimize sales and profits. Logistics management plays an important role in place decisions as firms determine how products will physically move from manufacturer to customer. Issues of cost and efficiency, timeliness, freshness, customer service, customer access, and control all affect the choice of distribution channel by an agribusiness.

All elements of the marketing mix are critical decision areas for the marketing manager. In the end, the marketing mix must deliver a consistent message—the position—to the target market.

DISCUSSION QUESTIONS

1. Many retail food stores also sell flowers. Why do you suppose they do this? What marketing decision does this illustrate? Besides food products and flowers, what other products or services do retail food stores offer? Why?
2. Pick a food product or an agricultural input. Use the total product concept to break the product down into the generic product, the expected product, the value-added product, and the potential product. In what other ways could the agribusiness add value to the product you have chosen?
3. Draw a product life-cycle curve for large, four-wheel drive farm tractors. Where would you estimate this product's life cycle is currently? Why? What might manufacturers do to prolong its life?
4. If you were introducing a new electronic dairy feeding system that would reduce feeding costs by 20 to 25 percent, what pricing policy would you suggest? Why? Would you stay with this policy indefinitely? Why?
5. You are the marketing manager for a retail food company that has a new 16-ounce fruit juice product, packaged in a reusable squirt-type bottle. Your target market is the active 18- to 25-year-old consumer group. Develop a promotion plan for this new product. What market communication tools would you use?
6. You are the marketing manager for a large animal health company. Your firm has just developed a very effective anthelmintic (wormer) for cattle. Your target market is large feedlot operations in the western United States. Develop a promotion plan for this new product. What market communications tools would you use?
7. What are the advantages of a manufacturer's selling directly to a farmer rather than to a distributor? Why can a dealer sometimes do a better job marketing to farmers than a manufacturer can?
8. Some retail food stores offer a service where they sell products via the Internet. Here, using communications software, the customer accesses the firm's inventory and price list through their website, chooses what they want, and the firm delivers the choices to the home at a time convenient to the customer. Why would food firms do this? What do you see as the advantages of this system? What do you see as the disadvantages?

CASE STUDY — PecsBake[1]

PecsBake is a medium-sized bakery company that supplies fresh baked breads and traditional Hungarian pastries to retail groceries in southwestern Hungary. Pecs (pronounced "paytch") is the fifth largest city in Hungary and a popular tourist destination near the Croatian border. On a recent management training and business development trip to the United States, the company's top management visited many

---

[1]This case study developed with the assistance of Jesse Lyon and Carmi Lyon.

American supermarkets. In the refrigerated goods sections they saw products not currently present in the Hungarian market—refrigerated breakfast pastry dough ready-made for microwave preparation, packaged in special cardboard/can containers. There were two common package sizes: 2 servings per container and 6 servings per container.

PecsBake's managers noted the increasing popularity of microwave ovens in Hungarian homes, and decided that they would like to introduce sweet Western-style pastry products (such as frosted cinnamon rolls) in Hungary. Their production experts were confident that PecsBake could also produce refrigerated, microwave-ready dough for traditional Hungarian pastries such as *turos taska* (pronounced "tourosh tashka"— a roll filled with sweetened cottage cheese), *lekvarosz batyu* (pronounced "leckvarosh bawtuyu"—a roll filled with jam), and *kakaos csiga* (pronounced "cock-a-osh cheega"— a roll filled with cocoa). On a per-serving basis, the Hungarian pastries would require 1.5 times the volume of the American products.

In general, Hungarians don't like sweet breakfasts. Their morning pastries are only semisweet at most. If PecsBake decides to sell American-style cinnamon rolls, they may need to be positioned as dessert products unless the firm can educate consumers about this new breakfast taste.

Microwaves are just emerging in Hungary. At present they are in only three or four out of every ten homes in the capital city of Budapest, and with even fewer in the countryside. "Food-in-a-hurry" isn't a widely held concept in Hungary. However, this is changing quickly as many Hungarians are working longer hours and sometimes two or three jobs.

U.S. cardboard/can manufacturers are currently reluctant to produce containers sized for the proposed Hungarian products unless they are paid higher prices to cover the cost of manufacturing the special containers. At this time they are only offering prices that would be economical for the Hungarian bakery on the same cardboard/can containers that they are currently supplying to large American companies.

PecsBake's managers are not certain that the proposed products will be well received by Hungarian grocery stores that already have limited refrigeration space. Alternatively, one manager suggested that they provide stand-alone refrigerated displays and sell their products in appliance stores in areas near the microwave displays.

PecsBake has enough capital to begin producing and marketing refrigerated, ready-made microwaveable pastry dough on a limited scale. They would like your advice about what products to introduce, how to package their products, and how to reach their customers.

Questions

1. What target market should PecsBake pursue?
2. What position should they take with this target market?
3. Outline the marketing mix you would suggest for PecsBake as they attempt to communicate your suggested position with the target market. What product/service/information strategy would you employ? How should they price their products? What promotion or market communications strategy should they use? What place strategy or distribution channel would you use? You may want to consider both the short term and the long term in your strategy.

REFERENCES AND ADDITIONAL READINGS

Manning, Gerald L., and Barry L. Reece. 1992. *Selling Today: An Extension of the Marketing Concept.* 5th ed. Weedham Heights, MA: Allyn and Bacon.

Nagle, Thomas T. 1987. *The Strategy and Tactics of Pricing: A Guide to Profitable Decision Making.* Englewood Cliffs, NJ: Prentice Hall.

Rogers, Everett M. 1995. *Diffusion of Innovations.* 4th ed. New York: Free Press.

CHAPTER 9

TOOLS FOR MARKETING DECISIONS IN AGRIBUSINESS

OBJECTIVES

- Outline the procedure for a marketing audit
- Analyze how sales forecasts are made and how they are used in the agribusiness firm
- Present methods for analyzing competitors
- Review techniques for studying customer attitudes and opinions

INTRODUCTION

Developing and managing a successful marketing program in an agribusiness can be a complex task, particularly in larger firms with many products. The needs of customers are constantly evolving. Competitors are continually introducing new products, services, and programs. Nearly all agribusinesses face highly seasonal demand that creates the possibility of critical bottlenecks in servicing customers. Unpredictable weather patterns further complicate market planning, and volatile agricultural commodity prices often cause the demand for farm supplies and services to fluctuate. Because of such complexities, agribusiness marketing strategies require a great deal of planning. Successful managers spend much time analyzing their market, assessing their own strengths and weaknesses, and mapping their strategic marketing plan.

Agrimarketers use a wide assortment of analytical tools and concepts—market planning tools to assist them in the market planning process. These tools play an important role in conducting the SWOT analysis described in Chapter 7. In addition, these tools play an important role in the evaluation phase of the market planning process. Some of these tools are highly sophisticated and involve complex mathematical models and extensive marketing information systems. Other tools involve far simpler approaches to data collection, while still others are simply methods of capturing subjective "gut feelings" for the market.

Agribusiness marketers stay on track by using a variety of market planning tools.

One point is clear regardless of which tools are used—agribusinesses using an information-driven approach to market planning are more likely to be successful than those who don't put data to work in their planning efforts. This chapter will look at four broad sets of tools that agrimarketers can use in the market planning process—the marketing audit, sales forecasting, competitor analysis, and customer analysis.

THE MARKETING AUDIT

Nearly every agribusiness and certainly all publicly owned companies have financial audits performed regularly by outside auditors to ensure that there are no discrepancies or oversights that could cause serious problems for the business and/or investors in the business. Similarly, a **marketing audit** is an objective examination of a company's entire marketing program. The marketing audit ex-

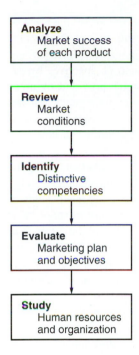

Figure 9-1
Marketing Audit Procedures.

amines the firm's marketing objectives and its plan for accomplishing these objectives in light of its resources, the market conditions, and its distinctive strengths and weaknesses (Figure 9-1). In the strategic market planning process outlined in Chapter 7, the marketing audit plays a key role in the evaluation stage of the process. It is a tool that firms can use to better understand past marketing efforts in order that future efforts will be even more effective.

The first step in a marketing audit is to carefully analyze how the agribusiness is doing with each product or service it markets. This step involves estimating total market potential, determining the firm's share of the market, and attempting to identify what factors are responsible for the firm's market position with each product or service.

Here the firm will reexamine its marketing goals for each of its products or services and assess what progress they have made toward achieving the goal. Areas where the firm falls short of the goal should be carefully explored to better understand why performance was poor. Did the firm miss the mark when assessing the needs of the target market? Did competitors introduce new products or programs that caused problems for the firm's marketing plan? Did the firm fail to execute the plan effectively? Or, was the marketing goal too ambitious? For example, a feed manufacturing company that has an objective of a 10 percent annual increase in poultry feed sales at a time when the poultry industry is under pressure from low prices might be better served by creating a new objective of holding poultry feed sales constant, while devoting more resources to expanding sales of swine premixes.

Answers to these types of questions will be provided in part by the remaining steps of the marketing audit. But, before moving to step two of the marketing audit, it is important to note that while much focus is given to areas where performance was poor, there is another, equally important, perspective. An effective marketing audit also attempts to capture the insights provided by areas where performance was exceptional. If sales of a new granola snack bar are growing twice as fast as expected, the food firm should dig deeply into the situation to better understand why this is occurring, if it will continue, and how they can apply this success to other products.

The second step of the marketing audit reviews general market and competitive conditions. Marketing factors external to the firm are evaluated, and expected trends explored. Sometimes probabilities are attached to these expected trends or market scenarios. This information helps agribusiness managers to shape their marketing strategies. The focus here is revisiting key assumptions about trends in technology, regulations and policies, general economic conditions, and the competitive environment. A marketing plan based on old forecasts of such areas is likely to fail. Marketers must challenge their previous assumptions, and generate new scenarios for the future they believe they will be moving into.

The third step of the marketing audit focuses on identifying the firm's distinctive competencies, that is, areas where the firm is particularly strong in relation to the competition. Distinctive competencies may include financial, human resource, marketing, and production advantages. As indicated earlier, these distinctive competencies often form the basis for the total marketing strategy. For example, consider an artificial-breeding company that is fortunate enough to have an extremely good line of bulls with exceptionally strong performance records. This measurable product quality distinction can form the basis for a quality-oriented marketing strategy with a premium pricing approach, a relatively small field sales force, and considerable emphasis on progeny testing.

While firms must give careful consideration to the idea of distinctive competencies during the market planning process, the intent here in the marketing audit is to reconsider the issue to make sure the areas identified earlier remain strengths. This part of the marketing audit may involve data collected during a competitor analysis or during a customer needs analysis.

The fourth step of the marketing audit is to objectively evaluate the agribusiness's overall marketing strategy, given a review of past performance in the firm's target markets, a reexamination of the firm's market environment, and a reassessment of the firm's distinctive competencies. In rapidly changing markets, strategies can become out of date and ineffective quite quickly.

This step of the marketing audit should also consider how the marketing plan is to be implemented. A good plan will be ineffective if it is poorly executed. Continuing to use the same channels of distribution because "that's the way we always do it" may be a poor choice. Do incentive travel packages offer the best type of sales promotion for dealers, or would it be best to return dollars in the form of a rebate—or perhaps just sell at lower prices in the first place?

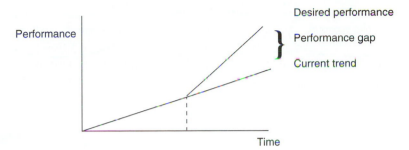

Figure 9-2
Gap Analysis.

Analysis of such implementation issues forces the marketer to think deeply about whether or not the current marketing strategy is still the correct one.

Given a thorough examination of where the firm has been and new forecasts for the future, the goal of this step is to make sure the firm's market strategy still fits. This is sometimes called **gap analysis** (Figure 9-2). The gap is the difference between desired performance at some point in the future and what performance will be if the firm does not change marketing strategies. Obviously, the bigger the gap, the greater the need for the firm to revisit its marketing strategy. Through this analysis, the next marketing plan can be made more effective in terms of the current and expected market conditions.

Finally, the marketing audit objectively evaluates the firm's human resources and organizational structure (see Figure 9-1). Given the critical role people and organizational structure play in the marketing process, this area deserves special attention. Are salespeople adequately trained to advise growers on technical matters? Should technical specialists in the field report to the area sales manager, or should they report to the home office? Do field salespeople have the right level of price authority to meet the marketing objectives? Does the organizational structure promote adequate cross-communications between salespeople, marketing staff, and the research and development department?

The firm's revised marketing strategy becomes the focal point for this final step. Can available human resources and the current organizational structure support what the firm wants to accomplish in the future? Evaluating human and organizational factors is one of the most critical facets of the marketing audit.

The marketing audit should be performed as objectively as possible. Because it is sometimes difficult for managers to evaluate their own program objectively (because they created it), managers should consider hiring an outsider to periodically review the program. A qualified consultant can evaluate a marketing program and raise important questions about the allocation of time, effort, and financial resources. In smaller firms with localized markets, a peer running a similar business in a different market can fill the role of the consultant and offer important, objective insights into the firm's marketing efforts.

Figure 9-3
Sales Forecasting Model.

SALES FORECASTING

Although agribusiness markets are well known for their volatility, agribusiness managers cannot escape the responsibility of forecasting sales. Nearly every management decision includes a critical assumption about the level and timing of sales volume. All production schedules are built around projected demand. Raw material purchases are based on expected sales. The personnel function of hiring and firing is greatly influenced by anticipated sales. Cash needs are based on sales forecasts. Capital investments in new facilities are determined by sales projections. The allocation of resources to research and development is based heavily on the expected sales of the resulting products. The fact is that a great majority of management decisions and much of the entire planning process rests squarely on forecasted sales.

Sales forecasting involves estimating sales in dollars and physical units as accurately as possible for a specific period of time. Many times this means forecasting sales for the total market, and then determining what share of the market can be captured with a specific product or service (Figure 9-3).

Both short-term and long-term sales forecasts are useful for agribusiness. Short-term forecasts, usually one season or shorter for agribusiness firms, are useful in formulating current operating plans. In food firms, such forecasts may be weekly or even daily. In a food service firm, forecasts may be hourly as the firm works to schedule food preparation to accommodate daily patterns. Longer-range sales forecasts are important for capacity and research and development decisions. In general, the longer the time period forecasted, the less accurate the forecast is likely to be. However, forecasting short-term volatility can be a major challenge as well.

Difficulty and complexity notwithstanding, agribusiness managers need relatively accurate sales forecasts. Such forecasts may be prepared in a formal manner using thorough economic and marketing analysis, or they may be more informal and based on assumptions from a variety of practical perspectives. Here, three types of forecasts will be considered—general economic forecasts, market forecasts, and sales forecasts.

General Economic Forecasts

General economic forecasts consider broad factors that affect the total economy. Government farm programs, inflation, the money supply, international affairs, exchange rates, interest rates, population demographics, and a host of other factors are included in such forecasts. A great many government and private economists spend much time tracking economic trends and making such projections. Agribusiness managers watch these opinions carefully as they formulate market plans. Forecasts of such general economic indicators have a fundamental role in an assessment of the market environment as part of the SWOT analysis and in the marketing audit.

Typically, these types of forecasts are generated by sophisticated econometric models of the domestic and world economies. Using such models, forecasts for a variety of indicators at different points in the future can be generated. This information is very important to agribusiness marketers. A firm selling capital equipment to farmers and ranchers knows that purchases are heavily influenced by interest rates. Long-term prospects for stable to declining interest rates mean this factor in the farmer's capital equipment purchasing equation is positive. A regional retail food chain may find that economists are predicting a harsh recession in the area served by their stores. Such information will have important implications for the firm's sales as some customers reduce luxury purchases, and focus more on staple products.

Market Forecasts

Market forecasts, or forecasts for specific industries or types of products, are based on general economic forecasts and a great deal of information about the specific industry involved. Agricultural economists continually monitor changing economic indicators that track or lead agricultural trends. Some use complex mathematical methods (econometric models) to predict demand for various products and commodities. In the food business, demographic data is carefully studied to forecast demand for specific products. Demographic factors such as family size, age, income level, and education tell market researchers much about what kind and how much of specific types of food products will be purchased in a region.

Factors determining farm market demand have been researched extensively. For example, agricultural economists have discovered various cycles in livestock production. Traditionally, hog production cycled every 4 years, while the cattle cycled about every 9 years. During these cycles, prices fell as the number of animals on feed increased, and profits declined. As a result, some less efficient producers exited from the market, livestock numbers dropped, and prices gradually increased. The cattle cycle was longer because of the length of the gestation period for cattle, and therefore it takes more time to increase production. Recent changes in the size and structure of hog and cattle operations and the technologies used by hog and cattle producers have lengthened and flattened these cycles. Still, information on where the industry is in the cycle is extremely important to feed companies, milling equipment companies, livestock marketing firms, and meat processing companies.

The demand for most farm inputs like feed is a **derived demand.** This means that the demand for a specific type of product or service depends greatly on the demand for another product or service for which it is used. The demand for fertilizer, for example, is a function of the demand for the crops it helps produce. When the

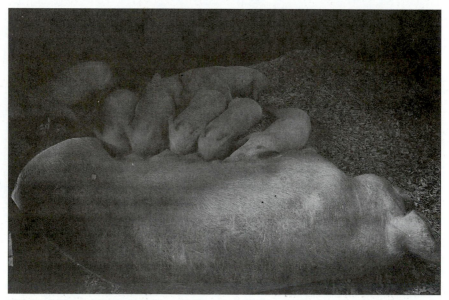

Market forecasts depend on an understanding of industry characteristics such as livestock production cycles.

manager of a fertilizer plant in the Midwest is projecting the demand for nitrogen, the corn market is carefully considered. The manager knows that the price farmers anticipate for corn in the next season will greatly affect the number of acres of corn they will plant and the amount of fertilizer they will want to buy. Consequently, agribusiness managers are careful students of the market for whatever products their supplies help to produce.

Not surprisingly, there is a great interest in current information on nearly every product produced by farmers. Summaries and analyses of market conditions for virtually every agricultural product are distributed widely each day. Important grain and livestock market information is instantly flashed to analysts and agribusinesses as transactions occur throughout the day via a variety of media—television, radio, satellite news services, and the Internet, to name a few. Government reports estimating current and projected acreage and livestock numbers are released frequently throughout the year to help agribusinesses monitor trends and adjust their plans.

Like agricultural input firms, food firms are also very interested in market demand. Food marketers pay special attention to two key areas when forecasting sales—regionality and seasonality. Food consumption patterns can vary dramatically from one region of the country to another. For example, grits sell well in the southeast United States, but few consumers may even know about this product in other regions of the country. Consumption of food products will vary dramatically by season as well. Summer fruit, vegetable, and ice cream sales, and holiday sales of turkey and ham are excellent examples. Combined with good demographic data, bringing regional and seasonal factors into the equation helps food marketers fine-tune their market forecasts.

Note that despite all the sophisticated techniques, forecasting market demand is far from an exact science. Random events such as unexpected weather, consumer fads, a food safety scare, or an international crisis can play havoc with the most elaborate market demand forecast. And, some firms argue that because of such events, most forecasting efforts are futile. However, firms that take a structured approach to forecasting market demand are much more likely to make good marketing decisions as compared to those firms who choose to be more reactive to market conditions.

Specific Product Forecasts

The actual sales forecast for a specific product within the firm is driven in part by the general economic forecast and the market forecast. Sales of a particular firm's low-fat yogurt are clearly influenced by general trends in the overall economy and trends in the market for dairy products. However, forecasting sales of individual products or services is complicated by the actions of competitors. It is one thing to forecast how much insecticide cotton growers in Texas will use. It is quite another to forecast sales of the specific insecticides offered by the firms competing in the Texas market. Here, because of uncertainty in forecasting the impacts of competitive marketing programs, predicting sales of specific products is quite challenging.

One sales forecasting method that is widely used involves projecting sales objectively based on past trends and then adjusting these projections subjectively to take into account the expected economic, market, and competitive pressures. This **trend forecast,** while simple, is reasonably effective in stable market situations. For example, a feed firm might start with a trend forecast of 50,000 tons for its catfish feed. However, based on information gathered about competitive programs, the firm has learned that a new competitor will be aggressively entering the market with a penetration pricing strategy. The firm expects about 20 percent of its accounts to at least try the new feed on a trial basis, and estimates that these trial purchases will represent about 25 percent of this group's purchases. The firm then adjusts its trend sales estimate by 5 percent (20% trial × 25% use) to 47,500 tons to account for the expected impact of the new competitor.

Sales forecasts are sometimes built-up from data collected by the sales force. A **build-up forecast** is constructed by asking salespeople to develop detailed sales forecasts for each of their major accounts. The accumulated sales estimates from all field salespeople then offer a grassroots estimate of sales expectations. This method is particularly valuable where competition is intense, as in the case of many farm inputs. Of course, build-up forecasts may have some inherent biases and depending on how the sales estimates are used, they can be under- or overinflated. For instance, if available production is allocated to sales territories based on initial sales estimates, salespeople may inflate their estimates to make sure they will have an adequate supply if production happens to be short. The reverse may occur when salesperson performance is evaluated based on actual sales relative to forecast sales. Here, the salesperson may be inclined to offer a modest sales estimate in order to help insure that actual sales will be relatively higher and the resulting performance review positive.

Consumer surveys and test panels offer a great deal of information about buying intentions. Several regional and national organizations, such as Doane's, Inc., and the *Farm Journal,* continually monitor farmers' attitudes and plans for the coming season through intention surveys. A panel of farmers may be paid a fee to give detailed reports about their intended farm plans on a regular basis, then report actual decisions made. Results of these private research studies are made available to subscribers on a fee basis. Farmers' planting intentions are also monitored by the USDA and released periodically throughout the year. Organizations like IRI (Information Resources, Inc.) perform these same tasks for food firms. In many cases, these services utilize scanner data purchased or acquired from retail food stores to show food manufacturers and retailers key trends in food consumption at the product level. The use of electronic scanners has made unparalleled amounts of information on consumer purchasing patterns available to food marketers. Such information can play an important role in their sales forecasting efforts.

The **Delphi approach** provides another useful tool for developing sales forecasts. Here, a panel of experts is asked to develop a forecast for the area of interest. These estimates are then pooled, reviewed, and any differences across the experts studied. The estimation process is repeated until a consensus forecast is achieved. This approach can be very useful when a forecast is needed quickly. The Delphi approach is also useful as a means to evaluate forecasts developed using other methods. Of course, experts can be wrong and the perspective of the panel members must be carefully considered when evaluating an expert opinion or Delphi forecast. This same method can also be used for forecasts of the general economic environment and the total market.

In some situations, especially with new products, a **test market** provides a useful way to collect data on potential sales. Here, an experiment is set up by selecting a test city or area with characteristics similar to the target market. The product is introduced, and the sales results measured. These results are then generalized to the target market of interest. Such information is valuable as it reflects actual consumer purchase decisions and not just attitudes toward purchase decisions as might be gathered through a survey.

On the other hand, test markets can be expensive, and it can be difficult to control all the outside factors that might influence sales of the product. For example, a juice maker might run a test market for a new 16-ounce juice product in Des Moines, Iowa. While the experiment is running, the area is hit by an unprecedented heat wave and competitors have production trouble, leaving them unable to meet market demand. The firm with the new juice product now has to decide which sales were the result of the heat and the lack of competition, and which sales were the result of an appealing new product. Market researchers work hard to control for these kinds of events during the design of the test market experiment.

COMPETITOR ANALYSIS

A thorough understanding of the competition is an important part of formulating any marketing strategy. Although it is usually unwise to simply react to competition, it is equally unwise to ignore competitors. Even the most dominant firms are often vulnerable to competitive encroachment from niche players, international competitors, or more traditional competitors with new technology or products. In fact, some marketing experts theorize that the larger and more successful a firm is, the more likely it is to fail. The argument is that large, successful firms often become complacent with their success, perpetuating the way of doing business that helped them become successful, and losing the edge that helped get them there. As markets and consumers change, old approaches become increasingly outdated, and smaller, more aggressive firms may begin to take over the market. Clearly, even successful firms must continually monitor competition.

One simple but effective method for analyzing competition is completing a formal strength and weakness analysis for each competitor. This tool simply involves listing the key strengths and weaknesses of a competitor in the same way firm strengths and weaknesses were analyzed earlier. With the list of strengths and weaknesses in hand, the marketer then focuses on the opportunities or threats each competitor presents. The final step in using this information is to develop a marketing strategy that capitalizes on the competitor's weaknesses and takes advantage of any opportunity that competitor vulnerabilities present.

TABLE **9-1** | **Competitor Strength and Weakness Analysis**

Competitor: Red River Tractor and Implement Company

Date: January 15, 2001

Compiled by: Prairie View Equipment Company

Strengths

1. New store
2. Good location
3. Large parts inventory
4. Quality "short lines"
5. Favorable credit terms
6. Manufacturer-owned store has sound financial backing

Weaknesses

1. High overhead cost
2. No outside salespeople
3. Poor on farm repair service
4. Major equipment line not well accepted in local market
5. Manufacturer-owned store less flexible to meet local needs
6. New manager not well known in the area

Strategy for competing against Red River

- Emphasize importance of dealing with a local business, keeping money in community
- Upgrade emergency farm repair vehicle and promote this service
- Hold open house to familiarize customers with our location
- Promote local owner's long-term interest in community and understanding of local needs

In the example shown in Table 9-1, a local farm machinery company, Prairie View Equipment, has noted several key facts about Red River Tractor and Implement that can have important implications for Prairie View's marketing plan. Based in part on a careful assessment of Red River's business, the locally owned company Prairie View has developed a strategy that emphasizes its local ownership and flexibility in servicing farmers.

Information on competitors can come from a variety of sources. An Internet search will likely turn up a considerable amount of information on any major food or agricultural input firm. Annual reports, SEC (Securities and Exchange Commission) filings, company press releases, and speeches by company executives can all offer useful information. Information supplied by salespersons or dealers can help keep tabs on other firms. Customers, industry experts, and other players in the marketing channel can all help provide a more complete picture of competitor activity.

The need for such information may seem so obvious to field marketing people that this exercise seems unnecessary. But, formalizing it in this manner, discussing it thoroughly, and developing a competitive strategy often improves communication, even in small firms, and results in a more proactive approach to marketing decisions. When all significant competitors are analyzed periodically, there is a far greater likelihood that a logical marketing plan will be developed and executed.

ANALYZING CUSTOMER NEEDS

The case for understanding customer needs has been made in Chapter 7 and Chapter 8. How do agribusiness marketers learn what customers want from their firm? There are a variety of tools available to the marketer to help them better understand the needs of their customers and prospects. Some of the more important tools will be reviewed here.

Customer Surveys

One of the best ways to learn what customers think about a product or service is to ask the customers directly. By communicating directly with customers through surveys, interviews, and informal conversations, agribusinesses can gain much valuable information, which can help in formulating market plans. Although there may be reluctance among some customers who do not want to be bothered or who distrust any information-collecting activity, experience shows that food and agribusiness customers are willing to share their opinions with firms they do business with, and when the data collection effort has been properly designed. When such information is used effectively, it is invaluable in making marketing decisions.

Market research relies heavily on surveys of various types to track customer attitudes. Extensive **personal interviews** lasting an hour or more may provide a great deal of information about customer purchasing decisions, attitudes, and how the product is used. Such in-depth interviews must be developed and performed by trained interviewers; they are therefore quite expensive, often averaging $100 or more per completed interview. In many cases today—especially if the interviewee is a large commercial farmer or a professional such as a veterinarian—the person being interviewed also wants compensation for their time while being interviewed. This can easily add another $50 to $100 (or more) to the cost of the session.

Telephone interviews have gained in popularity because of their lower cost and the speed at which a telephone interview study can be completed. However, it is impossible to collect as much information by telephone as compared to a personal interview because telephone interviews seldom last more than 15 minutes. It is difficult to explore many issues very deeply in such a short period of time. Phone interviews by professional market researchers may cost from $15 to $30 (or more) per completed interview.

Written surveys are probably the most widely used method of studying customer opinions and attitudes. Written questionnaires can be short or long, but longer ones usually get a much lower response rate. A quick written survey delivered through the mail is considered successful when 15 to 25 percent of the questionnaires are completed and returned. More formal mail surveys involving

phone follow-up and multiple reminders will generate a substantially high response rate—50 to 60 percent is common. Some firms are now using the Internet to distribute written questionnaires, and this medium can help execute a very
quick, very focused survey of customers. (The Internet can also be used as a
communication tool with consumer/buyer panels.) While there are advantages to
written surveys, it is difficult to collect much detail or in-depth reasoning behind
responses from a written survey. Yet because they are relatively inexpensive,
written surveys are considered a good way of collecting some kinds of market
information.

For all types of surveys, careful **pretesting,** or evaluating the questions so
that what is being asked is clear to the respondent, is important. Poorly worded
questions can cause the results of a survey to be virtually meaningless. The
meaning of a question may be clear to the author of the survey, but totally confusing to the respondent—or worse, it may be misunderstood, with the result
that it elicits incorrect information.

Focus Group Techniques

Focus group interviews are another method of learning more about customer
attitudes and one that is widely used by agribusiness marketers. The focus group
is a quick way of obtaining a reading on a group of customers. Depending on the
situation, focus groups can also be quite cost-effective.

A focus group interview is simply a discussion among six to ten customers
or other individuals of interest, guided by a skilled moderator. In many cases
this discussion is recorded so that careful analysis of the comments can be made.
It is not a formal interview so much as it is an informal discussion where people
get to know each other well enough to talk freely. As barriers to discussion
break down, participants begin to react to each other and the conversation unfolds, hopefully yielding much valuable information about how the group feels
about important marketing issues.

The key to success in a focus group interview is the synergistic effect of the
informal discussion. The group chosen for the interview is extremely important.
It must fairly represent the population of interest and it must be a group where
free-flowing discussion is possible. The moderator plays a very important but
extremely subtle role, making sure the right subjects are covered and that no one
individual dominates the discussion. The moderator must also be careful not to
bias responses in any way.

Focus group interviews can provide many new ideas and perceptions from
the customer's viewpoint about advertising programs, packaging, quality of
products, performance, and relative comparisons of competitors. Group interviews also add a great deal of support and help to clarify more quantitative surveys because they can suggest how strongly impressions are held.

However, focus group interviews can also be misused. They will likely generate very little factual or quantitative data. As such, results are very difficult to
tabulate or quantify. Some group members may tend to dominate and to influence the thinking of others, perhaps creating a "bandwagon" effect. Therefore,

any results must be interpreted with the nature of the group in mind—it is criti-cal not to extend the results of a focus group interview beyond their designed use. Still, focus group interviews offer a highly productive and relatively inex-pensive method for gaining insights into customer attitudes and for generating marketing ideas.

Internal Data Analysis

Agribusiness marketers can often get so involved in developing and implement-ing a marketing plan that they have little time for detailed market analysis. This is particularly true in smaller or local agribusinesses, where marketing people may have a number of responsibilities and have limited time to devote to careful exploration of what is going on in the market. One typically underutilized re-source for better understanding customer attitudes and preferences is internal customer and transactions data. By looking at the data a firm has on its own cus-tomers and their transactions with the firm, a marketer can learn much that will be helpful in developing or refining a successful marketing strategy.

Transactions data can provide a wealth of insights into a market. Sorting customers by their sales volumes can begin to help a firm understand the com-position of its portfolio of customers. Looking at sales across product lines may point out important cross-selling opportunities. Exploring the timing of pur-chases may allow a firm to improve customer retention. If a firm knows that the Hendricks Vegetable Growers always place their seed orders by January 15, then any delay past that date becomes a warning sign that the customer may be looking elsewhere for seed. In many cases, firm transactions data are obtained and stored solely for accounting purposes. However, the investment it may take to make these available for market planning purposes is likely time and money well spent.

Another tool for analyzing internal data is **market mapping.** It is common, for example, for much of the sales volume in a small agribusiness to be concen-trated in limited geographical areas or among a relatively few accounts. At the same time, other geographic areas have relatively little market penetration with many potential accounts left for someone else to serve.

In any given market, there may be physical reasons for an imbalance in market penetration. An interstate highway may make access to some areas diffi-cult. Certain types of farms may be concentrated in some parts of the market area. More often however, friends, relatives, or just well-established customers seem to predominate in one local area. Salespeople tend to spend less time in areas with which they are less familiar. In short, many agribusinesses have undiscovered market potential near them, if they were just able to discover and cultivate this potential.

One way of discovering these areas of untapped potential is through a market-mapping technique. This exercise can be done by hand in smaller firms. All cur-rent customers are pinpointed on a map of the market area (Figure 9-4). It is a good idea to color-code customers by volume of business, type of products pur-chased, or other informative characteristics when developing this map.

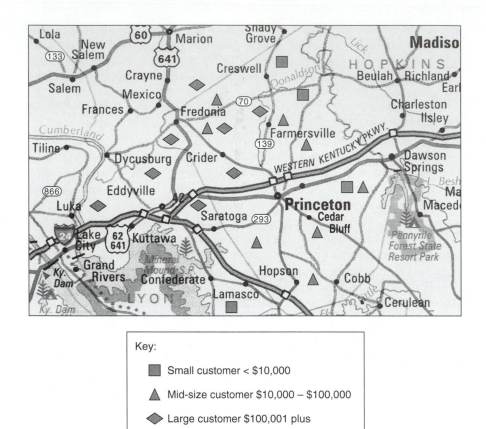

Figure 9-4
Market Map.

A market map can also be developed using any one of a number of computer software products designed specifically for this purpose. Recent developments in computer mapping software and geographic information systems (GIS) make it possible to download a variety of data on market characteristics such as number of pigs on feed, number of households with at least two children and family income of $60,000, and so on. Then, these data can be overlaid with store locations or competitor locations to get a graphic picture of the market situation.

Once this mapping has been completed, visual observation provides a great deal of information about weak and strong areas. When this mapping technique is first used, it is often a real revelation for marketing and management people. The market map suggests areas in which sales efforts should be targeted. When effort is concentrated in a new area and new customers are obtained, that market usually grows rapidly.

SUMMARY

Planning a marketing strategy requires a variety of techniques and tools to analyze markets and make decisions. Some tools are simple applications of logic, while others are highly sophisticated analytical models.

A marketing audit is a thorough and objective examination of a firm's marketing program. In a marketing audit, the firm's market performance and potential for each product are analyzed; current and future market conditions are reviewed; distinctive competencies (i.e., areas of unique strength relative to competition) can be recognized; current marketing programs are evaluated and execution of the marketing program is considered; and finally, the human resources available and their organization are studied.

Sales forecasting is a highly useful marketing tool that is especially valuable in agribusiness because of the volatility of agricultural markets. It usually begins with general economic forecasts and becomes more and more specific as the forecaster moves toward forecasting sales for an individual product or service. A variety of tools are used for forecasting sales of individual products including trend forecasts, build-up forecasts, analysis of scanner data, the Delphi approach, and test markets. Each of these techniques has advantages and disadvantages, which agribusiness marketers must understand to know when each may be the appropriate tool to use.

Marketing managers must also analyze competition and monitor their marketing programs. Formal competitor strength and weakness analysis is a useful tool that suggests strategic marketing approaches for capitalizing on the vulnerabilities of a competitor, or addressing a competitive threat.

Various marketing research methods provide invaluable information for developing marketing plans. Customer surveys and personal and group interviews are popular methods. Focus group interviews are a commonly used technique for collecting information on customers. Internal transactions data can yield many insights into customer behavior. Simple tools, such as pinpointing customers by type or size on a map, can also provide much information.

In the end, marketing managers must synthesize information from a variety of sources to assist them in developing a successful marketing program.

DISCUSSION QUESTIONS

1. What is the purpose of a marketing audit? Why is it important to have outsiders involved in marketing audits?
2. Why are sales forecasts particularly difficult to make in food and agribusiness firms?
3. What is meant by derived demand? Explain why the demand for animal health products is a derived demand.

4. Your cooperative is evaluating a new package of precision agricultural services they may offer to patron-members. This package includes an extensive field-mapping program, linking soil fertility information with yield monitor information, crop scouting in the summer months, and personal consultation in the winter. The package will require a significant investment of both financial and human resources. What data will the cooperative need to collect to make the "go/no-go" decision on this package? How should they collect this data?

5. What are the benefits of preparing strengths and weaknesses analysis on competitors?

6. What are focus group interviews? Are they of much value to small local firms? What are some important limitations of the focus group interview?

7. Give some of the reasons customer maps often show customers clustered in particular areas. What can be done about this situation?

CASE STUDY—A Big Assignment

"Yes sir, I have had an agribusiness course," Steve Turner said as he looked across the large desk. His new boss was sitting on the other side of that desk, and, frankly, Steve wondered what was coming next. He had just accepted a summer internship that would begin in a few weeks, but he had no idea what his assignment would be at Valley Co-op Producers.

"Do you know anything about marketing, Steve?" asked the manager.

"A little, sir," answered Steve nervously, still not sure what to expect.

"Well, we'd like you to prepare an analysis of our feed marketing program this summer, Steve. As you know, we have three divisions: feed, fertilizer, and general farm supplies—all managed out of our headquarters here. Although we want you to spend some time working in each division this summer, we'd like you to spend 2 days each week talking to people, looking at our records, or whatever else you feel would be helpful, to give us a good idea of where we stand in the feed market. Our feed sales have been dropping off lately, and I think it's because our equipment is getting old and our service is poor."

"The board of directors is considering expanding our feed business with a new feed mill, but we just aren't sure right now. There are a lot of cost factors that I'm working on, but I'd like you to take a look at the market side. Then we'd like you to make a report on what you find at our August board meeting. It's a big job, but we'd like to know what you can come up with."

Steve swallowed hard. He wanted to make a good impression—and it did seem like a great opportunity to learn a lot about the business. Yet he wasn't sure he was up to it. He wasn't even sure how to begin.

Questions

1. What kinds of information do you think Steve needs to find before he can complete his assignment successfully? How might he get this information?

2. What market analysis tools should Steve consider using to collect the data he will need for his report?

3. Outline a procedure that Steve might use to make his analysis and report.

10

SALES AND SELLING IN AGRIBUSINESS

OBJECTIVES

● Describe the role of personal sales in the marketing strategy of an agribusiness
● Discuss the unique challenges of selling in the food and agribusiness markets
● Identify the different categories of selling in the food and agribusiness markets
● Describe the primary activities of a professional salesperson
● Understand the different types of support that firms provide for salespeople
● Discuss some of the major characteristics of successful agribusiness salespeople
● Delineate the specific rewards and potential disadvantages of a career in sales

INTRODUCTION

When nobody sells, a terrible thing happens: Nothing. This poster on the office wall of an Indiana feed manufacturer aptly describes the importance of selling in the food and agribusiness markets. Personal selling is one of the most critical elements of a firm's promotion strategy (part of the marketing mix discussed in Chapter 8). An agribusiness firm's personal sales strategy involves the use of people to manage the business relationship with the customer. Sales are all about helping customers buy. This role may include such activities as communicating the value of the product/service/information bundle, executing the business transaction, and follow-up after the sale.

Note these types of activities can be accomplished in other ways—a salesperson does not need to be involved. A wide variety of media such as newspapers, radio, and direct mail can be used to communicate with customers. Transactions can be executed and orders taken over the Internet. Follow-up could be done via an auto-fax system that will send faxes with the desired product information to customers who call and request information. However, a personal sales strategy involves accomplishing these tasks, and others, in whole or part through people working directly with customers, prospects, and key influencers.

249

SALES AND SELLING

Personal sales strategies are developed at all levels of the food and agribusiness markets. When a salesperson visits a farmer to discuss a new line of herbicides, the firm is using a personal sales strategy. Likewise, when a salesperson talks to a local golf course superintendent about a new grass seed variety or a new piece of fairway mowing equipment, the firm is using a personal sales strategy. When a key account team from Kellogg's meets with the procurement group of Kroger, Kellogg's is using a personal sales strategy.

Personal selling clearly takes place throughout the many levels of the marketing channels for food and agribusiness products, not just at the final consumer or retail level. And, as customers, whether they are farmers, processors, or food companies, get larger and more sophisticated, the importance of personal selling increases. Likewise, as products and programs become more complex, personal selling becomes more important. Let's take a closer look at this important area in agribusiness management.

Sales-Profit Relationship

In its simplest form, selling is the act of transferring ownership of goods and services. Selling involves the trading of goods and services for money or business income. Professional salespeople help facilitate this process. The income statement emphasizes the importance of sales in business. This financial statement begins with sales, and then subtracts all costs associated with those sales, to determine the resulting profit (see Chapter 12). Profit, which begins with sales, is the primary (though not necessarily the only) motivation for the existence of the business. If the selling effort does not generate enough sales to cover overhead expenses as well as the expenses associated directly with sales, the business will be unprofitable and, over time, may fail.

As pointed out in Chapter 8, agribusiness firms use a host of strategies and tools for generating sales. Product development, pricing strategies, advertising, sales promotions, distribution strategies—all are aimed at helping a firm generate sales. But, for many agribusinesses, the professional salesperson bears direct and primary responsibility for sales, and therefore profits. This is particularly true in smaller firms, where the sales force may be the sole or primary means of promotion.

Selling, pricing policy, and firm profitability are also closely linked. In small firms, those who are selling are often authorized to conduct price negotiations with customers; therefore, they have direct responsibility for generating a gross margin that is sufficient to cover expenses and generate a profit. In larger organizations, salespeople may wield little or no pricing authority. In these cases, higher-level management assumes pricing authority, leaving salespeople responsible only for sales. In firms where responsibility for sales and profit become separated, or where those responsible for sales become enamored with simply generating more volume, it is possible for significant profit problems to arise. Successful firms are those that carefully coordinate the pricing and selling functions within their businesses.

What Is Selling?

Profits may begin with sales, but sales begin with activity! To sell is an active verb, which is to say that it involves doing something. Sometimes the selling activity simply "enables" the sale to occur, because the customer was already aware of, needed, and recognized the value of a particular product or service. On other occasions, potential customers may be unaware of the product or service's existence or unaware of their own needs, in which case the selling function begins at the fundamental point of first identifying the customer's needs and then helping the customer see that need. A cattle feeder may fail to recognize that the daily rate of gain of his or her livestock could be increased by 5 percent by changing the feed ration slightly and incorporating a particular feed additive. A salesperson for an animal health company may diagnose the problem, then present the cattle feeder with a product that will help them solve the problem.

In still other cases, a potential customer may recognize the basic need for a product, but may not understand how the value of a particular product or service can exceed its cost. One Washington (state) wheat producer feared that a new $150,000 combine would "cost too much" and so had not seriously considered trading equipment. But an aggressive implement dealer, focusing attention on the individual circumstances, showed the producer how the combine would save money by reducing labor costs and grain losses, and by saving precious harvest time. The salesperson made the sale and the farmer gained as well because he received the benefits of the new equipment.

It is important to recognize that selling does not create needs among customers. Rather it focuses customers' awareness of existing needs and then shows them how those needs can be met with the firm's product/service/information bundle. Selling must also arouse "wants"; that is, the customer must want the product or service in order for the sale to be made. Some people think of selling as being beneficial primarily to the seller, whose profits from the sale are obvious. Yet in a very real sense, both the seller and the buyer must profit from the transaction for the food or agribusiness firm to develop a long-term relationship with the customer. The buyer only trades dollars for products or services that are believed to be of greater value to the buyer than the cost (money, time, etc.) of the products or services.

Professional salespeople always work with the customers' best interest in mind. If the product or service will not genuinely improve the customer's situation, solve a problem, make them money, or increase their satisfaction level, then the sale should not be pursued. This is not only an ethical matter, but it is also a practical issue because if the customer does not believe they are better off after the purchase, there is little chance for future business. Or, even worse, the negative attitude of an unsatisfied customer may be vocalized to other customers and prospects, leading to a very difficult situation for both the salesperson and the firm.

In the end, the final decision of whether or not to buy is the customer's. After all, it is their money. But, professional salespeople are reluctant to sell products when they personally do not believe it is in their customer's best interests even

A professional salesperson is always
focused on the customer's best interest.
Photo courtesy of Pioneer Hi-Bred International, Inc.

though it might benefit the salesperson in the short run. The bottom line is that
the most successful food and agribusiness salespeople always work for their customers' best interests. When this genuine concern for the customer plays out, the
salesperson is rewarded with more sales in the long run.

Even in the best of circumstances, however, customers do not always benefit as much from a purchase as they had anticipated. When customers are disappointed, businesses often agree to make refunds or adjustments because they realize that the benefit to the customer must be greater than the costs. If this
doesn't happen, the business is likely to lose the customer. A good salesperson
helps to avoid this situation by not creating unrealistic expectations and by fully
educating the customer on what to expect. Most agribusinesses are well aware of
the fact that it is cheaper to keep existing customers than to lose them and then
attempt to find a new customer.

The benefit to the customer, especially for agribusiness firms, is often measurable in dollars and cents, just as it is for the seller. A Wisconsin dairy farmer
who spent $175,000 to purchase and install some new milk-handling equipment
expected to recoup her expenditure through cost reductions. Her calculations
showed an expected reduction of 40 worker-hours per week and a significant reduction in disease problems. These reductions would allow her to recoup the initial outlay within 4 years and show a return on investment of 10 percent over the
life of the equipment. A good salesperson will illustrate these benefits to a customer. Likewise, good customers will solicit this type of information from their
salesperson. Hence, the savvy salesperson is alert to these needs and has the
needed data available.

Sometimes the benefit to the customer cannot be expressed in dollars. Customer satisfaction or pleasure may take a much less tangible form than revenue increases or cost reductions. However, the idea is no less important. A consumer of a new high energy nutrition bar may be after a number of benefits—losing some weight, having more energy, enhancing their image as a person focused on fitness. A good salesperson knows that these benefits are just as real here as they were for the dairy farm example. And, even when the customer intends to use the product as an input for business purposes, the sale often hinges on some important element of anticipated satisfaction that is highly intangible. Pride of ownership, ease of use, safety, and a great many other intangibles are important to most customers, both farm and food. As pointed out in Chapter 8, such intangibles are an important part of the customer's perception of total value.

Marketing and Sales

While sales and marketing are often talked about together, they are very different activities. Reflecting on the description of the selling activity, we can begin to see how sales and marketing are related, and how these activities differ. As discussed in Chapter 7, a firm's marketing activities are longer term in nature. Agribusiness marketers focus on assessing the marketplace, considering what segments of the market to pursue, making decisions about all of the elements of the marketing mix. The focus of the marketing strategy is developing and supporting the firm's unique position in the market.

Selling is an important element of marketing—and a key dimension of the agribusiness firm's promotion strategy. In many companies, the sales force has the primary job for implementing the marketing plan. And, the sales force may be the most important part of the promotion strategy. **Selling** is defined as the act of transferring ownership of goods and services. It is the process of helping people buy. Sales is a short-term activity, looking for results this week, this month, this quarter. The sales force has to make things happen today. And, the sales force must act within the guidelines established by the company's marketing plan. In addition, the selling activity will be supported by other elements in the firm's marketing plan such as the product development effort, the pricing strategy, and other promotional activities. Sales and marketing can be very closely aligned in some food and agribusiness firms, and can be almost two distinct activities in others. However, the most effective agribusiness marketers align sales activities with the direction established by the marketing plan.

A recent development in the management of the agribusiness sales force is the use of marketing tools by the salesperson. Here, the salesperson still has the basic responsibility to make sales happen. But, the salesperson is asked to treat his or her sales area as a mini-marketplace, and use the marketing tools developed in Chapters 7, 8, and 9 to assess opportunities, and to develop and implement a sales plan that is very focused on the needs of the specific territory served. This approach is called field marketing, and represents a move toward even closer integration of sales and marketing activities by agribusiness firms.

Selling as a Process

Professional agribusiness salespeople are keenly aware that selling is an endeavor that is highly individualized. They will concede that there is no one way to sell. Instead, each sale is affected by four highly variable factors: (1) the salesperson, who has a unique personality, experiences, and skills; (2) the firm, which supports the sales activity in a very wide range of ways; (3) the customer, who has a unique personality, needs, and beliefs; and (4) the general situation, which will have its own unique background and circumstances. Any attempt to apply one rigid set of rules to every salesperson, firm, customer, and situation vastly oversimplifies the sales process, can be misleading, and often is just plain ineffective. The professional salesperson must be skilled and flexible enough to evaluate the diverse factors affecting each sale. The individual must use a highly developed sense of judgment in adapting to the changing situation throughout the selling process.

Selling is a process, and though each sale is different, there are general procedures and techniques that have been tried, tested, and proven by thousands of successful professional agribusiness salespeople over the years. Knowledge of these procedures forms the basis of successful selling. It is only after mastering the basic techniques that the professional salesperson can learn how to choose the most productive selling approach and adapt it to each unique selling situation in a creative, highly individualized way.

As indicated above, selling skills can be learned. While some people seem to have a greater affinity for selling than others, nearly everyone can learn the basic steps and skills involved in successful professional selling. Agribusiness professionals have found these selling skills helpful not only for actually selling products and services, but also for a much wider variety of responsibilities. One example of such responsibilities is the motivation and supervision of other employees through the process of "selling" them on ideas. Consequently, basic selling skills are a necessary tool to help the agribusiness manager perform more effectively.

Selling in Agribusiness: A Unique Challenge

Selling in agribusiness offers its own unique characteristics, which make it a distinctly rewarding kind of challenge. First of all, many agribusiness salespeople sell to farmers and ranchers, either directly or through sales to third parties such as dealers who then work directly with the agricultural producers. In either case the customer is likely to have a strong rural or agricultural orientation and value structure. Although it is difficult to pinpoint the precise nature of this "agricultural ethic," understanding it is crucial to the selling process in agribusiness when production agriculture is involved. This ethic tends to value practical experience and a familiarity with the language of farming.

Secondly, selling in the food and agribusiness markets typically revolves around long-term repeat business that is founded on high quality customer service. Most rural market areas can boast only limited numbers of customers, many of whom are members of a close-knit community. This is true whether the area is a local farming district or a larger dealer market that covers several hundred miles. Under these conditions, customer-salesperson relationships commonly span several years of steady service, and tend to change only when serious problems arise. (This is not to imply a low level of competition. Quite the opposite is true: agribusiness salespeople must work extremely hard to attract new customers.) It is not uncommon for a strong personal relationship to develop between the salesperson and the customer, once the customer experiences a sense of "being taken care of."

In food markets, the number of buyers is relatively small, the communication across the industry constant, and mistakes and successes are communicated quickly. So, the premium on exceptional customer service is present in these markets as well. And even though the accounts may be large and complex in food industry sales, personal relationships are still very important.

Third, selling in agribusiness generally demands a high degree of technical skill and knowledge of food and agriculture. The agribusiness salesperson often plays an integral part in customers' business decisions. The dairy marketing field technician, for example, provides farmers with crucial information about disease-control and milk-quality problems on the farm. The person selling processing equipment to a wheat miller is the expert on the design and implementation of automated milling systems. The selling strategy of countless food and agribusiness firms centers on this all-important aspect, the provision of state-of-the-art technical information and service by well-trained field personnel.

Fourth, most agribusiness sales jobs are extremely varied and seasonal. Each new season in the crop or livestock year brings a completely different set of responsibilities in serving the customer. For example, at various times the agricultural chemical salesperson may actively solicit new business, make early-order sales presentations, check out distribution problems, handle in-season technical product-use problems, work with end-customer complaints, collect payments on accounts, make market projections for next year, check inventory levels, and hold farmer-dealer information meetings. Likewise, much selling in the food industry must address seasonal peaks and valleys. The food firm salesperson may be involved in planning seasonal promotions, evaluating data on past promotional activities, monitoring inventory availability, and scheduling shipments.

Despite some unique characteristics, however, selling in agribusiness involves many of the same procedures and techniques that are common to the selling process in any field. These will be discussed in more detail in Chapter 11. But, understanding the specific challenges that the food and agribusiness markets bring allows the professional salesperson to deliver at an even higher level of performance.

TYPES OF SELLING IN AGRIBUSINESS

Agribusiness selling can be categorized either by the type of product or service being sold or by the level of the marketing channel at which the sale occurs. The latter method of classification is particularly useful, since the level at which a sale occurs (manufacturing and processing, wholesaling and distribution, retailing, farmer and rancher, or consumer) exerts a tremendous influence on the type of selling involved.

Selling to Manufacturing and Processing Firms

Agricultural manufacturing and processing firms generally take raw materials and manufactured inputs and combine them into products to be used by farmers, ranchers, growers, or the consumer of food and fiber products. Large numbers of industrial and manufacturing companies produce industrial inputs for the manufacture of agricultural products. Our study of agricultural selling at this level will concentrate only on those firms with unique characteristics which are the result of the firm's focus on agriculture and which are reflected in the firm's operations. Feed manufacturing firms, seed processing companies, meatpackers, and wheat millers are good examples of these types of firms. Such firms, though industrial in nature, require the salesperson to possess a unique understanding of agricultural practices, technology, and trends to be effective.

Selling at this level tends to involve complex products and services, and complex relationships between the buyer and the seller. Successful salespeople at this level often possess excellent technical skills that facilitate their frequent communication with product users, often engineers. Very often the manufacturer or processor will maintain very close agricultural ties and must continually evaluate the production agriculture sector as they make management decisions. As a result, a specialty equipment manufacturing company producing citrus processing equipment for packinghouses, for example, uses salespeople who demonstrate a keen understanding of engineering mechanics, fresh fruit physiology, and general agriculture.

Salespeople who sell to these firms usually deal with purchasing agents, engineers, and other highly specialized personnel. They often get to know several people at the firm, especially those who actually use their product. In many cases, because of the complexity of the relationship, the salesperson may be part of a multiperson team that represents the firm with the customer. The team might include the salesperson, a technical specialist, a credit manager, and someone from upper management. This team selling makes for a far more complex relationship building activity by the salesperson and the total sales team.

Salespeople who call on agricultural manufacturing and processing firms usually cover a wide geographical area, perhaps several states. They often travel by air because of the broad dispersion of potential customers, and they may be on the road for several days at a time. Sales are often large but infrequent. Manufacturing and processing firms commonly agree to contract sales for a lengthy period of time, to guarantee consistent supply shipments. The salesperson calls as frequently as required to service the account; solve any shipping, billing, or technical problems; encourage the use of the product; and maintain the relationship.

Selling to Distributors and Wholesalers

The agricultural input manufacturer may sell to distributors or wholesalers, who in turn move the product to dealers, or directly to farmers. Agricultural chemicals, for example, are generally sold to a distributor or directly to a dealer. Greenhouse and nursery supplies are usually sold to a wholesale distributor who carries a wide variety of products for use by nurseries and greenhouses. The food processor or manufacturer may sell to a broker, to a wholesaler, to a distributor, or directly to a food retailer. Complex relationships exist between major food manufacturing and processing companies and major food retailers. And, food manufacturing and processing firms will serve smaller retail chains through distributor networks. As indicated in Chapter 8, while direct relationships between manufacturers and processors and the retail level are increasingly common, almost all major agricultural input firms and food manufacturers will use distributors, wholesalers, and brokers to serve some market segments.

Salespeople who sell to distributors usually work with the buying agents who actually place the orders, but often spend considerable time with dealers and even the final users. They assume the responsibility of pulling the product through the marketing channel, which means that they funnel field orders through an appropriate distributor. In some firms, the salesperson never actually takes an order; instead, all orders go directly from the buyer to the manufacturing plant. Logistics is typically a priority focus of selling at this level. Shipping costs, inventory management, product availability, and timeliness are all issues of central focus to distributors.

A common feature of the salesperson's job at this level is building and staffing exhibits and displays at trade shows or conventions, or perhaps in a food store. In some product lines the salesperson may sponsor, arrange, and conduct final-user and dealer informational-promotional meetings. They usually spend a great deal of time working with their customer's clientele—solving problems and generating demand for their product.

Salespeople who work with distributors and wholesalers usually have offices in their homes. Some may travel a great deal, but because their geographical area is rather limited, they are seldom gone for more than one or two nights in a row. Others may have very intense travel schedules that take them across the United States working with their key distributor accounts.

Selling to Retailers

In agricultural input markets, generally the distributor or wholesaler sells to retailers, who in turn sell to the farmer. Agricultural pesticide distributors, who usually have their own sales force, are a good example of this. For food markets, the distributor/broker may be selling to the food retailer. The account representative of a grocery wholesaler is an example.

In other cases, the manufacturer/processor sells directly to the retailer. A good example of this is in the fertilizer industry. Farm equipment manufacturing representatives also generally sell directly to the local retailer. Likewise, many smaller food companies employ a sales force that sells and services retail accounts, taking orders, as well as assisting with inventory management and merchandising. In each of these examples, salespeople visit retailers in their territories on a fairly regular basis, and usually work closely with dealers to service the farmer-customer or the food consumer. They will often help the dealer or retailer by sponsoring local promotional programs, perhaps developing advertising materials, helping to train dealer-salespeople in technical matters, troubleshooting product complaints with farmers, and generally providing services that facilitate a better dealer-customer or retailer-customer relationship.

Similar sales jobs exist in larger manufacturing companies that own their own retail outlets. Company-owned stores are usually serviced or supervised by an area representative of the parent company, which serves much the same function as the salesperson calling on independent dealers. In some cases, area representatives may also have line management authority over several company stores, but their main responsibility is to assist sales through the retail outlet.

Salespeople who call on retailers usually do not travel large geographical areas, depending, of course, on the concentration of agriculture or the nature of the area, rural versus urban. Generally, these types of salespeople are home most nights (at least at some hour). They likely have offices in their homes and report to supervisors who may be located hundreds of miles away.

Selling to Farmers and Ranchers

In the retail food market, the sale to the final food consumer typically does not involve personal selling. However, in serving the production agriculture market, personal selling plays a fundamental role in building and maintaining the relationship between the input supplier and the producer. Selling at this level usually involves a great deal of close contact with farmers in a local area, and it generally entails on-the-farm sales calls and visits, which keep the salesperson very close to farming and the rural economy.

At this level of selling, sensitivity to specific farm problems that are relevant to the product or service is critical. As farms and ranches have grown in size, selling at this level has become much more like selling to dealers/retailers. Farms and ranches are increasingly complex organizations that are quite demanding of their suppliers. Successful salespersons develop strong business-oriented relationships with these accounts. It is common for the salesperson to develop close business and social ties to both farm and ranch customers, ties that often extend over an entire professional career. The technical and business expertise required of the salesperson at this level has continued to escalate. Large national companies and locally owned independent dealers pursue essentially the same set of sales responsibilities at this point.

The salesperson can expect to only rarely spend nights away from home. A company vehicle may or may not be assigned. The work is usually very seasonal, which means that peak agricultural seasons require very long hours.

Selling to Final Consumers

In some agribusinesses, a significant part of the business is conducted with non-farm consumers. Some examples include lawn and garden shops, nurseries, greenhouses, "pick-your-own" produce farms, and the like. These firms are an increasingly important part of agribusiness as interest in a rural lifestyle has blossomed. Such firms offer a variety of opportunities for those who are interested in agriculturally oriented selling. In most cases, consumers come to the business as they would to any other retail outlet, and the salesperson works at the retail location.

Successful selling generally requires a fairly technical yet practical knowledge of the product line. The hours are normally quite regular, though they must conform to customer convenience. Although much of the interaction with customers may be focused on their problems or questions, this represents a challenge to the salesperson that likes developing solutions, and enables such salespeople to reap the personal and financial rewards of accomplishment.

THE SALESPERSON

The food and agribusiness salesperson is directly responsible not only for generating revenue through sales, but also for providing the service that is the mainstay of a successful long-term relationship with customers, and for representing the company in the market area. The different facets of the selling responsibility are generally divided according to whether they require direct or indirect selling skills. Table 10-1 presents a job description for a typical entry-level feed salesperson position at a local cooperative.

Direct Selling Responsibility

Direct selling is the traditional and most basic function of the salesperson. The direct selling process involves prospecting for new customers, pre-call planning, getting the customer's attention and interest, making presentations, handling objections, closing the sale, and servicing the account. These steps may be carried out in formal or very informal ways. The direct selling responsibilities of a salesperson will be explored in detail in Chapter 11.

Indirect Selling Responsibility

Interestingly enough, **indirect selling** may take more of the agribusiness salesperson's time than direct selling. Indirect selling includes the service functions and follow-up provided by the salesperson. A long list of activities falls in this category, as shown in Table 10-1.

Market Intelligence   Market intelligence is a common responsibility of the salesperson. It may be a formally stated responsibility or an informal expectation, but in either case the company recognizes that the salesperson is in a position to know what is going on in the field. The salesperson is expected to keep the company informed about competitors' actions, market prices, product performance, the "mood" of customers, weather and crop conditions, key customer trends, and product inventory levels. Many companies even require a weekly marketing intelligence report to be prepared by their salesperson for the supervisor. This information is extremely valuable for supervisors and managers when they are making company-wide or division-wide decisions.

Technical and Product Information   Being well versed on technical and product information is a fundamental part of a salesperson's job. Customers expect professional salespeople to be fully knowledgeable about the agribusiness's products and services. Therefore, most companies provide much detailed information to their sales force about the products and services they sell. The salesperson must

T A B L E 10-1 | **Job Description for Feed Salesperson**

Job title:

Feed salesperson

Purpose:

Responsible for feed sales in four-county trade territory.

Supervision:

Works under direct supervision of general sales manager and is held responsible for sales results in feed department.

Major areas of responsibility

1. Responsible for feed sales to livestock farmers in four-county trade territory.
2. Helps farmers increase their profits by advising them on livestock production techniques and practices.
3. Promotes co-op products and co-op image to improve customer relations.
4. Sells and supervises livestock contacts.
5. Responsible for feed sales at a profit for co-op.

Duties

1. Sells complete feeds, supplements, premixes, and animal health products to livestock farmers in four-county territory.
2. Works with elevator manager in sales territory.
3. Sells co-op benefits, services, and feeding programs to customers.
4. Follows up on sales to make sure farmer is satisfied.
5. Helps farmer maintain proper records.
6. Plans promotions to increase feed business.
7. Consults with customer on sanitation, medication, waste management, and other special problems.
8. Develops, promotes, and supervises all contacts.
9. Makes sales calls daily and reports results to sales manager.
10. Prepares and maintains an up-to-date prospect list.
11. Handles any feed complaints.
12. Checks competition for prices and sales practices and reports results to sales manager.
13. Reviews monthly sales and tonnage figures with manager.
14. Helps develop annual sales budget.
15. Follows pricing and credit policy established by general manager.
16. Knows and interprets any regulatory issues.
17. Assists other departments by referring leads immediately.
18. Plans and schedules use of time.
19. Promotes and develops good customer relations.
20. Controls company car and cell phone expense.
21. Attends sales and product meetings as directed by sales manager.
22. Studies and recommends ideas to improve internal operations and profits.
23. May perform other duties as directed by immediate sales supervisor.

Goals

1. Increase sales over last year by 20 percent.
2. Increase feed profits by $10,000 over last year.
3. Make five sales calls per day.
4. Hold four feeder meetings during year.
5. Develop new sales and promotion ideas by January 15.
6. Keep sales manager informed of results.
7. Pick up 10 new feed customers this year.

become an expert on the company's products, and understand how the product brings benefits to the customer, regardless of level in the marketing channel. Many companies provide intensive product information training for new salespeople during the first few months on the job and then regular ongoing training throughout their careers. Typically, salespersons are provided a wide array of resources in this area. And, many firms now provide such information via a company intranet so salespersons can access current product information using their computers.

Handling Complaints   Handling complaints is often the salesperson's responsibility. Though this function is rarely a favorite part of the job, it is usually not a major problem. Only occasionally does a salesperson encounter a really difficult customer. Most customers are quite reasonable, even when they have a complaint, and most companies have well-developed procedures for handling complaints. A well-handled complaint often turns into a real "plus" for the salesperson because it helps establish a deeper level of credibility and rapport with the customer.

Collections   Collecting accounts receivable may, infrequently, be part of the salesperson's responsibilities. An old agribusiness saying warns that: "The sale is not complete until the money is collected." A salesperson who sells indiscriminately without concern for when and how payment will be made can create real problems for the company. One account that cannot be collected can wipe out all the profits for an entire region. Even when customers pay late, the cost to the parent company is usually significant. It is not uncommon for firms to counsel sales representatives on the importance of collecting payments from customers who are late paying their accounts. Others often handle this activity within the firm, but the salesperson that handles the particular account is often at least indirectly involved.

Trade and Public Relations   Trade and public relations are also an important part of the overall selling effort. How the general public and potential customers perceive the company is often very much a function of how the salesperson represents the company. The salesperson may work to build the company image through involvement in some community activities, and frequently takes part in trade shows, exhibits, field days, product field tests, and educational programs. Participation at county fairs through livestock auctions remains a popular way for many agribusinesses to show community support. In food firms, trade shows may be an important part of the salesperson's responsibility. Someone other than the salesperson typically handles formal public relations in food firms. However, even here, a professional sales representative is always sensitive to this part of his or her position.

   To most people, the salesperson is the company, because he or she is their only direct contact with it. Every action or lack thereof by the salesperson is part of this overall impression or image. The more local the company, the greater the significance of this impression. Even the salesperson's conduct in personal and business affairs has an important reflection on her or his effectiveness as a salesperson. Whether or not this is "fair," professional salespeople must realize the situation for what it is, and adapt to it.

Recordkeeping   Recordkeeping, budgeting, and administrative duties are a part of most professional salespeople's job responsibilities. Most salespeople submit weekly reports on their activities, the number of calls they made, and market conditions. Usually, they must fill out order forms, take inventory, and collect special information for company surveys. Most salespeople are asked to prepare annual sales forecasts for their area and to develop their expense-budget requests. Salespeople must keep accurate expense records and fill out expense statements for reimbursement. All this is typically not as time consuming as it sounds, and when handled properly it can clear the way for more immediately productive activities.

Certainly the wide adoption by most companies of the personal computer has greatly facilitated this planning and reporting process. Both sales representatives and the company benefit by having more information in a timely manner. In many agribusiness firms, the laptop computer is a fundamental tool for the sales representative. Reports and records may be kept on the computer, orders may be placed over the Internet, and customer records may be accessed to prepare for sales calls. Salespeople are generally expected to be "computer literate"—using computers and the Internet to communicate with the company *and* customers.

Public Agency Contacts   Public agency contacts are frequently part of the salesperson's job. It is usually considered important to maintain a positive relationship and good communication with county extension educators, university researchers, or other public agency representatives, who may well be opinion leaders in the market area. These people are often an excellent source of current information and can be very influential with at least some customers.

Support for Salespeople

Most agribusinesses do not expect their salespeople to "go it alone"—there is usually a team of specialists to support the sales effort in a variety of critical areas. The team's primary purpose is to assist the salesperson in the field and to ensure an effective total marketing program. The actual number of specialists supporting the sales force varies, of course, with the size and nature of the company. Some large multinational agribusiness manufacturing companies have literally hundreds of support people involved in highly specific functions. Smaller, local, independent agribusinesses may have only a few support persons, whose duties tend to be less specific.

The salesperson's relationship with the support team (Figure 10-1) is not all one-way. The salesperson is constantly feeding information to the support staff as part of a regular working relationship. This information is an important input for increasing the support staff's ability to carry out its own servicing responsibilities.

The Supervisor   The supervisor is the most common and often the most valuable support person available to the agribusiness salesperson. Supervisors often carry such titles as "district manager," "territory manager," or "sales manager" and are ultimately accountable for all sales in their area of responsibility. Conse-

Figure 10-1
Relationship of Salesperson to Key Support Staff.

quently, they typically exhibit a great deal of interest in the success of their salespeople and work closely with them to enhance their chances of success. Salespeople usually maintain contact with their supervisors at least weekly, and even more often if the need arises.

The supervisor generally passes along a great deal of information to new salespeople, especially in the first few months on the job. Detailed procedures and product information are spelled out through much close contact. It is not uncommon for a sales supervisor to spend several days in the field, calling on customers and introducing the new salesperson to them, as well as providing the new salesperson with helpful information about selling skills and methods.

A Training Program   A training program is another common support service for salespeople. It is becoming increasingly common to give salespeople considerable formal training in selling skills, product information, technical training, and a general company orientation. Sometimes training is accomplished through self-study materials. At other times, it requires a few days to several weeks at headquarters and in the field with experienced salespeople before being assigned a territory. However, all companies still rely very heavily on on-the-job training.

Increasingly, business firms, especially larger ones, are utilizing programs of regular periodic training, even for experienced salespeople. These sessions are often conducted by professional company trainers or outside consultants, and they usually provide very useful re-tooling for the experienced salesperson.

Technical Support   Technical support is frequently available in the form of special technical experts who are on-call at regional or home office locations to troubleshoot for the sales force. Most farm equipment companies assign agricultural engineers to help their field representatives solve special customer problems. Agricultural chemical companies usually have entomologists and agronomists on-call to help with customer complaints or special use product questions. A cotton producer concerned about a specific disease problem may contact the salesperson who in turn calls on the company agronomist to visit the farm and assess the problem. Food companies may have logistics experts or merchandising experts available to support a complex customer relationship.

Marketing Department   The marketing department often develops special promotions that support the activities of salespersons and special incentives that salespeople can use as aids to selling. For example, horticultural supply and agricultural seed production firms frequently develop "early-order programs" and special price incentives to increase business. Investments in advertising and promotion support the salesperson by creating customer awareness of products or services. National and local advertising programs often open the door for salespeople and make their jobs easier. In food companies, one of the salesperson's primary jobs may be coordinating national promotions with in-store activities. The task is to help the customer maximize the value of the investment in marketing made by the manufacturer.

In some industries—animal health products, for instance—the advertising department even develops special advertisements that can be personalized to fit a particular dealer and be placed in local newspapers, or on radio or television. These ads are often educational in nature—or may simply promote the product or the dealer. It is not uncommon for the parent company to split the cost of local advertising with the dealer. In this way, advertising can become an important sales tool for the salesperson.

Research and Development   Research and development, especially in larger manufacturing firms, provides important indirect support for the selling effort. It is the responsibility of the research and development staff to find and develop products and services that will meet customer needs. Their efforts provide the salesperson with the basic product; and the better the product, the easier the salesperson's job. The research department that makes a breakthrough in the development of a new package for displaying case-ready meat can make the job of a salesperson for a branded meat company very exciting and productive. Likewise, a steady stream of innovations for a feed ingredient firm gives the salesperson new opportunities to bring to their customers.

Credit Department   The credit department is often viewed by salespeople as more of a hindrance than a supporting resource, primarily because a salesperson who does not have to be concerned about credit approval for the customer could probably sell more. Yet, because collecting the money for a sale is an essential part of selling, the credit department is an essential part of the sales team. The policies established by the credit department are usually valuable in keeping the salesperson on solid ground financially. The successful salesperson understands that the job of selling isn't done until the customer is pleased and the bill is paid.

What It Takes to Be a Professional Salesperson

Many studies have attempted to determine what it takes to be a good professional salesperson, but none is really conclusive. For every characteristic that good salespeople appear to have in common, there are many exceptions. The personal qualities needed seem to be highly individualized, depending on how each person uses personality characteristics in selling situations. For example, contrary to popular assumption, some excellent salespeople tend to be rather introverted, that is, not particularly talkative, which is far from the stereotype of a typical salesperson. Here, even though their personality types are often extremely careful, dependable, conservative, they may over time develop strong relationships with customers. While it is impossible to define precisely what it takes to be a good salesperson, there are some general characteristics that successful salespeople seem to have in common.

Determination and Desire   Determination and desire, which represent an internalized commitment to accomplish an objective, seem to be common traits among successful salespeople. Desire, along with hard work, often accomplishes wonders in overcoming deficiencies in sales technique. Many of the traits of good managers discussed in Chapter 2 also apply to successful salespeople.

Self-Motivation   Self-motivation, or the ability to actually pursue an objective through self-discipline, is critical to successful salespeople, and greatly complements determination and desire. Most sales jobs allow the salesperson a great deal of freedom without close supervision. Much depends on the individual salesperson. Seldom is there a specific time schedule or routine to follow, or someone to tell the salesperson specifically what to do and when to do it. A good salesperson must decide what to do and then aggressively pursue the fully internalized goal.

Enthusiasm   Enthusiasm is an important element in successful selling. Probably nothing is so contagious as enthusiasm (except apathy). Salespeople who are excited about their product or service and what it can do for the customer have a real advantage. Enthusiasm does not necessarily mean extreme, outgoing behavior; even quieter people can display a sincere belief in what they are doing. Customers are quick to perceive how the salesperson really feels about the product or service.

Ability to Work with People   The ability to enjoy working with people is another characteristic common to successful salespeople. Selling is a people-oriented business, and a salesperson must work with a wide range of personality types and value systems. While it may be natural to like some people better than others, it is also necessary for the salesperson to feel comfortable with many kinds of people.

Self-Improvement   Constant self-improvement is necessary for successful salespeople. The job frequently changes and requires new and different approaches. Technological changes can quickly outdate the knowledge of the

Enthusiasm, determination, and confidence are all characteristics of a successful salesperson.

salesperson who is unwilling to spend time keeping up. Effective salespeople must stay ahead of their customers, and must continually search for ways to perform more professionally and skillfully.

Sensitivity to People   Sensitivity to people is critical to success. A good salesperson is able to hear what people feel as well as what they say, that is, to pick up on the small clues that are silent but clear messages, to empathize with the customers, and to identify with their needs. A good salesperson has a genuine interest in and cares for others as individuals, not just as prospective buyers.

Intelligence   There is no type of selling that does not require good problem-solving skills—analyzing, synthesizing ideas, and making logical decisions. Successful selling often combines technical expertise with a large dose of good sound problem solving.

Honesty   Honesty and a sense of ethics that is in tune with that of the clientele are essential to survival as a salesperson. Nothing can undermine a salesperson's success as quickly as questionable integrity. Not only is it important to be honest and ethical, but the salesperson must strive to appear honest and ethical as well. Reliability is another part of this characteristic: can the salesperson that makes a promise be trusted to follow through? In fact, research studies show that honesty and integrity are the top characteristics that customers want from their salesperson.

Ability to Communicate   The ability to communicate ideas clearly to others is invaluable. Salespeople do not need to be great public speakers. What is important is being able to convey ideas clearly so that others can understand them, and to listen to and understand what others are saying. Communication must flow both ways in the selling process. Good sales representatives understand their customers.

Neat Appearance   A neat appearance also carries some importance. Fairly or unfairly, many customers' perception of a product or service is influenced by their perception of the salesperson's appearance. Appearance is especially important in a customer's "first impression." If appearance is sloppy or ill kept, it is hard to convince the customer that this is really a professional person. Maintaining a neat appearance does not require buying expensive clothes or following the dictates of high fashion—rather, it means making a consistent effort to dress to fit the occasion. Given the wide range of definitions of "business attire" or "business casual," this means being sensitive to the specific dress code of the customer. The important thing is not how the salesperson perceives himself or herself but how he or she is perceived by others.

REWARDS AND CHALLENGES OF SELLING AS A CAREER

In the past, selling has suffered from an unfortunate stereotype, that of the fast-talking, pushy, unethical salesperson who will do anything to make a sale. The old joke said that an agricultural salesperson was a person who would "sell a farmer a milking machine and then take their cows as down payment." But today's professional food and agribusiness salesperson is a different breed. They are well trained, technically competent, concerned about long-term relationships with their customers, and have a high level of business and management savvy. As a result, they can look forward to a number of rewards.

Financial Rewards

Because of the importance of sales to any organization, professional salespeople, especially productive ones, generally earn good incomes. Young people who graduate from a university, college, or technical agricultural program with a record that indicates skills in leadership, dealing with people, and superior communication skills are usually considered to have selling potential. Opportunities and starting salary potential for these people are often considerably above average compared to other employment alternatives. Technical training and work experience with agribusiness is especially valuable in agribusiness selling. For more experienced salespeople, performance records are the primary factor in determining income.

Many agribusiness selling jobs feature some type of direct financial incentive, either a commission on each sale or a year-end bonus based on sales. Incentive programs appeal to many professional salespeople, because they feel such programs are a direct reward for their successful efforts. For some types of selling jobs, however, direct incentive programs are difficult to set up. For example, when the salesperson's job is primarily to assist and work with dealers who actually buy through a distributor, as is the case with most agricultural chemical companies, salespeople are generally paid a straight salary. In such cases, salespeople's salary increases or merit pay are usually based on their supervisor's evaluation of their performance. They may also share in the company's profits at the end of the year. In nearly every case, salespeople find their financial rewards tied very closely to their performance.

Advancement

Closely associated with the financial rewards of sales is the opportunity for advancement. Salespeople who are aggressive and productive are usually offered opportunities for increasingly more responsible sales positions (Figure 10-2). Initially, this may mean a larger territory or a greater number of important customers, but it may eventually mean moving into sales management and supervising less-experienced salespeople. Commensurate pay increases usually accom-

Figure 10-2
Professional Selling Career Path.

pany such advancement. In larger organizations, the advancement path is usually well defined and measurable, and requires several geographic relocations by salespeople as they move up through the sales organization. In smaller, more local companies, advancement is often not so formal in terms of title and peer prestige. The opportunity for professional growth in the smaller, local organization comes in the form of broader market assignments, community respect, development of close working relationships with customers, and added responsibility in the company.

Successful experience in professional agribusiness selling can also lead to opportunities for advancement into general management or other areas of the business. Most firms value field selling experience highly. Some firms require at least some field selling experience as a prerequisite for any important management position, except for positions in highly technical areas. Consequently, field selling is often a good stepping-stone into many areas of staff and line management, especially in marketing firms. In fact, some marketing oriented agribusinesses insist on sales experience before promotion into administrative or marketing management positions.

The tendency (especially among young people) to view sales as only a stepping-stone to something else is becoming less common. Many companies even build several different levels or designations into their field sales force structure to ensure that successful, experienced career salespeople carry titles that give them appropriate status and allow them to reap the financial rewards they deserve.

Personal Rewards

Many professional salespeople argue that the greatest rewards from selling are personal. They place a high value on the lifestyle associated with their jobs, and derive great satisfaction from the challenges of making a sale and closing the deal. The number of satisfied customers and the dollars generated by their efforts are viewed as tangible evidence of a job well done. They feel comfortable with the flexibility of their schedules, their freedom to organize their own time as they believe is most efficient, and their essentially autonomous operation as business managers.

They often speak of the close customer-salesperson relationships that have developed over time. Helping customers to solve their problems in areas that relate directly or indirectly to their company gives them a great deal of satisfaction, and being regarded as an expert in an important part of agribusiness builds their self-esteem. But to many, the ultimate satisfaction comes from the opportunity to work closely with food and agriculture and all the intangibles therein, an advantage that can offset all but the most extreme challenges.

Challenges

No description and discussion of a career path can be complete without some indication of the career's potential challenges, and professional agribusiness selling is no exception. It is, for example, not uncommon for a salesperson to invest a great deal of time and effort on a proposal or the preparation of a sales presentation, only to have it flatly rejected. Rejection is part of the job, and though no one asserts that it is pleasant, the true professional has developed skills and techniques for handling it graciously. It is important that these rejections not be internalized and taken personally. A successful salesperson accepts this as part of the sales career and goes on.

Some people who enjoy sales find the time required to do a good job is in direct conflict with their personal and family needs. Much of agriculture is a highly seasonal business and it is not uncommon for some salespeople to spend 60-plus hours per week or more on the job during peak season. Some sales jobs require considerable travel. While this may be an exciting opportunity for some, others find it difficult to balance the travel with their personal life.

Likewise, customer complaints are completely avoidable only if a salesperson never sells anything. Because most agribusiness salespeople are not hourly employees, they often have to rely on self-motivation, which is far easier to muster when the outlook is rosy. Though salespeople work closely with others, in many cases they are geographically isolated from their peers, other salespeople in their own company, and even from their supervisors.

In short, some aspects of the job can be considered serious drawbacks, depending on the individual's needs and lifestyle. Those who find selling rewarding tend to place the majority of their emphasis on the many positive aspects, and recognize that all jobs present some unique challenges.

SUMMARY

Personal selling is an integral part of the marketing strategy of most food and agribusiness companies. All food and agribusinesses depend on the sales process to generate income. Without adequate sales, a business cannot survive.

Selling is the process of helping people buy. It links the customer to the business. Selling has an important educational dimension because it helps customers become aware of alternatives that can benefit them. Whenever an exchange takes place, both the seller and the buyer expect to benefit.

Professional selling in food and agribusiness seldom means superficial door-to-door contacts. Professional selling is often focused on building a long-term business relationship that is based on a high level of technical ability. In some situations, it is a highly individual job in which each person is responsible for his or her own performance. In other situations, selling is a team-based activity with each member expected to bring his or her unique talents to the sales team.

There are many levels of selling in food and agribusiness. Some people sell only to manufacturers and processors, others to distributors and wholesalers, or to dealers or food retailers, while others call on farmers. Still others deal directly with the general public. Each level has its own unique characteristics, but selling at any level utilizes professional selling skills that involve close working relationships with customers.

The salesperson has a variety of responsibilities in addition to actual selling. Salespeople must monitor the market for information required by their company for decision-making purposes, provide technical support for their customers, handle complaints, collect accounts, keep records, and maintain good relationships with public institutions in their area. But salespeople have a lot of support in successfully fulfilling these responsibilities. Most salespeople receive special training, have technical specialists to call on for help, can utilize company advertising and promotional programs as selling aids, and can count on their supervisor to assist them when problems arise.

There are a great many personal attributes that are important to successful selling, but there is no one set of characteristics that is necessary for success. Some people seem to be naturally inclined toward sales, but nearly everyone can develop the skills necessary to sell successfully.

There are many rewards associated with selling. Financially, the selling profession can be especially rewarding. It is also an excellent route for advancement and personal satisfaction. But selling is hard work, and it can be discouraging when things do not go well.

DISCUSSION QUESTIONS

1. Before you read this chapter, you probably had an opinion on some of the characteristics of a "typical salesperson." List at least five of those characteristics. Now with a better view of the true nature of agribusiness salespeople, what do you believe are three of the most important characteristics of today's successful agribusiness salesperson?

2. Based on your personal experiences and observations, how do salespeople vary from industry to industry? Why do we see these differences? Discuss some of these differences for:
   a. A salesperson at an electronics superstore
   b. A field salesperson at a local agricultural cooperative
   c. A car salesperson
   d. A loan officer at a bank
   e. A salesperson for an advertising agency

3. Compare and contrast the role of a salesperson for a large food or agricultural input manufacturer with that of a local salesperson for a dealer or retail organization. If possible, interview these two types of salespersons. Where do they spend their time— that is, what types of activities make up their week? Who do they report to? What type of records are they required to keep? Without getting specific, how are they paid?

4. This chapter identifies some of the important personal characteristics of professional salespeople. Select a local food or agribusiness salesperson and discuss the list of personal characteristics with them. Can this individual help you identify additional characteristics for this list? Which characteristics do they feel are most important? Which do they feel are least important? Why?

CASE STUDY — **An Opportunity for Brad**

Brad Johnson had just returned to school after working as a summer intern for a large farmer-owned cooperative near his hometown. It had been Brad's first real exposure to agribusiness, except as a customer, when he would pick up supplies for his family's farm. He had spent most of the summer filling in for people on vacation, which had given him the opportunity to take a good look at the total business. Of course, it had been hard work and long hours, especially early in the season, when the bulk of the work had been fertilizer delivery and application. He had not been particularly interested in the retail fertilizer business before, but the overtime hours had allowed him to save the money he had needed for the fall term. In fact, the money had been his primary reason for taking the job.

The internship had made a real impact on Brad's career plans. He had always thought he wanted to farm. His father had said he could come back to the family farm, but Brad knew that would mean acquiring additional acreage. He felt his father wanted him to come back, but his father would never really say so. Brad's younger brother also liked farming. Brad had enjoyed the summer experience so much that he was pretty confused regarding his career choice.

Then, at the end of the summer, the general manager had given Brad some real encouragement. He had made it clear that they were pleased with Brad's performance and were interested in talking further about developing a full-time field technician position for Brad in fertilizer and crop protection chemicals. If it worked out, Brad would be responsible for sales to farmers, making recommendations about crops and fertility, taking grid soil samples, working with the firm's site-specific agriculture program and their GPS (global positioning system) equipment, holding grower meetings, and helping with custom fertilizer application during the busy season.

But Brad was uncertain. Although he had taken several general agriculture and agronomy courses, he did not feel he was ready to tell a farmer who owned a large tract of land what his fertilizer-chemical program should be. Another concern was selling. Brad had always had a poor impression of salespersons. Even if he did take the job, he was not sure he had the skill to communicate with experienced farmers. He hadn't even turned 21 yet and the prospect of a sales call on an experienced farmer was pretty intimidating. He became apprehensive when he thought about standing in front of farmer customers at a meeting. Still, it might be exciting, and the salary the manager had mentioned sounded good.

Brad was keenly aware that he had only one more year of school, and that some of his courses were locked in. But he had some flexibility, especially in his last term, to take several elective courses that might more adequately prepare him for this position. It was a big decision, determining the best course of action.

Questions
1. What are the pros and cons of Brad's possible agribusiness opportunity?
2. What things could Brad do during his final year to prepare him more fully for the job opportunity? (Consider both academic and nonacademic possibilities.)
3. What suggestions could you make to help Brad keep his options open?
4. What are some important questions Brad might discuss with: (a) his father; (b) his manager; and (c) his academic advisor, to make his decision easier?

11

THE SELLING PROCESS

OBJECTIVES

- Understand all of the steps in the selling process, from preparation and backgrounding to follow-up
- Develop methods of locating and qualifying prospective food and agricultural customers
- Develop specific objectives and strategies for sales calls
- Develop a strategy for approaching customers on a sales call, building rapport, learning what a customer's needs are, and generating interest in the product or service
- Learn the steps to deliver a sales presentation that will help the customer understand the product's benefits
- Identify common objections customers have when buying, and techniques for handling such objections
- Understand the critical task of closing the sale, and techniques for accomplishing this task
- Understand the importance of follow-up service and how this effort may increase future sales

INTRODUCTION

Professional selling is an exercise in individuality, expression, and personal interaction as well as a process with logical, measurable goals and results. It is the human element of selling that may mean that each interaction is unique—even repeated interactions with a single client. Because of its seasonal and biological nature among other unique characteristics, selling in the food and agribusiness industry offers individuals both challenge and opportunity. The products and services offered are highly varied, ranging from a highly technical piece of harvesting equipment to a new beef snack. But, through careful examination of some of the basic principles of selling, one can determine strategies to capitalize on the unique selling environment offered in the food and agribusiness industry.

Selling involves a process that can be learned by applying certain principles. Today's food and agribusiness sales professionals use these special selling skills in a variety of circumstances to improve the efficiency and probability of sales.

Careful and thorough preparation for a sale includes knowledge of the market, the product, and the prospective client. The sales process includes discovering customer needs and wants and communicating to customers that the product or service offered can meet those needs. This chapter examines some of the key concepts in planning a sales call strategy that will improve the salesperson's chances for success.

GETTING STARTED

Bob Gray was quite pleased that his youngest son, Mark, had decided to join the family seed business upon graduation from college in December. It was a tough decision for Mark. He had completed two successful internships while in college. One summer work experience was with a national agricultural chemical firm and the other with a leading farm equipment manufacturer. In both cases Mark was able to work in the sales and follow-up service area. The dilemma presented to Mark recently was the very attractive job offer he had received from the equipment manufacturer. Should he accept this job, gain additional experience, and join the family business later, or decline the offer and join the family business now? He chose the latter.

Bob, although very pleased with Mark's decision, realized this meant the business would have to expand to support three families—his own, Mark, and Mark's older brother Matt, who had joined the family business about four years ago. Matt's background was more technical and his degree was in agronomy. His skills complemented his father's, so their areas of expertise dovetailed nicely. The firm had adopted the strategy of slowly expanding over the last 4 to 5 years. Mark, however, didn't bring many new technical skills to the firm, except his proficiency with computers.

Bob Gray decided to have a planning session with his two sons to discuss the future of the business. Bob had started the firm about 25 years ago and had earned a reputation of producing high quality seed and being fair with his independent growers and honest with his customers. Although there was significant competition in the area, the intensity of crop production made this a profitable market. The firm was in good shape financially, but the three partners would have to construct a new marketing plan that would provide the foundation to allow the firm to grow.

After lengthy discussions, it became clear to Bob and his two sons that the firm must put greater emphasis on personal selling. It was clear that Mark had the personality and communication skills, some gained during his summer work experiences, to take on this new responsibility.

Up to this point Gray Seeds had relied primarily on word-of-mouth advertising to generate new business. Bob was always bothered by expenditures for advertising. He was never convinced this could do much to expand business in his local market area. Thus, he relied on satisfied customers and a few special "dinner" meetings the firm sponsored to bring in additional sales.

Now, however, it was clear the firm must be much more aggressive in the selling area. Gray Seeds had the facilities available to support additional sales. However, Mark, along with his father and brother, was not sure of the proper strategy the firm should use to increase sales.

THE DIRECT SELLING PROCESS

It is a common misconception (especially among those not involved in sales) that the selling process involves nothing more than talking to customers and getting an order signed. Reality is far different: the selling process often includes thorough preparation before a customer call is ever made. In addition, food and agribusiness salespeople devote a substantial portion of their time to servicing their accounts. There are some food and agricultural industries in which salespeople never actually take an order but spend their time assisting others who actually complete the sale, while other salespeople spend most of their time actually preparing and making sales presentations. Irrespective of how a salesperson's time is divided, he or she must never lose sight of the primary objective: to generate sales and increase revenue for the firm. This means that every contact with the customer is designed to maintain or expand business with that customer in some direct or indirect way. Not surprisingly, the ultimate objective of all selling activity is to sell!

The selling process embraces the various elements involved in preparation for the sales call, the actual sales call, and the follow-up activity after the sales call. Figure 11-1 shows the major steps in the sales process. Although these steps are not always followed in the exact order shown, this selling process is logical and one that is commonly used by effective food and agricultural salespeople.

Preparation

Many different kinds of preparation precede the salesperson's first contact with a customer. The salesperson prepares by becoming quite knowledgeable about the product or service, about the market, and about the customers being served. Without a thorough understanding of these important areas, the salesperson usually shows a lack of confidence that hampers effective selling.

This need for information is why many food and agricultural businesses budget considerable time and money to train new salespeople. Even the new salesperson who has just completed a successful academic program in an appropriate technical area of agriculture or food science still has to absorb a lot of practical and highly specific information about the individual company and its products. Many firms prefer to begin with only a brief orientation, after which

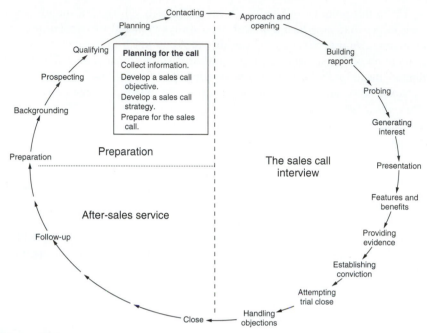

Figure 11-1
The Selling Process.

they send the new salesperson to the field, under close supervision—often working with a veteran salesperson. After a few months, when the new salesperson has had enough experience to put the additional training into perspective, the firm may follow up with more detailed formal instruction. There is nothing like a few complex technical questions from a challenging customer to bring home the importance of product and market knowledge.

Such training is only part of the salesperson's preparation for selling. Hours of the salesperson's time are frequently devoted to prospecting for customers, qualifying the prospects as they develop, and then making the initial contact that may result in a sale.

Backgrounding

Backgrounding implies much more than simply understanding the firm's product. It involves developing a keen awareness of the ways in which the market works, the ins and outs of competitive strategies, the strengths and weaknesses of competitors, the many customer needs and problems to be anticipated, the business and technical practices that are common among customers, and the unique characteristics of key regular and potential customers. Some of this background knowledge can be gained through formal study of company literature, but most of it has to be gained through experience. Consequently, most firms structure the first few weeks or months of work for a new salesperson to include frequent contact with senior sales personnel or supervisors.

But background knowledge is not exclusively the concern of new salespeople. It is an appropriate continual concern of all professional salespeople. Changing technology is a regular element of the selling challenge. Constant efforts to improve personal, social, and business skills offer another important dimension for professional growth. Professional salespeople expect to upgrade their knowledge regularly through independent self-study, company seminars, and daily experiences in the field. Salespeople who ignore the importance of this continual upgrading process usually end up with mediocre performance records because they are incapable of servicing their customers adequately.

Successful salespeople also keep abreast of their rapidly changing markets. The strategies of competitors can have a drastic impact on selling success; therefore, when a competitor initiates a new credit policy, alert salespeople quickly adjust their own selling strategies to counteract any negative effects. Similarly, salespeople keep informed about changes among their own customers, where the changes are occurring, and how these changes may alter customer needs. This is not quite so easy as it sounds initially, since many salespeople may have 200 accounts or more.

One useful technique for better understanding a salesperson's territory involves developing a detailed written analysis of market strengths and weaknesses, both for the firm and for key competitors. This written statement focuses attention on numerous areas of comparative advantage and disadvantage, and thus has important implications for sales presentations. Food and agribusiness salespeople should not overreact to the competition or allow the competition to determine their market behavior. Yet all salespeople need to be keenly aware of the relative strengths and weaknesses of their products, so that strengths can be emphasized and the effects of weaknesses minimized.

Database programs and contact management software can allow tracking of such detailed information at the customer level. It is typical for a small number of customers to account for a large proportion of a salesperson's sales (and profit). So, tracking extensive data on such important, individual accounts (customers) is important for successful selling. Other tools like market maps (discussed in Chapter 9) are also used by effective sales personnel to better understand their market territories.

Prospecting

Prospecting is the process of identifying and locating potential customers. Many different ways of prospecting have been devised, and their effectiveness depends largely upon the type of food or agricultural business and upon the skills of the salesperson involved. These methods generally fall into three broad categories: (1) cold calls, (2) cool calls, and (3) leads.

Cold Calls   Some highly successful companies rely exclusively upon **cold calls,** which involve simply stopping by farms or agribusinesses in the market area without any previous knowledge about the prospect. This method of prospecting can be quite productive even for companies that do not rely on it exclusively, since the law of percentages suggests that cold calls will eventually

Effective prospecting for new customers requires careful review of market data.
Photo courtesy of Pioneer Hi-Bred International, Inc.

turn up at least some prospects. Cold calls turned out to be very productive for one local certified soybean seed dealer in the Mississippi Delta, for example, and they were equally productive for a grain bin representative in central Iowa, suggesting that the method is particularly effective where there is widespread demand for the product or service. Although this kind of prospecting can be very inefficient, it can be very helpful in establishing new accounts or opening new territories. Thus, many firms strongly encourage at least some cold calling on a regular basis.

Cool Calls **Cool calls** involve calling on those you have little firsthand information on, but who have particular characteristics that are common to good customers and consequently qualify them as prospects. A feed salesperson once reported that canvassing her market area by first thinking carefully through the characteristics of her best customers, then looking for farms with similar characteristics, greatly improved her odds of identifying viable prospects.

This method of prospecting places the responsibility for identifying potential customers squarely on the salesperson's shoulders. One method for identifying potential customers relies upon visual observation. Machinery salespeople are alerted to observe the condition of certain types of equipment as they drive past farms and ranches in their market area. A good salesperson makes a mental note of what acreages different growers have planted.

Some commercial organizations sell lists of addresses for all the subscribers to certain magazines. These subscriber lists then yield names of potential users of specific products or services. These lists may contain details about the farm business that can generate excellent cool call opportunities. Noting attendance at local or regional education meetings that are sponsored by the Cooperative Extension Service, universities, or trade associations frequently helps to identify prospects.

Leads   **Leads,** or prospects that have been provided by knowledgeable sources, form one of the single most productive methods of prospecting. **Cool leads** are those potential customers whose need for a product or service may or may not be current. **Hot leads** are those potential customers whose need is believed to be current. Such customers may be definitely planning to buy, or they may simply have a current need and can be convinced to fill that need with the salesperson's product or service. Because hot leads are by far the most productive method of prospecting, most salespeople choose to follow up on such leads as soon as they discover them.

Good leads develop from a variety of sources. Successful company advertising usually results in numerous requests for information from potential customers. Trade shows and exhibits, which are common in many areas of food and agribusiness, often generate leads by identifying interested customers. In some agribusinesses, salespeople from one product division may learn that their customers are in the market for other company products, and may alert the other product divisions to the lead. A farm supply firm that already sells feed to a farmer should recognize the potential for acquiring that farmer's fertilizer business as well and aggressively explore these opportunities. This calls for close cooperation by different salespeople in the firm.

However, current customers remain one of the very best sources for leads. Satisfied customers often are willing, and even anxious, to share their satisfaction regarding the firm's product with their peers. Many customers feel that the wisdom of their own purchase decision is reinforced when their friends reach the very same decision. The sharp agribusiness salesperson can capitalize on this tendency by asking current customers about others who might be able to benefit from the company's products and services. The salesperson then uses this lead as an opener when contacting the new potential customer.

Qualifying Prospects

Common sense dictates that salespeople should make sure their prospects are qualified to buy before exerting much selling effort on them. Who to spend time with in the customer's business is also an important question. Agribusinesses or food companies with a purchasing agent or with one owner-manager may have a single clear-cut decision maker, although this is not necessarily so. For years, farming and ranching were thought to be male-oriented businesses, in which the decision was made by one person, but salespeople have recognized that the decision to buy often rests in the hands of husband-and-wife partnerships, or includes other family or nonfamily employees. Spending time with the wrong individuals in a customer's business is a waste of important sales resources.

Other factors that determine whether a prospect is qualified include the prospect's ability to pay. A prospect that wants and needs the product may not necessarily be able to afford it. Any customer whose financial condition may jeopardize collection of the account should be considered unqualified, or approached with a program and price that manages and reflects the risk of the account. It is usually the salesperson's responsibility to avoid potential account collection problems. If there is any doubt, the salesperson should avoid the account or consult with his or her supervisor.

It is also the salesperson's responsibility to determine whether the potential customer's need is well suited to the company's product or service. If not, then it is bad policy to sell to that customer because the resulting dissatisfaction may create bad publicity among more qualified customers.

Planning for the Sale

The **planning** step, or *preapproach,* involves the salesperson's preparation prior to actually making the sales call. This step is absolutely critical to the success of the sales call. The preapproach includes collecting information about the customer, developing a sales objective and strategy, and preparing for specific aspects of each sales call. The extent and formality with which the planning should be carried out varies widely with the situation. Where the customer-salesperson relationship is well established and the salesperson is highly experienced, the planning may be less formal. But underestimating the importance of the planning step can create very embarrassing situations, even for the seasoned salesperson. Some veterans relate stories about calling on customers and asking them what brand they are using, only to discover that it was their own!

Some people think all this preparation is unnecessary and wastes time. "Just go talk to them," they argue. But the best salespeople feel strongly that preparation pays off—especially for significant prospects that could generate a lot of business. A mistake on the first call can ruin the chances for a long-term relationship.

Sales are seldom made in the first call. Usually, selling in the food and agricultural markets involves a series of calls, each going one step further until the sale is actually made—and then ongoing calls to service the account and secure repeat business. So the sales plan is considered to be a learning process—a series of steps. The overall process begins with a long-term sales plan for getting a customer's business. This plan usually includes an unspecified number of sales calls and activities. Then, each sales call is planned—including a sales call objective and a sales call strategy. The key elements of a sales call strategy are presented next (see Figure 11-1).

Information   Much information about a potential customer goes into building a sales presentation. The exact nature of the information depends upon the circumstance, but almost invariably it includes the following:

1. **Information about the prospect organization.**

   Name

   General size of the business or farm

Brand of product or service now in use

Relationship with current suppliers

Problems or dissatisfactions with current suppliers

Past experiences and relationships with the firm

Name of buying authority (the buyer)

Ownership structure of the firm

Power structure in the organization

Image that the business promotes among its customers

Company philosophy, objectives, and strategy

Credit rating and ability to pay

Operating strengths and weaknesses

Company policies that may affect purchase decisions

Physical facilities that may affect purchase decisions

2. **Information about the buyer for the prospect organization.**

Age, education, and background

Social and professional interests

Personality type

Personal peculiarities and value structure

Hobbies and interests

This information is essential for developing a personalized sales strategy to be utilized with each customer. It can be used to tailor the opening, the interest building, and the type of presentation that is most likely to be productive. For example, one salesperson found that a large agribusiness customer had a buyer who was extroverted, saw herself as a community and industry leader, had never tried the salesperson's firm before, and had recently had shipment problems with her present supplier. At the same time, this salesperson handled a small account with a husband, wife, and two sons who ran their own business, were not engaged at all in community activities, were slow to adopt new techniques, had been doing business with the same supplier for the last 7 years, and saw no reason to change. Needless to say, the salesperson did not approach both potential customers in the same way.

Personality type, or social style as it is sometimes called, plays an important role when selling to a particular customer. There are a number of different models or frameworks for helping a salesperson better understand a prospect's values and buying behaviors. One framework classifies prospects/customers into one of four categories depending on how they make decisions and communicate with others. Figure 11-2 shows these four social style categories along with some typical characteristics.

Many people display characteristics of all four categories; however, there is typically one area of the grid in Figure 11-2 where most people fit, given enough

	Analytical (How is the problem solved?)	**Driving** (What will the solution do for me?)
	Traits: Deals with facts Logical Chooses words carefully	**Traits:** Direct Tells you how it is Challenges others' ideas Very professional Makes decisions quickly
	Wants: Service	**Wants:** Results
	Traits: Relies on what you think Easygoing, very agreeable Supports others' ideas	**Traits:** Expresses ideas openly Seen as more casual A dramatic communicator
	Wants: Support	**Wants:** Attention
	(Why is this solution the best?) **Amiable**	(Who else has used this solution?) **Expressive**

Figure 11-2
Typical Characteristics of Social Styles.

knowledge of their personality or background. There isn't a good or bad category, but determining the social style of your customer will allow you to present meaningful information during the sales call. For example, with an analytical prospect, you would not want to spend a lot of time telling the prospect that Rod Gumz, the number one farmer in the area uses your product, when the primary information the analytical wants is technical specifications and trial results conducted by a major university.

A buyer for a major food company who is a heavy user of price charting techniques, supply and demand figures, and other economic data might be classified as an analytical. A farmer who frequently asks What's the return? and consistently makes very quick decisions might be considered a driver. The farmer who bases his buying decisions solely on the recommendations of his chemical sales representative would be an amiable. Likewise, the food wholesaler who tells you his life story when you ask how things are going is probably an expressive. These examples are oversimplified, but should help make the point: knowledge of the "type" customer or prospect you deal with is critical in making an effective sales call.

Obtaining information about the customer can sometimes be a problem. Most people are somewhat averse to salespeople who appear nosy and ask too many questions. Some people may be quite uncomfortable when a salesperson seems to know a great deal about them, especially before a definite relationship has been established. So even obtaining the necessary information can be hazardous to the development of a potential relationship.

As with prospecting, much information can be obtained by observation. As the salesperson passes the place of business, he or she should simply note such things as the extent of storage space, the available equipment, and the brands and programs that are being used or advertised. Current customers with whom the salesperson has a good relationship are good sources of information about potential customers. Different impressions should be gleaned from different people, and a composite of the potential customer built from different sources. Other, noncompetitive professional salespeople who call on the customer may also be good sources. Still other sources include trade association directories, which often provide limited information about members, county agents, and sales supervisors. The World Wide Web (WWW) can be a superb source of information on business customers. Firm web pages can be literal gold mines of information, and a web search may turn up interesting and useful press releases.

Finally, the customers themselves should not be neglected as a source of information. Since so much of food and agribusiness selling revolves around technical problem solving, it is quite natural to go directly to the customer for the information necessary to make a sound proposal. Although the term preapproach refers mostly to preparation before the call, the ongoing nature of agribusiness selling suggests that calling on the potential customer for some technical information to help in the preparation of a major presentation may be quite appropriate.

Sales Call Objective   Before the salesperson can prepare a sales call, a sales call objective must be developed. In fact, a sales call objective should be established for each call made. The **sales call objective** is a tangible, measurable objective of what the salesperson expects to accomplish during the sales call. The salesperson that does not have a specific objective clearly in mind will find the planning process frustrating, and is likely to appear disorganized and unprepared to the customer. It is a good idea to actually write down the sales call objective. This technique may bring to mind ideas that the salesperson might overlook. And after the call has been completed, the written objective facilitates evaluating the success of the call. Many find that listing the call objectives along with the daily itinerary of sales calls is an efficient way to track performance.

The sales call objective need not be so straightforward as getting the customer to sign an order. Although the ultimate objective of all professional salespeople is to generate sales, a particular call may represent only one small part of that total process. The salesperson that visits a veterinary supply distributor to check current inventory levels for a new canine vaccine has a perfectly acceptable sales call objective. Convincing a prospective dealer to attend a special dealer management seminar where he or she will be exposed to many satisfied regular customers is a tangible sales call objective and a very practical step in establishing the relationship and ultimately getting the business. Occasionally, the sales call objective is simply to create goodwill. Although goodwill calls are acceptable, failure to develop sales call objectives for such calls sometimes results in a majority of calls that are little more than social in nature. Written sales call objectives help to increase the productivity of professional salespeople.

Sales Call Strategy   Each sales call should have a well-defined **sales call strategy.** This plan of action is designed to maximize the potential for success with the customer and make the most of the call. Early in the selling process the purpose may be to get more information, so the strategy may simply be to get the prospect talking. But later, when it is time to actually make a presentation or present a solution to the prospect's problem, the call should also emphasize product characteristics that are more advantageous than those of competitors. The strategy should include such things as the timing, location, and circumstances of the call, all of which should maximize the call's effectiveness. Is it best to set up an appointment or just to stop by? Should the call be at the customer's place of business or in neutral territory? What time of year, or what month or week would be best for the call? The successful salesperson is able to pose and answer these and many other similar planning questions.

The strategy should incorporate a degree of subtlety, so as not to antagonize the customer, especially in the early stages of the selling process. Sometimes there is a fine line between knowing enough about the customer to focus on key issues, and exhibiting so much knowledge about the customer's problem that he or she feels you are moving into areas that are "none of your business" or appear presumptive. The sales call strategy should capitalize on available information by orienting the sales call toward the most productive areas rather than simply trying to impress customers with the depth of information that the salesperson has acquired about their business. The discovery process should be allowed to flow naturally and smoothly. The information that has been gathered before the first call should be used to guide this process initially, and as the sales process moves along, additional information will suggest how to proceed.

Preparing the Sales Call   The process of actually developing specific plans for the call, including any materials that may be needed, is known as **preparing the sales call.** The more experienced the salesperson, typically the less time required to plan the details of the call. Yet this can be a very productive process, even for seasoned salespeople. It is a matter of selecting the proper approach and opening, planning transitions from rapport building to the presentation, selecting the most important benefits, anticipating objections and deciding the best way to handle them, and finally choosing a probable close. It is a matter of thinking ahead and planning.

Especially important calls should be mapped out in writing because this greatly enhances the quality of the planning process. The suggestion about formally outlining a sales call objective and strategy may seem like an academic exercise, but thousands of dollars may be riding on the effectiveness of the sales call. The tendency for inexperienced salespeople is to discount the value of planning. Conversely, even highly experienced salespeople spend considerable time planning for important calls. To ignore the potential value of proper preparation is unwise and can have very negative impacts upon success of the salesperson.

Making Contact

All the planning in the world may be wasted unless the salesperson calls on a prospective customer at precisely the right time and under the right conditions. Making contact can be very difficult. Food and agribusiness salespeople sometimes encounter strong prejudices against sales personnel; others find that prospective customers are just plain busy. Customers sometimes protect their privacy by surrounding themselves with such barriers as secretaries, receptionists, junior executives, assistants, hired workers, and spouses. Selling these "buffers" on the importance of taking up the buyer's time can be a difficult task. These intermediaries should be treated with respect and courtesy, since their importance can be difficult to assess.

Delay is an unavoidable part of any sales job, since buyers and important executives frequently make salespeople wait. Yet waiting too often, or for too long, can become habitual. A salesperson's time is also valued highly, particularly when there are other calls to make. Alerting the customer to this fact by rescheduling an appointment after a brief wait may partially solve the problem. Incidentally, waiting time can be used effectively rather than wasted. Productive salespeople spend their spare moments reviewing preparation for either the upcoming call or for later calls.

Appointments have become standard procedure in most areas of food and agribusiness selling, and are increasingly important with farmers. By making appointments, the salesperson shows that he or she recognizes the value of customers' time. Appointments also help salespeople employ their limited time more productively. Many agribusiness salespeople who work out of their own homes set aside the first hour of the morning for making or confirming appointments before they begin their actual travel.

The telephone, e-mail, and faxes have become great time savers. The best salespeople utilize these sales tools efficiently. It is quite common today for food and agribusiness salespeople to utilize cellular phones and portable computers with modems. This allows the salesperson to maintain constant contact with their customers as questions or problems arise. Cell phones also allow salespeople to answer a key customer's question immediately versus receiving the message on the home office answering machine and contacting the customer a day or two later. Finally, office contact is facilitated and orders are sent, inventory is checked and other tasks can be performed with the computer—versus relying on written order forms and reports and phone calls. These new technologies have been great time savers for the salesperson willing to invest a little time to learn how to utilize these systems properly. The reality is that the effective use of these tools has become mandatory in many companies.

Working with clerical and administrative support is vital to almost any salesperson's success. Experienced salespeople have learned how to respect the position of such intermediaries, and to work with them rather than trying to get around them. A sincerely friendly and respectful attitude is often the key. Most need to know why the salesperson wants to see their supervisor. If the salesperson has prepared for the sales call properly, the simplest thing is to refer to some specific plans that must be discussed. If the call is in any way a follow-up call, that fact can be pointed out. Relationships with receptionists and support personnel, as with the customers themselves, are often long term and repetitive, and so require continued maintenance.

Once contact with the customer has been achieved, the salesperson moves out of the preparation stage and into the heart of the selling process, the sales call interview.

The Sales Call Interview

The **sales call interview** is the act of establishing customer contact, developing or renewing a positive relationship, discovering the customer's needs, convincing the customer that your product/service can meet those needs, and securing the customer's commitment. The actual sales call is the image that most everyone envisions when selling is mentioned. This step relies heavily on interpersonal skills, and is best described as being primarily a communication process.

To an untrained observer, a sales call by an effective food or agribusiness salesperson may appear to be a casual conversation among business friends, during which pleasantries, industry gossip, and even occasional good-natured barbs are exchanged. Yet these seemingly innocent conversations can have important undertones for the alert salesperson that skillfully seeks information about the

Some customers like to make contact with their salesperson when taking a break from daily activities.
Photo courtesy of Deere & Company.

customer's concerns and needs, about the current market conditions, and about competitive activity. The fertilizer dealer who teases a chemical sales represen-tative by asking whether "this year's incentive trip to the Bahamas will be any good" may be transmitting important messages that the sales representative can use to advantage in a later visit.

On the other hand, the sales call interview may adopt a decidedly formal air, where each of the basic steps from rapport building to the close is distinct. In agribusiness selling, the more formal and complete sales call may occur only when major new products or programs are introduced, or the complete process could occur in almost every sales call, as is more common when selling agricul-tural equipment. The point is that the sales interview, whether formal or infor-mal, is a systematic communication process that can be productive both for the salesperson and for the customer. It involves several well-defined stages of ac-tivity, such as the approach, rapport building, asking questions, generating inter-est, the presentation itself, the trial close, handling objections, and, finally, the close. The style varies with the salesperson, the customer, and the situation.

Approaching Customers

The method of approaching a customer or prospect is absolutely critical to the success of the sales call. This brief greeting sets the stage and tone for the entire interview. Both the customer and the salesperson create impressions and draw conclusions about each other. Each consciously and unconsciously perceives and makes judgments about the other. Spoken and nonverbal messages from both par-ties bear upon the outcome of the interview. Salespeople can gain or lose control of the sales call simply on the basis of how they handle themselves during the ini-tial contact. And it is during that contact that any awkwardness in the relationship can surface, especially when the salesperson has minimal experience.

There are five basic points to consider when approaching customers:

1. *Dress in good taste.* A professional appearance is basic to professional selling. Personal grooming, including hairstyle, must fit within the tolerance limits of customers' tastes. It is understandably frustrating to realize that a customer's personal tastes may influence a sales relationship, but the simple fact is that they do. A salesperson must either remain within the boundaries of his or her customers' normal values or risk alienating them. Whether or not this is fair is not the relevant point. The fact is that a value structure does exist and must be dealt with by the professional agribusiness salesperson. Because the standards for what defines "professional appearance" and "business casual" vary so much from firm to firm, customer to customer, it is important for the salesperson to do their homework in this area.

2. *Greet customers with friendly confidence.* The consensus among professionals is that it is best not to be overly aggressive or assuming until definite rapport has been established. Although a warm smile and a firm handshake make for a good beginning, some customers may signal a discomfort with the handshake as overly aggressive, so care should be taken. In most cases, however, nothing surpasses the personal warmth of a sincere handshake for beginning the sales call on a productive note.

3. *Introduce yourself and your company.* The salesperson should introduce himself or herself clearly by stating his or her name and company. This idea has merit even in situations where the salesperson has called on the customer before, unless both have come to know each other very well. The introduction generally places customers more at ease. Many salespeople wear some type of identification where it can be easily seen to reinforce their name with the customer and to spare the customer the embarrassment of not remembering a salesperson's name. An alternative would be to reinforce the introduction by handing the customer an accompanying business card. Many experienced salespeople also have a file of their customer's business cards. This can be an extremely effective way to collect substantial information—name, title, phone, and address. Notes can be made on the back of the card regarding customer needs, personality, hobbies, follow-up requested, etc. As mentioned earlier, it is increasingly common for salespeople to capture such data in an electronic database.

4. *Use the customer's name, but don't overuse it.* The salesperson must be sure to state the customer's name exactly during the introduction. Many salespeople are reluctant to clarify matters by asking a customer to repeat the name, especially in cases where the two parties have previously met. Yet it is perfectly acceptable to ask the name of your customer to make sure you have it correct. However, in most cases the professional salesperson will have done their homework and have the customer's name well entrenched in their mind before the call. The salesperson should be on guard against overusing the customer's name, since the customer may become suspicious of a too-friendly salesperson who appears to have overstepped the bounds of the relationship.

5. *Have an icebreaker in mind.* Getting the conversation started is the hardest part, especially for new salespeople. Once the ice is broken, however, things usually go more smoothly. Knowing something about the customer and about the situation can help. Some salespeople say that it is useful to have an opening statement firmly in mind, as long as one is ready to adapt to shifting circumstances.

While the most appropriate opener depends greatly on the circumstances, possible openers include:

1. Asking a question about an area of mutual interest.
2. Commenting about some known customer interest or hobby.
3. Referring to an issue covered in the last meeting.
4. Complimenting the customer genuinely.
5. Giving a small gift that is somehow related to the sales call.
6. Providing a copy of technical material that may be of interest.
7. Indicating that a mutual acquaintance suggested the visit (if that is indeed the case).
8. Offering to help the customer with some task.
9. Simply stating the purpose of the sales call.

Building Rapport

The first few minutes of the sales call are usually devoted to building rapport with the customer, which helps both the customer and the salesperson feel comfortable with the situation and helps develop mutual trust. This period also gives the salesperson an opportunity to learn more about the customer's temperament and personality, to size up the situation, and to adapt the sales call strategy developed before the call to the particular situation. Building rapport demands that the social skills of the salesperson be highly developed.

Another important benefit of the rapport-building process is the discovery of interest areas and of inferences that can be drawn about the customer's needs and values. The alert salesperson makes a mental note of a farmer's obvious pride in having finished planting early, and prepares to highlight any timeliness benefits of the company product during the latter part of the call. Or, a salesperson for a food ingredients company makes a mental note of an offhand comment about a delivery problem with a shipment from a competitor.

Although socializing can be enjoyable, salespeople must remember that busy customers may react negatively to interruptions that are lengthy. Excessive use of time for socializing also subtracts from the salesperson's precious selling time. As with so many other delicate questions in establishing customer-salesperson relationships, proper balance must be achieved and maintained.

One of the main objectives of rapport building is to encourage the customer to talk. This is accomplished by asking questions and listening carefully to customer responses. Many salespeople mistakenly believe that their job is to keep talking themselves. A successful call, however, usually finds the salesperson interacting and directing the interview, while the customer does the majority of the talking. As the interview progresses, the salesperson looks for "bridges" or transitions from general socializing to the true purpose of the call.

Probing

Throughout the interview, one of the salesperson's most indispensable tools is the question. **Probing** is the art of asking questions. The salesperson that tactfully asks questions is able to focus attention on the customer and learn about the customer's needs and interests. It is amazing how often carefully designed questions can lead customers to their own conclusions, which is far more persuasive than being "told" outright by a salesperson. Customers often sell to themselves, and the skillful salesperson simply facilitates the process—especially when the product is a natural fit with their needs.

The effective use of questions in a sales interview is a skill that can be learned. The manner in which a question is asked may make a big difference. Tact demands that the salesperson avoid appearing nosy or controlling while asking questions. When questions flow appropriately within the conversation and are asked with genuine interest, they are usually more successful. Care should be taken to phrase questions in a nonthreatening manner so that they can be answered in a positive way. It is of little value to embarrass a customer or to force the customer to admit ignorance. Try not to make the inquiry in such a way that it forces the customer to adopt a polarized position, particularly when that position is counterproductive to the sales call objective. Open-ended questions that are designed to get the customer talking are usually better than questions that can be answered with a yes or no or a specific number.

Salesperson (Processing Equipment Firm): Looks like you had a good processing season! Any bottlenecks?

Customer (Vegetable Processor): No—things went pretty well—except for some days we fell behind. My people just couldn't keep up—then they'd hurry, and that's when we'd make some mistakes.

Salesperson: Things kind of got cramped sometimes, huh?

Customer: No more than every year I guess, but we did have a few down days due to equipment problems.

Salesperson: John, it sounds like your biggest limitation on business is your workforce at peak season. Is that right?

Customer: You got that right! I just can't get enough people we can trust on the equipment nowadays.

Salesperson: How many days are you short people in most years?

Customer: Usually 3 to 4 weeks during the peak season, I guess.

Salesperson: About a month? If I could show you a way to handle the same amount of business with two less people—and end up with higher processing yields—would you be interested?

Notice that this salesperson has successfully probed into an area of concern to the customer and then moved the conversation to the product. The questions were designed to encourage customer response and to identify a need that the salesperson could meet.

Generating Interest

As the interview progresses, the salesperson must move gradually from introductory, socially oriented conversation to the main subject of the sales call. To accomplish the sales call objective, the salesperson must generate the prospect's interest in the item/service that the seller has to offer.

Interest in a product or service is generated when the customer recognizes the possibility that a need might be met by using your product or service. When the customer is keenly aware of a need, the interest is correspondingly high. The customer may even request a sales call to discuss the need. This happens, for example, when a grain farmer who plans to expand his grain storage and handling system seeks to actively explore alternatives with a knowledgeable salesperson. Or, a new quality standard has forced a processing organization to look for new equipment designed to meet the standard.

In other cases, the customer may lack awareness of a particular need. Here the interest step is much more complex and may involve an educational process. For example, one nursery manager was unaware that the nursery's current pruning equipment and practices made their trees more susceptible to disease, and that new pruning equipment could reduce the incidence of the disease enough so that yields and profits would be increased.

Methods of generating interest include:

1. Making reference to cost savings or to potential increases in profit.
2. Telling a story or anecdote about the product.
3. Sharing an experience that relates to the product or to the need.
4. Making a startling statement about the product's benefits.
5. Showing or demonstrating a beneficial feature of the product.
6. Stating openly what one believes the problem to be and why the customer may be interested in exploring possible solutions.

The interest step is designed to direct the prospect's attention to a genuine need for the product, and to prepare the prospect for learning about the product itself. Interest building is a bridge to the main part of the sales call, the presentation.

The Presentation

The **presentation** is the very heart of the sales call. The primary objective of this step is to present the product or service so effectively that the customer will see it as satisfying a particular need. In many respects, the sales presentation can be likened to relating an interesting and convincing story, with the purpose of motivating the customer to a positive buying decision. Through it, the salesperson systematically presents important selling points or reasons for buying in such a way that they can be understood and accepted easily.

A proper presentation also encourages customer questions. The salesperson should be able to answer them in a manner that strengthens the message. The presentation may be highly technical, requiring the salesperson to have a thorough understanding of the product and of all its features. In most cases, it includes helping customers to identify and understand their need more clearly as a prerequisite to accomplishing the sales call objective. In any case, the presentation should be designed to help the customer respond favorably to the proposal.

A good sales presentation requires thorough preparation.

The length and manner of the presentation depend greatly on a host of factors that relate to the customer, to the salesperson, and to the situation. The most effective presentations are tailored to the specific needs of the particular customer. In most areas of agribusiness selling, the salesperson knows far more about the technical aspects of the product and its use than the customer wants or needs to know. Care should be taken not to give customers more information than they desire or need to make their buying decision. Nor should the product be oversold through implications that it will perform better than it actually will. Customer discontent is related to expectation level. Unrealistic expectations can lead to serious problems in the future.

As a general rule, the salesperson should strive for customer agreement on each selling point as the interview proceeds. That is, the salesperson should work toward getting the customer to believe and accept each piece of factual information as it is presented. This systematic acceptance of selling points leads logically to a positive buying decision on the part of the customer.

Although the presentation needs to be led by the salesperson, it should be far from a monologue. The more dialogue, the more successful the call is likely to be. The salesperson should use questions freely, and check to see whether there is customer agreement. He or she should also encourage customer questions because they reveal how far along in the buying process a customer may be.

Throughout the sales interview, but particularly during the presentation, alert salespeople will heed customer clues when specific points spark keen interest. Areas of particular customer interest are known as **hot buttons.** Clues, such as a series of questions about some specific area or careful examination of some aspect of the product, may indicate areas of important concern to the customer. When customers lean forward and look interested, they are providing nonverbal clues that may pinpoint high interest areas. The effective salesperson capitalizes on these clues by adding detail at crucial points and by reemphasizing interest areas at summary points in the interview.

Explaining Features and Benefits

The technique called *features and benefits* is a systematic presentation of factual information about a product given in such a way that customers can clearly recognize the potential gain from purchasing the product. This technique concentrates on the benefits that the customer will derive rather than on the physical characteristics of the product.

A **feature** is a descriptive, measurable fact about the product. Often features are technical, such as nutritional characteristics, weight, density, shipping schedule, and billing procedures. A **benefit** is an interpretation of what the feature is likely to mean to a particular customer or the business. Benefits are often stated in terms of personal gain (typically time and/or profit), and always in terms of value received by the customer.

Customers buy to obtain benefits; that is, they don't buy the product, they buy what the product can do for them. When a product is sold in terms of the value that the buyer may derive, the buyer's motivation for purchasing may be enhanced. A farmer does not simply buy seed. Farmers buy hybrid XP400, which will give them enough additional yield to finance a vacation trip with their families. A purchasing agent does not buy soybean oil, they buy a food ingredient that will deliver the right end product, the assurance that quality standards will be met, and that delivery will be timely—things that will ensure their next performance review is a good one.

It is naive to assume that the customer is aware of the benefits, even though they may seem perfectly obvious to the salesperson. Each benefit should be carefully pointed out, or at least clearly implied, in terms that are crystal clear to the customer. It is no accident that most commercial advertisements show the customer happily enjoying the benefits gained from the purchase. An advertisement is likely to show a producer proudly operating the new piece of equipment in one scene, and enjoying more leisure time with the family in the next.

The features and benefits method of scoring selling points works best when it is integrated into the conversational tone of the presentation. Initially it may be helpful to use the following lead-in phrases: "Because of . . . (feature) you will be able to . . . (benefit)." This general pattern can be expanded and adapted extensively to fit the circumstances, but basically it serves as a logical way of helping customers to understand what benefits they may expect (see Figure 11-3).

It is generally useful to make sure that the customer has accepted each benefit as it is explained. This leads the customer to a logical buying conclusion. A salesperson will sometimes use a brief rhetorical question, such as "and that's important, isn't it?" The seller will then watch for some sign that the customer has accepted and internalized the point before moving ahead to other points. If any one point seems to spark a particular interest or raise a question, the salesperson will reinforce or clarify the point before proceeding. Otherwise, confusion and/or resistance may creep in, which will make successive points more difficult to sell.

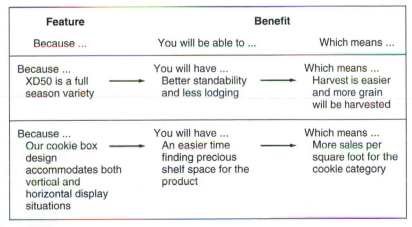

Figure 11-3
The Relationship Among Features and Benefits.

Skillful salespeople often refer back to an earlier moment in the sales interview when the importance of a particular point was established and agreed upon. Perhaps the customer had indicated a need or concern whose resolution is an obvious strong point of this product. This becomes a golden opportunity on which to capitalize tactfully. For example, early in the interview the alert salesperson may have extracted a customer comment about safety concerns resulting from the frequent use of inexperienced operators at harvest time. When the exclusive slip clutch system is brought up, the salesperson can suggest, "Now this feature is a particularly important safety device for your operation, isn't it, Pete?"

Providing Evidence

Professional salespeople must be ready and willing to provide evidence to support everything they say, even though that proof is usually not requested. If the salesperson's claims are questioned and cannot be supported, the salesperson quickly loses credibility. The customer will regard subsequent points, which may be quite accurate, as suspect.

Features can usually be proven easily because they are highly technical and measurable. Company manuals and brochures as well as invitations for personal inspection serve as proof of a feature's existence. They may be supported by technical information, by personal observation, or simply by logical deductions. Benefits may be obvious and measurable, but they are sometimes not completely tangible. No matter how obvious the proof may seem, the customer may need to have it spelled out.

Evidence is not only a means of supporting selling points as they occur in the presentation; it is also a highly valuable selling tool in itself. Providing evidence provides a convincing way of driving home important sales points in dramatic ways that make lasting impressions on customers.

Demonstration is one of the most powerful forms of evidence. There is nothing quite so convincing as seeing, feeling, tasting, hearing, and sensing the product or service in action. Experience is the most forceful kind of demonstration. It not only proves selling points but also often just about sells a product. Getting the farmer into the driver's seat to operate the equipment on his or her own farm is a costly, but powerful, form of selling. Calculating the specific costs and benefits of owning and operating an on-farm feed mill is highly productive. A detailed proposal showing the profitability of changing to bulk shipments can be very convincing. In the food industry, the proof may well be in the taste. Of course, there is nothing more damaging as a demonstration that fails; therefore, demonstrations must be planned very carefully.

Illustrations are another form of convincing proof as well as being a useful selling tool. General models developed by the company support staff can vividly illustrate how a proposal can work. Multimedia computer software can provide graphic video and audio support for the features and benefits of a product.

Comparisons that are well documented and practical are also useful tools. The comparisons must be closely related to the customer's situation or they will lose impact.

Testimonials are used widely in food and agribusiness selling, particularly when the salesperson is selling directly to producers. Farm magazines are full of advertisements that draw heavily on the testimony of other producers. These advertisements simply relate other customers' experiences, and they are often effective and convincing. For the salesperson, testimonials take on an even more personal tone—drawing on experiences of their own satisfied customers, often with photos, quotes, or references.

Test results supplied by independent researchers or even by the company itself are useful forms of evidence. Although manufacturer-sponsored tests are considered suspect by many customers, they can still be used to illustrate many points. Public and private performance tests of chemicals, seed varieties, and livestock breeding stock are widely used.

Company-supplied literature can be useful. This literature is particularly valuable for documenting a product's features. Engineering performance information can be detailed in a way that is usually quite acceptable to most customers. Note that the definition of "literature" here could include videos, websites, CD-ROM/DVD, brochures, and so on.

Establishing Conviction

Conviction is the customer's belief and decision that a product will be of benefit. It is the stage at which the customer has inwardly accepted the proposal as offered, and begins to think in terms of actually using the product. This belief may occur even though the customer has not yet verbalized a definite intention or made a commitment.

One effective technique for achieving conviction involves helping customers see themselves actually using the product or service and gaining the benefits from that use. Gradually, as the customer comes to accept and internalize the ideas presented by the salesperson, the sale nears completion. Skillfully and subtly, the professional salesperson guides the customer through the conviction stage by wording questions carefully, clarifying ideas, and helping the customer to visualize reaping the benefits of the product:

> "And that's what our full service policy is about, Ms. Valdez. We know how many things you have to do in April, and we feel we can help you be more effective by custom-applying your fertilizer and crop protection chemicals, professionally and expertly. This would leave you free to concentrate on ground preparation, logistics, and careful planting. You won't need to worry about information/recordkeeping either—we will take care of it. And, in a year like last year, that would be a big help, wouldn't it?"

It is important not to push too hard or too fast. The customer must come along at their own pace. Even when the benefits may seem perfectly obvious to the salesperson, pushing too aggressively can easily cause the customer to rebel. Once customers begin to feel pushed, resistance builds rapidly. If this occurs, the best tactic is usually to back off quickly before the customer's mind is made up. Then take a fresh approach, perhaps by getting the customer talking again.

Trial Close

The **trial close** is a technique designed to find out whether the customer is ready to buy. It is a "trial balloon" to find the customer's position in the buying process. If the customer indicates that a decision to buy has been or is about to be reached, the salesperson will attempt to close the sale. If not, then the salesperson should continue the presentation. Essentially, the trial close sends out "feelers" that determine the next course of action.

Trial closes often appear in the form of questions. These questions fit neatly into the normal flow of conversation, but they yield important information about the customer's thinking processes. For example, salespeople will ask, "Well, what do you think of that plan?" "When could you stop by to see the demonstration?" "How much credit would you need?" "Which model would work best in your operation?" Note that each of these questions is designed to probe an issue related to the completed sale. A positive response to any of these or similar questions would indicate that the customer has accepted the basic proposal and is well on the way to a positive buying decision. The next obvious step is to clarify final details and close the sale.

A trial close should be made whenever the customer gives a **buying signal.** Buying signals are verbal or visual clues that the customer has decided to buy the product. When a customer asks, "And what about the guarantee?" it is usually an indication of serious interest. When the customer leans forward and examines the label with obvious interest, attempt a trial close.

Another good time to try a trial close is immediately following the strong points of the presentation. If there have been several points, summarize them very briefly and make a trial close. Similarly, after a detailed plan or illustration has been presented, the salesperson should attempt a trial close. Whenever a question or objection has been raised and proof successfully supplied, the salesperson should use a trial close. If the customer has been satisfied, the trial close may show readiness to complete the sale. If not, the salesperson will undoubtedly find himself or herself handling objections.

Objections

An **objection** is a negative reaction or concern expressed by the customer about the product or about any selling point that has been made. It constitutes a reason for not buying. If the customer's concern is legitimate, the salesperson must deal with it constructively in order to complete the sale. Because of their negative character, objections can upset the balance of inexperienced salespeople. But objections offer valuable insights into the customer's thought processes, and present crucial issues for completing the sale. Rather than meeting objections with fear and trembling, salespeople should view them as a challenge. Sometimes objections result from a simple misunderstanding, and the barrier can be broken down easily. At other times, the expressed objection is really an excuse that is designed to hide the real reason for not buying and which must be overcome if there is to be a sale. Most of the time, however, it is a legitimate concern that, if dispelled, may well result in a sale.

Experienced salespeople seldom hear new objections; most concerns have already been handled numerous times. This factor results in preparedness to deal with objections from a position of strength. Less-experienced salespeople can counter by anticipating probable objections and preparing responses or strategies for handling each. Preparation and practice hold the key for most salespeople to properly handle objections.

Objections may arise at any point in the sales interview, even at the very beginning. Generally, they should be dealt with immediately to prevent them from becoming barriers to accepting subsequent ideas. Only when the objection involves technical information that will be developed later in the interview should the discussion be postponed. Even in those cases, the objection should be acknowledged, and deferred until later in the presentation only with the customer's approval.

The objection that is suspected of being an excuse should be dealt with differently. Some customers may be reluctant to admit that money is a problem for them; thus they may ask to "think it over for a while." The first step is to check out excuses thoroughly to ensure their validity. Asking questions to clarify the objection is usually helpful. Sometimes the most appropriate course of action is to ignore the excuse as if it had never been heard. This approach may be particularly useful when the customer is simply delaying the decision for the third time, and a critical date is approaching. If it was indeed a real objection it will certainly be voiced again. Another suggestion is simply to confront the customer with, "Mary, if we find a solution to that problem, can we complete this transaction?"

Handling Objections   The basic technique for handling objections includes four steps:

1. *Listen!* Be sure you allow the customer to express their objections. It is important to give the customer the opportunity to articulate their concerns as clearly as possible to make certain you can focus clearly on the key issue. This also lets the customer know you respect them.
2. *Restate!* Second, restate the objection in your own words. One of the most effective techniques for handling objections is to simply reword the objection and feed it back to the customer. For example:

> "You feel that hydraulic system isn't powerful enough to handle your situation. Is that right, Pat?" or
> "From the questions you raise, you seem to have concerns about how this variety withstands moisture stress, right?"

This step ensures that the salesperson understands the objection correctly. It is surprising how often salespeople think they understand the nature of the objection when, in fact, they do not. Restating and clarifying removes the obstacle and embarrassment of dealing with the wrong objection. Additionally, this technique reassures the customer that he or she has communicated successfully with the salesperson. The customer is usually flattered to know that a pressing objection is being taken seriously. The professional agribusiness salesperson should attempt to reword an objection very carefully to capture the customer's true feeling. Sometimes a judicious choice of words softens the objection slightly, thereby making it easier for the salesperson to handle. And finally, restating and clarifying gives the food and agribusiness salesperson more time to determine exactly how the objection should be handled. This is particularly important for less-experienced salespeople.

3. *Ask!* Customers' explanations of their objections may well disclose key issues in meeting the objections. Like restating, questioning creates extra time for the salesperson to decide how to deal with the objection. Even more importantly, questions start the customer talking, explaining more about individual needs and concerns, and providing more essential information. It is not uncommon for a customer to actually develop a solution to her or his own objection as the conversation proceeds. Frequently, customers convey additional information about the specifics of their situation as they answer the salesperson's questions. This information could be highly useful in the later stages of the selling process. For example:

> "Brian, as a grower, why do you feel that we are not as convenient as the other vegetable processors in the valley?" or
> "Betty, delivery seems to be an issue. How many deliveries each week will you need to keep our product on your shelves?"

4. *Act!* Fourth, handle the objection using any one of several techniques. There are countless ways of handling objections, and no one method is best for all situations. However, there are several commonly used techniques that can be adapted to varying circumstances and people. One technique involves agreeing with the objection, then adding information that either addresses the objection, or minimizes it. For example:

> "You're right, our price is higher. However, remember that our price includes delivery and no payment until June. In addition, I know you like the quality of the recommendations our field agronomist has been providing."

Another technique involves turning the tables on the objection, and showing the customer how the objection is actually a customer benefit. This technique works well when there was some misunderstanding about the product. For example:

> "Yes, Marla, that dual-switch system does take more time to disengage—and we designed it that way for a reason. That extra step helps prevent engaging the grinding mechanism accidentally. Other customers have found this feature really helped them address some serious safety concerns."

Sometimes, it is necessary to directly deny an objection that a customer has raised. When a customer has based an objection on misinformation, the best approach may be to politely and tactfully, but firmly, declare that the information is incorrect. The salesperson must be ready to back up an opposing position, but the risk involved in frankly contradicting the customer may have a high payoff. When using this approach, it is imperative for the salesperson to help the customer save face by offering a possible reason for the misinformation or by suggesting that it is a common misconception.

Occasionally, a customer will raise an objection that is quite valid and cannot be countered in any reasonable way. In this case, simply agree that the customer has a valid point, leave it to the customer to evaluate, and move on to the next point. For example:

> "Yes, that's right, Jim. But as with anything, you must weigh the pluses and minuses to see what is best for you."

The best way to regain momentum is to make sure that the customer has enough information to put the objection into proper perspective. Do not dwell too long on the issue, or the extra attention will give the issue undue importance.

It would be quite uncommon to find an agricultural or food sales situation where there were no valid objections. Often the objection or its importance is a matter of judgment and the customer may not be easily convinced his or her objection is not valid. But all is not lost. Most food and agricultural sales involve rather complex technology and purchasing situations. In a given situation, products have both pros and cons. And the salesperson's job is to make sure the pros outweigh the cons. So the overall strategy in handling objections is to build a strong and convincing case for buying the product up front. Note that if a customer cannot afford or does not need the product after having been given a reasonable opportunity to buy it, these objections must be accepted. A salesperson only wastes valuable time by concentrating too long on a single customer.

The Close

If objections have been handled successfully or the positives outweigh the negatives, the salesperson is ready to close the sale. The **close** is the act of securing a commitment from the customer. A successful close accomplishes the sales call objective. In many cases, this means a signed order or a verbal commitment to buy, but it may be simply a customer agreement to do something specific that is an important step that will lead to the sale. In any case, a successful close provides tangible evidence of accomplishment.

Closing is very difficult for many salespeople. Basically, it involves nothing more than asking for the order or for some form of commitment. Yet many professional salespeople find this step quite uncomfortable and are often less than effective in executing it. For some, the fear of being refused invokes a fear of failure so they are reluctant to attempt the close. Others fail to recognize the importance of a formal effort to close, assuming that closing will somehow take care of itself. Still others lack skills in closing techniques. Obviously, this is a very important step in the selling process, and one that deserves additional discussion.

A confident attitude is critical to closing successfully. The salesperson that sounds apologetic or unconvinced often conveys signals that destroy much of what has been built up in the presentation. The close should not be treated as a joke but as serious business. The salesperson should act positively and confidently to convey the impression that completing the sale is the only logical next step.

There is no one time at which a close should be attempted. The alert sales-person is *always* ready to close. It is not necessary to wait until all the sales points have been covered thoroughly. The professional salesperson always maintains sufficient flexibility to close whenever the customer is ready, even though all the prepared material may not have been covered. Certainly, an appropriate time to attempt to close is immediately following an obvious buying signal from the customer, or after a positive response to a closing feeler. Good judgment is required to ascertain the right time to attempt a close. And good judgment comes from experience. However, even those with little experience can learn to use the five widely accepted, basic techniques for closing.

1. *Direct close.* The direct close is the most straightforward and obvious method—simply ask for the order—"OK, may I go ahead and send in the order?" Food and agribusiness salespeople commonly use this close when the sales call has been very candid and positive throughout. It is also appropriate when there is a clearly defined relationship in which it is presumed that the salesperson will continue to service that account. This closing method simply verifies the continuation of that assumption, but wisely does not take the account for granted. For example:

   "OK, may I go ahead and send in the order?"

2. *Summary close.* The summary close is also popular among agribusiness salespeople. Used most often when the presentation has been long or complicated, it summarizes the major selling points and then asks for the order or for some other commitment. The summary close concentrates mostly on the anticipated benefits. It subtly takes advantage of the agreement that has been reached on each major selling point, and suggests that completion of the sale is the logical consequence of all the selling points. Often these selling points are summarized in sequence so that the customer can also visualize them as the close progresses. For example:

   "OK, Lisa, now let's take stock of what we've said. First, we showed how the early-order program locks in lower prices. Then we calculated an interest savings of almost 1.5 percent. And, we have a solution for the warehousing issue you raised. Now, it seems clear that this is a program designed to fill your needs. Can we get it started today?"

3. *Choice close.* The choice close offers the customer a choice between something and something, not something and nothing. It assumes that the customer is buying, and offers no obvious alternative. The choice close is a powerful close, particularly when some sales resistance has been dealt with satisfactorily. For example:

   "Would you like to begin with 500 or 1,000 pounds?"
      "Should I have them ship the full line of inventory, or should we begin with only the limited line?"

The less-experienced salesperson may initially feel that the choice close is manipulative or too aggressive, and so may be reluctant to use it. Yet when the customer has responded favorably to the presentation and to the salesperson's handling of objections, the choice close fits smoothly into the flow of conversation and is often quite effective.

4. *Assume close.* The assume close never asks for the order, but assumes that the customer has already made a positive buying decision and propels the customer toward enacting that decision. It is used primarily when little sales resistance has been encountered and the salesperson believes that the customer has responded favorably to the proposal. Although there are many forms of the assume close, each features a statement implying the positive buying decision, then initiates the process of finalizing the purchase. For example:

> "OK, then I will stop by Monday with the papers you need to sign," or
> "Then you need our frost-tolerant, early maturing variety. I'll make sure to hold 100 units for you."

The aggressiveness of the assume close is awkward for some food and agribusiness salespeople, who feel that it pushes the customer too fast. Its success depends on its application to appropriate circumstances. Most professional salespeople can easily tell whether the customer has found solid agreement at each step. And, when the relationship is good, the assume close is a natural extension of the presentation. In many cases, this close represents the continuation of a well-established relationship. If the customer has any thoughts of not continuing that relationship they will most certainly make it known. If this occurs, there is an objection that must be uncovered and dealt with before another close is attempted.

5. *Special feature close.* The special feature close describes some special feature of the product that has not been previously described. It is appropriate only when not all the selling points have been covered in the presentation or when the salesperson intentionally holds back special benefits to use at just the right time. This technique is often used when there have been many objections during the presentation and the customer needs additional incentive to make a buying decision. This may be an inducement or a special feature that is available to everyone who purchases now. It offers an extra incentive for immediate purchase. For example:

> "And the reason I'm calling on you now is that August is a special bonus month. You get one free with every five ordered when you purchase in August."

Or, it may be a concession or something extra for a specific customer. Care should be taken in using concessions, since legally the same "deal" must be available to all customers. It is not necessary to publicize the deal, but the deal must be available to other customers should they request it. Often, the concession takes the form of a special service by the salesperson. For example:

> "And if we can complete this deal today, I can just drop the point-of-purchase display and collateral materials by here tomorrow on my way home."

Ending the Sales Call

After the sales call objective has been accomplished, finish any necessary administrative details, compliment the customer on the buying decision, thank the customer for his or her business, make sure that the customer knows what to expect next and when to expect it, and leave quickly. There is usually little good to be accomplished from remaining for any length of time. Lingering too long presents an opportunity for the customer to change the decision.

Sometimes, perhaps often, the customer may refuse to commit to the proposal, even after a good sales presentation. When this occurs, try not to push so hard that the customer refuses to reconsider the proposal at a later time. If possible, get an invitation to call again sometime. In food and agribusiness selling, it often takes many calls for a sale to be completed. Even when the sales presentation has not been effective, be sure to thank the customer for taking time with you, and then leave graciously.

The Follow-Up

In almost every type of food and agribusiness selling, the follow-up after the formal sale makes the big difference in successful long-term selling. The **follow-up** includes all the customer's experiences with the company or product after the sale has been made. It embraces delivery, billing, collecting, using, servicing, and maintaining the product. Since much agribusiness selling occurs in a tightly knit community or market area, or a marketplace where the network among buyers is quite close, and since agribusiness selling concentrates heavily on repeat business, servicing the account is critical to successful selling. In most areas of agribusiness selling, considerably more time and effort is expended in servicing the account than in the formal sales presentation. It might even be a good idea to consider the follow-up as the first step in the next sale.

In most companies, the majority of the postselling functions, such as delivery, billing, and maintenance, are the direct responsibility of someone other than the salesperson. However, if anything goes wrong, it will be the salesperson that will ultimately suffer the greatest penalty in terms of difficult or impossible future sales because the salesperson is the one who is most visible in the selling process. Although this may not seem fair, the salesperson must accept working responsibility for the way in which the company services a customer. The professional salesperson recognizes and accepts this responsibility by acting as a liaison between the customer and the company. It is of great importance to long-term success for the customer to feel his or her needs, have been satisfied.

Servicing the customer actually begins with the sales call itself. Customers should understand completely what will happen next and when it will occur. The old adage that all discontent is relative to expectation is certainly true for the relationship of the customer and salesperson after the sale is completed. The salesperson should make absolutely sure that the customer knows when to expect delivery and billing, and the terms of payment. It is also a good idea to reassure the customer that the salesperson is willing and anxious to handle any questions or problems that might arise.

One effective technique for promoting customer satisfaction is to recheck the situation directly after the projected delivery date by revisiting the customer. This is often a critical time period. The salesperson is in a position to deal with the problem of late shipments. Confusion about merchandise and its use can be clarified at this time. Generally, an indication of concern gets things off to a good start. This callback is most critical for newer customers, for sales of highly technical products, or for unusually large sales.

Billing and collection procedures often present critical difficulties. In a great many food and agribusinesses, the total billing process is the responsibility of a central office. The salesperson does not become involved in the situation unless a problem surfaces. When a problem with a collection does occur, it usually places the salesperson in an uncomfortable position. However, as mentioned earlier, there is a great deal of truth to the statement that "a sale isn't complete until the money is collected."

Many companies make the salesperson that made the sale responsible for initial contact regarding collection efforts on problem accounts. Only a very small percentage of accounts become problems, but those few accounts can develop into uncomfortable situations. Most collection problems can be avoided by making sure customers understand the credit policy at the time of the sale, and by working closely with the company credit manager in the initial extension of credit. Those problem situations that do arise require the utmost in tact on the part of the salesperson. The objective is to make the collection without alienating or losing the customer.

Servicing the account can mean a great many things, depending on the nature of the product or service involved. However, it is usually valuable to make a brief call on larger accounts during or just after the peak use period, to confirm the customer's satisfaction with the product. If there is any dissatisfaction, it is imperative for the salesperson to know it. Often dissatisfaction results from some misunderstanding that can be corrected easily if caught at this time. If customers are unhappy, it is far better for them to complain to the salesperson than to other customers.

This follow-up call makes a very good impression on the customer by establishing that the salesperson really cares how the customer is doing. It sets the stage for continued sales. In fact, regular "service calls" to see how the customer is doing can bring the customer new ideas, check out emerging needs, and ensure ongoing satisfaction in critical key account situations.

SUMMARY

Professional selling is a highly individual process. Each salesperson, each customer, and each situation is unique. Yet selling is a process that can be learned by applying certain principles. Professionals then use these special selling skills in a variety of circumstances to improve the efficiency and probability of sales.

Preparation is the first step in the process. Food and agribusiness salespeople must be knowledgeable about their product, the prospect, and the market. They, therefore, spend time prospecting for and qualifying new customers and planning a sales call strategy that will improve their chances for success.

The first few minutes of the sales call are often the most critical because they set the tone for the duration of the presentation. During this period the salesperson must build rapport, ask questions to learn more about the customer, and arouse interest in the product. There are various techniques that can be utilized, each of which can be productive, depending on the circumstances.

The presentation is often thought to be the core of the sales call, but it can be done well only if the planning and opening have been well executed. In the presentation, the salesperson describes the features and benefits of the product, makes demonstrations, and offers evidence where necessary. This is a conversational period during which the salesperson attempts to instill conviction in the customer. Sometimes the customer may raise objections (reasons for not buying) that must be handled properly. When the customer is satisfied, the sale will be closed.

The days, weeks, or months following the sale are also a critical part of the sales process. In this follow-up period, the customer should be serviced to ensure satisfaction, thereby hopefully creating the possibility of future sales.

The sales process is a logical approach to discovering customer needs and convincing customers that the product or service offered can meet those needs. Each salesperson approaches the problem differently, but the same principles are helpful to every professional salesperson. The challenge of the sales process depends both on the endless variety of products or services and on the unique characteristics of the customer, the salesperson, and the situation. Yet the basic principles of selling can be applied to agribusiness selling in a way that effectively manages the unique characteristics of the agricultural environment.

DISCUSSION QUESTIONS

1. Figure 11-1 shows the steps in the selling process. Which two steps do you feel would be the most difficult for a new salesperson? Why?
2. Salespeople in agribusiness must be good time managers. What are three ways a salesperson might improve time efficiency?
3. Select a food firm or a farm input business and list the information needed to properly prepare for a sales call on the business and/or product you choose.
4. Compare and contrast features and benefits. How is each utilized in the selling process? For the food or agribusiness product of your choice, develop a list of the product's most important features. Then, for each feature, write a statement of the benefits the customer will obtain from that feature.
5. What are ways to handle objections? Give an example of the use of each (different than the examples presented in the chapter).

| CASE STUDY — **Diane Cooper: Building a Sales Strategy**

Frank Bradley was just leaving the milking parlor after finishing the evening milking as Diane Cooper drove past on her way home. "Boy, I sure would like to have Frank's business," she thought. "Selling Frank feed for his 400-cow herd would really help us get the sales increase we are shooting for this year."

Diane Cooper was one of 15 salespeople for Pro-Dairy Feeds. She had joined the firm right out of college. While Diane was recognized as the leading salesperson in the company, she was always looking for ways to increase her business.

The next day she called a local extension agent to ask some questions about Frank's dairy operation. The agent knew Frank, but had never dealt with him much. Diane did learn that Frank had used G & H Feed products for the last several years.

"Frank is really quite a guy, and one of the best dairymen in this area, some say," the agent told her. "I understand he's an officer in the DHIA and still works with the 4-H Club. His daughter showed the champion 2-year-old at the fair last year."

According to the agent, Mrs. Bradley was the most outgoing member of the family and was fairly assertive. He also thought she had a strong voice in many of the farm decisions. In fact he was certain that nothing happened on the farm without Mrs. Bradley's approval.

Diane was an acquaintance of Frank's son, Jim. They met at the state university, but had only infrequent contact. Jim was now living at home and helped out as needed on the farm. He worked full-time as a loan officer at the local Farm Credit Service office.

"Oh, by the way, Diane," the agent had told her, "I just heard that Bud Peters over at G & H Feed is retiring in two months. And, you know that Bud has always worked closely with Frank."

Diane thought about this conversation. She knew this prospective change in salespeople presented a significant opportunity. She wanted to begin developing a strategy to get Frank's business. And, she knew she needed to map out a plan.

Questions

Map out a plan for Dianne's first sales call.

1. Who should she contact first, Frank, his wife, or Jim? Why?
2. How should this contact be made with the individual you selected in question 1?
3. What information should Diane have in hand before she makes the initial sales call?
4. How aggressive should Diane be during this first call? If you feel you need more information about Frank's personality, list the assumptions made to defend your answer.
5. What are some potential features and benefits Diane might use in this or some future sales call with this dairy?
6. Should Diane make this sales call alone or with assistance from outside (a technical person, the extension agent, etc.)? Why?

PART I

FINANCIAL MANAGEMENT
FOR AGRIBUSINESS

Cash is king. Money talks. Time is money. Put your money where your mouth is. All are clichés, but each expression holds an element of truth when it comes to effective management. This section explores the financial aspects of agribusiness management.

12

UNDERSTANDING FINANCIAL STATEMENTS

OBJECTIVES

- Describe the importance of a financial information system in any agribusiness
- Develop a working knowledge of how financial statements aid the agribusiness manager's decision-making process
- Understand financial statement terminology as it is used by the agribusiness manager
- Describe the balance sheet and illustrate how the agribusiness manager uses it
- Describe the income statement and illustrate how the agribusiness manager uses it
- Understand the accrual basis of accounting
- Describe the statement of owner's equity and illustrate how the agribusiness manager uses it
- Describe the statement of cash flows and illustrate how the agribusiness manager uses it
- Understand how managers develop and use pro forma financial statements to make decisions affecting the future of the agribusiness

INTRODUCTION

"The bottom line." Probably no term is used more often among managers than the word *profit*—or the "bottom line." For any business, the bottom line on the income statement is crucial. This figure represents a composite of how the firm, and its management, has performed over the past year. It is a guideline to measure the relative success or failure of the firm over this period of time. **Profit,** the amount remaining from a sale after cost of the product and operating expenses have been paid, is often used as a historical benchmark to provide evidence of the skill and ability demonstrated by decision makers within the organization. The firm's CEO (chief executive officer), corporate board members, and management team are all enamored with this word "profit." The importance of this figure over time can be related to the ability of the firm to grow, enter new markets, and introduce new products.

Although the bottom line is of great importance in itself, other types of financial data are also needed to properly evaluate firm performance. The successful agribusiness manager must understand what the firm did (or did not do) that led to the resulting bottom line, and the successful manager uses this understanding to improve the bottom line in the future. This is what the study of financial management is all about. The successful manager understands the financial operations of the firm well enough to use this information as a tool for improving firm performance. Although the bottom line provides a benchmark for comparison, successful agribusiness managers must understand the interrelationships of other accounts on the balance sheet and the income statement to properly manage the business.

THE IMPORTANCE OF FINANCIAL STATEMENTS

Financial management requires a working knowledge of how to interpret financial information from a firm's records. Such information is used to satisfy two distinct needs of the agribusiness firm. First, and perhaps more importantly, is the need for information that can be used internally by managers in decision making. Second, information is also needed for financial reporting, or reporting financial performance to stockholders, entities outside the firm such as lenders, and others who have interests in the firm.

Without financial information, agribusiness managers at any level find it difficult to successfully pursue the goals and objectives of the organization. Each agribusiness enterprise therefore must accumulate historical records of financial information that are vital for its continued success. The importance of financial information and records is evidenced by the tons of paper, the billions of forms, the millions of computers, and the thousands and thousands of people who are employed in recording business activities throughout the country.

Modern financial recordkeeping had its beginning some six centuries ago in Italy. The growth of commerce in Venice and Genoa, which were great commercial centers, created an accompanying need for business records. As a response to this need, a system of records and bookkeeping was developed that is still widely used throughout the world today. This system summarizes the records of a firm by dividing them into two basic documents. These documents are called the balance sheet and the income statement. Together, they make up the primary financial statements of the firm. The remainder of this chapter will discuss the importance of these financial statements and illustrate how agribusiness managers may use these statements for decision-making purposes.

Correct Information at the Proper Time

Agribusinesses seek to generate the greatest possible returns from the resources they possess. Associated with the profit objective are several other objectives, such as producing a quality product or service, rewarding the employees of the business, helping the business grow, striving for environmental stewardship and promoting a positive public image for the firm as a "citizen" of the community. These types of objectives are common to all business organizations, from the huge food company to the rural, small town feed dealer. Such objectives are usually not accomplished by means of a single, brilliant tactical maneuver, but rather through the consistent use of resources to their greatest potential over a long period of time. The consistent use of resources to meet the firm's objectives requires managers to have useful and timely financial information and records. Business records are tools to help managers guide the operation of the business intelligently and make good management decisions that are consistent with the needs, objectives, and goals of the company. The collection and use of information reported for these purposes is often referred to as **managerial accounting.**

In addition, firms must also satisfy the reporting requirements that are established by various outside entities. For instance, every agribusiness must keep track of sales, purchases, expenses, and profit or loss. Records that document these areas are necessary to meet the reporting requirements of governmental units, lending institutions, investors, and suppliers to the business. The collection and use of information reported for these purposes is often referred to as **financial accounting.**

Although a records system must be designed to meet both the managerial and financial reporting needs of the firm, its design should be guided by the following criteria:

1. It should be simple and easy to understand.
2. It should be reliable, accurate, consistent, and timely.
3. It should be based on the uniqueness of the particular business.
4. It should be cost-effective to implement and maintain.

When designing the records system for an agribusiness, it is usually advisable to secure the services of competent professional advisers or consultants who are not members of the firm. These professionals can help to determine objectively which system of records best fits the firm's needs. In addition, all but the smallest agribusinesses need trained and competent bookkeepers and/or accountants to maintain their managerial and financial reporting system.

An effective accounting system is at the heart of good financial management.

USES OF A GOOD ACCOUNTING SYSTEM

The accounting system also functions to prevent errors and to safeguard an agribusiness's resources. To do this, records must be maintained accurately and honestly by competent personnel. As the business grows larger, a system of checks and balances should be instituted to ensure that no one person has complete control over any transaction. For example, employees who are working as cashiers or collecting monies should not be engaged in bookkeeping. The person who is performing purchasing duties should not be keeping the books or writing the checks.

Whenever and wherever possible, the responsibilities of records, reports, and controls should rest with at least two people. A retail salesperson normally has any customer cash refunds verified by another employee. Outside auditors are typically used to verify the accuracy and integrity of the organization's financial records. All this is not to imply dishonesty so much as to confirm the credibility and integrity of all involved, and to verify that records are being kept properly. When proper checks are built into an accounting system, there is little reason to suspect impropriety.

Good financial records should provide the basis for:

1. Determining the success of the business in terms of profitability during specific time periods or cycles.
2. Determining the general financial condition or health of the enterprise at a given moment.
3. Predicting the future ability of the business to meet the demands of creditors, of change, and of expansion.
4. Analyzing the trends in performance as they relate to management's abilities and to the success or failure of the past decisions and achievements.
5. Choosing among the various possible alternatives for future use of resources within the firm.

Working with Accountants

Some agribusiness managers, such as Chuck Altman, the owner and president of a sizable wholesale fertilizer company, confess to a fear of financial management and accounting simply because they are unfamiliar with the way the system works. Chuck's unfamiliarity stems from the fact that his company started small, and his financial knowledge and skills have not grown with the company. Many other managers have come up through the ranks in larger companies from the areas of sales, production, and operations, and they also have not been exposed to the financial component of the business. It is not necessary for the manager to know how to do the accounting personally or to maintain the records of an organization. Instead, the agribusiness manager should concentrate on understanding how the system works and what information the recordkeeping process produces that can be useful for decision making and forecasting, and for financial reporting.

The agribusiness manager who works with controllers, accountants, and bookkeepers is likely to find many that view the world about them in a conceptual and abstract manner. The approach of such specialists can produce information, which, though accurate and interesting, may not be understandable or meaningful to those controlling and managing the organization. Determining which information is needed and what form it should take is the responsibility of the manager.

Each manager reviews the information supplied by the records of the organization, and determines if the information is to be used solely for managerial decision making. If so, it should be tested against the criteria provided previously. He or she must then ask such questions as: "Will this information or data allow management to make better decisions?" "Is the information prepared in such a way that managers can easily understand and interpret it?" "Is it information that the particular agribusiness truly needs?" Records that do not meet these tests should be altered or dropped. However, information that will be used in financial reporting to governmental units, lenders, and others generally is much more standardized in terms of criteria (i.e., profitability, solvency, and liquidity) and measures (i.e., rates of return on assets and equity, solvency ratios, liquidity measures, etc.). The reporting of this information must be in accordance with generally accepted accounting principles and practices.

THE ACCOUNTING PROCESS

The recordkeeping process derives largely from the original documents of the organization. These documents include such things as sales slips, receiving tickets, checks, invoices, employee time cards, and bills. One could say that such original documents are the building blocks for the entire recordkeeping system.

If we look closely at Chuck Altman's fertilizer business, we will see that each day the firm incurs more expenses and acquires more revenue. These day-to-day transactions are recorded in a book or computer file called the journal. The journal is also referred to as the book of original entry for the business. Here are recorded, in chronological order, all the transactions of the business. The journal, then, provides a running account of the day-to-day transactions or activities of the business. In a small business there may be only one book, computer file, or general journal in which all the transactions of the business are recorded. As a business such as Chuck Altman's grows larger in size, it is necessary to have several specialized journals or computer files for particular areas of the business to record such separate categories as sales, purchases, and available cash. While the journal records the transactions or activities of the business chronologically, it does not put them into any meaningful form by which the manager can interpret or use the information presented.

Business Information Must Be Organized

In looking at Chuck Altman's fertilizer business, it is apparent that he should have a record of its property or assets. Such a financial statement is called a **balance sheet.** The balance sheet shows the financial makeup and condition of a business at a specific time by listing what the business owns, what it owes, and what the owners have invested in the business. It shows a balance between assets and claims against them. **Assets** are those things of value that are owned by the business. Those who are interested in a particular agribusiness require such information as the amount of available cash, the amount of money that customers owe to the business, the amount of inventory or merchandise available, and the resources needed to carry on business activities, including buildings, land, office and sales space, and equipment.

Those who are interested in the business are also concerned with the **liabilities** of that business. Liabilities are the sums of money that have to be paid to creditors at specified dates in the future. To put it another way, liabilities represent the sums that are owed to people outside the business. Thus, liabilities represent claims by "outsiders" against business assets. In addition to information about assets and liabilities, the full financial picture requires information about the amount of money that the owners have invested in the business. This sum is called owner's equity or **net worth.**

Then, too, there is the need for records that detail revenue and expenses in such a way that the manager may meaningfully measure the success of the business, in terms of profit or loss, for decision-making purposes and by various public entities for tax purposes. These records are used to organize the information into ledger accounts for each source of revenue and for each expense category. These accounts are summarized in a financial statement called an income statement. Ledger accounts, then, provide for the separating, categorizing, and recording of related transactions or activities within the business, and they do so in an organized manner. Maintaining financial information in separate ledger accounts not only makes this data more usable, but provides information that is more easily understood by the agribusiness manager.

The accounts in the ledger are summarized on a predetermined regular basis called the **accounting period.** The accounting period typically represents a one-year time span in the operation of the business. This may be a calendar year (January–December) for some firms or a different fiscal year for other firms (i.e., July–June). Many agribusinesses are seasonal in nature. A calendar year accounting period may not make sense for a seed corn firm since this really cuts across two different production and marketing seasons. Hence, firms tend to report financial information for one year that includes a single production period. Management in medium and large firms typically request updated monthly and/or quarterly financial statements. These updates represent summaries of the accounts in the ledger and help generate the key financial statements for the firm.

FINANCIAL STATEMENTS

Financial statements are summarized statements of the financial status of a business, usually prepared as a summary of accounts from the ledger. The primary financial statements usually consist of the balance sheet and income statement. Essentially, a balance sheet shows what a business owns, what it owes, and what investment the owners have in the business. It can be likened to a snapshot that shows the financial makeup and condition of the business at a specific point in time. The **income statement,** on the other hand, is a summary of business operations over a certain period, usually the period between the dates of two balance sheets. It could be compared to a video of the firm that details the financial activities as they occurred over a period of time. Generally, the balance sheet illustrates the current financial position of the business. The income statement, alternatively, demonstrates financial changes since the last time a balance sheet was prepared. The income statement is also known as the operating statement, the profit and loss statement, or simply the P&L statement.

The length of the accounting period and the dates for issuing financial statements are particularly important in agribusinesses that are highly seasonal in nature. In most parts of the country, fertilizer and chemical operations conduct a very large portion of their annual business in the 3 to 4 months before and during the planting season. For example, a balance sheet constructed just prior to the rush season is likely to show large amounts of fertilizer on hand and little money owed by customers. If this same statement is drawn up just after the busy season, it is likely to show less fertilizer on hand and larger amounts of money owed by customers. Such financial statements can be very misleading, if they are not interpreted in light of the seasonal characteristics of the business. Consequently, the agribusiness manager must understand why variations in financial statements are inescapable in highly seasonal agribusinesses. Previous years' financial statements often provide a more valid comparison than a simple month-by-month comparison.

Brookston Feed and Grain Company: An Example

The Brookston Feed and Grain Company (BF&G) typifies many smaller agribusiness firms. BF&G is a retail agribusiness firm located in the Midwest. It sells a wide range of fertilizers, chemicals, feed, seed, and farm supplies. BF&G also merchandises grain—primarily corn and soybeans. The firm provides a variety of services to its customers such as feed delivery, fertilizer application, grain drying, and grain storage. BF&G purchases fertilizer from national manufacturers and blends products to fit individual customer's needs. BF&G is located on a rail line and ships unit trains of corn to the Gulf and Southeast. The firm has a grain market area of about 25 miles. Table 12-1 and Table 12-2 summarize BF&G's operations during a recent year.

THE BALANCE SHEET

As outlined, the balance sheet represents a financial summary of what the business owns and owes and of the investment that the owners have made in the business (Table 12-1). Financial resources that have monetary value are called assets (*a*). Assets are usually listed at the top or the left-hand side of the balance sheet (Table 12-1, *a* to *s*).

The amount that the business owes to creditors is called liabilities (*aa*). Liabilities are generally located in the middle section or on the right-hand side of the

TABLE **12-1** | **Balance Sheet for Brookston Feed and Grain Company December 31, 2001**

(a)	**Assets**		
(b)	**Current Assets:**		
(c)	Cash	$115,000	
(d)	Accounts Receivable	240,000	
(e)	Inventory	310,000	
(f)	Prepaid Expenses	20,000	
(g)	Other	5,000	
(h)	Total Current Assets		$690,000
(i)	**Fixed Assets:**		
(j)	Land		115,000
(k)	Building	490,000	
(l)	Less: Accumulated Depreciation	85,000	
(m)			405,000
(n)	Equipment	660,000	
(o)	Less: Accumulated Depreciation	135,000	
(p)			525,000
(q)	Total Fixed Assets		1,045,000
(r)	**Other Assets**		10,000
(s)	**Total Assets**		$1,745,000
(aa)	**Liabilities**		
(bb)	**Current Liabilities:**		
(cc)	Accounts Payable	210,000	
(dd)	Notes Payable	125,000	
(ee)	Accrued Expenses	70,000	
(ff)	Advances	27,000	
(gg)	Total Current Liabilities		$432,000
(hh)	**Long-Term Liabilities:**		
(ii)	Mortgages	200,000	
(jj)	Other	138,000	
(kk)	Total Long-Term Liabilities		$338,000
` (ll)	**Total Liabilities**		$770,000
(mm)	**Owner's Equity**		
(nn)	**Owner-invested Capital:**		
(oo)	Common Stock	195,000	
(pp)	Retained Earnings	780,000	
(qq)	**Total Owner's Equity**		$975,000
(rr)	**Total Liabilities and Owner's Equity**		$1,745,000

balance sheet (Table 12-1, *aa* to *ll*). Legally, the creditors of the business would have first claim against any of its assets. The value of the assets over and above the firm's liabilities is the owner's claim against the assets, or owner's equity. Owner's equity (*mm*) is also referred to as net worth and represents the summation of several specific accounts. The owner's equity section usually appears just below the liability section on the balance sheet (Table 12-1, *mm* to *qq*).

This brings us to the dual-aspect concept of the balance sheet. The balance sheet is set up to portray two aspects of each entry or event recorded on it. For each resource of value, or asset, there is an offsetting claim against that asset. The recognition of this concept leads to the accounting equation:

$$\text{Assets} = \text{Liabilities} + \text{Owner's Equity}$$

An examination of the following situation will clarify this concept.

When Jody Snyder decided to start her landscaping business, she deposited $2,000 of cash in the bank. This sum represented an investment of $1,000 from her own funds and $1,000 that she had borrowed from the bank. If Jody had prepared a balance sheet at that time, it would have shown assets of $2,000 cash balanced against a liability to the bank of $1,000 and an owner's claim of $1,000. Using the balance sheet formula:

$$\text{Assets} = \text{Liabilities} + \text{Owner's Equity}$$
$$\$2,000 = \$1,000 \quad + \$1,000$$

As this formula indicates, there will always be a balance between assets and the claims against these financial resources (liabilities plus owner's equity). The balance sheet is so named because it always balances. The various components are presented in the following section.

Assets (*a*)

It was stated earlier that assets represent resources of value that are controlled by the business. Of course, the business does not legally own anything unless it is organized as a corporation (see Chapter 4). But regardless of whether the business is organized as a proprietorship, a partnership, or a corporation, all business bookkeeping should be tabulated as a separate entity and kept apart from the personal funds and assets of its owner or owners. Small agribusinesses where personal and business assets are mixed have difficulty in sharply defining genuine business performance. Business assets are typically classified in three classes: current assets (*b*), fixed assets (*i*), and other assets (*r*).

Current Assets (*b*)   For accounting purposes the term **current assets** is used to designate cash or assets that will be converted to cash during one normal operating cycle of the business (usually 1 year). The distinction between current assets and fixed and other assets is important because lenders and others pay much attention to the total value of current assets. At a minimum, this figure gives an approximation of the cash generation potential of the firm.

The value of current assets bears a significant relationship to the stability of the business, because it represents the amount of cash that might be raised quickly to meet current obligations. For example, a nursery may discover that it is necessary to fumigate a large acreage prior to planting nursery stock. Owners might experience great difficulty raising the funds for this emergency because their current assets are limited.

The major current asset accounts may include the following:

Cash (c)   **Cash** funds are those that are immediately available for use without restriction. These funds are usually in the form of checking account deposits in banks, cash register money, and petty cash. Cash amounts should be large enough to meet any obligations that fall due immediately. BF&G has cash balances of only $115,000 as of December 31 (see Table 12-1). As carloads of fertilizer or deliveries of seed arrive soon after the first of the year, BF&G may have to borrow additional short-term funds to meet its purchasing obligations.

Accounts Receivable (d)   **Accounts receivable** represents the total amount owed to the company by customers for past purchases. Essentially, these accounts result from the granting of credit to customers. They may take the form of charge accounts on which no interest or service charge is made or they may be of an interest-bearing nature. In either case, they are a drain on the cash position of the firm. The larger the outstanding amount in accounts receivable, the less money the company will have available to meet current needs. The amount of money in this account depends on the firm's credit policy, that is, how much credit is extended to customers and how efficient the business is at collecting these outstanding accounts.

Some agribusinesses depend heavily on credit as a selling tool. Agricultural production is a seasonal business, and many customers prefer to postpone payment until after harvest and/or selling of crops. BF&G is in an industry in which selling on credit is common. As Table 12-1 indicates, on December 31 Brookston's customers still owed a total of $240,000 from sales made during the past year or earlier. This is a figure that BF&G's management and creditors will watch closely, since borrowing money to pay the firm's obligations is an added expense that reduces the business's profitability. The challenge is for BF&G to offer enough credit so as not to hurt sales, while at the same time keeping its credit policy tight enough not to jeopardize its own cash-flow position. (This is described in more detail in Chapter 15.)

Inventory (e)   **Inventory** is defined as those items that are held for sale in the ordinary course of business or that are to be consumed in the process of producing goods and services to be sold. Inventory items are usually valued at cost (actual funds expended) or market value (what they are worth), whichever is lower. BF&G has $310,000 worth of inventory (at cost) as of December 31. The firm may already be building some inventory for spring sales. The objective of its managers is to keep inventory as low as possible in order to minimize the cash investment while still maintaining an adequate supply of products to meet their customers' needs. Control of inventory and related inventory expenses is one of management's most important jobs, particularly for retailers. Good accounting records are particularly useful in controlling inventory. As BF&G becomes better at matching supplies of product with customers' demands, the firm will lower its inventory investment and increase its return to investment in inventory.

Prepaid Expenses (f)   **Prepaid expenses** represent products or services that have been paid for in advance but not yet utilized. A good example would be prepaid insurance. A business often pays for insurance protection for as much as 3 to 6 months in advance. The right to this protection is a thing of value (an asset), and the prepaid or unused portion can be refunded or converted to cash. BF&G has $20,000 in prepaid expenses, insurance in this case. Other examples that fit this category would include prepaid rent and prepaid taxes.

Other Current Assets (g)   A firm may have various other assets that can easily be converted into cash. For example, BF&G has $5,000 in "marketable securities." An example of **other current assets** could be an outside investment in another company's stocks or bonds, an investment that can be converted to cash during the accounting year if needed.

Fixed Assets (i)   **Fixed assets** are those items that the business owns that have a relatively long life. Fixed assets are typically used to produce or sell other goods and services. If they were to be held for resale, they would have to be classified as inventory (current assets), even though the assets might be long-lived. Normally, fixed assets are categorized into the accounts: land, buildings, and equipment. However, some managers may list types of buildings and equipment separately on the balance sheet.

BF&G's balance sheet could be made more specific if its equipment (*n*) were divided into three main categories: fertilizer equipment, grain storage and handling equipment, and rolling stock (trucks, spreaders, etc.). Some companies even list large, especially expensive equipment, such as "floaters" (self-propelled fertilizer applicators with large floater tires, which may cost $100,000 or more), as separate entries on the balance sheet.

One other important aspect of tracking fixed assets for the balance sheet that should be considered is depreciation (*l* and *o*). Generally, all fixed assets, with the exception of land, depreciate (decrease in value) over time. For example, BF&G's largest floater, which is 3 years old and showing wear, may be worth only $50,000, much less than its original $90,000 purchase cost. For a balance sheet to show the true value of the company's assets, it must reflect the asset's loss in value over this 3-year period. This loss in value due to use, wear, and tear on the fixed asset is called depreciation. The net fixed asset value on the balance sheet reflects only the "accounting" value of the asset in its present state. The real value of fixed assets may be considerably different. The actual market value could be accurately established only by selling the asset, but an ongoing business cannot sell its assets.

Several methods may be used to determine the amount of depreciation. Annual depreciation is allowed as an expense item on the income statement (discussed later), and so it significantly affects total expenses, profits, and, consequently, taxes. Internal Revenue Service regulations typically determine the method used to calculate depreciation. But for both tax and accounting purposes the business can deduct the loss in the value of an asset each year during the useful life of the asset, until it reaches the point where the total purchase price (plus transportation and assembly costs) has been fully depreciated.

Land is usually entered at its purchase cost, even though its current value may be much greater. Accountants argue that because it is impossible to accurately determine its real market value without selling the land, the balance sheet should reflect its original purchase cost so as not to overstate the value and mislead the reader. Obviously, when agribusiness firms have significant land assets whose value has increased greatly, the balance sheet may vastly understate the real market value. For example, Jody Snyder's landscaping business is located on 10 acres of land that her father purchased in 1941 for $10,000. A short time ago a large food store chain offered her $300,000 for the same 10 acres. If Jody could revalue her land, her owner's equity would be much greater. Accountants are working on new methods of reporting fixed assets whose value has been drastically increased by rapid inflation, but most businesses still report the purchase cost, unless otherwise noted. If Jody were seeking additional bank funding for her business, she would be allowed to revalue the land account on her balance sheet to a more realistic level.

Land is a unique fixed asset in that it is not depreciated over time the way other fixed assets are.

Photo courtesy of Deere & Company.

Other Assets (*r*)   A miscellaneous category called other assets accounts for any investment of the firm in securities, such as stock in other private companies and bonds that are held for longer than one year. The category also includes intangible assets, such as patents, copyrights, franchise costs, and goodwill. Goodwill is the extent to which the price paid for an asset exceeds the asset's physical market value, usually because of the value of the reputation established in the market area by the previous owner. Items in the other assets category usually have a longer life than current asset items and are generally nondepreciable in nature. In many cases, they cannot easily be sold within the operating year or represent financial resources that the firm would not sell.

Liabilities (*aa*)

Liabilities consist of money that the business owes to "outsiders" (other than capital invested by the owners). Liabilities are claims against the business's assets, but they may not be claims against any specific asset, except for some mortgages and equipment liens. This means that unless a creditor holds a lien or mortgage (legal claim) against a specific fertilizer truck, spreader, or parcel of land, that creditor has no claims against individual assets; his or her only claim is against a specific dollar portion of the total value of the company's assets. Essentially, liabilities are divided into two classes: current liabilities and long-term liabilities.

Current Liabilities (*bb*)   The term **current liabilities** describes those outsiders' claims against the business that will fall due within one normal operating cycle, usually 1 year. Some of the more important current liabilities entered on the balance sheet are the following:

Accounts Payable (cc)   **Accounts payable** represent the amount that BF&G owes to vendors, wholesalers, and other suppliers from whom the business has bought items on credit. This category also includes any items of inventory, supplies, or capital equipment that have been purchased on credit and for which payment is expected in less than 1 year. When BF&G purchases 100 tons of hog supplement from their supplier on short-term credit at $150 per ton, accounts payable immediately increases by $15,000.

Notes Payable (dd)   **Notes payable** are sometimes labeled as short-term loans or liabilities. This category represents those loans from individuals, banks, or other lending institutions that fall due within a year. Also included in this category is the specific portion of any long-term debt that will come due within a year.

In highly seasonal agribusinesses, short-term loans are a very important part of financial management (see Chapter 14). Cash needs intensify during the peak season, when the agribusiness must pay for increased inventory and for financing accounts receivable. This situation requires careful cash management. BF&G shows $125,000 in short-term credit as of December 31.

Accrued Expenses (ee)   The **accrued expenses** account represents the aggregation of several individual accounts such as wages payable and taxes payable.

Each may be reported separately or one account may include all of these obligations. They include those obligations that the business has incurred for which there has been no formal bill or invoice. An example of this would be accrued taxes. BF&G knows that the business has the obligation to pay $5,000 in taxes, an amount that is accruing or accumulating each day. The fact that the taxes do not have to be paid until a later date in the operating year does not diminish the daily obligation. Another example of accrued expenses would be BF&G's wages of $2,000. Although they are paid weekly or monthly, they are being earned daily or even hourly, and they constitute a valid claim against Brookston's assets. An accurate balance sheet reflects these individual obligations.

Advances (ff)   Sometimes firms receive payment for goods in advance. BF&G shows a $27,000 balance for this account. This means that the firm owes its customers $27,000 worth of product/services. A proper balance sheet will illustrate the payments from customers as advances or deferred income.

Long-Term Liabilities (*hh*)   Outsiders' claims against the business that do not come due within 1 year are called long-term liabilities or, simply, other liabilities. Included in this category are bonded indebtedness, mortgages, and long-term loans from individuals, banks, or others. Any part of the principal for a long-term debt that falls due within 1 year from the date of the balance sheet is recorded as part of the current liabilities of the business. BF&G has both a mortgage (*ii*) and other long-term debt (*jj*).

Owner's Equity (*mm*)

The **owner's equity** or net worth section (*mm*) details the claims of the owners against the business's assets. It is a summation of the accounts that report owner's equity (i.e., earnings retained in the business, contributions, paid-in capital, etc.). The balance sheet balances because of the inclusion of all applicable accounts and their respective amounts for all three sections of the balance sheet, including the owner's equity section. Viewed a different way, the balance sheet formula becomes:

$$\text{Assets} - \text{Liabilities} = \text{Owner's Equity}$$

In an incorporated business, the owners' original or contributed investment to the business is listed as a separate entry called common stock (*oo*). This does not necessarily represent the current market value of the common stock, but rather its original value. (Market value of stock is a totally separate issue, determined solely by buyers' and sellers' perceptions of the value of the business.)

The category of retained earnings (*pp*) represents the net gain on the owners' original investment. If no profits were ever drawn from the business, the retained earnings figure would reflect the total amount of profit that the business has made since its inception. Of course, most owners expect to remove profits regularly from the business as a return on their investment. Thus, retained earnings represent whatever net profits the owners have chosen to leave in the business as additional contributed capital. BF&G owners have left $780,000 of earned profits in the business to combine with their original $195,000 investment. For most companies, retained earnings are an important source of capital for growth.

When the business is a sole proprietorship or partnership (see Chapter 4), it is customary to show owners' equity as one entry with no distinction between the owners' initial investment and the accumulated retained earnings of the business. However, in the case of the incorporated business, there are entries for stockholders' claims and also for earnings that have been accumulated and retained in the business. Of course, if the business has been consistently operating at a loss, the owners' claims may be less than the initial investment; and, in the case of a corporation, the balancing account could be an operating deficit rather than retained earnings.

The complete combination of all the entries outlined creates a full balance sheet. This financial statement provides a great deal of information. It illustrates what the business owns and what claims exist against these resources. Managers need this information to help them decide what actions they should take in running their business. Outsiders seek this information to determine, among other things, the creditworthiness of the firm. Owners utilize this information to judge the performance of the management team.

INCOME STATEMENT

The **income statement** (Table 12-2) summarizes revenue and expenses during a specific period of time and demonstrates the profit or loss that results from the deduction of expenses from revenue. For these reasons, this financial statement is most commonly known as the income statement. Titles like the operating statement, profit and loss statement, or statement of earnings are also used for the income statement.

The income statement provides insight into management efficiency; therefore, it is a key financial statement for operating managers. Its very format emphasizes the basic profit formula, and, hence, holds the key to many operating management problems:

Sales

– Cost of Goods Sold

Gross Profit

– Expenses

Net Profit

Because this financial statement summarizes the activities of the company over a specific period of time, it lists only those activities that can be expressed in terms of dollars relating to that specific period of operations for the firm. While comparing two balance sheets can indicate the changes in the position of the company for a particular accounting period, the income statement details how this change took place during that same period.

TABLE **12-2**	Income Statement for Brookston Feed and Grain Company Year Ending December 31, 2001

(a)	**Sales:**		
	Grain and Soybeans	3,120,000	
	Seed	260,000	
	Fertilizer and Chemicals	780,000	
	Feed	780,000	
	Miscellaneous Supplies	156,000	
	Service Income	104,000	
	Total Sales	5,200,000	100.00%
(b)	**Cost of Goods Sold:**		
	Grain and Soybeans	2,995,200	
	Seed	202,800	
	Fertilizer and Chemicals	694,200	
	Feed	624,000	
	Miscellaneous Supplies	131,040	
	Service Expense	0	
	Total Cost of Goods Sold	4,647,240	89.37%
(c)	**Gross Profit**	552,760	10.63%
(d)	**Operating Expenses:**		
(e)	Salaries and Benefits	81,200	1.56%
(f)	Full-time Wages	57,400	1.10%
(g)	Part-time Wages	10,400	0.20%
(h)	Commissions	15,600	0.30%
(i)	Depreciation	52,400	1.01%
(j)	Maintenance and Repairs	38,980	0.75%
(k)	Utilities	20,800	0.40%
(l)	Insurance	26,000	0.50%
(m)	Office Supplies/Expense	10,400	0.20%
(n)	Advertising/Promotion	13,000	0.25%
(o)	Gas and Oil	17,740	0.34%
(p)	Delivery and Freight	57,200	1.10%
(q)	Rent	7,800	0.15%
(r)	Taxes, Licenses, Fees	12,000	0.23%
(s)	Miscellaneous	8,300	0.16%
(t)	Payroll Tax	10,900	0.21%
(u)	Bad Debt	3,640	0.07%
(v)	**Total Operating Expenses**	443,760	8.53%
(w)	**Net Operating Profit**	109,000	2.10%
(x)	Other Revenue	18,200	0.35%
(y)	Interest Expense	35,000	0.67%
(z)	**Net Profit Before Taxes**	92,200	1.78%
(aa)	Taxes	27,400	0.53%
(bb)	**Net Profit After Taxes**	64,800	1.25%

The income statement identifies the dollar volume of business during a specific period and then matches to it, as precisely as possible, the expenses incurred to perform these business operations. Not all the cash expenditures (dollars spent) during an accounting period can be attributed to business transacted during that period. For example, BF&G may purchase a truck in one year but use it for several years. It would be inappropriate to charge the entire cost of the truck (an expenditure) to the business during the year in which it was purchased, since the truck will last for several years. Therefore, only that part that is used during the operating period is reported as an expense (previously referred to as depreciation). This loss in value is reported on the balance sheet (Table 12-1, *l* and *o*). Similarly, BF&G could have purchased seed (inventory) near the end of an accounting period without selling it until the following period. This amount of inventory should not be charged as an expense for the period since it was not sold.

Therefore, only as an asset is used up or sold does it become an expense to the business. That is, as any asset becomes part of the operation and is directly or indirectly sold to a customer, it becomes an expense. As the BF&G truck is used, it wears out. For business purposes, one might even say that it is being sold to customers bit by bit as this wearing-out process occurs. These "bits" going to the customer are listed as depreciation expense for the current period.

The primary purpose of the income statement is to match precisely the expenses and the revenue from the business generated during that period so that management can accurately measure business profits and, thus, performance. This profit can then be compared to the asset base of the firm to begin to evaluate management performance and financial progress.

The accounting process helps in distinguishing between expenses and expenditures. An **expenditure** is incurred whenever the business acquires an asset, such as a truck, building, or fertilizer, whether it is used immediately or years later. **Expenses** are expenditures that are incurred by the business during the accounting period being reported. Expenses directly affect owner's equity—higher expenses mean lower profit, which, in turn, dictates a lower addition to the firm's owner's equity. Whenever an asset is acquired, the business must make arrangements to pay for it, either immediately or later. The date of cash payment is extremely important to the firm's cash flow, but it has little to do directly with profits or losses.

Accounting on a Cash Versus an Accrual Basis

The **cash-basis approach** to preparing an income statement implies that revenue and expenses occur when cash is received or paid. Many agricultural producers and some small agribusiness firms use this approach. The **accrual approach** says that revenue and expenses truly exist whenever they are earned or incurred regardless of *when* the cash transaction occurs. The accrual approach is a much more accurate approach for reporting net profit for a firm, because it more closely matches the expenses incurred during a period to the revenue generated.

If a cash-basis income statement is used, the net profit reported can be misleading. Cash net profit can *understate* true accrual net profit to the extent that:

- Revenue is generated but not converted to cash—such as an increase in accounts receivable or an increase in the amount of products stored or held in inventory.
- Expenses are incurred during prior or later years but cash is paid this year. Examples are a decrease in accounts payable, a decrease in accrued expenses, an increase in prepaid expenses, or an increase in purchased supplies.

Cash net profit can *overstate* true accrual net profit to the extent that:

- Revenue generated in prior years is converted to cash this year—as in the opposite of that discussed above.
- Expenses are incurred this year but paid in cash in a prior or later year—as in the opposite of what was discussed above.

Hence, the more accurate approach to reporting net profit is to use the accrual approach. This approach will be used in the discussion that follows. (*Note:* Cash record systems can be used to generate an "accrual adjusted" measure of net income by systematically adjusting the cash revenue for changes in certain inventory accounts, and by adjusting cash expenses for changes in accrued expenses. These adjustments reflect the difference between the amounts shown in these accounts on the beginning and ending balance sheets for the year.)

Income Statement Format

The format for the income statement varies somewhat from business to business, but such statements generally begin with sales and subtract the appropriate expenses, with profit showing as a remainder.

Sales (*a*)   **Sales** represent the dollar value of all the products and services that have been sold during the period specified on the income statement. These may be either cash or credit sales. Sometimes customers return products after the products have been purchased. The dollar value of all returns is usually subtracted from the dollar value of all sales. In some cases, the returns are shown as a separate entry. Some customers may be given price discounts for the goods or services they buy. Either the discounted price or the full price may be shown on the income statement, with a special entry indicating how much discount has been given. The format to follow in such reporting might look like the following:

Gross Sales

– Returns

– Discounts and Allowances

Net Sales

In Table 12-2, the income statement for BF&G illustrates five accounts for product sales revenues and one account for service income revenue. Formats for the sales figure vary, but more information is contained in this income statement versus one that might report total sales of $5,200,000 under one general account.

Cost of Goods Sold (b) **Cost of goods sold** represents the total cost to the agribusiness of goods that were actually sold during the specified period. In the case of retail firms, whose purpose is to resell a previously purchased product, this category is a rather straightforward accounting of the actual purchases plus any additional freight charges. In BF&G's case, the grain that was purchased and resold during the year actually cost the firm $2,995,200, the seed cost $202,800, fertilizer and chemicals cost $694,200, etc.

Many types of agribusiness firms are involved in processing or manufacturing. In these cases, determining the cost of goods sold is considerably more complicated because it involves not only the costs for raw materials but also many internal, direct manufacturing costs. In such cases, the cost of goods sold section becomes more complex, and it is recommended that the cost of manufactured goods be detailed to show important cost breakdowns. A seed corn production firm would represent a good example of a manufacturing firm. They are assembling inputs to produce and market seed.

To provide an accurate cost of goods sold figure, current inventories of raw and finished products must be balanced against those of the previous accounting period. In a manufacturing or processing business, the cost of the raw materials and direct labor that have been used during the accounting period are usually included in the cost of goods sold, and are subtracted from sales to establish the gross margin or gross profit. Decreases or increases in inventories from one accounting period to another reflect consumption of or additions to inventories, so that the balancing of changes in inventory is intended to reflect the actual costs incurred during the current accounting period.

For example, if a firm consumed a large amount of raw materials, which naturally would reduce inventories from the previous accounting period, the net change in inventory would be reflected in the cost of goods sold. BF&G had an inventory of $330,000 in the previous period; now it shows an inventory of $310,000 (Table 12-1). The decrease in inventory of $20,000 would have to be added to new purchases of $4,627,240 to equal the total cost of goods sold of $4,647,240 (b). The formula is:

Beginning Inventory	$ 330,000
− Ending Inventory	−310,000
Net Inventory Change	20,000
+ Purchases	+4,627,240
= Costs of Goods Sold	$4,647,240

In addition, if BF&G paid any freight or transportation expense in receiving inventory at their facility, this expense would also be included as part of cost of goods sold.

Gross Margin (Gross Profit) (c)   **Gross margin** or gross profit represents the difference between total sales and total cost of goods sold. The gross margin is the money that is available to cover the operating expenses and still leave a profit. If the gross margin is not large enough to cover operating expenses of the business, losses, and not profits, will be the result.

Gross margins are particularly important to retail agribusinesses because such businesses have relatively little control over cost of goods sold. The prices of the goods that an agribusiness purchases are a critical factor affecting its gross margin. Different products usually have different individual gross margins, so the total gross margin for the business will also depend on the particular combination or mix of products and their sources. Often management can affect the cost of goods sold through careful purchasing. BF&G has a gross margin of $552,760, which is enough to cover its expenses and still leave a profit. Note in the example that this account is actually labeled as "gross profit." Since service income revenue has been included under the net sales figure, BF&G utilizes the term gross profit to reflect gross margin from the five product lines plus all service income revenue.

To demonstrate the importance of pricing decisions on gross margin and bottom line profit, let us assume that BF&G was able to raise its price by 1 percent on current sales of $5.2 million. This would have increased sales to $5.252 million. With the cost of goods sold remaining constant, gross profit would have increased to $604,760; but after operating expenses have been subtracted, net operating profit would have increased by 48 percent to $161,000, and net profit before tax would have increased by over 55 percent. The same kind of effect on the bottom line would be seen if the cost of purchased fertilizer or seed (cost of goods sold) were reduced slightly. This points to the fact that business success is largely centered on relatively small but important changes, and illustrates the tremendous importance of using relevant financial data to explore such changes. (These kinds of changes and their analysis are explored further in Chapter 16.)

Operating Expenses (d)   **Operating expenses** represent the costs that are associated with the specific sales transacted during the time period designated on the income statement. It is easier to interpret these expenses if they have been divided into major divisions such as:

Marketing expenses, including:

Sales, wages, salaries and commissions

Transportation

Advertising and promotion

Administrative expenses, including:

Auditing fees

Directors' fees

Management salary

Office expenses

Travel expenses

General expenses (overhead), including:

Depreciation

Insurance

Taxes

Rent

Repairs

Utilities

(Manufacturing costs, where incurred, are included in cost of goods sold.)

In BF&G's income statement, the various expense categories are simply listed (e through u). The returns and allowances have already been excluded. Net sales for the year ending December 31 were $5.2 million. This detailed listing of operating expenses in the above categories will make the analysis and interpretation of this valuable financial information much easier for BF&G management.

Net Operating Profit (w)   Also called the net operating margin, the **net operating profit** is the amount left over when total operating expenses (v) are subtracted from the gross margin or gross profit (c). The net operating profit is affected by the same factors that influence gross margins, plus the factors associated with business operating expenses.

Net Profit Before Taxes (z)   **Net profit before taxes** is the amount that remains after taking into account any nonoperating revenue or expenses. Nonoperating revenue would include any revenue derived from other sources, such as interest or dividends earned on outside investments (x). Local cooperatives may include patronage refunds from their regional cooperative in other revenue. BF&G generated $18,200 in other nonoperating revenue from interest on investments and the sale of some very old equipment. On the other hand, BF&G incurred interest expenses of $35,000 ($y$). This $35,000 is interest on money that BF&G had borrowed from various sources (as they appeared on the balance sheet), and so is not directly a part of the operation.

Net Profit After Taxes (bb)   **Net profit after taxes,** or net income as it is sometimes called, simply takes into account the federal business profits tax (aa). The rate of tax depends on many factors, including the size of the profits, profit levels in previous years, type of business organization, and several complicated tax regulations. In larger corporate organizations, tax rates often reach nearly 40 percent of the profits. In the case of proprietorships and partnerships, federal profit taxes are not levied against the businesses themselves, but are assessed against the individual owners. Profits are taxed as personal income (see Chapter 4). Cooperative businesses often have lower entries either for taxes or for after-tax profits due to the nature of their organization and its respective operating objectives (see Chapter 5).

T A B L E 12-3 | Statement of Owner's Equity for Brookston Feed and Grain
Year Ending December 31, 2001

Owner's Equity, December 31, 2000		$910,200
Retained Earnings, December 31, 2000	$715,200	
Net Profit After Taxes, 2001	$64,800	
– Dividends	0	
– Withdrawals	0	
Increase (Decrease) in Retained Earnings	$ 64,800	
Retained Earnings, December 31, 2001		$780,000
Common Stock, December 31, 2000	$195,000	
+ Increases	0	
– Decreases	0	
Increase (Decrease) in Common Stock	$ 0	
Common Stock, December 31, 2001		$195,000
Owner's Equity, December 31, 2001		$975,000

STATEMENT OF OWNER'S EQUITY

The **statement of owner's equity** is usually the shortest and least complicated of the financial statements. It details the changes that affect the owner's equity accounts. The primary change is usually a change in retained earnings resulting from a net profit or loss. Other changes could occur due to contributions made or distributions withdrawn from the owner's invested capital. Also, stock could be issued by the agribusiness.

The statement of owner's equity for Brookston Feed and Grain Company for the year ending December 31, 2001 is provided in Table 12-3. As can be seen, the common stock equity account did not change from beginning to end of year. However, BF&G made a net profit after taxes of $64,800, as reported on its income statement for 2001, and all of the net profit was retained in the business. So the retained earnings account increased from $715,200 on December 31, 2000 to $780,000 on December 31, 2001. Consequently, owner's equity increased by $64,800, from $910,200 on December 31, 2000 to $975,000 on December 31, 2001.

STATEMENT OF CASH FLOWS

This statement is basically a way to show the cash inflows and outflows of the firm for a period of time. It can be tabulated at the end of a period, like the balance sheet and income statement, to help interested parties see what happened to cash. Or, more importantly, it can be constructed at the beginning of a period in an attempt to budget appropriate changes over the period.

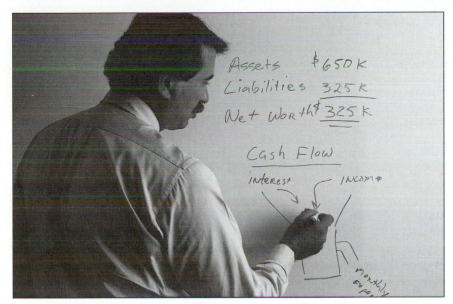

Assets $650K
Liabilities 325K
Net Worth $325K

Cash Flow
Interest   —— Income

monthly expo.

The statement of cash flows is used to manage that most critical firm resource: cash.

The statement of cash flows, like the balance sheet, always balances; that is sources of cash (dollars flowing into the business) always equals uses of cash (dollars flowing out of the business) over the period. The primary categories for this statement are: (1) cash flows from operations, (2) cash flows from investments (disinvestments) made by the firm, and (3) cash flows from financing transactions, such as contributions from owners, borrowing, and repaying debt. The statement of cash flows for BF&G at the end of the accounting period is shown in Table 12-4.

PRO FORMA STATEMENTS: A CASE STUDY

Each of the four financial statements discussed in this chapter provides a way to report historical accounting information. Historical information provides information needed to analyze past financial performance, but it does not provide the information needed to assess future plans. To provide such information, financial statements can be prepared for a future period. Such statements are called **pro forma statements.** The following sample case study is used to illustrate two pro forma statements—a balance sheet and a statement of cash flows.

The Aggieland Landscaping Company provides landscaping services to a wide region of east central Texas. This company has done reasonably well since its inception 5 years ago. However, the owner, Ted Henderson, has a problem. He needs to replace $50,000 of depreciated equipment, but he does not have the cash to pay for the new equipment. He thinks he can sell the depreciated equipment for $5,000, but it will cost $55,000 to replace it. Thus, Ted visited his banker and discussed the possibility of a loan. Ted constructed the balance sheet shown in Table 12-5 for the visit to his banker.

TABLE 12-4	Statement of Cash Flows for Brookston Feed and Grain Company Year Ending December 31, 2001

Cash Flow from Operating Activities		
Net Profit After Taxes	$64,800	
+ Depreciation	52,400	
– Increase in Accounts Receivable	17,000	
– Decrease in Accounts Payable	15,200	
– Decrease in Accrued Expenses	15,000	
Net Cash from Operating Activities		$ 70,000
Cash Flow from Investing Activities		
Cash Proceeds from Sale of Equipment	$10,000	
– Cash Purchase of Equipment	55,000	
Net Cash from Investing Activities		$(45,000)
Cash Flow from Financing Activities		
+ Increase in Notes Payable	$35,000	
– Decrease in Mortgages	30,000	
Net Cash from Financing Activities		$ 5,000
Net Change in Cash		$ 30,000
Beginning Cash Balance		$ 85,000
Ending Cash Balance		$115,000

TABLE 12-5	Balance Sheet for Aggieland Landscaping Company Year Ending December 31, 2001

ASSETS			LIABILITIES:	
Current Assets:			Current Liabilities:	
Cash		0	Accounts Payable	5,000
Accounts Receivable		14,000	Accrued Expenses	2,000
Inventory		21,000	Current Portion of Long-Term Debt	10,000
Fixed Assets:				
Equipment	100,000		Long-Term Liabilities:	
Buildings	50,000		Long-Term Debt	40,000
– Accumulated Depreciation	80,000			
Net Buildings and Equipment		70,000	TOTAL LIABILITIES:	57,000
Land		80,000		
			OWNER'S EQUITY	128,000
			TOTAL LIABILITIES	
TOTAL ASSETS		$185,000	AND OWNER'S EQUITY	$185,000

Ted, after a discussion with his banker, decided he would purchase the $55,000 of equipment in order to be able to replace the worn-out equipment and be in a position to handle more business. Ted wants to estimate the impact this will have on his firm in the short run, so he wants to develop a pro forma statement of cash flows. He believes that the inventory, accounts receivable, accounts payable, and accrued expense accounts will stay the same. Depreciation for the next year will be $25,000. In addition, during the next year Ted has payments of $10,000 to make on his existing long-term debt and he will need to make a principal payment of $10,000 on his new debt. He estimates next year's profit will be $50,000, which he normally withdraws for family living. With this information and these forecasts, Ted can now develop a pro forma statement of cash flows (Table 12-6). In addition, he can construct a pro forma balance sheet (Table 12-7).

Ted's estimated statement of cash flows balances if he requests a loan of $50,000, of which $10,000 plus interest is due by the end of 2002. This is an absolute minimum for the loan request, however. As can be seen by examining the pro forma statement of cash flows, Ted will need all of the profits for personal living expenses, and he has made no allowance for contingencies. In addition, at the end of the period the first payment on the new loan will be due. In reality,

TABLE **12-6**	**Pro Forma Statement of Cash Flows for Aggieland Landscaping Company Year Ending December 31, 2002**

Cash Flow from Operating Activities		
Net Profit After Taxes	$50,000	
+ Depreciation	+ 25,000	
± Change in Accounts Receivable	0	
± Change in Accounts Payable	0	
± Change in Accrued Expenses	0	
– Withdrawal for Family Living	– 50,000	
Net Cash from Operating Activities		$25,000
Cash Flow from Investing Activities		
Cash Proceeds from Sale of Equipment	$ 5,000	
– Cash Purchases of Equipment	– 55,000	
Net Cash from Investing Activities		($50,000)
Cash Flow from Financing Activities		
+ Increase in Equipment Notes	$50,000	
– Principal Payments on Long-Term and Equipment Notes	– 20,000	
Net Cash from Financing Activities		+ $30,000
Net Change in Cash		+ $ 5,000
Beginning Cash Balance		0
Ending Cash Balance		$ 5,000

TABLE **12-7** | **Pro Forma Balance Sheet for Aggieland Landscaping Company Year Ending December 31, 2002**

ASSETS			LIABILITIES	
Current Assets:			Current Liabilities:	
Cash		$ 5,000	Accounts Payable	$ 5,000
Accounts Receivable		14,000	Accrued Expenses	2,000
Inventory		21,000	Current Portion	
			of Long-Term Debt	20,000
Fixed Assets:				
Equipment	$105,000		Long-Term Liabilities:	
Buildings	50,000		Long-Term Debt	60,000
–Accumulated Depreciation	55,000			
Net Buildings and Equipment		100,000	TOTAL LIABILITIES	87,000
Land		80,000		
			OWNER'S EQUITY	$133,000
			TOTAL LIABILITIES	
TOTAL ASSETS		$220,000	AND OWNER'S EQUITY	$220,000

Ted's request to the bank must be a figure that both he and the banker can live with. A conservative approach to estimating profits and other relevant changes in the firm's financial statements is probably the best approach.

SOME IMPORTANT ACCOUNTING PRINCIPLES ▬▬▬

It is important that agribusiness managers understand the following principles and ideas about financial accounting.

1. Only facts that can be recorded in monetary terms should be reported on the balance sheet and income statement.
2. Records or accounts are kept for business entities. Personal and business transactions must be carefully separated.
3. Accounting methods assume that the business will continue to operate indefinitely.
4. Resources (called assets) owned by the business are ordinarily recorded at their cost or market value—whichever is lower. This practice is called the cost basis of valuation. The amounts listed on the financial statement do not necessarily reflect the true market value of the assets.
5. Every accounting event is composed of two transactions: changes in assets and changes in liabilities and/or owner's equity. All assets are claimed by someone; therefore, claims against assets must always equal the assets listed for the firm. These claims can be found among banks, suppliers, owners, and stockholders.

6. Most accounting is handled on the accrual basis of reporting. The objective of the accrual method is to report revenue or the income statement for the period during which it is earned (regardless of when it is collected), and to report an expense in the period when it is incurred (regardless of when the cash disbursement is made). This procedure more clearly reflects the profits of the business.

7. The format for the income statement must reflect the unique needs of the organization. To do this requires the continued help of competent financial professionals.

8. Once a format is developed, it is not sacred and should be changed as necessary to meet changing conditions. However, a degree of consistency should also be maintained to allow comparisons on a historical basis to previous financial statements.

9. Finally, one of the major purposes of records and financial information is to provide the necessary fuel for informed decision making on the part of agribusiness managers.

SUMMARY

Profit is the focal point of any business, but it is the net result of many factors, many of which are reflected on the two primary financial statements. Agribusiness managers rely on financial statements to evaluate how the business is doing, to suggest the direction it should take, and to report its financial condition and performance to outside entities. It is important that this financial data be current, complete, and accurate to facilitate financial decision making within the firm.

All financial statements rely on a systematic recordkeeping system. Accountants have developed highly refined sets of rules for handling every conceivable financial transaction. The daily transactions are summarized in balance sheets, income statements, and other financial reports for owners, managers, and other concerned parties.

A balance sheet shows assets, or what the business owns, at a certain point in time. It also shows liabilities, or claims of creditors against those assets, and the value of owner's claims against the assets. Assets always equal liabilities plus owner's equity as all of the assets must be claimed by someone. Both assets and liabilities are broken down into several classes to provide a useful financial picture of the business.

The income statement summarizes the revenue over a period of time and then carefully matches the appropriate expenses incurred in generating that revenue. Any excess of revenue over expenses is profit. The format of the income statement emphasizes gross margin, or revenue left after the cost of goods sold has been covered. The gross margin represents dollars left to pay operating expenses, which are generally itemized on this report, and make a profit.

The statement of owner's equity and the statement of cash flow are useful in helping understand changes in these two key areas. Pro forma statements are projections, based on careful forecasts of key variables. Such statements help managers use financial data to make decisions that will impact the future of the organization. Agribusiness managers must be familiar with the intricacies of all these statements, both actual and pro forma, since they become the core of many management decisions.

DISCUSSION QUESTIONS

1. Why is financial data important to the agribusiness firm and the agribusiness manager?
2. List and discuss three uses for the balance sheet and three uses for the income statement for agribusiness managers.
3. What are the major sections of a balance sheet? Why does the balance sheet equation always balance?
4. What are current assets? Why is the management of current assets so important to food and agribusiness firms?
5. Why do accountants usually value assets at their cost or market value, whichever is lower?
6. Define the difference between current and long-term liabilities. Why is this distinction important?
7. What are the major sections of the income statement?
8. Discuss the meaning of depreciation and describe how it is treated on the balance sheet and income statement. How is the amount of depreciation determined?
9. Define the difference between an expenditure and an expense from an accounting point of view.
10. What is the difference between gross margin and net profit, and how are they related?
11. Managers may have other business objectives besides making a profit. What are three examples of these objectives? Discuss how these objectives compete with the profit objective.
12. For the case study on Aggieland Landscaping Company, assume that two new landscape firms have opened in Ted's trade territory. Ted now believes that instead of a profit of $50,000 in the coming year, his business will only generate a profit of $30,000. If none of his other assumptions change, how much will he need to borrow to buy the new equipment? Is this a good idea, given his new profit forecast? Why or why not?

CASE STUDY — **Julie Rowe**

Julie Rowe graduated from Chalmers High School 8 years ago with a background in vocational agriculture. Julie had worked for the Ford Motor Company as a truck driver until a year ago, when she decided to start her own business. Her father was an established poultry producer, and through him she had learned that the Payless Grocery Company was looking for someone to haul eggs to their egg processing plant.

Julie had saved $26,000. She decided to invest $20,000 of her savings in a proprietorship to haul eggs for Payless. She decided to keep $6,000 in her personal account for personal expenses and emergencies.

Julie purchased a number of items to start the business. A refrigerated truck that she needed cost $42,000. The truck was 2 years old, and she estimated that it would last for another 5 years. She also purchased her own storage tank for gasoline so that she could save on fuel for the truck, and this cost another $800. She spent $1,000 for gas, small supplies, and miscellaneous inventory. Julie calculated that she needed at least $1,000 to make ends meet before she received the first check from Payless. (Payless pays for hauling eggs on a monthly basis.)

Julie's sister agreed to keep the books. Julie's father let her use the family garage for an office and agreed not to charge rent the first year. She purchased a computer and some other office equipment for a total of $5,200.

Julie's father was willing to lend her $10,000. The principal is to be repaid in 18 months. The interest rate is 8 percent per year. At the end of 12 months Julie will repay her father the interest for 12 months, but no principal. At the end of 18 months Julie will pay her father interest for 6 months, plus principal.

Julie borrowed the balance of what she needed from Chalmers National Bank. The loan was to be repaid in five equal, annual principal payments, with interest at an annual rate of 10 percent.

At the end of the first year Julie's expenses were as follows: gas, oil, repairs and tires, $5,200; telephone, $400; office supplies, stamps and utilities, $630; and miscellaneous supplies and expenses, $890. Her revenue from the business was $41,600. She had been making $26,000 per year when working for Ford, and she calculated that she should make at least that much to justify the time spent in the business. Her sister spent about 1 day a week doing the office work, and Julie felt that she should be reimbursed at a rate of at least $40 a day for her efforts. Assume that her tax rate is 25 percent.

Questions

1. Construct a balance sheet as of the first day of business for Julie. The balance sheet is to be constructed after she purchased the assets needed to start the business. The date for the balance sheet is 12/31/00. How much did she have to borrow from the bank?

2. Construct an income statement for Julie's first year of business. Make the following assumptions:
   - Depreciate all fixed assets over 5 years and assume a salvage value of $0.
   - Julie makes her payment to the bank. She also pays her father interest only on the loan from him. She makes both payments on 12/30/2001. Julie does not pay Social Security tax on her sister's salary.

3. Construct a balance sheet for the end of the first year for Julie's business. Use the information provided on Julie Rowe's beginning balance sheet and income statement to complete an ending balance sheet for Julie Rowe. The ending balance sheet date is 12/31/2001. *Note:* You will have to calculate Julie's cash balance on 12/31/2001, because it is not $1,000. Make the following assumptions:
   - The ending inventory for Julie's supplies is the same amount as her beginning inventory, $1,000.
   - Julie lives with her parents so that reduces her need for living expenses. She decided to invest one-half her net profit after taxes back into her business as retained earnings. Net profit after taxes for 2001 is reported on the income statement.
4. Construct a statement of owner's equity for Julie's business for her first year.
5. Construct a statement of cash flows for Julie's business for her first year. Which activities resulted in a positive cash flow and which resulted in a negative cash flow?

13

ANALYZING FINANCIAL STATEMENTS

- Discuss how different viewpoints are possible depending on who evaluates agribusiness financial statements
- Explain the importance of establishing benchmarks for financial relationships in the firm
- Calculate and interpret financial ratios in the four major areas of analysis
- Illustrate how financial ratios can aid the decision-making process for an agribusiness manager
- Outline and discuss the limitations of financial ratio analysis
- Describe the profitability analysis system and understand its role in decision making in the agribusiness firm

INTRODUCTION

The financial statements of any agribusiness provide a wealth of information for managers, owners, lending institutions, and the government. Some of this information requires little interpretation, but much of it is meaningless unless it is put into proper perspective. Many managers who look at a financial statement are like the explorer who sees an iceberg for the first time and fails to realize that what is seen is only the tip and that 90 percent of the iceberg is hidden under water. Highly useful information can be found in financial statements, but this is not always apparent to the casual observer. Without proper analysis, financial statements can be meaningless scraps of paper.

Financial analysis might be compared to a person's medical checkup, where much more is expected than a superficial glance from the physician. For physicians to understand if or where a problem may exist, a series of tests must be performed and many questions must be asked. Those who are interested in the

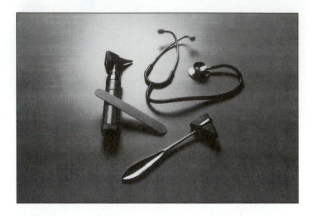

Like a physician, the agribusiness manager has a set of tools that can be used to assess the financial health of the firm.

financial well-being and progress of the business must follow the same procedure. Agribusiness firms that survive and prosper must have managers who use the tools of financial analysis to check the vital financial functions and health of the firm and then prescribe the changes needed to keep the business on course to meet its goals and remain viable and competitive in the future. Tools that allow those who are interested in the business to use financial analysis in determining how successful the business performance has been, what problems or opportunities exist, and what alternative or remedial courses of action might improve performance in the future are discussed in this chapter.

FINANCIAL STATEMENTS: THEIR USE IN EVALUATING PERFORMANCE

Because they can be interpreted from differing points of view, a company's financial statements should provide enough perspectives to satisfy all interested parties. The community looks at what the agribusiness is spending on being a good citizen. Members of a cooperative are interested in the efficiency and savings that result from patronage. Suppliers are interested in the firm's ability to market and pay for products. Customers are interested in the long-term viability of the firm and its ability to provide products at a reasonable price and in a timely manner. Employees are interested in reaching firm goals which, in turn, should increase their compensation, while the board of directors is interested in the effectiveness of the management team in using the firm's financial resources in the most profitable manner.

A manager such as Barry Meade, president of the Meade Food Brokerage is inclined to approach the business financial statements primarily from two points of view—first, to evaluate how the business has performed, and to determine ways financial information can be used to improve decision making in the future. Profit is the primary gauge for success or failure. Some firms will organize the firm into separate "profit centers" to facilitate this analysis. Second, the manager must also keep in mind that the financial statements and their analysis should provide complete, accurate, and timely information for lenders, investors, and the government, since success requires the satisfaction of these interested parties.

When Pilar Fernandez invested in ConAgra, she had a high degree of interest in this firm's profits and general health. However, her primary interest was centered on the rate of return generated on the funds she had invested in the stock of this company. She expected her investment to equal or surpass alternative investment opportunities with similar risks. Members of cooperatives, on the other hand, concentrate their interest on the efficiency of the firm and the savings accruing to them through their patronage. These represent the implicit returns for patron members as discussed in Chapter 5.

Meanwhile, Mary Hartman, who heads the commercial loan department of Farmers' Bank and Trust, is interested in not only a firm's profits, but also its cash flow, because it is an important barometer of the firm's ability to repay loans. Because she is interested in protecting the loan against loss, the firm's assets versus liabilities are a primary focus for her, as for all lenders.

Governmental units are interested in financial statements from several points of view, depending on their particular needs. For example, the Internal Revenue Service is interested in profits, the tax assessor is interested in asset valuation, and the Labor Department is interested in wages and employee information. Each of the many governmental units will require certain kinds of information, and supplying it in proper form is mandatory for the business.

The agribusiness manager must be aware of each of these perspectives in designing and analyzing the financial statements of the business. Construction of these statements in a format useful to the agribusiness firm is the first important step. This chapter outlines how these statements may be used to evaluate agribusiness performance—the second important step in effective financial management.

Timely Analysis

A successful agribusiness operation depends greatly on managerial planning. However, since planning involves the consideration of an uncertain future, even the best management plans can go astray. Therefore, the management team must monitor the firm's progress, or lack of it, toward previously established goals. Many agribusinesses fail because management discovered too late that goals were not being realized. Through continual evaluation of financial records, management might have recognized the problem as it was developing, which would have allowed sufficient time for corrective action. This assumes the records were complete and accurate. Proper and timely evaluation allows management to begin to implement contingency plans, when necessary. Such a combination of financial and marketing evaluation tools was discussed in Chapter 9.

Use of financial data in firm evaluation is analogous to an automobile's oil level. There is a pressure gauge to indicate when the oil level is dangerously low. However, the sensible motorist will check the oil periodically when filling the gas tank to prevent potential problems. It is the same with business records. These records should be timely enough to prevent serious problems from developing. A spot check of the business's health should be taken at regular intervals. In most businesses, the financial statements should be prepared and analyzed on a monthly basis, or at the very least on a quarterly basis, so that problems or opportunities can be determined before they develop or pass by. The cost of the process is small compared to the risk of discovering too late that problems or opportunities have been missed.

What Areas Require Analysis?

The collection of data can become an unending process with increasing costs and diminishing returns. Generally, data should be relevant to the decision-making process to be useful for management.

Agribusiness managers can use financial ratios to help monitor financial position and performance. Four areas are normally explored when financial ratios are used to analyze a firm. These areas are profitability, liquidity, solvency, and efficiency. The information provided on the firm in each of these areas is presented below.

Profitability

- Trends in revenues and expenses, such as sales, operating expenses, overhead costs, and wages.
- The firm's success in terms of actual profitability and also in terms of trends in profitability.
- The firm's use of assets to generate a return on equity sufficient to satisfy current and future investors.

Liquidity

- The firm's cash position; that is, its ability to meet current commitments, such as payrolls and supply purchases.

Solvency

- The firm's capital structure; data on sources and uses of funds and the firm's ability to meet future change and expansion.
- The firm's ability to repay long-term loans and debts.

Efficiency

- Trends in production and performance, measured in accordance with previously established standards of efficiency.
- The competency of the management team in terms of the proper financial and operating use of the resources of the firm.
- The firm's ability to be globally competitive in a constantly changing marketplace.

Begin by Establishing Benchmarks

The interpretation of financial information need not be a complex process. It is centered largely on developing benchmarks or points of reference. Much information can be secured by very simple procedures. One of the easiest ways to determine trends and identify problems or opportunities is simply to compare the current period with similar periods from the past. Comparisons between last year's

TABLE 13-1	Dixie Manufacturing Company Comparative Balance Sheets

	December 31, 2001	December 31, 2000
Assets		
Current Assets:		
Cash	$ 272,000	$ 210,000
Marketable Securities	380,000	395,000
Accounts Receivable	349,000	270,000
Inventory	550,000	610,000
Total Current Assets	$1,551,000	$1,485,000

numbers and the current period or with an average figure, or comparisons between the current month and the same month last year, or to trends, may be useful.

At this point, an important concept should be noted. Financial analysis and records do not solve problems or create opportunities—people do. All the analysis of records can do is to help define and identify problems and opportunities. Alternative courses of action or constraints to action may be suggested and identified, but these records alone do not have a cognitive value. The manager must take specific actions based upon his or her financial analysis for this process to be useful.

Suppose that Neil Gray, controller of the Dixie Manufacturing Company, notes that accounts receivable for December stands at $349,000. That figure in and of itself may have little significance. However, if he notes that last year's December accounts receivable was $270,000, he can be alerted to a potential problem or opportunity (see Table 13-1).

There are many possible reasons for this change. More questions (and answers) may well be needed to provide sufficient information for the manager. Product prices could simply have gone up 29 percent, thereby raising the value of accounts receivable, or the total sales for the year could have been 29 percent higher.

The critical issue is that a simple comparison with a past period gave the figure added significance. This simple and easily prepared comparison is one of the best methods for identifying points where the manager should raise questions. To evaluate this properly, Neil needs the answers to several questions. Has the firm changed its credit policy? Have sales increased significantly over this time period? Which of Dixie's accounts represent the bulk of the $349,000 in receivables due their firm? Has this changed in the past year? Answers to these and similar questions will help Neil determine if any action is necessary. So Neil would find the answers to the questions posed, identify the cause of the deviation, and then, if needed, take corrective action(s).

Performance measures can also be compared to industry averages. Various trade organizations, universities, governmental agencies, consulting firms, etc., collect, compile, and analyze industry data. Such reports can also be used for comparative purposes.

The financial statements and the budgeted forecasts provide another excellent comparison. This is particularly true if the former is an income statement. Table 13-2 presents income statement information in a useful manner for comparative purposes. The first column contains the previous year's actual figures for sales, cost of goods, gross margin, and expense data. The second set of columns contains this year's budgeted figures. In other words, these data represent Pacey's financial goals for the current period. The third set of columns of financial information illustrates the actual amounts for the current year. The final two sets of columns present similar information—budgeted and actual—for the most recent quarter.

Management of Pacey Farm Store can compare actual operating results to the budget and previous year's figures to indicate relative performance in cost control (expenses), buying (cost of goods), and pricing (sales). Cursory analysis indicates Pacey has a significant gross margin problem. Currently, gross margin is about $4,600 under the budgeted goal for this point during the period ($42,810, year-to-date, actual, versus $47,439, year-to-date, budget). Both the pricing decisions and buying practices must be reviewed to determine what specific problems exist.

Common-Size Analysis

Common-size analysis is another useful method in effectively evaluating a business's financial situation. The greatest benefit of this tool is to put things in perspective. Common-size analysis simply expresses the balance sheet and income statement figures as percentages of some key base figure, which could be derived from similar businesses or from the firm's own total sales, total assets, budgets, or forecasts. They provide the manager with benchmarks that give figures an added dimension.

For example, the Pacey Farm Store (Table 13-2) may show a figure of $5,374 for sales promotion in the "This year to date" column. When one realizes that this accounts for 2.9 percent of net sales, the magnitude of the sales promotion figure has some meaning. Comparison with similar figures from the past yields even more insight. Realization that last year's promotion expense was only $4,725 might elicit a hazy reaction, but comparison with a promotion expense last year that constituted only 2.7 percent of net sales will trigger an immediate response. When Heidi McClain, the sales manager, looks at this comparison, she can see that promotion expenses are 19 percent above the budgeted figure, [($5,374 − $4,500) ÷ $4,500] × 100. A similar analysis relative to other expenses indicates that "other general expenses" are more than triple year-ago levels. Action, or at least an investigation, is called for at this point.

Common-size analysis should include raw data from the original statements to prevent distortion or masking. This is particularly true when comparing an

TABLE 13-2 | Pacey Farm Store Expense Budget and Control Report

	Last year to date (Actual)		This year to date (Budget)		This year to date (Actual)		90-day (Budget)		90-day (Actual)	
	Dollars	Percent of Sales	Dollars	Percent of Sales	Dollars	Percent of Sales	Dollars	Percent of Sales	Dollars	Percent of Sales
Net sales	175,502	100.0	180,000	100.0	185,310	100.0	81,889	100.0	85,305	100.0
Cost of goods sold	131,566	75.0	132,561	73.6	142,500	76.9	62,185	75.9	66,847	78.4
Gross margin	43,936	25.0	47,439	26.4	42,810	23.1	19,704	24.1	18,458	21.6
Full-time labor	19,764	11.3	20,500	11.4	19,764	10.7	9,882	12.1	9,527	11.2
Part-time labor	4,392	2.5	5,000	2.8	4,392	2.4	2,196	2.7	1,929	2.3
Overtime labor	0	0	0	0	0	0	0	0	0	0
Management fee	4,500	2.6	4,500	2.5	4,569	2.5	2,000	2.4	2,031	2.4
Total property expenses	1,800	1.0	1,800	1.0	1,800	.9	900	1.1	900	1.0
Warehouse expenses	550	.3	500	.3	580	.3	250	.3	290	.3
Sales promotion	4,725	2.7	4,500	2.5	5,374	2.9	2,000	2.4	1,388	1.6
Interest	3,750	2.1	3,750	2.1	3,583	1.9	1,875	2.3	1,592	1.9
Other general expenses	365	.2	500	.3	1,244	.7	165	.2	411	.5
Total expenses*	39,846	22.7	41,050	22.9	41,306	22.3	19,268	23.5	18,068	21.2
Net operating margin	4,090	2.3	6,389	3.5	1,504	.8	436	.6	390	.4

* Some totals for percents may not equal sums of categories due to rounding.

individual firm's percentages with those of similar firms. One firm may be very large and have sales in the hundreds of millions, while a smaller firm's sales might be in the millions or even thousands. The actual dollar figures and not just percentages would help amplify and clarify differences.

RATIO ANALYSIS

One of the most important tools for financial analysis is **ratio analysis.** The strength of financial ratio analysis lies in the fact that the relationships used between data on financial statements eliminate the weaknesses of dollar comparisons, which are sometimes not only confusing but may be misleading. Ratio analysis permits relative comparisons of important financial data and relationships, and such relative comparisons can be very insightful.

For example, we may consider the following relationships for two firms:

	Firm A	Firm B
Current Assets	$200,000	$1,000,000
Current Liabilities	100,000	900,000
Net Working Capital	$100,000	$ 100,000

Note that the difference between current assets and current liabilities is called net working capital. Both firms have the same net working capital, $100,000. However, Firm A, all things being equal, is in a healthier financial condition than Firm B because the current ratio, an indicator of liquidity (current assets ÷ current liabilities), for Firm A is 2:1; whereas, the current ratio of Firm B is only 1.1:1. The current ratio is the number of dollars of current assets available to meet obligations due during the upcoming year. In this case, it is the ratio of current assets to current liabilities that provides the more accurate reading of the situation, rather than the straight dollar comparison.

Why Use Financial Ratio Analysis?

Financial ratio analysis is often used to measure the performance of the many facets of the business mentioned in the introduction to this chapter. When properly used (with its limitations understood), this financial management tool can be a very useful management aid. Many agribusiness managers use ratio analysis extensively because it not only provides better indicators for managerial decisions but also is useful in the following ways:

1. *Easy to calculate.* Most ratios compare two accounts that are normally provided by the income statement or the balance sheet. Because these data are readily available, not much time or expense is required to compute a financial ratio.

2. *Easy to make comparisons.* Ratios facilitate comparison of past with present performance, as well as comparisons with firms of a similar nature. Ratios are of particular value to corporate boards of directors.

3. *Easily understood.* Not all members of the management team are financial sophisticates, and ratios provide overview information simply and clearly for all management personnel.
4. *Able to communicate a firm's financial position and performance to outside interested parties.* For example, financial authorities and stockholders, cooperative patrons, or investors may rely on ratios to determine a firm's creditworthiness and success in the use of its resources.

Agribusiness managers and analysts are likely to develop their own favorite set of financial indicators. Because lenders may use certain ratios to evaluate a firm's creditworthiness, those ratios need to be monitored by management. It would not be productive to present an exhaustive set of ratios nor pretend that we can produce a selective set of ratios that is of use to *every* agribusiness firm since these firms vary so greatly in size and type. We present below some of the more commonly used and popular financial ratios that cover the four areas for evaluation.

Selecting the Proper Ratios

Agribusiness managers should exercise care in selecting those ratios that provide useful information for decision making in their own unique businesses. Professional advice from outside the business can often be of great help. The first consideration to be made in selecting key ratios is whether the ratios cover those areas of the business where knowledge is critical to good business decisions.

For example, Brookston Feed and Grain (Chapter 12) would have a high degree of interest in ratios that relate to accounts receivable and inventory because its fertilizer and grain business is so seasonal. Loss of control over either of these important accounts could be fatal to maintaining the firm's liquidity. Large variations in different periods are therefore an important criterion for determining which accounts to analyze. Another factor is the sheer size of an account. A corn processor like National Starch would be extremely aware of changes in costs for direct labor and raw materials, because they account for the major portion of production costs. Agribusiness managers must study their own operations and use those financial ratios that are best suited for the unique business involved.

Just a few of the more commonly used and most helpful ratios are discussed below. These ratios will be related to the balance sheet and the income statement for Brookston Feed and Grain (review Table 12-1 and Table 12-2 in Chapter 12). From a manager's point of view, there are four key areas to monitor, and hence four types of ratios to explore:

1. Profitability ratios
2. Liquidity ratios
3. Solvency ratios
4. Operating or efficiency ratios

Profitability Ratios

Profitability ratios include several different indicators that help assess a firm's profitability and record of performance. The net sales figure is used in three of the five ratios rather than the gross sales figure because sales, after returns, allowances, and discounts provides a more accurate measure of the true sales for a firm. Net sales is not inflated by amounts that were ultimately returned to customers in the form of discounts and allowances, or returned merchandise.

Earnings on Sales: Net Operating Profit Divided by Net Sales   The relationship between net operating profit or earnings that Brookston Feed and Grain generated and sales (the **earnings on sales ratio**) is shown here:

$$\text{Net Operating Profit} / \text{Net Sales} = \text{EOS (Earning on Sales)} \quad \text{(Equation 1)}$$
$$\$109,000 / \$5,200,000 = 0.02096 \text{ or } 2.1\%$$
(From Table 12-2, income statement, w and a)

The earnings on sales ratio (equation 1) focuses on management decisions that reflect operating efficiency and pricing policy. Earnings may be increased by changes in pricing policy. Low prices may generate increased sales, but could produce zero profits, while higher prices might reduce sales significantly. The pricing strategy must be developed after careful consideration of the competitive environment. Sales forecasts can be helpful in this area, and ratios may be used to improve sales projections. Likewise, a low EOS ratio may indicate costs are out of line. Brookston Feed and Grain can attempt, through management decisions, to reduce such things as labor, administrative expenses, and selling costs.

Return on Sales: Net Profit Divided by Net Sales   Managers often find it helpful to look at income after all business expenses, including interest and taxes. The **return on sales ratio** is:

$$\text{Net Profit} / \text{Net Sales} = \text{ROS (Return on Sales)} \quad \text{(Equation 2)}$$
$$\$64,800 / \$5,200,000 = 0.0125 \text{ or } 1.25\%$$
(From Table 12-2, income statement, bb and a)

Equation 2 indicates how profitable Brookston was when all costs and other income of the business have been included and compared to overall sales. This ratio is one on which all those interested in Brookston will place a priority, and any deviations will be monitored carefully. ROS is a key ratio for management as it addresses virtually every aspect of their job: marketing and setting prices, purchasing, and cost management.

Return on Equity: Net Profit Divided by Owner's Equity   This ratio is probably the most widely used profitability ratio and takes the investor's point of view of the firm. The **return on equity ratio,** which determines the return on investor ownership, is particularly valued by owners when comparing investment opportunities.

Net Profit / Owner's Equity = ROE (Return on Equity)   (Equation 3)
          $64,800 / $975,000 = 0.0665 or 6.65%
(from Table 12-2, income statement, *bb,* and Table 12-1, balance sheet, *qq*)

This ratio is helpful in determining the wisdom of investment in BF&G. It can be invaluable in encouraging additional investment of equity capital if greater cash flow is needed in the business. Stockholders also use this ratio as an indicator of the relative value of their stock.

Return on Assets: Net Profit Plus Interest Expense Divided by Total Assets   A slightly different profitability ratio is called the **return on assets ratio.** This ratio measures the net profit generated on the total investment in the business. The return to the total investment would include not only the return to owners, but also the return to creditors' investment in the business. Interest costs must be added to the net profit. Thus, this ratio is more inclusive than the ROE ratio discussed above. It measures net returns relative to both outsider and insider investment in the business.

(Net Profit + Interest) / Total Assets = ROA (Return on Assets)   (Equation 4)
     ($64,800 + $35,000) / $1,745,000 = 0.0572 or 5.72%
(from Table 12-2, income statement, *bb* and *y,* and Table 12-1, balance sheet, *s*)

Gross Margin Ratio: Gross Margin Divided by Net Sales   Depending on how the firm's income statement is set up, this may also be called the gross profit ratio. The **gross margin ratio** shows how much BF&G has left from each dollar of net sales to pay operational and other business expenses plus make a profit.

          Gross Margin / Net Sales = GM (Gross Margin Ratio)   (Equation 5)
($5,200,000 – $4,647,240) / $5,200,000 = 0.1063 or 10.63%
          (From Table 12-2, income statement, *a* and *b*)

Here Brookston has 10.63 cents of each sales dollar left to cover expenses and make money. As prices and BF&G's product mix (combination of products sold) change, gross margin will also change. More detailed analysis may be needed here if the gross margin is too low. Costs of goods or raw product and shrinkage may be pinpointed for further study. This ratio is crucial. Any drop in gross margin is a signal for immediate managerial action. Small changes here have a great impact on the bottom line.

Liquidity Ratios

Liquidity refers to the ability of the firm to pay bills as they come due. Hence the focus of **liquidity ratios** is on assets that can be easily converted to cash, and liabilities that must be paid in the short term. Liquidity ratios are another comparison tool in financial ratio analysis that may help agribusiness professionals more clearly identify short-term cash-flow problems. Let's consider some of the most common liquidity ratios that are helpful in determining Brookston's ability to meet its short-run obligations. Often firms refer to the amount of working capital needed in their businesses.

Net Working Capital: Total Current Assets Minus Total Current Liabilities

Total Current Assets – Total Current Liabilities =   (Equation 6)
NWC (**Net Working Capital**)

$$\$690,000 - \$432,000 = \$258,000$$

(From Table 12-1, balance sheet, *h* and *gg*)

However, this is a dollar figure and, for reasons discussed earlier in this chapter, a ratio of two numbers may provide a better measure of liquidity.

Current Ratio: Total Current Assets Divided by Total Current Liabilities

The **current ratio** is probably the most popular liquidity ratio. It is calculated as shown below:

Total Current Assets / Total Current Liabilities = CR (Current Ratio)   (Equation 7)

$$\$690,000 / \$432,000 = 1.60$$

(From Table 12-1, balance sheet, *h* and *gg*)

This ratio indicates BF&G's ability to meet its current obligations. In this case, there is $1.60 of current assets for every $1.00 of current debt. In most cases, a current ratio somewhere around 2.0 signifies ample liquidity for the firm.

There is concern among many financial analysts that the current ratio may have some limitations as an indicator of a firm's ability to meet current obligations. The need to quickly liquidate inventories, accounts receivable, marketable securities, etc., to raise cash may cause a sharp decrease in their value. Therefore, managers often use a second ratio that more clearly delineates a firm's ability to meet immediate cash needs.

Quick Ratio: Total Current Assets Minus Inventory Divided by Total Current Liabilities

(Total Current Assets – Inventory) / Total Current Liabilities = QR (Equation 8)
(Quick Ratio)

$$(\$690,000 - \$310,000) / \$432,000 = 0.88$$

(From Table 12-1, balance sheet, *h, e,* and *gg*)

For most firms a **quick ratio** between 0.8 and 1.0 signifies sufficient strength in liquidity. Consequently, at first glance Brookston would appear to be in good shape. However, the ratio would need to be monitored because any difficulty in collecting accounts receivable could cause problems. This ratio is of great interest to lenders of short-term funds. These funds may be needed because any change in value for marketable securities and for accounts receivable could result in their immediate cash values being lower than the ratio indicates.

A new or rapidly growing firm with immediate cash needs for labor, supplies, and goods must be more aware of this ratio than older, more established businesses. The ease of converting inventory to cash must also be considered. It is possible for a firm to be making a profit, and have a strong owner's equity position in relation to total assets, and still be so starved for available cash that it is unable to take advantage of discounts or quantity buying, meet emergencies, or even pay current bills. Bankruptcy results when a firm is unable to pay its bills as they come due, so the liquidity area is one of fundamental concern for agribusiness managers.

Solvency Ratios

A third challenge to management is keeping the firm solvent. **Solvency** is related primarily to a firm's ability to meet long-run obligations or total liabilities. These **solvency ratios** pinpoint the portions of a business's capital requirements that are being furnished by owners and by lenders. Evaluation of solvency ratios gives an indication of the likelihood that lenders will incur problems in recovering their money. These ratios can have a real effect on the amount of long-term money a firm can borrow. They also affect alternative sources of outside capital. If creditors supply a greater portion of the total business capital, and, thus, assume a greater share of the risk, they normally would demand more financial documentation and would more closely monitor management's decisions. These ratios can also indicate when a firm should consider borrowing more of its capital needs, with consequent opportunity for increasing return on its own investment.

Debt-to-Equity Ratio: Total Liabilities Divided by Owner's Equity   This ratio (equation 9) indicates the relationship of BF&G's owner's equity to the total liabilities of the firm:

$$\text{Total Liabilities / Owner's Equity} = \text{Debt-to-Equity Ratio} \quad \text{(Equation 9)}$$
$$\$770,000 / \$975,000 = 0.79$$

(From Table 12-1, balance sheet, *ll* and *qq*)

In Brookston, we find a ratio of 0.79 to 1, or total liabilities that are equal to 79 percent of owner equity. Interpreted another way, this ratio suggests that for every $1.00 of owner investment in BF&G, there is 79 cents of outsider investment. Lenders tend to get nervous when their investment is greater than that of owners, which would result in a **debt-to-equity ratio** greater than 1.0.

Here we can see the ownership interest in contrast to the creditor interest in the firm. Some lenders require the 1-to-1 ratio mentioned above as the absolute upper limit. Changes in this ratio over a period of time can be of great significance in planning long-range financial programs for Brookston. As BF&G is able to continue generating healthy solvency figures, lenders may begin viewing the firm as a good credit risk and offer "preferred customer" status, and, potentially lower interest rates.

Solvency Ratio: Owner's Equity Divided by Total Net Assets   This **solvency ratio** (equation 10) shows the relationship between what the owners are contributing toward supporting the firm and the total net assets of the firm.

$$\text{Owner's Equity / Total Net Assets} = \text{Solvency} \quad \text{(Equation 10)}$$
$$\$975,000 / \$965,000 = 1.01$$

(From Table 12-1, balance sheet, *qq, h, gg, q,* and *kk*)

Note that in Equation 10, Total Net Assets = Net Working Capital + Net Fixed Assets: that is, Total Net Assets = $(h - gg) + (q - kk)$. Net working capital is calculated by subtracting current liabilities from current assets, and net fixed assets are calculated by subtracting long-term liabilities from fixed assets.

There is no exact standard for every business, but usually if owners contribute less than 50 percent of the total net assets of a firm, one is more likely to find solvency problems developing and to experience difficulty in securing more long- and short-term credit. So this ratio typically needs to be greater than 0.5.

Debt-to-Asset Ratio: Total Liabilities Divided by Total Assets   The third solvency ratio is used to determine the proportion that lenders are contributing to the total capital of the firm:

Total Liabilities / Total Assets = Debt-to-Asset Ratio   (Equation 11)
$770,000 / $1,745,000 = 0.44
(From Table 12-1, balance sheet, *ll* and *s*)

Note that the **debt-to-asset ratio** (equation 11) conveys the same information as the debt-to-equity ratio (equation 9). Both ratios are used to report the solvency position of a firm.

Changes in this ratio can signal danger if the lenders' portion becomes excessively high. By the same token, a low percentage invested by lenders could signal the opportunity for expansion or additional borrowing potential. It would appear from this ratio that Brookston could expand its use of debt or liabilities. Expansion plans and current interest rates would be important factors to consider in making this decision. The upper limit that lenders assign to this ratio is usually 0.50 or less, which corresponds to a 1-to-1 upper limit to the debt-to-equity ratio. At 50 percent both outsiders and insiders have an equal investment in the firm.

Efficiency Ratios

The final area to be discussed in relation to the use of financial ratios is in the area of firm efficiency. This area, in particular, offers the greatest opportunity for BF&G's management to develop unique, meaningful ratios that will be of greatest value to its business in the efficiency or operations area. Here, ratios tend to be highly tailored to fit the specific type of business being evaluated, and the efficiency ratios needed to monitor a retail farm equipment dealership are very different than those used by a soybean processor.

Asset Turnover Ratio: Net Sales Divided by Total Assets   The **asset turnover ratio** can be used to determine the intensity with which Brookston's assets are used, as measured by the number of times assets "turn over" in a period:

Net Sales / Total Assets = Asset Turnover   (Equation 12)
$5,200,000 / $1,745,000 = 2.98
(From Table 12-2, income statement, *a;* Table 12-1, balance sheet, *s*)

This figure shows that Brookston turned its assets about three times during the year. For this ratio to take on meaning, Brookston managers would have to compare it with the situation in previous years and to the turnover ratio in similar businesses. If last year's turnover ratio had been only 1.8, Brookston would know that it has significantly improved asset utilization. A retail grocer would be very unhappy with this ratio because of the necessity for rapid turnover in that kind of business, while an agricultural producer would be extremely happy with this ratio, since production agriculture tends to use a huge amount of assets relative to sales.

The asset turnover ratio (equation 12) suggests that the more BF&G can sell with a given set of assets, the higher its return on investment will be. Many large agribusiness firms use this as a primary measure of effectiveness or efficiency of the management team. Management strategies that affect this ratio are primarily focused on increasing sales volume, using assets more effectively, increasing prices, reducing ineffectively used assets, reducing accounts receivable or inventory, or choosing better alternatives for use of available cash.

Inventory Turnover Ratio: Cost of Goods Sold Divided by Ending Inventory   Another similar, yet more specific measure of efficiency is focused on inventory. In most firms, the **inventory turnover ratio** is of great concern. A significant portion of a firm's assets may be tied up in inventory:

Cost of Goods Sold / Ending Inventory = Inventory Turnover Ratio   (Equation 13)
$4,647,240 / $310,000 = 15.0
(From Table 12-2, income statement, b; Table 12-1, balance sheet, e)

The turnover ratio of 15.0 means that Brookston was able to sell its inventory 15 times over the year. In a different view, BF&G was able to generate $15 in sales for every $1 invested in inventory.

Here, caution must be used, since the measure of inventory must be meaningful. This is a static reading that is taken. In Brookston, with an extremely seasonal business, the ending inventory is not representative, and could provide a meaningless figure. The manager may need to average beginning and ending inventory; he or she may also take monthly or estimated average inventories to arrive at an acceptable figure. In a firm like Brookston, the typical monthly average would probably be a valid figure to use in an analysis of inventory efficiency. When comparing the firm's sales level and inventory level, the correct figures must be used. In most cases, firms value inventory on an "at cost" basis. Thus, the cost of goods sold figure must be used as an indicator of sales activity.

The rate of inventory turnover indicates how successfully working capital has been managed. If capital is being tied up in inventory, a higher margin will be required on sales, because too much stock is on hand. Poor inventory management can also add a severe burden of interest expense if short-term financing at a high rate of interest is being used to finance the investment in inventory. Alternatively, a high rate of inventory turnover could indicate a lost opportunity for sales because of out-of-stock conditions or inability to meet delivery requirements. Agribusiness managers must also consider discounts for early delivery and match these against current interest rates. If Brookston has storage facilities for fertilizer and the discount received from a supplier is greater than its interest cost, profit can be enhanced by adding to inventory levels.

Days Sales in Accounts Receivable Ratio: Accounts Receivable Divided by Net Sales, and the Result Multiplied by 360 Days   Another measure of efficiency is found in the average collection period of accounts receivable. An extended period for collection of accounts receivable could indicate that profits might be reduced because of added collection costs, interest on funds needed to support the accounts, and bad debt losses. On the other hand, too low a figure could indicate that an overly strict credit policy was causing lost sales. Two important criteria should be used; generally, the calculated collection period (1) should match that of others in the industry, and (2) should be at least equal to the time extended by suppliers or vendors of the firm. This important area can be monitored by means of the **days sales in accounts receivable ratio:**

(Accounts Receivable / Net Sales) × 360 days =   (Equation 14)
Days Sales in Accounts Receivable
($240,000 / $5,200,000) × 360 = 16.6 days
(From Table 12-1, balance sheet, *d;* Table 12-2, income statement, *a*)

A firm's credit policy and standing with creditors, as well as changes or trends, should be indicators of efficiency in managing working capital. One of the best rules of thumb in relation to accounts receivable is that the collection period should not exceed the regular payment period by more than one-third. If, for example, the credit policy states that accounts are due in 30 days, the calculated collection period (days sales in receivables) should not exceed 40 days.

Again, caution must be used in evaluating the receivables ratio. Accounts receivable may vary from time to time or seasonally in a firm because of the nature of its business or general economic conditions, and these factors should always be taken into consideration when interpreting the ratio.

Wage Efficiency Ratio: Labor Cost Divided by Net Sales   A number of efficiency ratios can be developed by exploring relationships between costs and sales. These should be unique and meaningful to the individual business and to the type of industry. In this area, management should be operating on predetermined control standards of efficiency. Only one specialized efficiency ratio will be examined here, the **wage efficiency ratio** (equation 15). However, numerous other ratios can be used. Two other popular ratios are sales per square foot and administrative costs per dollar of sales.

Labor Cost / Net Sales = Wage Efficiency Ratio   (Equation 15)
$175,500 / $5,200,000 = 0.034 or 3.4%
(From Table 12-2, income statement, *e, f, g, h, t,* and *a*)

It is particularly important for management to monitor labor costs, especially if the business is expanding through new product lines or if new equipment is being added to the business.

Summary of Ratio Analysis

1. Many ratios can be expressed in reverse order; that is, "sales to receivables" to one person may be "receivables to sales" to another. Either is correct, but the interpretation should also be reversed.
2. All ratio comparisons should be based on similar data from similar businesses; to compare the "sales to fixed assets" ratio of a breakfast cereal manufacturing firm with the same ratio for a small veterinary clinic would be meaningless.
3. In addition to comparisons of ratios between companies, comparisons of ratios from period to period in the same agribusiness are justified. Seasonality, for example, in many agribusinesses dictates that a variety of ratios must be evaluated monthly or quarterly with different goals for each period.
4. Monitoring trends is an important aspect of financial ratio analysis. For example, gradually improving ratios are more impressive than static, declining, or highly variable ratios.
5. The development and use of a large number of financial ratios may be confusing. Hence, ratios should be specifically selected with regard to the nature of the problems for which solutions are desired within the firm.
6. In addition to ratios from balance sheet and/or income statement accounts, ratios related to the physical nature of the business are often used as measures of efficiency. For example, inventory throughput ratios, that is, tons spread per day or per truck, might be used by a firm like Brookston. These may be called management ratios, physical efficiency ratios, or general efficiency ratios.
7. At their best, ratios are aids to better decision making within the agribusiness, not substitutes for it. They should be used in tandem with other financial tools to properly assess performance.

Limitations of Financial Ratio Analysis

Because financial ratio analysis is being used extensively today, there is a potential for misuse of the technique. In spite of their advantages, financial ratios are merely indicators of performance. If a particular facet of the business is headed for trouble, a change in a ratio can only sound a warning. Even a drastic change in a given ratio may not isolate and identify the actual cause of the problem. More often than not, additional analysis is required before the appropriate corrective action can be taken. Care must be taken to ensure that all comparisons are between genuinely similar elements. Some of the more general limitations of financial ratios are discussed below. In addition, each specific ratio may have limitations that are unique to its use, and many of these were highlighted above.

Changes in the accounting methods of the agribusiness itself, or differences between the firm's accounting methods and those of similar firms, may limit the use of ratios. If Brookston Feed and Grain were to change the way it values inventory, assets, and/or change its depreciation method, comparisons of its ratios with ratios of previous periods would lose their validity.

Time factors also pose a significant constraint on ratio analysis. Financial statements are indicative of one period of time. Comparing one monthly period to a different monthly period or to a yearly statement can present a distorted view of the business, because of the seasonality aspects of some businesses. Comparing Brookston's January accounts receivable as a percentage of current assets to its July accounts receivable percentage would provide a comparison that would be virtually useless, because of the seasonality aspects of the business. Bad information or analysis can be worse than no information at all.

Also, extraordinary items can distort the financial ratios for a firm. A one-time revenue or expense is an example of such a distortion. The gain or loss resulting from the extraordinary sale of a capital item can distort financial ratios, since this sale is not part of the routine business for the firm.

As mentioned in Chapter 12, the use of income statement data using the cash basis of accounting can result in tremendous variation and distortion in profitability and efficiency ratios. Data from income statements prepared using the accrual basis of accounting should be used whenever possible for calculating financial ratios.

THE PROFITABILITY ANALYSIS MODEL

Three fundamental measures of how well the business is being managed are (1) return on sales, (2) return on assets, and (3) leverage. The combination of these very important financial relationships into a single ratio, **return on equity (ROE),** provides one of the most useful financial tools available to measure the performance of the agribusiness and to assess the skill and ability of the management team. Many sophisticated managers, bankers, investors, and boards of directors depend on this conceptual measure as the primary gauge of a business's success.

ROE in Perspective

Figure 13-1 shows the step-by-step process of developing an ROE analysis model. This model is called the profitability model or profitability analysis system. Sometimes, this type of analysis is also referred to as the DuPont model. ROE is measured here in terms of return on equity. The diagram illustrates the three key component ratios which individually and together affect ROE. Management can use this chart as a tracking system or early warning system to determine when action is needed if the firm is to continue to move toward its ROE goals.

Figure 13-1 indicates that ROE is directly affected by (1) the earnings from sales (net profit divided by net sales), (2) the intensity assets are used, measured by asset turnover (net sales divided by total assets), and (3) the use of outsiders'

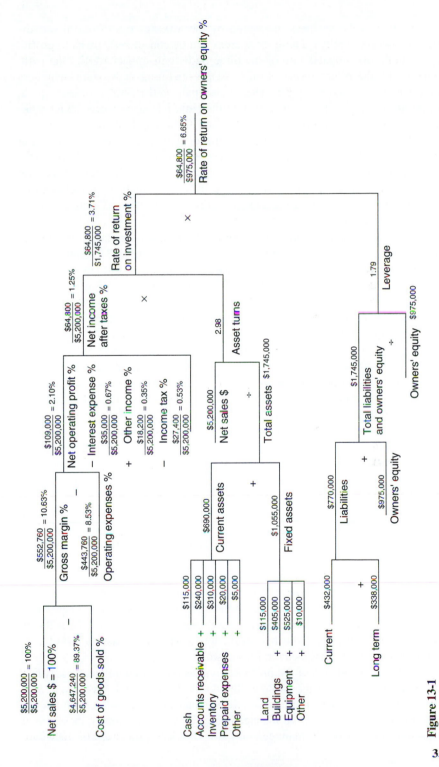

Figure 13-1
Profitability Analysis Model for Brookston Feed and Grain Company.

funds to expand the business, measured by the leverage ratio (total assets divided by owners' equity). These three areas can be considered "paths to profit" and anything that impacts one of the three paths will impact ROE. One path, measured by the return on sales ratio, is the operating path. Here managers worry about sales and expenses. The second path is the asset efficiency path, measured by the asset turnover ratio. On this path, managers worry about making efficient use of the assets invested in the firm. The final path is the debt path. This path is measure by the leverage ratio. Here, managers focus on using debt in a profitable way. So, for a manager who wants to improve ROE, there are three possible paths to pursue—operations, asset efficiency, and debt. The step-by-step process for constructing the profitability analysis model is illustrated using the financial statements for Brookston Feed and Grain Company (Table 12-1 and Table 12-2).

Each of these paths to profit is important to reaching the firm's profit goal, and should be carefully considered. The top part of Figure 13-1 relates to the earnings picture and helps the manager focus on operating problems related to personnel, pricing, product mix, etc. Careful control of expenses is essential, and monitoring systems must be implemented to deal with internal as well as external changes over time—that is, internal personnel and management decisions versus decisions driven by external factors such as competition. As can be seen in Figure 13-1 cost of goods sold (89.37%), operating expenses (8.53%), interest expense (0.67%), other income (0.35%), and the tax expense (0.53%) are all divided by net sales to calculate each as a percentage of net sales.

The asset turnover concept, the second part of the profitability analysis model, suggests that the higher the sales volume produced with a given set of assets, the higher the firm's ROE will be. For example, as more feed and grain are sold for BF&G, the associated assets are being utilized more effectively, up to a point. A firm may attempt to push volume to a level at which physical processing or handling capacity is reached. However, the cost of squeezing out the marginal volume may be prohibitive as these capacity limits are approached.

In the asset utilization area, managers must frequently evaluate their investment in assets: when should equipment be replaced, what fixed facilities are needed and what should be rented, whether inventories can be reduced without reducing sales, how extra cash should be invested, etc. All of these decisions relate directly to asset efficiency and indirectly back to the ROE ratio.

When net profit after taxes (1.25%) is multiplied by the asset turnover ratio (2.98), the result is the return on investment (3.72%). Note that the **return on investment (ROI)** used in this model differs slightly from ROA. ROI has the interest expense subtracted from net profit, where **return on assets (ROA),** includes the interest expense as a return to assets.

The third critical area of the profitability analysis system relates to solvency via financial leverage, or use of debt. Should the firm expand, given an acceptable solvency ratio? What implications does expansion have for the firm's future earnings? Can a larger firm be managed as efficiently? Obviously these are all tough questions faced by management, stockholders, bankers, etc. However, careful evaluation of the solvency relationship helps managers better understand how use of debt will impact the profitability of the business. For BF&G, total liabilities and owner's equity ($1,745,000) is divided by owner's equity ($975,000) for a leverage ratio of 1.79.

Improvements in any of the three individual ratios, or paths, will improve profitability as measured by ROE. It is important for the manager to understand that these ratios are interrelated; changes in one may affect performance in the others. Thus, the final impact of a given change in the firm's ROE must be viewed as a dynamic relationship among these three separate ratios. For example, using more debt will raise the leverage ratio, which would tend to increase ROE. However, more debt also means more interest expense, which will reduce ROE. Both the leverage effect and the ROS effect must be considered to determine the final impact on ROE.

Price Is an Important Consideration

When a price increases, some customers will shift to other firms and some will remain. If it were foreseen that most customers would stay despite the price increase, the earning power would rise because any reduction in asset turnover would be offset by an increase in the firm's gross margin.

If the firm lowered prices, hopefully an increase in the asset turnover would offset the decline in gross margin. But the question is, how much will sales be increased, or what is the "elasticity" of demand (see Chapter 3)?

Successful use of a low margin and high asset turnover strategy is evident in the rapid increase in the number and patronage of discount firms and warehouse stores. Generally, businesses with a low turnover will have a high margin, and firms with a high turnover will have a low margin, which is the case with the discount firms.

ROE, or the earning power of the business, offers the advantage of bringing together in a single figure the complex relationship of asset turnover and gross margin, and allows managers additional insights into how decisions and changes made in either area affect the other. Decision making in any agribusiness does not occur in a vacuum. An attempt to control inventory or reduce assets is likely to affect sales. Certain items may be out of stock, for instance, and some sales will be lost. The use of ROE and the profitability analysis model will help the agribusiness manager gauge the effects of such decisions on overall earning power.

While ROE is certainly not a ratio that can be compared from agribusiness to agribusiness on an unqualified basis, it does offer a more easily compared measure of different firms' resources than almost any other ratio. For this reason, ROE is considered the most accurate measure of effective resource use by managers, boards of directors, lending institutions, investors, and others interested in the business. The ROE is sensitive enough so that even small changes in trends should cause the manager to be pleased (if it is improving) and try more of the same, or to be concerned (if it is declining) and begin searching for methods of improvement.

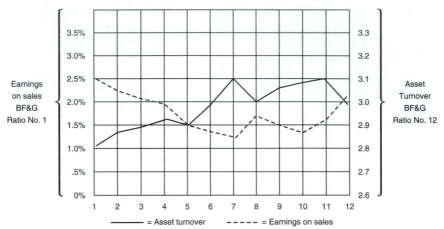

Figure 13-2
Graph of Key Ratios for Brookston Feed and Grain Company.

Changes in figures on the turnover, earnings, or solvency ratios can also point to or help identify places where problems or opportunities arise. For example, Brookston Feed and Grain might note a drop in its ROE figure and see that the asset turnover figure was up slightly, but the ROS ratio was down somewhat. Here, the firm could concentrate its management efforts on such areas as sales expenses or administrative costs or the selling price. Careful use of ROE through the profitability analysis model can be the single most helpful indicator of complete firm performance available to the agribusiness manager. By classifying expenses into categories on the income statement (sales, administration, etc.), and by combining income statement and balance sheet data into an easily understood format, problems and opportunities can be more easily located to take remedial action.

GRAPHING TO INCREASE UNDERSTANDING

Simple graphs of financial ratios often enhance the ability of managers to interpret financial data and information. Graphing of important ratios, for example, allows many more periods and time spans to be compared and the more time or periods or ways in which something can be compared, the more meaningful data become (Figure 13-2).

The graph illustrates the impact of a period when BF&G lowered prices from period 5 to period 11. Asset turnover increased (the solid line in Figure 13-2), but at the same time, earnings on sales (the dashed line in Figure 13-2) decreased as a percentage. If another factor, ROE, was compared, Brookston could measure visually the net effect of price changes, and plan its sales strategy accordingly. Monthly reassessment by means of these ratios would allow BF&G to see whether its plans were still on target.

Graphing financial information can help the manager see underlying trends and patterns.

Graphing increases comprehension by presenting a visual picture and it allows for comparison of separate aspects of financial information in a very clear manner. A graph also can be widely distributed to less sophisticated members of the management team for use in decision making, since it is easier to conceptualize than a set of numbers.

SUMMARY

Financial statements provide a wealth of information for a variety of persons and organizations interested in evaluating the agribusiness. Owners, managers, lenders, and government agencies all look at financial statements from a different perspective.

Analysis of financial statements is often comparative: over time, against budget, or relative to similar firms. Ratios show important financial relationships in the business. These ratios must be understood and used carefully or they will be misleading—used properly, they are indicators of strengths and weaknesses in the business.

Ratios generally measure the important areas of profitability, liquidity, solvency, and efficiency. There are important ratios for each area that provide useful indicators of performance. Return on the owner's investment is the most fundamental measure of profitability. The profitability analysis model shows the relationship of profit, sales, assets, and owner's equity. Sometimes graphing financial ratios over time is a highly valuable management tool.

Financial analysis is a fundamental tool for the agribusiness manager. Although it is not a substitute for good judgment, financial analysis gives the manager important insights for improved decision making.

DISCUSSION QUESTIONS

1. Discuss the major value of financial analysis to agribusiness managers.
2. Discuss the value of common-size analysis to the agribusiness manager. What insights do these statements provide?
3. Describe what financial ratios a general manager, the president of its bank, a prospective investor, and a government official might track in an agribusiness. Explain your reasoning.
4. What are the major types of financial ratios? Do all areas need to be evaluated? Why?
5. What are two reasons why an agribusiness firm might need to evaluate ratios monthly? Under what circumstances would a less frequent financial evaluation be sufficient?
6. List the limitations of ratio analysis for the agribusiness manager. What key issues must the agribusiness manager be aware of if she is going to use financial ratios effectively?
7. What criteria should the agribusiness manager use to select the financial ratios the firm will monitor?
8. What are the advantages of ROE as a tool of analysis for agribusiness managers? What specific ratios are used to develop ROE?
9. What specific actions might an agribusiness manager take if the earnings side (ROS) of the ROE was down? If the asset turnover side was down?
10. How is the leverage ratio (debt path) related to the ROS (operating path) and asset turns (asset efficiency path)? If the use of debt increases, how is ROE impacted?
11. Discuss the merits of graphing ratios. What insights does graphing provide to an agribusiness manager?

CASE STUDY—Altman Dairy Company

Use the following condensed financial statements (Table 13-3 and Table 13-4) from the Altman Dairy Company to develop a profitability analysis model for the company.

Questions
1. Follow the step-by-step formula shown in Figure 13-1 and develop a profitability analysis model for Altman Dairy. What is the firm's ROE?

TABLE **13-3** | **Altman Dairy Company Condensed Income Statement**

Sales	$1,700,000
Cost of goods sold	1,450,000
Gross margin	$ 250,000
Sales expenses	60,000
Administrative expenses	30,000
Other expenses	25,000
Total operating expenses	115,000
Net operating profit	135,000
Interest expense	15,000
Net profit before taxes	120,000
Income tax	45,000
Net profit after taxes	75,000
Dividend paid	15,000
Retained earnings	$ 60,000

TABLE **13-4** | **Altman Dairy Company Condensed Balance Sheet**

Current Assets	
Cash	$ 60,000
Accounts receivable	220,000
Inventory	230,000
Total current assets	$510,000
Net fixed assets	240,000
Total assets	$750,000
Current Liabilities	
Accounts payable	$220,000
Notes payable	110,000
Total current liabilities	$330,000
Long-term liabilities	130,000
Total liabilities	$460,000
Owner's contribution	180,000
Retained earnings	110,000
Total liabilities and owner's equity	$750,000

2. When finished with the profitability analysis model, calculate and interpret at least one profitability, liquidity, solvency, and efficiency ratio, using the information in Table 13-3 and Table 13-4.
3. What are some strengths and weaknesses of the Altman Dairy Company identified by your ratio analysis? What other information would be helpful to you in answering this question?

FINANCING THE AGRIBUSINESS

- Discuss the reasons an agribusiness may choose to increase its financial resources
- Describe the alternative types of capital available to the agribusiness
- Outline the various types of loans available to the agribusiness and the situation that would cause the agribusiness to select each specific type
- Explain how specific variables impact the annual percentage rate (APR) paid on a loan
- Discuss the relationship between the agribusiness firm's tax rate and APR
- Show the usefulness of the cash budget in making loan requests
- Understand the usefulness of pro forma financial statements in financial planning
- List some of the important external sources of financing
- Describe what is needed to choose a lending institution and to apply for a loan
- Describe the means of generating capital funds internally and the importance of this financing alternative
- Discuss leasing alternatives used in agriculture, including land, operating, custom hire, and capital leases

INTRODUCTION

Cash is king. Proper cash management is the lifeblood of any agribusiness. Although money itself is not capital, it represents the amount of capital a firm could control. Cash is needed for financing such assets as machinery and equipment, accounts receivable, materials, and supplies. This is why the managers of an agribusiness are expected to be experts in cash management.

Managers must make sure that sales on credit (accounts receivable) turn into cash for the firm, that inventories are maintained and sold to generate cash, and that cash is available to meet short-term financial obligations—both planned and those unforeseen. Whenever and wherever financial resources are secured, money is used in the enterprise with the full expectation that it will be returned with a profit. This profit-making ability is essential for the viability of the firm, and the manager who lacks this expertise will find the agribusiness hard-pressed for funds.

There are three sources from which the manager may raise the funds needed to operate an agribusiness: (1) investment by owners, (2) borrowing, or (3) funds generated by profits and funds available due to depreciation expense. In most medium-to-large businesses, the major source of funds (over 50%) is the owner's equity of the firm (also called "owned capital"). The larger the company, the more it depends on owner's equity as a source of funds. One major reason for this is that larger companies usually enjoy access to public offerings of their stock or equity, and possess the ability to attract investors, a situation that is not shared by smaller companies. Whatever the kind or size of a business, its ability to generate profits will ultimately determine the amount of funds that are made available for its use.

REASONS FOR INCREASING FINANCIAL RESOURCES

The ultimate reason for increasing the financial resources of an agribusiness is to increase its revenues, and ultimately its profits, by generating additional business. Extra funds are used for general purposes, to increase liquidity or the cash position, or for expansion and growth. An agribusiness may find that its funds are tied up in current or fixed assets and that it is unable to meet its day-to-day obligations. Bills cannot be paid with such nonliquid assets as accounts receivable, inventory, new orders, or a piece of equipment. Consequently, an agribusiness requires working capital in the form of cash. The principal source of cash must be the revenue generated by the business itself, but in short-term situations additional cash may be required to meet the day-to-day obligations of the business. This is particularly true of the many agribusinesses that are highly seasonal in nature. In this case, cash funds can become tied up in inventory and in accounts receivable, which will not be turned into cash until some later date. As a rule, many agribusinesses find it advisable to keep on hand enough cash to equal 20 to 25 percent of the amount of their current liabilities. This helps ensure the firm's ability to make short-term payments as well as any unexpected obligations.

The most important use for additional financial resources is for expansion. Expansion can require either short- or long-term commitment of funds. Short-term expansion involves such factors as increased labor, increased inventories, and increased accounts receivable. Long-run expansion encourages more ambitious projects, such as the purchase of new equipment, land, and buildings. The objective of increasing the capital of an agribusiness is to increase its sales volume and revenues, and consequently its profit, through the shrewd application of increased assets. Capital, or the financial resources of a business, comprises in its broadest sense all the assets of that business, and represents funds provided by the owners of the firm and the contributions by outsiders.

Determining When Financial Resources Should Be Increased

As an agribusiness manager considers the possibility of acquiring additional financial resources, several questions should be asked and carefully answered:

1. Why are additional funds needed?
2. How does the use of additional funds fit with the overall mission of the business?
3. What increases in revenue, profit, and/or net cash flow will be generated by the additional funds?
4. When is this increase in revenue flow expected?
5. When will these additional funds be needed?
6. For what time period will these additional funds be needed?
7. How much is needed in the way of additional financial resources?
8. Where can these additional funds be obtained?
9. How much will these additional funds cost the agribusiness?
10. If the funds are borrowed, how will the indebtedness be repaid?
11. How will this indebtedness affect profit?
12. How will this indebtedness affect solvency and liquidity?
13. What risk is involved that may delay the time period that funds may be repaid?

The manager who is seeking additional financial resources for the agribusiness should use the foregoing as a checklist to select the one alternative that is likely to be the most beneficial to the firm.

DAVIES FARM STRUCTURES, INC.

Doug Davies had started his farm-building company some 20 years ago with $1,000 in cash that he had saved plus $1,000 that he had borrowed from the bank. Today, Doug's company is a corporation that is owned primarily by members of the family. Doug is currently investigating an expansion opportunity, and his opportunity will be used to illustrate some of the decisions involved in financing an agribusiness.

Doug's present annual sales exceed $2 million, and although he now erects some commercial and public buildings, his major source of revenue remains farm buildings of various kinds. His workforce often exceeds 50 people. Recently, Doug was offered the opportunity to purchase a small lumberyard in Roland, a town of 6,500 people. The lumberyard had a yearly gross of over $750,000 and was earning 15 percent on owner's equity and 5 percent on sales in net after-tax profits. Doug had used the financial tools discussed in Chapters 12 and 13 to analyze the business, and felt that the lumberyard would be a great opportunity to expand and integrate into his own business. He knew that he had two expansion alternatives: either to expand his current operation or to diversify his business. He saw the lumberyard as an opportunity to do both at once.

Doug analyzed a number of advantages. His analysis of the lumberyard revealed that it could benefit from an increase in inventory turnover. By combining his building company's needs for lumber with the lumberyard's need for increased sales, he could increase the sales volume of the lumberyard by about 70 percent. The added purchasing volume for the construction company would allow him to buy lumber more cheaply because he could buy it in greater quantities and in full carloads. He could also consolidate his storage and handling operations and operate more efficiently because he would be located on a rail siding. His offices and entire operation could be moved to one location, and he figured that combining the operations would result in lower administrative costs.

With savings in cost of goods sold and administrative expenses, along with more operational efficiency, he foresaw the potential of lowering the prices for both businesses and being more competitive in both marketplaces. His figures showed that even if he lowered prices by 5 percent and assumed existing sales volumes for the two operations, he could maintain existing gross margins. But he hoped that the lowered price would help sales eventually climb as much as 50 to 75 percent because of his strongly competitive price position and his positive image in the farm structure business. At this point, the future looked very bright for Doug and his family. They felt they had successfully addressed questions 1 and 3: "Why are additional funds needed?" and "What increases in revenue will be generated by these funds?" While Doug's own company was in a very comfortable position financially, he knew that he would have to secure added financial resources to purchase and operate the lumberyard. The balance of this chapter will relate to the principal tools and alternatives that Doug Davies should consider in making this expansion decision.

DEBT OR EQUITY FINANCING

The next question that Doug had to ask himself was, "What kind of capital do I need?" Basically, there are four types of capital:

1. Short-term loans: 1 year or less
2. Intermediate-term loans: 1 to 5 years
3. Long-term loans: more than 5 years
4. Equity capital: no due date

Short-Term Loans

Short-term loans are generally defined as loans for 1 year or less, and they are used whenever the requirement for additional funds is temporary. Davies recognized the need for funds to increase inventory for the spring and summer, when the building business is at its seasonal peak. Some of these funds would also be needed to support accounts receivable as inventory is sold to customers. An important characteristic of short-term loans is that they are usually self-liquidating; that is, they often start a chain reaction process that results in their repayment:

Loan => inventories => receivables => cash => repay loan

While short-term loans may be made on an unsecured basis to firms that are well established, there is often a requirement for collateral, or for the loan to be secured by some of the firm's assets. Collateral can take many forms, but for short-term loans it most often takes the form of current assets. Some of the most common kinds of collateral are inventory, accounts receivable, warehouse receipts, and marketable securities. A personal guarantee of the loan by the owners is also common. In other words, the owner or owners endorse the obligation to repay the debt, and become personally liable if the firm is unable to meet the payment.

Short-term loans may be regular-term loans, with a specific amount due at a specific time, or they may be revolving or line of credit loans. Managers who anticipate a need for short-term funds often apply for a line of credit in advance of their needs. A **line of credit** is a commitment by the lender to make available a certain sum of money to the firm, usually for a 1-year period and at a specified rate of interest, at whatever time the firm needs the loan. Usually, the loan must be repaid during the operating year of the firm.

With a line of credit a manager is assured of protection in the form of cash that is available as it is needed; there is also the added advantage of not incurring interest on the borrowed funds until they are actually used. Lenders who make a line of credit available to an agribusiness often require that a monthly copy of the firm's financial statements be furnished to them so that they can monitor the financial health of the firm. Doug Davies wanted to avail himself of a line of credit for his short-term cash needs for seasonal purposes. He did not feel that he would have a problem securing these funds because he could pledge his inventory and accounts receivable as collateral against any outstanding loans.

It is important for the agribusiness manager to recognize that short-term borrowing is only appropriate for temporary uses. When, for example, funds are borrowed to increase inventories in order to accommodate increased sales volumes, and the loan is expected to remain in force for some time, a more permanent form of funds is needed. Such a permanent loan will increase the total working capital of the firm. Likewise, short-term loans should not be used to acquire fixed assets whose repayment term will likely be much longer than one year.

Intermediate-Term Loans

Intermediate-term loans are typically used to provide capital for 1 to 5 years. Such a loan is almost always amortized; that is, the process in which the amount of a loan is reduced by equal periodic (i.e., monthly, quarterly, annual, etc.) loan payments. These payments include both interest and principal. A simple illustration is provided in Table 14-1.

The purpose of the intermediate-term loan is to provide the agribusiness with a source of capital that will allow growth or modernization without forcing the owners to surrender control of the business. These loans provide for additional working capital, which can be used to increase revenues and sales; the funds generated by the increased revenues will, in turn, help to retire the loan.

In many respects the intermediate-term loan is similar to the short-term loan. Most require some sort of collateral and/or security against fixed assets, if that is

TABLE **14-1** | **Example of the Repayment Schedule for a $5,000 Loan[a]**

End of Year	(1) Payment	(2) Interest[b]	(3) Principal (1–2)	(4) Balance[c]
0				$5,000.00
1	$2,010.57	$500.00	$1,510.57	3,489.43
2	2,010.57	348.94	1,661.63	1,827.80
3	2,010.58	182.78	1,827.80	0

[a]Amortized with three equal annual payments and a 10 percent interest rate.
[b]Calculate by multiplying the balance at the start of the year by the contractual rate of interest (10%).
[c]Balance at the start of the year minus the principal payments (column 3) for the year.

the purpose of the loan. Intermediate-term loans provide permanent increases in capital for the agribusiness whenever larger inventories, larger accounts receivable, new equipment, and/or modernization are essential to the growth and profitability of the firm. Doug Davies foresaw a need for an intermediate-term loan. He wanted to increase both his accounts receivable and his inventory as he acquired the lumberyard. He also needed funds to pay for moving and consolidating his operation at the new central site.

Long-Term Loans

In general, **long-term loans** have duration of more than 5 years. The time distinctions among these loans are somewhat arbitrary, and there is some overlap in the functions of intermediate- and long-term loans, depending on the philosophies and policies of lender and borrower. But the real difference between intermediate- and long-term loans usually rests with the planned use for the funds, as well as with the long-term prospects for the existence of the firm and its solvency. The purpose of the long-term loan is most often for real estate, that is, for land and buildings. As the lender examines requests for long-term loans, he or she becomes deeply concerned with evaluating the past track record of the firm, the skill and ability of the management team, and the stability of the business enterprise. The security for long-term loans is usually a mortgage or claim on the fixed assets of the firm, and the longer the period of the loan, the riskier it becomes for the lender. There is always the chance that an unstable enterprise will be forced to dispose of fixed assets in a forced sale, where these assets may bring only a fraction of their true value.

Long-term loans are nearly always amortized over the loan period and secured by a mortgage or claim on a specific fixed asset. Sometimes bonds are used to secure long-term capital, but a small firm seldom has the size or financial strength to sell a bond issue. Because Doug Davies plans to build new storage facilities at his new site as well as enlarge the office building to accommodate the consolidated business, he will also need a long-term loan.

Equity Capital

If the agribusiness is not in a strong financial position (solvency is discussed in Chapter 13) or cannot meet the stiff collateral requirements of lenders, it may have to turn to equity capital to meet its long-term needs. **Equity capital** can be used for the same purpose as borrowed funds, but there is an important difference: equity capital does not have to be repaid. It becomes a permanent part of the capital of the business. Equity capital is secured either by reinvesting profits from the business or by finding investors who are willing to invest additional funding in the business.

Lenders pay particular attention to equity when they are making long-term loan commitments, and they may insist that a larger percentage of the owner's money be invested in the capital of the agribusiness than the lender's. This is particularly true of new businesses, where risks are more difficult to calculate. Some owners do not wish to increase their equity for various reasons, but it may be the only prudent way of securing long-term capital funds. Doug Davies will strongly consider expanding his equity base. Inasmuch as his business is already organized as a corporation, it will be easier for him to make the move should he so desire. (A more detailed analysis will appear later in the chapter.)

THE COST OF CAPITAL

When a business borrows money, it incurs special costs that are paid to the lender. One of these is interest, but interest is not the only cost of borrowing money. Several other factors affect the net cost of borrowed capital:

1. Repayment terms and conditions.
2. Loss of control; that is, compensatory balances, points, and stock investments.
3. The income tax bracket of the firm.

Repayment Terms

The repayment terms and conditions directly affect the rate of interest that is actually paid. If Doug Davies borrowed $100,000 for one year at the stated interest rate of 8 percent, the amount of interest paid would be $8,000. At the end of the year Doug would pay the lender $108,000, and his interest rate would have been the same as the stated rate of 8 percent because he used the entire $100,000 for the entire year. This type of interest is called **simple interest.** The formula for simple interest is:

(Amount of Interest Paid / Amount of Available Capital) × 100 = Annual Interest Rate

($8,000 / $100,000) × 100 = 8%

Sometimes, however, loans are **discounted,** which means that the amount of interest to be paid is deducted from the amount the lender provides the borrower. If this method had been used in Doug's case, the $8,000 interest to be paid, or

0.08 × $100,000, would have been deducted from the loan amount, and Doug would have had the use of only $92,000 in capital. The discounted loan formula is as follows:

Amount of Loan − Amount of Interest Paid = Amount of Available Capital

At the end of the year, $100,000 would have been repaid to the lender; but because Doug had the use of only $92,000, the "effective" interest rate was not the stated rate. The "effective" interest rate for a discounted loan is calculated as follows:

(Amount of Interest Paid/Amount of Available Capital) × 100 =
Effective Annual Interest Rate
($8,000 / $92,000) × 100 = 8.7%

The cost of interest on this discounted loan was then 8.7 percent and not the stated rate of 8 percent.

Banks often require that borrowers have a **compensating balance** in their accounts at the lending bank. To obtain a $100,000 loan, Doug might be required to maintain a minimum balance of $20,000 in his company bank account while the loan is outstanding. This means that he would have the use of only $80,000 in additional capital from the loan. The formula for calculating the effective rate of interest in this case is:

(Amount of Interest Paid / Amount of Available Capital) × 100 =
Effective Annual Interest Rate
($8,000 / $80,000) × 100 = 10%

If Doug normally carries a cash balance, this amount could be deducted from the compensating balance to lessen the increase of the effective interest cost.

Sometimes lending institutions also require that a certain amount of **points** (service charges based on the face value of the loan) be paid to secure the loan. These charges for risk and for loan servicing are made in advance, and the amounts so charged are usually deducted from the total capital at the time the loan is made. If Doug borrowed $100,000 at 8 percent with 2 points, his total loan cost would be:

$100,000 × 0.08	$ 8,000	interest
+100,000 × 0.02	2,000	points
Total Cost	$10,000	

Thus, a point amounts to one percent of the value of the loan.

Another demand that lenders sometimes make is that the borrower purchase a certain amount of stock in the lending institution, an amount that is determined by the value of the loan. The lender might require the purchase of one share of stock, valued at $10, for each $1,000 that is borrowed. In reality, this is a form of discounting that increases the effective cost of the loan. Actual or effective interest rates must be revealed to individual borrowers by commercial lenders under federal truth in lending laws, but this law applies to consumer loans and time purchases. It does not apply to most commercial or business transactions.

If the loan were repaid in monthly installments, the effective rate of interest would increase substantially. This rate of interest is called the **annual percentage rate** of interest or **APR.** The formula for calculating the APR on an installment loan is as follows:

$$APR = \left(\frac{2 \times P \times F}{B \times (T + 1)} \right) \times 100$$

where:

APR = annual percentage rate of interest
P = payments per year
F = dollars paid in interest
B = amount of capital borrowed
T = total number of payments

If the terms on Doug's loan included an 8 percent interest rate and monthly installment payments, the APR calculation would be:

$$APR = \left(\frac{2 \times 12 \times \$8,000}{\$100,000 \times (12 + 1)} \right) \times 100$$

$$APR = \left(\frac{\$192,000}{\$1,300,000} \right) \times 100 = 14.8\%$$

The simple interest loan gives the borrower the greatest amount of capital for the longest period of time. The discounted and compensating balance loans result in a slight decrease in available capital, so the borrower pays a higher annual percentage rate. The installment loan results in the highest APR since the full amount of the loan is available for only one month. At the end of the first month, part of this capital is repaid to the lender.

Other Restrictions

Lenders will often place restrictions on the management prerogatives of an agribusiness during the loan period. These restrictions vary from requiring monthly and annual financial statements or other financial information regarding inventories, accounts receivable, and accounts payable to actual restrictions on expending capital funds without the approval of the lender. Often banks require that firms maintain certain ratios during the loan period, such as a current ratio not below 2.0. Agribusiness managers must be sure that they can live comfortably with these restrictions before they agree to them. Otherwise, they may find themselves severely handicapped in decision making and in their flexibility to meet changing conditions and capitalize on new opportunities.

Interest Rates and Taxes

One of the things that agribusiness managers often overlook is that they can deduct from the business's taxable profits every dollar of interest paid, because the interest is a business expense. To know the effective cost of borrowing funds, the manager must know the after-tax cost of interest. This effect can best be seen by looking at net profits after taxes, before and after borrowing.

Using Doug Davies' company (a corporation) as an example, and assuming that he had borrowed the $100,000 at 8 percent interest, consider the following tabular information.

	Before Loan	After Loan
Net operating profit	$50,000	$50,000
Interest charges	-0-	-$ 8,000
Net profit before taxes	$50,000	$42,000
Income tax (assume 25% rate)	-$12,500	-$10,500
Net profit after taxes	$37,500	$31,500

The difference between the two situations is $6,000. The effective interest rate for decision-making purposes is then only 6 percent. The formula is:

$$\text{After-Tax Cost} = \text{Before-Tax Cost} \times (1.0 - \text{Marginal Tax Rate})$$
$$\text{After-Tax Cost} = 8\% \times (1.0 - 0.25) = 6\%$$

In this example, the firm paid $8,000 in interest on the original $100,000 loan. However, since these interest payments are tax deductible business expenses, the IRS really subsidizes the interest payments. The amount of the subsidy is equal to the firm's marginal tax rate, in this case 25 percent. So the firm paid "out-of-pocket" interest costs of $6,000.

The marginal income tax rate mentioned is the rate of income tax paid on the last increment of taxable income. Proprietorships and partnerships (see Chapter 4) must also be aware of this cost as they decide between investing their own funds and borrowing capital funds for their businesses.

The Leverage Principle

Leverage is the concept of financing through long-term debt instead of equity capital. Many managers like to use debt as a lever against equity as much as possible so that they can maximize the amount of assets or capital at their disposal. Several factors affect the leverage principle. First, it must be remembered that as the proportion of debt to equity increases, lenders are likely to increase the cost of supplying capital funds because of the deterioration in the solvency measures and the resulting increase in risk. It must be understood that risks for equity holders also increase as debt increases, because they hold a last-place claim on the firm's assets in the event that all does not go well. In general, equity capital is risk capital—in the event of financial problems, all other creditors are paid before equity owners are paid. Leverage, or increasing the proportion of debt to equity, can be either a profitable or an unprofitable decision.

As a rule of thumb, the after-tax rate of return on the capital of the agribusiness must exceed the after-tax cost of the firm's debt to increase profits. For example, if a firm has the ability to return 10 percent on the borrowed capital and the after-tax cost of borrowing that capital, or money, is 6 percent, borrowing more money should increase profits.

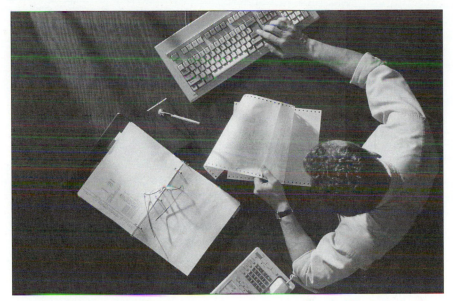

Careful analysis is required to determine the right type and level of financing for the agribusiness firm.

Determining How Much the Agribusiness Should Borrow

The question of how much an agribusiness should borrow is one that agribusiness managers frequently ask. Some answer by saying, "All I can get," while others say, "Let's pay off the mortgage and eliminate our long-term debt." These philosophical generalizations are not adequate for determining how much an agribusiness should borrow. The good manager always establishes criteria and a frame of reference for such decisions. This section will deal primarily with intermediate- and long-term debt, since it is assumed that short-term debt will be paid off from conversion of current assets to cash. The amount of debt that is most desirable depends on several factors, some of which have already been discussed. Many of these factors are easy to measure, but others are much more difficult.

The first factor to consider is the amount that the agribusiness will be able to generate for debt servicing (repayment of the loan). While available funds can be calculated from all sources of cash flow, generally two factors are considered the primary inputs for debt servicing: (1) net operating profits for the year, and (2) depreciation. Net operating profits must be further reduced by any interest that is due, income tax to be paid, dividends owed on owner's equity, or patronage refunds in the case of a cooperative (see Chapter 5).

For example, if Doug Davies has a net operating profit of $50,000 and depreciation of $25,000, he has a preliminary total of $75,000. Depreciation is added to net operating profit because it is a noncash operating expense that is subtracted from gross profit to calculate net operating profit. However, the depreciation expense is not paid to an entity outside the firm, so those dollars are available to pay debt obligations. For debt servicing purposes, he would have to deduct $8,000 in interest expense, $25,000 in taxes, and $5,000 for stock dividends, which would leave the firm with only $37,000 for debt servicing:

Net operating profit	$50,000	
Depreciation	+25,000	
Total available		$ 75,000
Interest	$ 8,000	
Income tax	25,000	
Dividends	5,000	
Total required		$ 38,000
Amount available for debt servicing:		$ 37,000

Doug would also have to consider other possible needs for these funds, such as boosting working capital and increasing stockholder dividends. When lending institutions look at these debt servicing figures, many use the rule that no more than 50 to 60 percent of the total should actually be counted as available for debt servicing, because of the possibility of missing budgeted figures or emergency situations.

If the additional capital that is to be borrowed will increase revenues and profits, thereby increasing debt servicing potential, then the amount borrowed can be increased accordingly. Forecasting such new earnings accurately is crucial. Many managers tend to be overly optimistic, especially in the short run. Remember Murphy's Law in this regard: "If anything can possibly go wrong, it will." The risk can be lessened considerably if the manager's profit forecast is understated. For example, if Doug feels that he will add $10,000 to his debt servicing capability through the loan, for the first year, at least, he would be advised to actually count on only half this amount, or $5,000, as available for debt servicing.

Several other factors must be considered as Doug analyzes the amount of money he should borrow. Debt servicing costs can be extended to the upper limits if:

1. Investors agree not to withdraw funds if the firm faces adverse market or competitive conditions.
2. The firm has a favorable debt-to-equity ratio (solvency), or large amounts of working capital.
3. The firm has fixed assets that can readily be converted to cash without incurring large losses.
4. There are redundant fixed assets that can be sold.
5. There is a low risk on the asset purchased, as, for example, with a new piece of equipment that will save labor.

If accelerated depreciation or special depreciation measures are used to increase the amount of depreciation that is taken each accounting period, then the amount available for debt servicing must be considered in that perspective, and the manager may want to increase the amount of this particular contribution to debt servicing. Finally, the manager will want to take a long, hard look at the overall stability and success of the firm and of its management team. Such factors as profits, control of inventories, accounts receivable, turnover, and efficiency will be the final elements in determining the amount of capital that the firm should borrow.

OTHER TOOLS FOR FINANCING DECISIONS

Two other techniques, or tools, play an important role in making financing decisions. These are the cash budget and pro forma financial statements, both of which can help the agribusiness manager to look ahead intelligently and can aid the decision-making process immeasurably. Both of these tools were introduced in Chapter 12, but will be explored in more detail here.

Cash Flow Statement

A **cash flow statement** is really a projection of the firm's cash inflows and outflows for a future time period (Table 14-2). It allows the manager to estimate the amount of cash needed to take advantage of cash discounts, to finance seasonal demands, to develop sound borrowing programs, to expand, and to make plans for debt servicing.

The length of time covered by a cash flow statement depends on the unique nature of the agribusiness. The primary considerations are the current supply of cash and the seasonality of cash inflows and outflows for the business. Highly seasonal agribusinesses will need to prepare their cash flow statements over longer periods of time than those whose business activity is fairly constant.

When Doug Davies prepares his cash flow statement, he will follow the budgeting steps outlined in Chapter 15. With the help of his planning committee, he will estimate, from both his existing business and the one he hopes to acquire, his cash receipts and cash payments. These estimates in reality become goals or budgeted figures, so careful and honest input is needed. Both cash receipts and expenditures are recorded on a month-to-month basis during the period—the end result being the cash balance at the end of the period. A goal must be set to determine whether the amount is sufficient (Chapter 13). For example, Doug might decide that the cash equivalent of a certain number of days' sales, or a certain percentage of current liabilities, would be the benchmark or goal.

If the cash balance is inadequate, then short-term borrowing or other adjustments might be needed. If the cash balance is larger than needed, the excess can be invested temporarily in marketable securities, other income-producing assets, or used to reduce current or long-term debt. The cash flow statement can help the manager to decide whether there is a need for short-, intermediate-, or long-term loans or equity capital. If the cash amounts are ample at certain times and short at others, short-term capital is needed. If there is a persistent trend on the short side, intermediate- or long-term capital is needed.

TABLE **14-2** | **Cash Flow Statement for Three Months Ending March 2002**

	January Budget	January Actual	February Budget	February Actual	March Budget	March Actual
Expected cash receipts:						
1. Cash sales						
2. Collections on accounts receivable						
3. Other income, investments called						
4. Total cash receipts						
Expected cash payments:						
5. Raw materials						
6. Payroll						
7. Other factory expenses (including maintenance)						
8. Advertising						
9. Selling expense						
10. Administrative expense (including salary of owner-manager)						
11. New plant and equipment						
12. Other payments (taxes, including estimated income tax; repayment of loans; interest; etc.)						
13. Total cash payments						
14. Expected cash balance at beginning of the month						
15. Cash increase or decrease (item 4 minus item 13)						
16. Expected cash balance at end of month (item 14 plus item 15)						
17. Desired working cash balance						
18. Short-term loans needed (item 17 minus item 16, if item 17 is larger)						
19. Cash available for dividends, capital cash expenditures, and/or short-term investments (item 16 minus item 17, if item 16 is larger than item 17)						
Capital cash:						
20. Cash available (item 19 after deducting dividends, etc.)						
21. Desired capital cash (item 11, new plant and equipment)						
22. Long-term loans needed (item 21 less item 20, if item 21 is larger than item 20)						

Pro Forma Financial Statements

Because the cash flow statement deals with only one account, cash, it is wise to go one step further and prepare a pro forma balance sheet and income statement. These statements really just project the best estimates of what the business will look like in the future (Table 14-3 and Table 14-4). The pro forma financial statements are typically prepared on a quarterly basis for an operating year. Again, the more seasonal an agribusiness, the more frequently it needs the pro forma statements prepared.

The **pro forma financial statements** will provide a look into the future of the business and will help the manager determine what the financial needs of the business will be during and at the end of the operating period. If Doug fails to use this tool, he may not recognize problems until they actually arise, and then it may be too late to take corrective action. The most important figure in the preparation of these pro forma statements is estimated sales (Chapter 9). This is where as many well-informed people as possible should become involved. Doug might well involve his family, the bookkeeper for the lumberyard, salespeople from both firms, supplier representatives, and his banker, to name some. Past experience and future price trends are key ingredients, along with expected competition from other firms in the area, and general economic conditions.

It is vital that these two tools be based on solid goals in the areas of cash balances, inventory turnover, accounts receivable collection periods, revenues,

TABLE **14-3** | Pro Forma Monthly Income Statement

Pro Forma Monthly Income Statement for the Month Ending _____

	Figures Based On
Revenue from sales	Sales forecast for the month
Cost of sales	Sales forecast, historical data
Gross margin	
Operating expenses:	
Selling expenses	Budget for the month
General expenses	Sales forecast, historical data
Total operating expenses	
Net income from operations	
Other expense:	
Interest expense	Outstanding debt, expected additional debt
Net profit before taxes	
Income taxes	Tax rate
Net profit after taxes	
Earnings withdrawn	Owner's intentions
Retained earnings	

T A B L E **14-4** | **Pro Forma Balance Sheet**

Pro Forma Balance Sheet for the Month Ending _____

	Figures Based On
Assets	
Current assets:	
Cash	Desired cash balance
Accounts receivable	Average collection period
Inventory	Monthly inventory turnover for season
Total current assets	
Fixed assets	Present figure adjusted for period's depreciation and any planned investments
Total assets	
Liabilities	
Current liabilities:	
Notes payable	Amount of borrowed funds needed to balance assets with equities
Accounts payable	Expectation of number of days' purchases on the books
Accrued liabilities	Same as preceding period
Total current liabilities	
Long-term debt	Expected and additional borrowings
Total liabilities	
Owner's Equity:	
Paid-in capital	
Retained earnings	Present amount plus period earnings to be retained
Total owner's equity	
Total liabilities and owner's equity	

and expenses. These goals help the manager in the budgeting and forecasting process. Specific goals also imply there is a management plan that can be objectively evaluated at the end of the period.

With documented goals, management can use the interim cash flow statement and financial statements to check their progress against their estimates and assumptions. If actual and planned performance vary widely at any point, the reason can be explored and the weaknesses corrected. As a professional manager, Doug Davies will avail himself of all these financial tools as he progresses in his new venture, prepared and as ready as anyone can be for the future.

EXTERNAL SOURCES OF FINANCING

There are a multitude of sources of capital available to any agribusiness. Some of these sources are used only in specific situations, while others are more common. The most important sources of financing for the agribusiness will be discussed in the following sections.

Commercial Banks

Commercial banks are the major source of borrowed funds for most agribusinesses. It is estimated that they provide over 80 percent of the funds loaned, excepting trade credit. Commercial banks usually offer a full line of banking services, including checking accounts, savings accounts, and loans. Banks make many kinds of loans, such as short-, intermediate-, and long-term loans, lines of credit, and special loans.

Banks also make personal unsecured loans to owners; many other kinds of secured loans, such as mortgages against real property; chattel mortgages against tools and equipment; and loans against the owner's life insurance policies, stocks and bonds, and so forth.

Often a bank will offer to help sell a product by buying sales installment contracts from the seller. Sales installment contracts are contracts made by the buyer to pay for the product in a specified manner over a period of time. When a financial institution buys the contract, payments are made directly to that institution. The business gets its money immediately. This procedure makes it easier for the customer to finance the purchase and helps the cash flow of the business. This is particularly helpful to retailers who are selling relatively large investment items, such as farm implements, tractors, and combines.

Accounts Receivable Loans

Accounts receivable loans are loans in which the bank lists a business's accounts receivable as collateral. This may be done on either a notification or non-notification basis. Notification means that the bank informs the debtors that it wishes to collect the part of the money that is owed. The bank receives the payment and deducts a service charge and interest, then credits the balance against the borrower's loan. Under non-notification, the borrower collects the receivables and forwards them to the bank. Recordkeeping and interest costs are usually high, and managerial flexibility is lost, so non-notification loans should typically be avoided.

Many bankers are reluctant to provide accounts receivable loans, or simply charge higher interest rates on these loans to discourage their use by borrowers. Also, since there is no guarantee that all of a firm's accounts receivable will be collected, the bank will limit the amount of accounts receivable that may be used as collateral against any loan. This limit will be a function of the firm's credit policy, record of collections, and current as well as forecasted future economic conditions. It is common for this limit to approach 50 percent of a firm's accounts receivable.

Warehouse Receipts

Warehouse receipts represent a means of using inventory as security for a loan. As inventory is stored in the warehouse, the borrower sells the inventory to the bank, and then buys back the receipts from the bank as product is sold. This type of loan is feasible only on nonperishable items, and allows the borrower to manage with limited working capital.

Warehouse facilities may represent a large storage building, a grain bin or shed, for example, or simply be a fence around a large site of coal to be used by a major processing organization to generate electricity. The key in warehouse receipts from a lender's point of view is to know reliably what is in the facility. Thus, if a borrower has financed $2 million worth of grain through a warehouse receipt, it is critical that a representative of the bank makes sure the grain is in the storage facility and has the ability to measure quantity and assess quality. This comes at a cost, borne by the borrower—hence, these loans are used less by most agribusiness firms compared to other financing alternatives.

Insurance Companies

Insurance companies are always looking for places to invest funds they have collected from policyholders. Most insurance companies are interested in intermediate- and long-term loans on fixed assets, such as equipment or real estate. They prefer large loans and mortgages for collateral. If the owners or the agribusiness itself has insurance policies with a particular company, that company will usually lend the agribusiness amounts that are equal to the cash value of the policy at very favorable interest rates.

Commercial Finance Companies

Commercial finance companies are those finance companies that specialize in business and commercial loans. They are not to be confused with personal finance companies, which make loans to individuals. Commercial finance companies often grant loans that are riskier than those that banks will accept; therefore, commercial finance companies normally charge higher interest rates than banks. Commercial finance companies may also demand a considerable amount of control over management decisions. This is particularly true if the loan involved is high risk. Sometimes commercial finance companies will pay off all firm debts in order to consolidate the firm's indebtedness into one loan held by the finance company. This can be of particular value if cash flow presents a problem, because payment schedules can be reconstructed within the constraints of the agribusiness's cash flow.

Factors

The **factor** is a very specialized source of capital funds. A financial institution or individual acting as a factor buys accounts receivable from the firm at a discounted price. They buy these accounts receivable without recourse; that is, factors assume the risk of bad-debt losses. Except when they are convinced that the accounts are sound, they may understandably pick and choose those they buy. Their procedure is to notify the individual accounts and collect directly from these accounts. Many agribusinesses do not want their customers dealing directly with the collecting factor. However, in some cases it is the only way for a business to raise quick cash.

Cooperative Borrowing

Cooperatives, of course, have all the conventional borrowing sources, but in addition agribusiness cooperatives can borrow from the Banks for Cooperatives, which are part of the Farm Credit System. In **cooperative borrowing,** the cooperative patrons, who are also its borrowers, own these special banks. The bank makes short-, intermediate-, and long-term loans to its members. To receive a loan, a cooperative must purchase an amount of membership stock that is equivalent to the amount of money being borrowed. This stock bears interest, and is revolved or repurchased whenever the debt is repaid and the firm has funds available for that purpose. Often the Banks for Cooperatives can offer better interest rates than some commercial banks because it is a nonprofit organization operated exclusively for its members' benefit. Because they specialize in cooperative loans, the bank's personnel are often able to offer management help and guidance to member-borrowers.

Trade Credit

One of the most neglected sources of capital is the credit advanced by suppliers and vendors of the agribusiness firm or **trade credit.** If the agribusiness is creditworthy, most suppliers and vendors will allow credit terms. The manager can often negotiate for longer credit terms than are usually extended. For example, Doug Davies was such a steady purchaser of treated lumber that he was able to get his supplier to extend his normal 30-day credit terms to 90 days from invoicing. In Doug's case he had often collected for the farm building prior to the 90-day credit period, so in a very real sense the lumber supplier also became a supplier of Doug's business capital, but without any cost to Doug. In other cases, a supplier may be willing to sell to an agribusiness on consignment. This means that the business does not have to pay for supplies until it is able to actually sell them. The agribusiness manager should make sure that suppliers and vendors are extending maximum terms, and that the procedure for paying accounts payable takes every advantage of all credit terms extended.

Leasing or Renting

An alternative to owning a durable asset is to lease that asset. A **lease** is a contract by which the control over the right to use an asset is transferred from the **lessor** (owner) to the **lessee** (person acquiring control) for a specified time in return for a rental payment. There are several types of leases, which are discussed on page 384.

Land Lease   Leasing land is a common method of acquiring control of a durable asset in agriculture, land. A **land lease** is a contract that conveys control over the use rights in real property from the lessor to the lessee without transferring title. The contract usually specifies the property's intended use and the conditions of payment for that use.

There are several types of land leases. The more common types are cash, crop-share, and livestock-share. A cash lease has a predetermined cash fee per unit (i.e., per acre, total farm, etc.) for the use of the land irrespective of the resulting yields or product prices. With a crop-share lease, the lessee pays a specified percentage of the crop to the landowner (lessor) for the use of the land. The share is set so both parties are compensated for the resources each provides. A livestock-share lease is usually an arrangement in which the landowner furnishes the land and buildings, while either or both of the parties own the livestock. Returns from the sale of livestock are divided between the two parties according to the share arrangements.

Operating Lease   An **operating lease** is usually a short-term rental arrangement (i.e., hourly, daily, weekly, monthly, etc.) in which the rental charge is calculated on a time-of-use basis. The lessor owns the assets and performs almost all the functions of ownership, including maintenance. The lessee pays the direct costs, such as fuel and labor. However, the terms may vary and may even be negotiated between the two parties.

Examples of operating leases are rental cars, trailers, moving trucks, etc. Agricultural equipment can also be leased. Such leases are especially useful for part-time producers who operate small acreages and cannot afford to own general-purpose pieces of equipment and for producers who farm larger acreages and have equipment breakdowns or have been delayed in planting, harvesting, etc. Also, specific-purpose types of equipment, whose demand varies by region or by type of service, are popular items for operating lease arrangements.

Custom Hire   **Custom hiring** is a form of leasing that combines labor services with the use of the asset being leased. Common examples in agriculture include custom harvesting for crops, custom applications of chemicals and fertilizer, and custom or contract feeding of livestock.

Custom hiring resembles other types of leasing since it avoids the financial drain of capital investments and the obsolescence risk of owned capital. At the same time, the specific functions are performed by the individuals or firms that control the equipment or facilities (lessor), who may have more training and experience with the equipment than the lessee.

Capital Lease   The **capital lease** is a long-term contractual arrangement in which the lessee acquires control of an asset in return for rental payments to the lessor. The contract usually runs for several years and cannot be cancelled. So the lessee acquires all the benefits, risks, and costs of ownership, except price variations of the asset; but does not have to make the usual investment of equity capital. It is comparable to a credit purchase that is financed by a loan with 100 percent financing. However, it should be recognized that lease prepayments have the same cash flow effects as down payments for credit purchases.

Leasing provides an opportunity for many agribusiness firms to extend their capital assets without having to borrow. However, leasing is the equivalent of borrowing, because it takes its place as a means of acquiring capital. Much of the money used in leasing comes from financial institutions and insurance companies. Special organizations have been set up to lease a business almost anything. Doug Davies is seriously considering leasing trucks for his business in order to conserve cash.

Leasing, a form of borrowing, is commonly used to finance equipment purchases.
Source: Photo courtesy of Deere & Company.

The typical lease, as would be expected, will cost more than interest on a loan. The lessor (person granting the lease) must charge an amount to cover profit for arranging the lease, the element of risk, interest on the capital involved, and the depreciation on the item leased. The longer the lease period, the lower the lease charges will be per period, which means that if Doug Davies leases a truck for 4 years, he will pay a lower yearly rate than he would if he leased it for 2 years. In many cases, the lessee (person renting the property) can arrange to purchase the property at the end of the leasing period for a predetermined amount. The popularity of leasing has grown significantly in recent years. This financing arrangement ten years ago was relegated to business use only. Today, however, leasing is one of the most popular ways for individuals to acquire the use of new and used automobiles.

Advantages and Disadvantages of Leasing

A business leases to avoid using its cash resources for purchasing assets. It will not have to resort to borrowing or selling equity. Many agribusinesses take the position that their funds should be used for expanding their operations rather than buying assets that can just as easily be leased. Leasing is also a deductible expense, and, in some cases, may be cheaper than borrowing, buying, and incurring the cost of depreciation. Another advantage is that leased assets can be turned back to the lessor and newer or better assets procured. This is of special value when change or new technology is pertinent to the asset. Today, for example, many firms lease computer equipment for a specific period of time, say three years, and then turn the hardware in for new equipment. This allows firms to utilize up-to-date technology at a reasonable cost.

But leasing is not without its drawbacks. For most businesses, leasing will cost more than borrowing. Leasing also commits the business to certain payments, whereas if the asset were owned it could be sold to at least minimize the financial commitment. Finally, leased property may increase in value, and such increases in value are only of benefit to the lessor or owner.

Other Sources of Capital

The agribusiness can draw upon many other sources of capital, including bonds, debentures, and promissory notes. These sources of capital are discussed below.

Bonds   **Bonds** are usually issued by corporations. They represent an obligation by the corporation to pay a certain sum at a specified time in the future. They are issued in series and are redeemed in the order in which they are issued. Usually, bonds carry a specified rate of interest that is paid on an annual basis. They are generally used for raising long-term capital, and the date of redemption is often 20 to 30 years from the date of issue. Bonds are not secured by any particular collateral other than the general assets of the firm, but they have a preferred risk; that is, they are redeemed prior to the common stock of the corporation or some other obligation of the firm. They may be traded on stock exchanges if they qualify. As a source of funds, bonds are of most value to large corporations.

Debenture   **Debenture notes,** which are issued in exchange for the loan of capital, usually create a security against the general assets of the firm, or a part of the company's stock and property, though not necessarily in the form of a mortgage or lien. Most debentures are restricted as to their sale and transfer among lenders, and usually have coupons attached to facilitate the payment of interest. They are generally issued and redeemed in a series, without any one debenture taking precedence over another.

Promissory Notes   A **promissory note** is a promise by the borrower to pay to the lender a particular amount of money and a particular amount of interest after a specified period of time. Promissory notes are common to agribusiness firms and banks, private individuals, and creditors. Agribusiness firms may also accept promissory notes from their customers. Such notes may be negotiable, that is, the holder can sell them and the new owner will have the same claim against the borrower as the original lender. When an agribusiness firm holding negotiable notes from customers needs cash, negotiable promissory notes can be sold to a bank or other person, usually at a discount. For example, Doug Davies might sell a farmer some materials to help build a new dairy barn, say $50,000 worth, and accept in return a negotiable note that is due in 6 months and bears a 10 percent interest rate. In the meantime Doug might need cash, in which case he could sell the note to a bank or other financial intermediary. If the farmer's credit is good, and the banker feels the interest rate is good, the note might be purchased by the bank for face value or at a very small discount that reflects a service charge. The farmer would then pay the bank when the note is due.

CHOOSING A BANK

The selection of the right bank can spell the difference between success and failure for the agribusiness firm. The right bank can be a lot more than just a place to deposit funds, service a checking account, and request an occasional loan. Convenience should not be the main criterion for selecting a bank. Many agribusinesses are located in rural communities and small towns. They feel that loyalty demands they do business with banks in their community, but this feeling has merit only if the bank can meet the needs of the agribusiness. Many small banks simply do not have the facilities or resources needed to make a growing agribusiness viable. One of the ways in which a small bank can expand its services is by facilitating working arrangements with larger banks. Here the best of both worlds can be combined; that is, the business can have the advantage of dealing locally and the advantage of having full service banking available.

Requirements for Selecting a Bank

There are five main points to consider when selecting a bank:

1. Is the bank progressive in its attitude toward agribusiness?
2. Are the credit services offered sufficient to meet the agribusiness's needs?
3. Is the bank large enough to meet the agribusiness's capital needs?
4. Does the bank have qualified personnel who are knowledgeable about agribusiness?
5. Are the bank's management policies in line with the agribusiness's objectives and financial strategy?

Is the Bank Progressive?   The progressive agribusiness manager needs a progressive bank, one that keeps up with changing times and with the community it serves. Physical facilities do give some indication of progressive thinking, but the real test is the organization and its people. Do the employees have a genuine interest in serving the agribusiness? Are they interested in the community and do they participate in civic affairs? Does the bank have personnel who specialize in agribusiness services, and have they appointed someone to handle the firm's account? Has this person called on the firm? Does the bank offer the firm innovative ways to meet its capital needs? These are but a few of the ways to judge how progressive a bank is or will be as it serves a business.

Check Credit Services Offered   The banker should understand the unique needs of the agribusiness. Can or will the bank make loans on receivables or warehouse receipts; provide lines of credit; or make short-, intermediate-, and long-term loans? Will it honor drafts, accept negotiable notes, and the like? Will it provide credit references for potential customers, help to solve capital problems, refer customers by building the firm's image? What does the bank charge for its various services? Are interest rates and fees competitive?

How Big Is Big Enough?   The size of the bank can be an important factor, especially if the agribusiness's credit needs are large and increasing. Banks are regulated by state and federal agencies that restrict the size of loans and the amount that a bank can lend to any one customer. Managers must make sure not to be caught in a situation where this legal limit will constrain the business. If a local bank cannot meet the firm's capital requirements, the agribusiness manager should make sure that it can make arrangements with other banks to fund the balance of the credit needed. Is the bank large enough to have a separate, well-managed trust department? If the agribusiness sells on installment loans, can the bank handle a working relationship with these buyers? Since size and prestige are often related, the agribusiness manager should make sure that the bank has a solid image in the financial community.

Qualified Personnel   Agribusiness is a unique form of business. It is important for the personnel at the bank selected to know the needs of the agribusiness firm. An understanding of agriculture, seasonality, and the like is indispensable. Do bank personnel also understand financial information, analysis, and the tools of financial management and planning? Are the bank personnel knowledgeable enough to provide sound input to the firm's capital and credit needs? A knowledgeable, interested, and dependable banker can become an important member of the management team, who is just as interested in the agribusiness's success as is the business. The banker's enlightened self-interest is to the agribusiness's advantage, too, because it leads to a healthy, growing, successful source of capital that will properly serve the business's unique needs.

The Bank Policies   The agribusiness manager wants a sound, well-managed bank. Banking relationships should be more or less permanent in nature. As the bank and the agribusiness work and grow together, knowledge, experience, and personal relationships take on greater meaning. Banks can be ultraconservative or ultraliberal in their lending policies. The needs of the business dictate that a bank be found that is willing to assume a risk when it makes capital available to the firm. Some banks will not lend money to a firm unless a need can be proven, while others may lend a customer into bankruptcy. A bank that has sound lending policies, knowledgeable personnel, and is able to advise as well as lend money should be chosen. The agribusiness manager should look upon the banker as a partner in the business, and should do the utmost to make the relationship a profitable and satisfactory one for both parties.

Getting Ready for the Loan

Often the agribusiness manager's success in securing capital from a lending institution will depend solely on how he or she goes about it. Actually, the best guide for preparing for a loan would be the 13 questions (listed earlier in the chapter) that the agribusiness manager should ask before adding capital to the firm. If the person who is negotiating the loan can answer these questions, then preparation is complete. These data should be well organized and presented clearly. Incompleteness and vagueness are deathblows, since they demonstrate the manager's ineptness and lack of financial acumen. Bankers are greatly impressed by managers who know what is going on and can prove it with the financial data to back up claims and requests.

Historical evidence of performance, trends, and future plans are all-important ingredients. Some of the most important information includes:

1. Balance sheet and income statements for at least the last 3 years.
2. Trends in sales, expenses, profits, etc. (financial ratios are helpful here).
3. Description of the market (customers), products and services, and suppliers.
4. Information about working capital, aging accounts receivable, turnover, inventory, and so forth (again, ratio analysis and the profitability analysis model are helpful).
5. Credit and character references, and the background of the management team.
6. Evidence of planning for the future; that is, cash budget, expansion plans, and pro forma financial statements.
7. The applicant's personal history and that of the agribusiness.

The applicant should be candid and honest. Facts should not be withheld, even those that could be damaging, since they are likely to surface anyway. Damaging facts that are brought out through the manager's initiative will be less damaging than they would be if the banker discovered this information was withheld.

A good question for the agribusiness manager to ask is, "If I were presented with this information, what would I think of it and would I make a loan on this basis?" Offering the banker a tour of the facility and an opportunity to meet other personnel from the agribusiness can also be beneficial in creating a favorable impression, and can be of benefit in building a better long-term relationship.

INTERNAL FINANCING FOR THE AGRIBUSINESS

One of the most important sources of capital is the money obtained from retained earnings. Some managers fail to fully understand the importance and role of retained earnings. It is not that managers lack awareness of these funds, but because they simply have not used the financial tools and techniques previously described, they have no idea how to use retained earnings to their fullest extent.

Equity Capital

Equity capital represents funds that are obtained by the firm through retaining the profits it has made, through the investment of more money by the owners, or through taking into the business additional people who are willing to risk their funds. In some small or new businesses, this may be the only alternative for securing capital funds. A specialized type of financing organization, the **venture capitalist,** focuses on providing equity capital to new and high potential businesses. These venture capital firms typically take an equity position in a firm that may be very risky, but be deemed to have major long-term potential. Many of the Internet start-up companies in the food and agricultural markets have venture capital backing.

Some owners are not anxious to sell equity to other people. These owners feel that the loss of control over the business is not worth the additional capital. Such owners should be aware that the borrowing of funds may place far more restraints on their control than the sharing of ownership. And if an owner defaults, or is slow in paying a term loan, he or she can completely lose control of the management of the business. Remember, equity capital does not have to be repaid at any certain time, and often there is no absolute need to generate funds for distribution to the owners. Equity capital should always be considered as an alternative, and weighed against other capital sources.

Common Stock

The most prevalent form of equity capital is the kind that is secured through the sale of **common stock.** For the small company, this may mean selling shares of stock primarily to people who are known to the present owners. There are always people in any community who have funds to invest in a promising business venture. The firm's banker can often be helpful in suggesting interested people. Employees of the firm are also a potential source of stock purchasers, especially if the firm offers a special purchase plan that grants employees preferential prices. Common stock is usually voting stock; that is, owners of common stock have a voice in the management of the firm. Sometimes common stock is divided into classes, of which only one kind carries the voting privilege.

Doug Davies has carefully studied the financial needs of his firm. He has decided on the particular mix of borrowed and equity funds that he thinks is optimum for the business. His intentions are to offer the lumberyard's owner, as part of the purchase price, a certain amount of stock. He also intends to offer stock to his employees and to members of the community on a limited basis. He determined this ratio of equity to borrowing by using the tools in Chapters 13 and 15 that relate to solvency.

Care must be exercised that laws relating to the sale of securities are observed when stock is offered to the public. All states have blue-sky laws, which regulate the sale of securities and stocks. In some cases, federal laws regulating the sale of stocks also apply. It is essential that a team of advisors be assembled before a public offering of stock is made. The business's banker, legal counsel, and auditor should be among those involved. If a firm wants to make a larger public offering, it will usually secure the services of an investment banker. These bankers perform a very special service by offering for sale the new stock offerings presented by companies. When an investment banker underwrites a stock issue, that banker makes an agreement with the corporation to market its securities by buying them and reselling them to the public. Investment bankers typically charge a commission of from 3 to 10 percent, depending on the difficulty of selling the stock and the quantity offered. It is unusual for even small investment bankers to underwrite less than $250,000 of stock.

Present owners of the business do not necessarily have to lose control of a business by sale of its common stock. They can retain control by keeping a sufficient amount of the stock issued themselves. Often bonds or debentures are sold as convertible issues. This means that they can be converted at some future time to a certain quantity of common stock. Companies offer this conversion feature as an inducement to secure buyers for their bonds and debentures, because if a company is successful, its stock may appreciate considerably in value.

Preferred Stock

Preferred stock is the stock to which a corporation shows preference. In the event of the liquidation of the corporation, the owners of preferred stock would be repaid before holders of common stock. Most preferred stock also has a definite annual dividend rate, which means that it would pay a percentage (e.g., 6%) of the face or issue value on an annual basis. Sometimes corporations will reserve the right to defer this dividend until a later date should the corporation have financial difficulties. In exchange for its preferred nature in the event of liquidation, most preferred stock does not carry with it any voting rights or control in the management affairs of the corporation.

Other Internal Financing

Partnerships may secure more capital by selling portions of their business to others who are willing to risk their money in the business. These others may be either general or silent partners. A general partner assumes the same rights and liabilities as other partners, while a silent partner has restricted rights and liabilities (see Chapter 4). Or an owner may simply lend money to the business just as any outside creditor would, if that owner does not wish to commit any additional funds on an equity basis.

SUMMARY

As the agribusiness manager strives to turn everything that he or she touches into gold, it is important to remember that not all sources of financial help are equally useful or equally applicable to all situations. The securing and managing of the financial resources of any agribusiness firm is a complex function, but careful attention to the tools, techniques, and principles discussed in this chapter will increase the agribusiness manager's chances of success.

The agribusiness manager needs to know the various kinds of loans, the cost of borrowing, and whether short-, intermediate-, or long-term capital is needed. The manager must explore all sources of capital to discover whether borrowing, equity financing, or some combination thereof is best for the particular agribusiness involved. But even more importantly, the manager's ability to assess the optimum amount to be borrowed, and to formulate a realistic plan for repayment, will make the agribusiness's financial strategy a firm and steady foundation for the future.

Financing the agribusiness is a necessary and important management responsibility. Money must be available to finance capital purchases and operate the business on a day-in, day-out basis. There are three primary sources of capital:

borrowing, funds generated from business operations, and additional investments from owners.

Borrowing can take many forms. Short-term loans of a year or less are normally used to finance seasonal business needs. Intermediate loans of 1 to 5 years are generally used to purchase equipment or finance increases in the volume of business. Long-term loans are usually for major business expansion, such as buying land and erecting buildings. Interest rates and repayment schedules vary according to a great many factors, including time, risk, amount of money borrowed, past experience, and the soundness of the firm's financial base.

Managers must consider many different sources of borrowing; for example, commercial banks, insurance companies, and commercial finance companies, to name three. Choosing the right lender is often a critical management decision. Larger agribusinesses borrow from the public by issuing bonds or debentures through the money and capital markets.

Some agribusinesses, particularly newer ones, rely heavily on equity financing. Equity capital results from retaining profits in the business rather than distributing them to owners. Other agribusinesses sell additional stock to attract additional dollars for the business. Each financing method has pros and cons and must be considered carefully by management.

DISCUSSION QUESTIONS

1. Suppose an agribusiness firm has decided it must increase capital. What are three ways to accomplish this? Would the method selected be different for: (a) a corporation, (b) a partnership, or (c) a proprietorship? Why? Defend the "best" selection for each business form.
2. What are the three general classes of loans? Give two examples of a firm needing each type of loan, and list the specific asset or type of capital acquisition you have in mind.
3. Why would a firm usually not finance a new truck with short-term debt? What problems could this cause?
4. What are the advantages of a line of credit as compared to a short-term loan?
5. Is there ever a situation when a borrower would prefer a discounted loan to a simple interest loan? Explain your answer.
6. You purchase a house for $180,000. The bank is willing to loan 90 percent of the value of your home plus 2 points. How much money do you need at the "closing" (i.e., when you take possession of your home)?
7. What financial information should you make available to the lender when you seek a loan? Be specific.
8. Explain how the IRS helps pay some of the interest on your business loan from the bank.
9. What are the major sections of the cash budget? For each section, outline the sources of information or the tools of finance you can use to forecast the section.

10. What is the value of pro forma statements?
11. Why is trade credit important to consider as a way to finance an agribusiness?
12. If you own a small agribusiness in a rural area, under what conditions would you prefer not to work through your local town bank? Be specific.

CASE STUDY — **Jamal Wood**

Jamal Wood, CEO of Wood Marketing Services, has decided to increase the firm's financial resources to acquire much needed new computer and information technology equipment. Wood Marketing is a full-line advertising and public relations service working with food and agribusiness firms. By using a cash budget he knew the amount of capital needed, when it could be repaid, and, after consultation with the board of directors, that these funds would be from external sources.

Jamal determined the firm's capital requirements amounted to $250,000. This investment would allow the firm to completely replace the existing information management system with a state of the art wireless network, and a database-driven customer relationship management system. Assume Wood Marketing's tax rate is 40 percent and the simple rate of interest is 8 percent. The loan period is one year.

Questions

Given this information, answer the following questions. Show the formula used for your answer. *Note:* the first four questions ask for two answers, one in dollars and the other in percent.

1. Calculate the amount of interest paid and the effective interest rate for a simple interest loan.
2. Calculate the amount of interest paid and the effective interest rate if this loan also required a compensating balance of $10,000.
3. Calculate the amount of interest paid and the effective interest rate if this loan was a discounted loan.
4. If this was a one-year installment loan with twelve equal, monthly payments, show the APR and the amount of interest paid on the loan.
5. What is Jamal's after-tax cost of borrowing? How would this change if his firm were in the 25 percent tax bracket? What does this say about debt versus equity financing?

15

FINANCIAL CONTROL IN AGRIBUSINESS

- Define control systems and explain their purpose
- Identify the areas of the agribusiness in which control systems are most often used
- Determine when control systems are needed
- Describe how the control process works
- Learn and apply the steps for developing budgets
- Understand qualitative and quantitative techniques for developing forecasts and budgets
- Understand why credit control is such an important factor in agribusiness, and the ways in which credit can be controlled
- Determine the costs of extending credit

INTRODUCTION

Central to the successful management of any agribusiness firm is the ability of both managers and employees to predict and monitor their progress and accomplishments. From a practical point of view, the general objective of the organization, "to maximize profits," does not provide the kind of guidance necessary for monitoring progress. No one knows what maximum profits are or can be. In addition, profit is the result of many, many actions taken by the agribusiness manager. Managers don't affect profit directly, they make decisions that affect sales, costs, margins, asset use, and so on. Progress must therefore be measured against more specific goals and in more focused areas to provide useful information to managers. If profits are to be maximized, then managers need an almost continuous supply of information on their progress toward a wide range of goals. Supplying and using such information is what financial control is all about.

In its simplest form, **control** means making sure that plans are on target and on schedule, and then taking remedial action as needed to ensure success. A basic control system involves three steps:

1. Establishing predetermined goals or standards of performance.
2. Measuring performance against these predetermined goals and standards through an information-gathering system.
3. Taking action to correct deviation from goals and standards.

In this chapter, we will look at this important area in some detail to better understand how agribusiness managers make use of financial control systems to improve decision making.

CONTROL SYSTEMS

While managers seek to inspire the most imaginative and motivated approaches from each employee, they also must prevent overlapping efforts and extension of authority beyond some reasonable limits. Control systems are designed to insure such guidance and to help managers and employees make sure their efforts are moving the firm toward its goals.

Consolidate Authority and Coordinate Efforts

Control systems help managers establish decision guidelines for employees. Such guidelines enable the manager to delegate decisions that need to be delegated, while allowing the manager to make the decisions they need to make.

Control systems provide guidance and insure activities are moving in the right direction.

One of the most familiar examples of this aspect of control would be the purchase order and the policies and procedures that are associated with it. For instance, the Hillsdale Farmers' Cooperative has a policy that the general manager must approve all purchases above $1,000. This control procedure requires the originator to submit a request for approval to the general manager. If the general manager approves the request, then it goes to the purchasing agent and accounting department for ordering and payment. In this way, unauthorized purchases of items costing more than $1,000 by Hillsdale employees is prevented. At the same time, employees have authority to make smaller purchases, freeing the general manager to focus on more important tasks.

Monitor Trends and Results

A control system should show trends, or the direction in which the business is going. Controls sound a warning when plans or programs are heading off purpose. Sales may be way up and inventory way down, or vice versa. Certain items may not be selling. Once this is recognized, the problem can be studied and appropriate action taken. A product may have become obsolete, pricing may be wrong, or a competitor's product may be cornering the market. Without controls to alert the agribusiness manager in time to take corrective action, the situation could become serious or even disastrous to the business.

Provide Information for Planning

The information from control systems becomes a critical part of the planning process. By tracking trends, identifying past successes and failures, and forcing the firm through the goal-setting process, a good control system provides important information for planning activities. An example would be the recognition that 75 percent of the sales of a certain item are made in a 3-month period. This information can be used to guide plans for future production, inventory, and storage decisions.

Provide Information for Decision Making

Good control systems can help provide answers to questions such as "Under similar circumstances, what happened in the past?" or "What past experience should guide our present and future activities?" For example, "When we raised prices before, what was the reaction of our customers, our salespeople, our competitors—and what did we do about it?" A good control system allows managers to access information useful in answering questions such as these, so that any lessons learned in the past are not lost. Information based on past experience can be an invaluable part of a control-information system.

Where Are Control Systems Found?

In most agribusinesses, control systems or procedures are found in the following general areas:

1. Financial control systems
2. Operational control systems
3. Marketing control systems

Agribusiness managers need control systems that are based on predetermined goals or standards of performance. Without measurement criteria the information or procedure will produce nothing of value. Control systems or procedures must also be used only in separable areas of business activity where reliable records are possible. There is an old adage that says "if you can't measure it, you can't manage it" and that is certainly a principle of an effective control system.

Even the smallest organization will require control systems that are related to finance, production of goods or services, personnel, and marketing. For smaller businesses, such as a local implement dealer, these control systems may be quite simple. However, a large organization, such as Monsanto, requires much more complex sets of key performance areas and a much larger number of control systems. Each food and agricultural firm must develop the unique kinds of information and control systems necessary to its success.

Key Elements of Control Systems

Control as a management task was discussed in Chapter 2. Now, some of the key elements of control systems will be described. While control systems come in a multitude of forms, such as budgets, forecasts, production standards, standard costs, personnel evaluation systems, planning programs, and financial plans, each has some elements in common. The principles for developing and using control systems are basically the same, regardless of the business area being monitored. Control systems and procedures are based upon the manager's plans, or what the manager wants or hopes will happen in the future. They also require ongoing monitoring of progress toward these plans. Finally, all control systems include an action phase where action can be taken as needed, based on feedback from the control system. They key elements of a control system are presented in Figure 15-1.

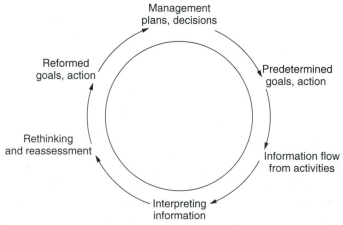

Figure 15-1
Key Elements of a Control System.

Developing Control Systems

The most effective control occurs when attention is directed to the areas that are most vital to the performance and survival of the organization. Pasteur said, "Chance favors only minds which are prepared." We might rephrase this to read, "Success favors only those organizations that are prepared"; in other words, those firms that recognize and concentrate on those few, most important areas. Various management writers have called these areas **KPAs (key performance areas)** or KIs (key indicators). We will use the KPA terminology. As an agribusiness manager decides just which areas are key performance areas, there are several questions to ask:

1. How does the area impact the performance of the organization?
2. Is there considerable risk or variation in the area?
3. What would be the impact on the organization if there were a major problem in this area?
4. Is this area new, or do we have considerable experience in this area?
5. What is the scope and scale of this particular area—is the area large or small relative to the total organization?

Such questions help managers isolate those areas where good management is absolutely essential to the performance of the organization.

After a KPA is identified, then managers must decide what information to collect about the area. Again, a series of questions is in order:

1. How will the information be used?
2. What will occur if the information is not known?
3. Can timely and accurate information be collected?
4. What is the cost (dollars and time) of collecting the information in relation to its value?

Many control systems, when tested, are found to lack sufficient importance to justify the effort involved in maintaining them. Too many controls can render an entire control system ineffective. Too few can be fatal, so the test for significance becomes vital to the success of the enterprise. The agribusiness manager should test each new control system, as well as review existing control systems on a regular basis.

Because the principles and steps for developing control systems are basic to all forms of control, we will focus the discussion on two areas. First, an important control tool, the budget, will be explored. Then, an important control area, credit management, will be explored and the fundamentals of a control system for the credit area developed. The agribusiness manager can adapt the principles and steps discussed for these two areas, the budget and credit management, to the development of almost any financial control system.

BUDGETING: A KEY TOOL FOR CONTROL

A **budget** is a specific forecast of financial performance that is used as a tool for control. An organization may have a cash budget, a capital budget, an advertising budget, and/or a research and development budget. The size and complexity of the organization will determine the kinds of budgets that are needed. A small business may need only an overall budget with different sections, such as sales, production, and finance. A large business may have budgets for departments, divisions, regions, products, etc. Budgets may be either long-range or short-term. A **short-term budget** is generally one that will be implemented within a year, and it usually requires shorter reporting periods. A **long-range budget** is two or more years in implementation and is usually reported on a semiannual or annual basis. The short-term budget, therefore, becomes a component in reaching the long-range budget objectives. Budgets may also be made for specific projects. Examples here might include constructing a new building, launching a new research program, or introducing a new product.

It is important to recognize that budgets help determine organizational direction. Investing in or reducing the financial and human resources devoted to the different areas of the agribusiness determines the long-run direction of the firm in terms of growth, new products, and size. With the allocation of resources, managers bring the company's vision and mission to life. Budgets should reflect the priorities and directions or goals of the business. It is only by changing the resource allocation that a manager can change or redirect the firm's focus. The budget is one of the most important vehicles for reflecting and directing such changes.

For example, if the management team of the Agrow Fertilizer Company were to decide to actively seek a larger share of the homeowner market, the capital, marketing, and cash budgets would need to reflect this objective. The Agrow capital budget might be revised to provide either more resources for production facilities or more resources for bagging and packaging equipment, to produce smaller packages for homeowner use. The marketing budget might well show expanded investment in advertising and promotion, added sales personnel, and the like. The cash budget might reflect the added drain of inventory that would result from the presence of a new product line. As this discussion suggests, budget control mechanisms are a basic part of the planning process.

Budgets and Forecasts

What is the difference between a budget and a forecast? Though it is true that some budgets are forecasts and some forecasts are budgets, in management a **forecast** generally refers to some prediction of the future. We may have a forecast for the general economic situation, or a sales forecast for a specific territory in west Texas. (Sales forecasts were discussed in Chapter 9.) As indicated, the **budget** is typically a specific forecast of financial performance that is used as a tool for control. For example, the Hillsdale Farmers' Cooperative would have a sales forecast or a prediction of the amount of business the organization expects to conduct during a certain period. To accompany this, Hillsdale would also develop a budget that is related to the advertising and promotional expenditures needed to secure the projected sales. Because the survival of the organization depends on the capacity of revenues to exceed expenses, the budget and the sales forecast are inseparable.

Tips for Developing the Budget

The ultimate goal of budgeting is to predict results and control inputs to increase the returns from resources available to the agribusiness. Preparing a budget for even a small firm is not a one-person job. Key people, advisors, and information sources should all be involved. Bankers, accountants and/or auditors, suppliers, consultants, and even customers may be involved in the full budgeting process. Within the agricultural firm there must be a sound organization and communication system that ensures responsibility, authority, accountability, and commitment to accomplishment. The major steps in preparing a budget (or almost any control system that the agribusiness manager might use) are shown below.

1. Set up a committee of those who are involved and schedule dates for the development process.
2. Review the organizational goals, objectives, policies, etc.
3. Review the present situation, where the business is now and how it got there.
4. Review and relate input (cost) factors to output (revenues) factors.
5. Determine why the budget is needed and the specific objectives or courses of action sought.
6. Designate the budget to be developed, and list its component parts.
7. Specify how and when the budget will be used.
8. Specify the information and records necessary to meet the intended uses.
9. Indicate the units of measure that will be used (dollars, tons, bushels, etc.).
10. Specify the records needed to secure the desired data and information.
11. Locate the sources of information or data needed to service the system.
12. Designate and identify the budgeting period.
13. Designate the frequency of issuance, and updating, of information and data from the system.
14. Specify who is to be involved in preparing those records that are pertinent to the budget.
15. Specify who will receive the information and when they will get that information.
16. Identify who is responsible for meeting the objectives and standards contained in the budget—budgets must be constantly monitored to ensure that the system is on target.

While many organizations develop budgets by simply taking last year's figures and adding or subtracting a fairly arbitrary amount, there are other, more thoughtful, approaches. Quick adjustments from last year's figures are certainly better than no budget at all; however, a more careful evaluation of changing conditions is likely to lead to a more effective budgeting process. Some of these more thoughtful approaches are presented below.

Qualitative Techniques in Budget Development

Qualitative approaches must always be part of budget development. The quantitative approaches discussed later in this chapter are only as good as the information on which they are based and must still be interpreted and implemented by people.

One qualitative technique that is often used successfully is consensus. In **consensus,** knowledgeable people who are associated with the business are asked separately for their opinions about a particular factor. For example, Agrow might ask the personnel manager, the operators' superintendent, a supervisor, and an outside consultant to predict or estimate the productivity of a second shift. Everyone involved shares and discusses their opinions, and all then report their (possibly revised) opinions separately. The aim is to derive an answer on which all, or at least a substantial majority, can agree. (This approach is similar to the Delphi approach for making sales forecasts discussed in Chapter 9.) The weakness of consensus is that it may simply encourage the pooling of a firm's ignorance. The technique's strength is that it provides an opportunity for people to present various points of view. Innovative ideas may result from this sharing opportunity.

Another qualitative technique depends on the intuition and experience of the leader. The owner-manager of the small market consulting and brokerage firm may, over a period of time, develop a thorough working knowledge of the key relationships in the business. Success lies in the manager's ability to generalize these past experiences into the future setting. The weakness here is that experience or intuition may be too narrowly based or out of date. If he or she is to be successful, the owner-manager of the market consulting and brokerage firm must be able to synthesize a large number of factors inherent to the local farm community in order to develop sound control systems. Most agribusiness managers hone this skill to a fine edge as they gain experience over the years.

Another technique is **logic,** or the combination of fact, induction, and deduction, with or without the aid of colleagues. Logic must be fortified by pertinent information and data to be successful. If a careful analysis is made, the control process can be substantiated and demonstrated so that others can evaluate the procedure and possibly improve on it. The manager of the market information and brokerage firm might gather information about the competition, the general economy, the productivity of the farms in the area, and the kind of year that farmers had. These factors might form the basis for developing the budget. They could be reviewed with other people who might be able to add or subtract or modify and improve the facts or reasoning that provided the basis from which the budget was developed.

Finally, we have the use of **scenarios,** which involves developing a series of flexible possibilities. These possibilities may vary according to such things as sales volumes, economic conditions, labor market trends, and so forth. The market information and brokerage firm might develop a budget and sales forecast based upon the several possible levels of farm income forecast for the marketing area. Then budgets would be developed for each scenario. Sometimes it is helpful for the manager to develop budgets based on an optimistic, a most likely, and a pessimistic scenario. This allows some contingency planning and allows the budget to be modified according to the actual developments in the market. This technique can be most helpful, but the manager should use great care in developing the scenarios.

Quantitative Techniques for Developing Budgets

Building on the qualitative approaches for developing budgets, there are several useful quantitative techniques for budgeting and forecasting. One of the most commonly used quantitative procedures is **linear projection.** To use this method, the manager first looks at past trends, then uses simple mathematics to project these same trends into a future time setting.

For example, if labor costs for the Hillsdale Farmers' Cooperative have increased by 5 percent during each of the past 4 years, then the general manager there might predict a 5 percent increase in labor costs for each of the next 4 years. This technique can be a useful starting point if the manager does not depend on it blindly. It is easy to use and draws on the actual history of the organization. However, linear projection does not capture any anticipated changes in the future market environment, except those reflected in the trend.

Another quantitative technique is to develop a model of the area of interest. A **model** is an abstract presentation, usually in mathematical terms, that captures all the factors believed to be pertinent, and demonstrates the relative influence of each factor on the entire situation. Models also show how change in any one or group of factors affects the rest of the factors or the whole.

This may sound complex, but an example will help to clarify it. To seek a larger share of the homeowner fertilizer market, the Agrow Fertilizer Company might have found it necessary to add a second shift to produce the extra fertilizer. This second shift would immediately double wage costs assuming the entire plant is in production. Adding a second shift would also increase wage costs by a given factor, perhaps 10 percent, since second-shift workers usually demand a wage incentive for working nights. Also to be taken into account would be the generally lower productivity of a second shift, usually 5 to 15 percent less than day-shift production. The independent variable of higher labor cost, because of the wages paid for the second shift, a wage differential for the second shift, and the lowered productivity of the second shift, could increase the total cost for the plant by 40 percent or more. Using this technique for budgeting in a particular area of the business, we must know:

1. What are the most important variables and how are they related?
2. What are the most likely future values for these important variables?
3. What are the most important constraints or restrictions?
4. How are these variables and constraints related to and integrated with the measure of performance under study?

The answers to these questions must then be formulated into a model of the situation, and then interpreted in relationship to the total problem or goal, which in this case would be to gain a larger share of the homeowner fertilizer market. It would be necessary then to relate the model of labor costs to the sales forecast to determine costs and revenues and to determine the profitability of the action being contemplated.

Finally, we have **simulation,** or a systematic trial-and-error approach to the problem of budgeting and forecasting. This approach is often combined with modeling to develop a more flexible and insightful quantitative assessment of the decision area. Simulation usually involves evaluating several possibilities that are derived from past operational experience and records. This is where industrial engineering techniques, such as standard costs and time and motion studies, become very important to a firm. The various simulations or possibilities can be run on a computer numerous times to determine what is most likely to happen in the future.

An example for Agrow might be to simulate the profit of the homeowner fertilizer market over a wide range of expected economic conditions such as the unemployment rate. Using mathematical techniques, it is possible to calculate the profit that the new business is expected to generate for a wide range of unemployment rate figures. The manager can then look at this distribution of profit figures and make an assessment of how risky the new venture might be given uncertainty in the key variable, the unemployment rate.

Whatever the method a manager uses, certain assumptions and/or constraints will be present. The agribusiness manager must form assumptions about how certain entities will react, about how a number of forces will behave, and about what conditions will or will not be present during the future period under consideration. For example, the market information and brokerage manager must assume that a possible drought will or will not cause increased price volatility in the coming year. Managers also operate within certain personal constraints, such as philosophies and beliefs, ethical standards, and attitudes toward people. Agribusiness managers must be sure that the assumptions and constraints on which they are operating are valid and do not simply reflect their own lack of knowledge or experience, biases, or prejudice.

The Control System/Budgeting Process: A Case Example

To gain a better understanding of the control system and budgeting process, consider the following case situation for Kalamazoo Farm and Industrial Tractor Company. This company is a family-owned corporation, with Jeff McDonald as general manager; his brother Josh as sales manager; his daughter Cassie as shop manager; and Laura, Josh's wife, as office manager. The organization is typical of many smaller agribusiness firms. As sales manager, Josh McDonald was in charge of setting up the sales budget and sales forecast control systems. As he looked over the steps for developing a control system, he saw that the first step was to determine who should be involved in the budget development process.

Josh decided to select his committee from salespeople in the organization, along with his wife, who was in charge of the financial records of the firm. He also decided to involve the firm's banker, his New Holland sales representative, and even two key farmer-customers. He thought that later he might also seek data and assistance from the county agricultural agent or an agricultural economist from the land-grant university. Once Josh had set up the committee, he wanted to work informally with some of the members on an individual basis. Others he wanted to involve in a formal meeting only.

Running a successful farm equipment dealership requires an effective control system.
Source: Photo courtesy of New Holland.

Before the meeting, Josh reviewed and put down on paper the organization's objectives and strategies regarding sales. He then reviewed the present situation and current market conditions, and the records of what happened last year. He spent some time understanding the key factors that really influenced last year's results. Josh committed his review to paper so that he could present it to the planning committee. At this point, he also developed some of the reasons he wanted to establish this control system. Explaining the reason for developing the budget and the goals he wanted to accomplish would be the first step in determining the committee's course of action.

When the sales forecast and budget committee met, Josh presented them with the data and information he had prepared. The committee discussed all the issues involved and added new information and insights of which Josh had not been aware. The next thing on the agenda was to decide the parts or components of the report system. Sales were divided into large-commercial tractor sales and small-garden tractor sales, combines and harvesting equipment, tillage equipment, and farm and industrial parts sales (Table 15-1). Based on committee feedback, it was believed that this breakdown of sales would be the most meaningful one for forecasting and control purposes.

Once the components of the budget had been determined, the next step was to identify how the budget was to be used and the people and parts of the organization that would benefit from the use of the sales forecast and budget. The report was then designed so that it would satisfy the needs of those who might be interested in using it. These included the members of the sales staff and the repair and service department. The report would also be helpful to the firm's banker and to New Holland, Kalamazoo Farm and Industrial Tractor's primary supplier.

At this point, the committee determined where the information would be gathered and how it was to be compiled. They decided that the original sales invoices to customers would provide the input information for the sales forecast control system. The invoices would be the source, and the measure would be dollars of sales. The budget period was designated as the company's fiscal year. Sales budgets and sales forecasts are usually made for one year, with intermediate steps or objectives on a monthly or quarterly basis. Frequent updating is necessary to make the report usable. As sales manager, Josh would determine the frequency of review and updating based on shifts and changes in the market. The preparation of the sales control report was assigned to the office manager, who would have access to the information required to develop the report.

The last step was identification of the person responsible for meeting the performance objectives and standards within the control system. As sales manager, Josh was assigned this responsibility. It is necessary to build into the sales forecast or budget, or any other control system, the evaluation criteria that will be used to determine whether the information prepared is actually serving the needs of the organization. The final sales control report is shown in Table 15-1.

Because of the interrelationships of all budgets in an agribusiness and the unexpected or unknown events that can occur, budgets should be assessed and evaluated whenever new facts surface. There should be a regular planned evaluation or review period. This review permits deviations or problems to be spotted quickly and the effects, positive or negative, to be evaluated before it is too late to take corrective action.

All budgets and all forecast and control measures may be important to the life of the company, but the sales forecast and budget are usually developed first because so many other areas are driven by sales activity. Those serving on the sales forecast and budget committee at Kalamazoo Farm and Industrial Tractor were able not only to add their own unique perspectives to the discussion and development of the system, but also to assume a positive responsibility for accomplishment because they were part of the development team. Good budgets and other control systems can aid the organization by promoting a goal-oriented team approach and by providing the specifications and controls required for a successful operation. Now, let's take a look at a control system for an area of fundamental importance to many agribusinesses: credit.

TABLE **15-1** | **Sales Control Report for Kalamazoo Farm and Industrial Tractor Company**

Sales Budget: January–March

Sales	Budget	Actual	Deviation from budget
Large commercial tractors	$2,540,000	_____	_____
Small garden tractors	540,000	_____	_____
Combines and harvest equipment	2,500,000	_____	_____
Tillage equipment	800,000	_____	_____
Farm and industrial parts	250,000	_____	_____
Total sales	$6,630,000	_____	_____

CONTROLLING CREDIT

Credit control receives special attention in many agribusiness firms because careful management of the credit function is so important to firm success. A common reason for business failure is poor credit control, and this is especially true for smaller firms. New businesses must be particularly wary of their credit control as they fight to establish themselves and attract customers. The extension of credit in many agribusinesses is an important and convenient way for customers to borrow money (because that, in fact, is what it really is). For the agribusiness and its managers, however, credit can create a costly, complex managerial headache.

Credit Is Part of Marketing

It is important to recognize that the least costly way of doing business is to sell for cash. Many large and successful firms have been built on a cash-only basis. If cash sales are the least costly method of doing business, then the next principle must state that credit extension can only be justified to the extent that it improves sales volume and profit. If the extension of credit by a business does not meet this test, it will cause profits to shrink. In almost every sector of agribusiness, credit is something that the purchaser expects or willingly takes. Buyers often look upon it as a right and may be unaware of its high cost, a cost that must ultimately be paid or passed on to the purchaser if the seller is to maintain economic viability.

Two grain and supply elevators in a small Michigan town were in close competition for the farm business in the market. Other agribusiness firms, basic manufacturers of fertilizer, pesticides, etc., were selecting the largest and most successful farmers and making them direct dealers. Credit terms became more and more loose. Finally, the management of one of the elevator companies took a long hard look at the cost of credit extension. As a result, they decided to sell on a cash-only basis and reduced prices by the amount of the cost of credit plus the expected increase in volume they thought a lower price would generate. One of their promotional slogans was, "Why pay more to subsidize the farmer who never pays?" No one likes to pay for another's debts, and yet, when credit is extended, someone must pay for the cost of bad debts, interest on the accounts receivable, and other costs of monitoring the credit function. As a result of their new policy, this company ended up with most of the successful farmers as customers. Actually, some farm customers bought from the elevator company when they had cash and from its competitors when they needed credit. This is not always the right way for a firm to go, but it is always to be considered as a viable alternative to credit extension.

Sales and marketing managers are typically the leading proponents of credit extension. One of the most successful ways of making salespeople responsible in the credit area is to make them accountable for credit costs, including bad debts, and putting this cost squarely into the sales profit and budget control system where it belongs. Credit is not an operational cost, but a sales cost in nearly all cases.

The Costs of Credit

Costs of credit come in many forms, some of which are not so easily recognized by agribusiness managers. Some of the most important of these are:

1. Interest costs on borrowed funds to support accounts receivable.
2. Lost earnings opportunities (opportunity cost) for the use of equity funds tied up in accounts receivable.
3. Liquidity problems, which can cause a firm's bankers and suppliers to raise interest rates or cut back on credit extension to the firm.
4. The costs of losses on bad debts or extremely slow accounts that may have to be discounted.
5. The costs of administering the credit program.
6. The indirect costs of a too liberal credit policy, which can actually allow some customers to get into financial trouble that results in loss of business.
7. The indirect costs of a too liberal credit policy, which can force a competitive reaction of lower margins or even more liberal credit policies.
8. The cost of legal and other activities needed to collect accounts receivable.
9. The work-time frustration and hard feelings of both managers and customers over disputed accounts.
10. The increased risk of accounts receivable to the existence and viability of the firm.

With all these costs it would seem unlikely that credit would be extended at all. And yet, credit remains an important benefit provided to customers by agribusiness firms. The highly seasonal and erratic cash flows of marketing crops and livestock is likely one reason why credit is so prevalent in the agribusiness sector. In addition, some traditional lenders have moved away from the food and agribusiness sector. Finally, competitive forces are such that most agribusiness managers must learn to administer credit programs skillfully.

Credit as a Profit Center

Since most agribusinesses are in the lending business in one way or another, one of the options open to managers is to use the extension of credit as a profit center. It is possible to minimize costs or, in some cases, actually make a profit on credit extension. It is common practice to charge interest on accounts receivable. Basic competitive conditions, the interest rate the firm is paying for credit, and the amount of funds available are some of the key factors determining the rate charged. It is not uncommon for agribusiness firms to charge from 1 to 2 percent per month on unpaid balances. Caution must be exercised to see that the disclosure of credit terms is made in compliance with the federal Truth in Lending Act. In addition, most states have uniform consumer credit codes that regulate credit terms and interest rates.

Another potential approach is to grant discounts for cash sales. These discounts can be absolute or can be extended for short periods of time, such as 10 days or 30 days from invoicing the customer. A very common discount is 1 percent for 10 days. This means that the purchaser may deduct 1 percent of the invoiced amount if it is paid within 10 days. When the firm's suppliers offer such terms, the agribusiness manager should take advantage of them, since the annual interest savings can be very high.

A cash discount can be converted to an effective annual interest rate. Say that Agrow Fertilizer Company wished to reduce its accounts receivable by offering a cash discount to its customers. Agrow offers a 2 percent discount on items that are paid in cash within 30 days, and the account is due in full in 120 days (2%/30, net 120 days). The following formula can be used to compute the effective annual interest rate that Agrow customers "earn" on the money they "invest" by taking the discount:

Effective annual interest rate (EAIR) = [(discount % / (100–discount %)) × 365 days] / (Account due date – Cash discount period)
$$\text{EAIR} = [(2/(100-2)) \times 365]/(120-30) = 0.0828 \text{ or } 8.28\%$$

Agrow has now defined the benefit that a customer taking advantage of the discount receives. (This same approach can be used by Agrow to evaluate its own options when paying suppliers.) Here, a customer paying during the discount period is effectively earning 8.28% on the money paid to Agrow. Customers taking advantage of this program buy at a discount, and Agrow gets its money quicker.

Retailers can often avail themselves of such credit card services as Visa and MasterCard. There is a charge for the firm to join and a charge for servicing the firm's accounts. Credit cards do, however, insure against bad debts and slow collections and reduce the cost of accounts receivable. Managers must weigh these service charges against their own credit-extension costs and select the most profitable alternative.

As in granting extensions of credit, the agribusiness manager should carefully study the use of credit terms available from suppliers and vendors. Strategy in this general area of financing the agribusiness, detailed in Chapter 14, can increase the profitability of a firm considerably.

Planning the Credit Program

Careful planning and development of the firm's credit program is the key to success. The first step is to develop goals and objectives for the credit program. These should be clearly stated in terms of increased sales, competitive factors, and profitability.

The second step is a careful analysis of the costs and benefits of a credit program. At this point, it is important that the costs of credit be considered. Benefits can include earnings through charges levied, increased sales volume, meeting a competitive threat, and taking early delivery with consequent storage cost savings.

The third step is to assign responsibilities for the administration and supervision of the firm's credit program. This is an important function. The person supervising the credit program should operate it within certain defined written policies and constraints developed by top management, such as maximum limits of credit, credit credentials or worthiness of purchasers, and collection policies.

Determining the creditworthiness of a purchaser is probably the top priority in avoiding bad-debt loss and slow payment. This requires a careful check of the purchasers against the "3 C's" of credit: character, capacity, and capital. **Character** refers to the personal habits, conduct, honesty, and attitudes of the credit customer. **Capacity** refers to the earnings and cash flow with which the customer can meet personal obligations. **Capital** refers to the solvency or owner's equity of the credit customer. It is a measure of the collateral or security that can be collected if the customer fails to pay. Capital represents the last resort for collection of a bad account, and if the manager expects to be forced to liquidate a customer's assets to collect the outstanding credit, that sale should not be made in the first place. Information on these factors is available from the applicant, banks, credit references, local and national credit rating firms, and credit bureaus. In larger firms careful evaluation of the 3 C's through a formal credit application process is common. In smaller firms, however, credit may be granted without such a careful review. The informality of the credit process in smaller firms is closely linked to credit problems in these firms.

Monitoring the Credit Program

Once the credit program is set up, the manager must carefully follow how well the program is operating. Several financial ratios and tools can be helpful in the monitoring and analysis of credit programs. A few of the more important ones are offered here.

The most important tool in monitoring a credit program and accounts receivable is the monthly aging of these accounts. Each account from the ledger is summarized in a report similar to the one shown in Table 15-2.

An increase in the percentage of accounts that are slow paying, or increases in the amounts of receivables, calls for a review of the credit program and credit policies before the situation becomes dangerous to the firm's financial health.

Ongoing monitoring is required to ensure a credit program is performing according to plan.

Source: Photo courtesy of Pioneer Hi-Bred International, Inc.

TABLE **15-2** | **Account Aging Schedule**

Period outstanding	Accounts receivable	Percentage of total this period	Percentage of total last period	Percentage change
Less than 30 days				
30–60 days				
61–90 days				
Over 90 days				
Total				

To assist in monitoring changes in accounts receivable, a ratio measuring changes in credit sales volume from one period to another is helpful.

(Credit Sales for Current Period / Credit Sales for Previous Period) × 100 = Percentage Change in Credit Sales

This simple ratio provides a quick indication as to how the proportion of credit sales is changing over time.

Bad debts are an agribusiness manager's nightmare. When the agribusiness sells and fails to collect, it is far worse than not making the sale at all. Not only does the business suffer from the loss in gross margin, but it also has to chalk up the loss from the cost of the goods or service sold, plus the aggravation and cost of trying (usually unsuccessfully) to collect the debt. One simple way to monitor bad debts is the formula:

(Bad Debts / Total Credit Sales) × 100 = Bad Debts as a Percent of Credit Sales

This figure can be tracked over time and checked regularly to insure it is within predetermined bounds.

TABLE **15-3** | **Sales Increases Required to Offset Bad Debt Losses**

Bad debt loss	Contribution margin		
	5%	10%	20%
	(You must increase sales by the amount.)		
$ 1,000	$ 20,000	$ 10,000	$ 5,000
$ 5,000	$100,000	$ 50,000	$ 25,000
$20,000	$400,000	$200,000	$100,000

The cost of bad debts is very serious for many agribusinesses. Table 15-3 provides a better understanding of just how seriously this cost affects business profits. The tremendous amount of sales increase necessary to offset bad debt loss is vividly demonstrated in this table. For example, $20,000 of bad debt requires an increase in sales of at least $100,000 at a 20 percent contribution margin on sales just to bring profits back to what they were without the bad debt. (The contribution margin is gross margin less variable costs. This figure is discussed further in Chapter 16.) Credit sales have to be monitored very carefully or they can quickly consume all the profits a business is generating. Because the stakes are so high, a carefully conceived and executed credit program is necessary for any agribusiness and its management team.

SUMMARY

As we have seen, financial control systems are the means by which an agribusiness manager coordinates the activities of the business, predicts trends and results, amasses data for planning, sets goals, and utilizes the lessons of past experience. Control systems can be focused on the marketing, production, personnel, and/or the financial areas of the agribusiness. The larger and more complex the organization, the more complex the control system.

There are a number of important steps in developing an effective control system; of these, the three most central are the review of organizational goals and objectives, analysis of the current situation vis-à-vis those goals, and the relation of projected costs (inputs) to projected revenues (outputs). Control systems must be constantly monitored and reevaluated. Budgeting is one important tool for control. Both short-term and long-range budgets are useful to the agribusiness manager. Most control systems will combine both qualitative techniques (consensus, logic, scenarios, etc.) and quantitative techniques (linear projection, models, simulation, etc.) when making forecasts.

One of the areas in which development of a strong control system is most important is credit control, where upward spiraling costs often cause serious financial problems for agribusinesses. Two possible solutions are to make the extension of credit a profit center or to offer a discount for cash sales. In any event, credit programs should be planned and monitored carefully in order to develop systems for determining a customer's creditworthiness and for minimizing accounts receivable or outright bad debts.

The agribusiness manager who incorporates these basic tools for control into the daily activities of the business will stand a good chance of developing and maintaining a profitable enterprise.

DISCUSSION QUESTIONS

1. Define control and explain why it is such an important management task.
2. Explain the purposes of control systems and describe the areas where they are most likely to be found in an agribusiness. How would your answer vary for (a) a manufacturer of grain drying equipment; (b) a large (1,000 cows) dairy operation; (c) a local organic food market?
3. Review each of the 16 steps for developing a control system and the purposes of the steps. Which steps do you believe to be the most important? Why?
4. How is the budget related to the firm's goals and objectives?
5. What is the basic difference between a forecast and a budget? How are these two concepts related?
6. Discuss the use of various qualitative and quantitative approaches to developing forecasts and budgets. You have been asked to develop a sales forecast for a local pizza restaurant. How would you make your forecast? What methods and data would you use? (Chapter 9 may be helpful here.)
7. Why do agribusiness firms offer credit? Why is credit considered a sales tool?
8. Discuss and explain the true cost of credit extension.
9. Under what circumstances can the extension of credit become a profit center?
10. Why is determining creditworthiness the most important step in extending credit? For an agricultural producer in your region of the country, develop a list of questions/data needs that an agribusiness would need to evaluate the 3 C's of credit for that producer.

CASE STUDY—Delaware Poultry Supply Credit Policy

The Delaware Poultry Supply is considering a change in its credit policy. Currently the firm offers a 2 percent discount on items paid in cash within 15 days, with the account due in full in 60 days (2/15, net 60). Under the proposed change, the cash discount would be 4 percent on items paid within 15 days, with the account due in full in 90 days (4/15, net 90). Assume a 365-day year.

Questions
1. What are some of the potential benefits to Delaware Poultry Supply from this change in credit policy?
2. What are some of the potential costs to Delaware Poultry Supply from this change?
3. Calculate the equivalent annual interest rate (EAIR) for the two credit policies. What do these figures tell you about the proposed change?
4. Outline a credit control system for Delaware Poultry Supply.
5. Should Delaware Poultry Supply adopt the new policy? Why or why not?

TOOLS FOR EVALUATING OPERATING DECISIONS IN AGRIBUSINESS

O B J E C T I V E S

- Understanding the decision-making process and how it may be used to solve management problems
- Explain fixed and variable costs and their relationship with business volume and profit
- Learn to calculate the breakeven point for a firm
- Apply volume-cost analysis techniques to important agribusiness operating decisions
- Discuss alternative strategies to reduce a firm's breakeven point
- Evaluate pricing decisions within the agribusiness firm and estimate the impact of a price change on the breakeven point

INTRODUCTION

Effective decision making is a critical talent of the successful food and agribusiness manager. Regardless of which commodities, services, products, and activities an agribusiness focuses on, the manager of that organization is faced daily with a myriad of decisions to be made at many different levels. From where to order lunch for the staff meeting, to determining where the next production facility should be built, agribusiness managers make decisions. And, it is the manager's responsibility to make effective, timely, and sound decisions for the ultimate good of the business.

Professional managers approach this decision-making activity systematically. The process of decision making involves identifying the problem, summarizing facts, identifying and analyzing alternatives, and taking action. This chapter examines some of those critical decision-making steps and the tools used to facilitate the process. Our focus in this chapter is operating, or shorter-term decisions. In the next chapter, we turn our attention to longer-term investment decisions.

415

PERFECT PALLET

The Perfect Pallet Company has grown steadily over the past decade and management has been pleased with this record of performance. Recently, however, the business had been sold to an international food company. As the new owners assumed control, they began to apply pressure for better performance and a higher return on their investment. Sandy Johnson, the newly appointed general manager at Perfect Pallet, is struggling with all kinds of questions, alternatives, and ideas for improving profits. Although she has some good accounting information, she is not sure how to best use this information to answer tough management questions including:

> "How much must Perfect Pallet sell to cover all costs?"
>
> "What volume of business is necessary to generate a 10 percent ROE?"
>
> "If prices are lowered by 5 percent to improve sales volume, will this be profitable?"
>
> "Can we afford to hire a new salesperson?"
>
> "Can Perfect Pallet afford new computer-controlled, electronic sawing equipment?"
>
> "How should the board of directors be advised about which of several alternative expansion investments is best for the business?"

Some of these questions frame operating decisions, or decisions that have a shorter (usually 1 year or less) time frame. Some of these questions frame investment decisions with impacts that will be much longer than a single year. Regardless of focus—operating or investment—professional agribusiness managers view decision making as a process and use analytical tools to help make decisions whenever feasible. Then they complement formal analyses and decision making with their personal experience, judgment, and intuitive feel for the situation. With the greater access to many sources of information—both financial and marketing—managers today rely much more heavily on computer-generated spreadsheets and database analysis systems to help them make more informed management decisions.

DECISION MAKING

An Overview

Decision making is the process of choosing between different alternatives for the purpose of achieving desired goals. In the following sections, we will look more closely at this definition.

Process   The first important idea here is the recognition that decision making involves some type of process. The word "process" implies activity, or doing something. It is important to recognize that good decision making is an active

Like ripples on a pond, many decisions made by agribusiness managers have consequences that expand over time.

process in which the manager is aggressively and personally involved. Of course, decisions can be made by default; that is, one can do nothing for so long that there is no longer a decision to be made. Putting off decisions until it is too late is a problem most people know something about. But, effective decision making includes an active participant whose actions are timely.

It is important to note that deciding to do nothing is *not* always a default decision. Deciding to "wait and see" may be a logical and correct choice. A default decision represents a failure to decide. The end result may be quite acceptable, but any positive result of a default decision is purely accidental. Any success is in spite of management, not because of it; therefore, most professional managers do not make default decisions.

Choosing   The second key idea in the definition of decision making is *choosing*. Choosing implies there are alternatives from which one must select. When one has no alternatives, there is no decision to be made. The alternatives must also be feasible. They must be realistic and reachable. For example, quitting or exiting is always an alternative, but it is seldom a realistic one. Choosing also involves selection—picking from among the available options. Such selection, picking the "right" alternative, is at the heart of decision making.

Goals   Finally, decision making is *purposeful*. Effective decision making requires that a clear goal be firmly in mind. If management refuses to set clear-cut goals at the beginning of the year, it is impossible to evaluate performance at the end of the year. Firm performance should be evaluated as you would evaluate employees of the firm. Goals or anticipated results should be clearly outlined as quantitatively as possible to allow assessment of performance at year's end. Likewise, effective choice requires that some criteria, such as goals, be used to guide the selection process.

Goals, like alternatives, must be feasible and specific. To say "My goal is to make as much profit as possible" is not much help in operating a large California vegetable grower-shipper operation. This is far too general a goal to be of much use. But a specific goal, such as "To generate a 15 percent after-tax return on investment, maintain an annual growth rate of 5 percent, and provide an opportunity for meaningful employment for family members" will be exceedingly helpful in guiding and evaluating day-to-day decisions.

The Decision-Making Process

The decision-making process is simply a logical procedure for identifying a problem, analyzing it, and arriving at a solution. This can be carried out in a formal manner, where many people are involved in its various aspects, work for many weeks or months on analysis, spend a great deal of money, and publish lengthy reports which outline proposed solutions. Or the process can occur informally over coffee in just a few minutes with no written report at all. The more important the issue, the more likely the process will be formalized. In any case, effective decision making is a systematic process that involves some key elements and several rather specific steps.

Three necessary elements are part of the decision-making process. First, decision making is built around facts. The less relevant, factual information there is available, the more difficult the decision-making process. Second, decision making involves analysis of this factual information. Analysis may be a highly rigorous statistical treatment using large computer spreadsheets, or it can be simply a logical thinking process. In either case, decision making requires careful examination of facts. Finally, the decision-making process requires an element of judgment, a subjective evaluation of the situation based on experience and common sense. Although it is theoretically possible to be completely mechanical in the decision-making process, seldom, if ever, are enough facts available or are there sufficient resources or time to completely analyze the situation. Human judgment is a necessary part of professional decision making.

With these three elements in mind, let's look at the steps involved in the decision-making process.

1. *Problem identification.* This is often the toughest part of decision making. It is easy to confuse symptoms with the real problem. The problem may seem to be low profits, when low profits are simply the result of an inefficient, high-cost distribution system. Once the problem is clearly defined, it can usually be more easily addressed.
2. *Summary of facts.* This step brings to the surface and highlights information pertinent to the problem and its solution. It may be critical to note overall

company goals, the impact of the problem on the business, environmental factors limiting possible solutions, or technical facts that affect the outcome.

3. *Identifying alternatives.* This step identifies and lists feasible alternative solutions, exploring various possibilities. Only feasible solutions should be considered.

4. *Analysis.* This step may require rigorous examination, weighing the costs and benefits of each alternative. Analysis considers both short- and long-term company goals. Although analysis should be objective, the final selection process should include some subjective evaluation of alternatives.

5. *Action.* One of the most critical steps in the decision-making process involves implementing the chosen alternative. Often this requires careful planning prior to the execution, but it is a critical step. Management responsibility goes much further than just deciding; it requires execution and results.

6. *Evaluation.* The final step in this process occurs some period of time after action has been taken. Management must evaluate whether or not the firm is better off by the action taken. If the situation has improved, no further action is necessary. However, if the action taken has not caused the desired results, management must go through this process again and seek new action alternatives.

DECISION TOOLS

There are numerous tools for analyzing alternatives and making management decisions, and the number is growing rapidly. Some of these tools are complex, while others are simple. The decision-making process just described is, in itself, a decision tool. But among the more important decision tools used by agribusiness managers is volume-cost or breakeven analysis. This is because most food and agricultural businesses are relatively capital intensive, requiring large investments in land, plants, and equipment. The food and agricultural industries are so seasonal that such large investments in many firms can be used only for very short periods of time—for example, during planting or harvesting. The capital-intensive nature of the industry emphasizes the importance of both investment decisions (Chapter 17) and the efficient use of fixed assets (operating decisions).

The remainder of this chapter is devoted to a discussion of this important management tool. Volume-cost analysis techniques are discussed and a format for utilizing this tool is presented. Numerical examples using data from the firm Brookston Feed and Grain Company, presented in Chapter 12, will illustrate the use of volume-cost or breakeven analysis and the many questions this decision tool can help management address.

VOLUME-COST ANALYSIS

Volume-cost or **breakeven analysis** is a tool for examining the relationship between costs and the volume of business generated by the firm. This tool analyzes differences in the kinds of costs encountered by every agribusiness and how the volume of business affects them. Volume-cost analysis (or breakeven analysis, as it is sometimes called) shows the level of business necessary to break even and/or to earn a specific amount of profit under various cost and price assumptions. Volume-cost analysis can show the impact of changes in selling price on the volume of business necessary to reach a certain profit level. It can reveal specifically how anticipated cost changes will affect profit levels. It can be useful in evaluating various marketing strategies, such as advertising and promotion expenditures, individual product pricing decisions, and the amount of sales a new salesperson must generate to cover salary and other costs.

The basis for volume-cost analysis is the separation of costs into two categories, fixed and variable. **Fixed costs** are those costs that do not fluctuate with the volume of business. Examples of fixed costs would be depreciation, interest, and insurance. **Variable costs** are those costs that change directly with the volume of sales. Examples would include the cost of goods sold, overtime, and commissions. The key question in classifying costs into these two classes is whether the cost is *directly* affected by how much is sold. Said another way, fixed costs are present regardless of the amount of firm sales. As soon as a business gears up for a particular level of sales, it incurs a certain amount of expense whether or not it makes any sales at all. These are fixed or sunk costs.

On the other hand, some additional expenses are incurred as product is sold. These incremental expenses are not charged to the income statement if the sale is not completed. These are variable costs. Note that the emphasis is on the sale. The actual sale of a product or service is the point of determination for this cost. Even in a manufacturing or processing plant, where costs are incurred throughout the production process, the crucial point is the actual sale. Until the sales transaction is completed, no costs are counted as expenses and are therefore not included on the income statement. Instead, they remain in inventory and show only on the balance sheet. If there are no sales during a period, by definition, there are no variable costs. Selling something actually causes the variable costs to be incurred.

Some people tend to confuse variable costs with controllable costs, but they are not the same things. While some variable costs are controllable by management, others are not. For example, fuel cost associated with delivering products is variable because it is incurred automatically when and only when product is actually sold and delivered. But management usually can do little to control it. On the other hand, advertising costs are generally controllable, but they do not vary directly with sales. Theoretically at least, advertising causes sales, which is the opposite situation from a variable cost. Once the advertising expenditure is committed, so is paying the advertising bill, whether or not sales result. Thus, advertising is a fixed cost, even though it is controllable.

Perhaps a graphical illustration can further clarify these important fixed and variable cost concepts. Assume that the Louisiana Crayfish Market, Inc. (LCM Inc.) has total fixed costs (or overhead, as it is sometimes called) of $200,000 per year.

Some costs of doing business for Louisiana Crayfish Market are fixed, and other costs are variable.

Fixed Costs

Fixed costs are constant regardless of the volume sold during the period. If LCM Inc. opens for business but has no sales at all, then its total fixed costs are $200,000. If its sales volume is $200,000, its fixed costs remain at $200,000. If its sales volume is $500,000, its fixed costs are still $200,000. The horizontal line in Figure 16-1 shows that regardless of sales volume, fixed costs will remain at $200,000.

Now, look at average fixed costs as a percentage of sales. If LCM Inc. sells only $200,000 worth of crayfish products during the year, its fixed cost per dollar of sales is 100 percent ($200,000 fixed costs ÷ $200,000 sales × 100). If sales are $400,000, fixed costs drop to 50 percent of sales. If sales are $500,000, the fixed cost percentage drops to 40 percent of sales. At $1 million of sales, fixed costs drop to 20 percent of sales. This fixed cost percentage continues to drop as long as volume is increased in the same physical plant (Figure 16-2). The fact that fixed costs as a percentage of sales falls rapidly as sales increase is extremely important to operating efficiency in any agribusiness. Firms often make expansion decisions based upon this economy of scale efficiency. As long as no additional assets are required to farm an additional 80 acres, the farmer expands and is able to lower average fixed cost. Of course, this can't go on forever—at some point limits to resources force additional increments of capacity to be added, and total fixed costs increase.

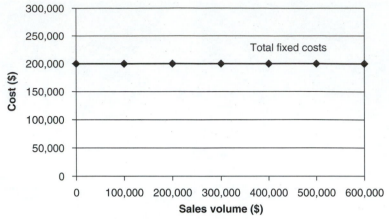

Figure 16-1
Total Fixed Costs, Louisiana Crayfish Market, Inc.

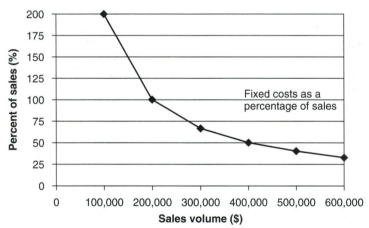

Figure 16-2
Average Fixed Costs, Louisiana Crayfish Market, Inc.

Variable Costs

Variable costs, on the other hand, behave quite differently. Assume that the variable cost per pound of sales for LCM Inc. averages 60 percent of each dollar of sales; that is, every time a dollar's worth of product is sold, a 60-cent cost is incurred (including the cost of goods). At zero sales, LCM Inc. incurs no variable cost, so the initial total variable cost is zero. If sales are $200,000 and each dollar of sales incurs 60 cents in variable costs, total variable costs would be $120,000 for the period (0.60 × $200,000). At $400,000 sales, total variable cost is $240,000 (0.60 × $400,000) (Figure 16-3).

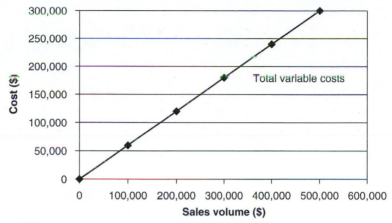

Figure 16-3
Total Variable Costs, Louisiana Crayfish Market, Inc.

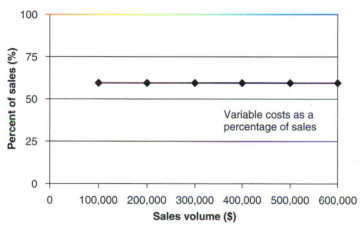

Figure 16-4
Average Variable Costs, Louisiana Crayfish Market, Inc.

Now, looking at average variable costs per dollar of sales, if LCM Inc. sells $200,000, its variable cost per dollar is 60 percent. If its sales are $400,000, its variable cost per dollar is 60 percent. In fact, no matter what the volume of sales, the variable cost per dollar sales remains essentially at 60 percent, or, in other words, constant (Figure 16-4) when calculated as a percentage of sales.

The key ideas presented in these graphs can be summarized as follows: (1) *As sales increase, total fixed costs remain constant but fixed cost as a percentage of sales falls,* and (2) *as sales increase, total variable costs increase but variable cost as a percentage of sales remains constant.*

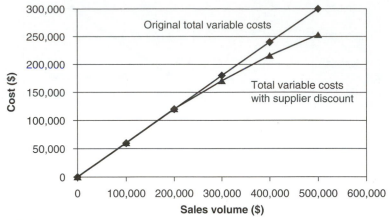

Figure 16-5
Total Variable Costs with Supplier Discounts, Louisiana Crayfish Market, Inc.

Semivariable Costs

Semivariable costs are a third type of cost. This is a cost that is partly fixed and partly variable, such as the electricity costs. There is a fixed, or minimum, charge per month, and then the usage charges increase with business activity, and most likely with sales. These types of cost behavior can and should be considered in volume-cost relationships. A simple way of handling semivariable costs is to carefully estimate the fixed portion, the amount that is part of the cost of just being open for business, and then estimate the portion that is incremental with each additional sale. With this allocation of semivariable costs, volume-cost analysis can proceed.

Special Problems in Cost Classification

There are certain situations that create special problems in classifying costs as fixed or variable. These are examined in the discussion below.

Incremental Costs Not Constant   One difficulty can be variable costs that increase incrementally with sales, but not along a straight line. That is, as sales increase to higher levels, the additional cost for each unit becomes increasingly smaller, or greater. This situation can be handled nicely, so long as the exact relationships are known. The graphical or mathematical analysis becomes more complicated, but the analysis can be performed and used in the same way. For example, consider what a firm can do to reduce per unit variable costs. One way to do this is for the agribusiness to secure volume discounts from suppliers. For LCM Inc. this may mean their cost per 1,000 bags is reduced by 1 cent per bag if the firm purchases over 5,000 bags from their supplier. As additional bags are used, variable costs go up, but not as much as before the volume discount. This would result in the curvilinear relationship between volume and variable cost depicted in Figure 16-5.

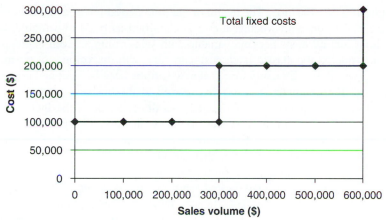

Figure 16-6
Changes in Total Fixed Costs, Louisiana Crayfish Market, Inc.

Lumpiness   Some costs are lumpy; that is, they are fixed within certain sales volume ranges, but once a certain point is reached, a whole new set of fixed costs are incurred. For example, when a delivery truck reaches its maximum capacity, sales cannot be increased further unless another truck is purchased. Purchasing an additional truck suddenly increases fixed costs, such as depreciation, insurance, and licenses for the new truck (Figure 16-6). This lumpiness can also be included in graphical or mathematical volume-cost analysis. Again, it just becomes more complicated—the principles do not change.

Allocating Costs   With the calculation of breakeven for the entire firm completed, many agribusiness managers desire to take this analysis one step further. If they run a diversified firm selling many products, they may want information about breakeven for each of these product lines or divisions of the company. Although it is clear where many costs will fit—that is, what division will be responsible for a particular cost—there are several expenses that must be allocated among several divisions.

As one can imagine, it may at times be quite difficult to allocate these costs. For example, in a diversified agribusiness operation like Brookston Feed and Grain, some of the manager's salary should be allocated to the grain division, some to the feed division, seed division, etc. The question is, how much? Some managers simply allocate costs based upon sales of the different divisions. Thus, if grain represented 80 percent of firm sales, the grain division would be "billed" for 80 percent of the manager's salary. This allocation scheme is often the worst method of dividing costs among divisions of the firm. The grain volume for the firm may be great, but upon further investigation we may find that the manager only spends about 25 percent of his/her time in this area. Thus, in deciding upon a proper cost allocation system, it is often important to ask many questions and make sure that the scheme selected is the fairest allocation method.

Length of Time Period   The length of the time period under consideration has a great deal to do with whether a cost is classified as fixed or variable. In the very long run, all costs become variable with sales volume. Even depreciation, often considered the classic fixed cost, can vary with the volume of business if the time period is sufficiently long that new facilities can be built and equipment purchased. For example, over a 5-year period, if the business grows, an entire new set of facilities may be constructed as a direct result of the increased volume of business. But in the very short run, say one day, all but the most direct costs associated with the cost of the product and the actual cost of the sale are fixed and cannot be changed within that period.

Therefore, it is important to define the time period that is under consideration. The time period selected depends entirely on the problems that are of concern and the decisions to be made. In attempting to utilize volume-cost analysis, time periods considered are usually reasonably short, such as a season or a quarter, or, most commonly, a year.

In practice, the fine points of separating fixed and variable costs are not that critical to making rough approximations of the volume-cost relationships. The analysis, of course, can be no better than the information used in making the analysis. But experience shows that so long as reasonable care is taken in classifying costs as fixed and variable and a realistic time period is assumed, close approximations of fixed and variable costs may be made without regard to semivariable costs allocations, nonlinear variable costs, or lumpiness. Volume-cost analysis can then proceed, providing highly useful insights into important management questions. Of course, if major violations of the simple assumptions are known to exist, caution should be used, and one should attempt to take these special situations into account in the analysis.

Volume-Cost Analysis Procedure

Let's take a typical operating statement from Brookston Feed and Grain (Chapter 12) and analyze the volume-cost relationships. There are four distinct steps in determining the breakeven point from the firm's income statement.

Step 1: Classify Fixed and Variable Costs   The actual classification of expense items from a firm's income statement depends on the makeup of each. Thus, it is necessary to be familiar with the operation and its accounting system in order to be accurate with the classifications. For example, rent for one firm may be strictly a fixed monthly amount, while another firm may have a rental contract that ties rent to the level of sales, making it a variable cost.

Take a look at Brookston Feed and Grain Company (Table 16-1) and classify its fixed and variable costs for the year. BF&G is an actual agribusiness firm operating in the Midwest; the determination of its fixed and variable costs is based on the actual situation as interpreted by its manager. Note that in most cases fixed costs are tracked in total dollars and variable costs are calculated as a percentage of sales.

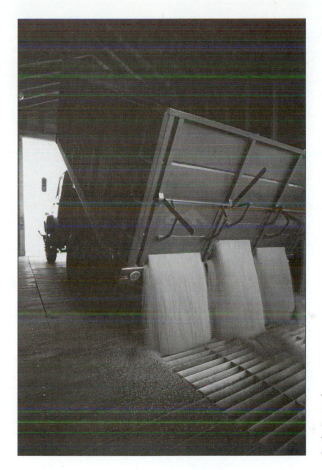

The breakeven point for Brookston Feed and Grain Company is that point where sales revenue just covers total cost.

Cost of goods sold—variable. For most production-oriented agribusinesses, this is the largest single variable cost. It, in fact, is a perfect example of what was depicted in the graphs earlier in this chapter for variable cost. No charge is made under "Cost of goods sold" until the product is actually sold. If product is purchased from a supplier but not sold by period's end, there is no charge against cost of goods—thus, no variable cost.

Salaries and benefits—fixed. BF&G pays its manager a set salary. This represents a fixed cost to the firm. Some weeks the manager may only put in 40 hours, but during the busy season BF&G's manager works in excess of 60 hours. The manager is paid a fixed amount monthly.

Full-time wages—fixed. Although these people are paid by the hour, this is still considered a fixed cost. If BF&G had paid any overtime, this part of wages should be considered a variable cost. Since it is hard to attract and

TABLE **16-1** | Income Statement for Brookston Feed and Grain Company

Year Ending December 31, 2001

Item	Amount	Percent of sales
Sales	$5,200,000	100.00
Cost of goods sold	4,647,240	89.37
Gross profit	552,760	10.63
Salaries and benefits	81,200	1.56
Full-time wages	57,400	1.10
Part-time wages	10,400	0.20
Commissions	15,600	0.30
Depreciation	52,400	1.01
Maintenance and repair	38,980	0.75
Utilities	20,800	0.40
Insurance	26,000	0.50
Office supplies/Expense	10,400	0.20
Advertising/Promotion	13,000	0.25
Gas and oil	17,740	0.34
Delivery and freight	57,200	1.10
Rent	7,800	0.15
Taxes, licenses, fees	12,000	0.23
Miscellaneous	8,300	0.16
Payroll tax	10,900	0.21
Bad debt	3,640	0.07
Total Operating Expenses	443,760	8.53
Net operating profit	109,000	2.10
Other revenue	18,200	0.35
Interest	35,000	0.67
Net profit before taxes	92,200	1.78
Taxes	27,400	0.53
Net profit after taxes	64,800	1.25

retain good help, BF&G pays workers a minimum of 40 hours per week, even though in slow times the manager must "find work" for them to do.

Part-time wages—semivariable, half fixed, half variable. BF&G employs some "permanent" part-time people. They are paid a lower wage than their full-time counterparts and receive fewer benefits. They also do not work a 40-hour week. As sales increase, part-time hours tend to increase, thus, the equal split between fixed and variable costs.

Commissions—variable. BF&G pays a commission to its salespeople as part of their compensation. Since commissions aren't paid until a sale, this represents a variable cost.

Depreciation—fixed. These costs are not a function of sales. Even with depreciation schedules that do not have the same depreciation charge each year, this amount is not tied to sales.

Maintenance and repair—semivariable, half fixed, half variable. Since equipment is on a scheduled maintenance program regardless of use, the manager has split this cost equally between fixed and variable. The fixed portion captures the regularly scheduled maintenance, and the variable portion captures those repair costs that are related to volume, or use of the facility and equipment.

Utilities—semivariable, two-thirds fixed, one-third variable. Since extra utility costs in the fall are caused by higher grain sales, the manager felt some of this expense should be classified as variable.

Insurance—fixed. This cost is not dependent upon sales.

Office supplies/expense—semivariable, half fixed, half variable. Since more sales result in higher supply costs, some of these costs are variable. Yet, some office expenses are borne by the firm regardless of sales. Hence, this cost is split between variable and fixed.

Advertising/promotion—fixed. These are expenses intended to create sales; they are not caused by sales. Only promotional programs like trading stamps, incentive gifts, and double coupons would be a variable cost.

Gas and oil—semivariable, half fixed, half variable. Since more sales result in higher gas and oil expenses, part of this cost is variable. But, some gas and oil expenses are borne by the firm regardless of sales.

Delivery and freight—variable. Here, there is no expense unless there is a sale.

Rent—fixed. BF&G makes a flat monthly payment to the owners for rent. Thus, this expense is not related to sales.

Taxes, licenses, fees—fixed. These are typically a flat fee and not related to sales.

Miscellaneous—variable. The manager of BF&G estimated that the majority of these costs did, in fact, vary with sales. This expense category represents the aggregation of several small expenses that do not need to be listed separately on the income statement.

Payroll tax—fixed. This cost is tied primarily to salaries and benefits and is not a function of sales.

Bad debt—variable. This expense tends to increase as sales increase. Firms often develop guidelines for bad debt costs. These benchmarks tend to be expressed as a percent of sales. For BF&G, the rule of thumb is that if bad debt loss is under 0.10 of sales, the firm is doing well in the credit and collection areas.

Interest—fixed. This expense varies with the amount of money borrowed, not sales volume. Interest on long-term loans is clearly a fixed cost. The status of interest on short-term, in-season loans is less clear. However, borrowing to finance inventory is not caused by the sale. It is to enable an anticipated sale. Only in circumstances where borrowing is caused by a sale would this part of total interest be classified as variable.

Step 2: Summarize Fixed and Variable Costs   The next step in the procedure to calculate breakeven is to summarize fixed and variable costs. Total dollar fixed costs are summed first. Then variable costs as a percent of sales are added together. Variable costs may be tracked in dollars per unit—that is, dollars per ton or dollars per bushel or as a percent of sales. The physical unit measure is acceptable for a firm that sells product in one physical unit—a fertilizer firm sells only fertilizer and prices it by the ton. However, most agribusinesses sell a variety of products in several different physical units. Thus, tracking variable cost as a percent of sales is usually the best option. Totals for the Brookston Feed and Grain Company example are included in Table 16-2.

The fixed and variable cost summary shows that Brookston committed itself to a cost of $348,327 the day it opened its doors for business. Additionally, every time it sold a dollar's worth of product, it incurred, on average, an incremental or additional cost of 91.87 cents.

Step 3: Calculate the Contribution to Overhead   CTO **(contribution to overhead)** is the heart of volume-cost analysis and an important figure for many management decisions. It shows the portion of each unit of sales that remains after variable costs are covered and that can be applied toward paying or covering fixed or overhead costs. Each time a unit of product is sold, the variable costs must be covered first. Anything that remains makes a contribution to overhead.

In this example, $1 of sales, of course, generates $1 of revenue. Of this $1 of revenue, 91.87 cents must go to cover variable costs. This leaves $0.0813 ($1.00 – $0.9187) as a contribution to overhead for each $1 of sales.

Revenue (Sales)	$1.0000
Less: Variable Cost	–0.9187
Contribution to Overhead	$0.0813

Thus, for every dollar generated in sales, BF&G has just over 8 cents left to contribute to the firm's overhead after variable costs have been paid. Thus, Brookston's contribution to overhead or CTO is 8.13 cents or 8.13 percent.

The breakeven relationships essentially answer the question, how many unit contributions at 8.13 cents per unit does Brookston have to make before its overhead is paid? One can think of overhead as a tank that must be filled over the

TABLE **16-2** | **Classification of Fixed and Variable Costs for Brookston Feed and Grain Company**

Expense	Fixed cost (Dollars)	Variable cost (Percent of sales)
Cost of goods sold		89.37
Salaries and benefits	81,200	
Full-time wages	57,400	
Part-time wages	5,200	0.10
Commissions		0.30
Depreciation	52,400	
Maintenance and repair	19,490	.37
Utilities	13,867	.13
Insurance	26,000	
Office supplies/expense	5,200	0.10
Advertising/promotion	13,000	
Gas and oil	8,870	0.17
Delivery and freight		1.10
Rent	7,800	
Taxes, licenses, fees	12,000	
Miscellaneous		0.16
Payroll tax	10,900	
Bad debt		0.07
Interest	35,000	____
TOTAL	$348,327	91.87%

operating period before any profits are made. In our example, this fixed cost tank has a constant capacity of $348,327.

Step 4: Calculate the Breakeven Point The question is, "How many units (dollars) of sales must there be, with each unit contributing just over 8 cents toward the $348,327 overhead, to completely cover all costs?"

If Brookston opens its doors but sells nothing during the period, it incurs its fixed costs of $348,327, but has no revenue, so the entire $348,327 is a net loss. If they manage to sell one unit ($100 per unit) of product, they incur the fixed cost of $348,327 and the variable cost of $91.87 for that unit. The total cost is $348,418.87 against an income of $100.00. The loss is $348,318.87, which is pretty bad, but this loss is not as great as if they sold nothing. The loss is reduced by $8.13 or 8.13 cents per dollar of sales. This represents BF&G's **contribution margin.**

The obvious and critical question then is, "How many times must this process be repeated before the firm reaches the zero-loss point?" Or, stated differently, "What is the firm's breakeven point?" Each time another $100 unit is sold, its revenue is used first to cover the variable cost of $91.87. The remaining $8.13 is used to cover the fixed cost or "overhead" burden.

The CTO in any breakeven example may be calculated as a percent of sales or in dollars (cents) per unit. In these examples, CTO is assumed to be SP–VC (selling price minus variable costs), in percent or dollars/unit. CTO may also be calculated as GM–VC (gross margin minus variable costs), again, as a percent or in dollars per unit. Either approach produces the same result since sales minus cost of goods sold equals gross margin. If variable costs are measured as a percent of sales, the breakeven amount is in sales dollars and is calculated as shown below.

$$\text{Breakeven in Sales Dollars} = \frac{\text{Fixed Costs}}{1.0 - \text{Variable Costs as a Proportion of Net Sales}}$$

If variable costs are measured in dollars (cents) per unit, then breakeven volume is in units and is calculated as shown below.

$$\text{Breakeven in Units} = \frac{\text{Fixed Costs}}{\text{Selling Price Per Unit} - \text{Variable Costs Per Unit}}$$

In our example, Brookston tracked variable costs as a percent of sales, so CTO is:

$$\text{CTO} = \text{SP} - \text{VC} = 100\% - 91.87\% = 8.13\%$$

Since we are using percentages, in this format, SP (selling price) is always 100 percent of selling price or it equals 1.0. So, CTO can be calculated as shown below:

$$\text{CTO} = \text{SP} - \text{VC} = 1.0 - 0.9187 = 0.0813$$

For BF&G the breakeven sales volume in dollars is:

$$\frac{\$348,327}{0.0813} = \$4,284,465$$

For Brookston Feed and Grain, breakeven has been calculated to be sales of $4,284,465. If BF&G generates sales greater than this amount over the operating period, the firm will make a profit. With sales above the breakeven level, the contribution margin continues to be generated, but the fixed costs have already been covered. Thus, CTO now becomes **CTP (contribution to profit).** The number is the same—8.13 cents—but now this "margin" contributes to profit for Brookston.

Brookston's actual sales were $5.2 million. Since actual sales are higher than the calculated level of breakeven sales, Brookston made a profit. Given our calculation of CTO, we should be able to estimate the net operating profit (including the interest expense as an operating expense) for Brookston that is shown in Table 16-1.

	Actual Sales	$5,200,000
−	Breakeven Sales	4,284,465
	Sales above Breakeven	915,535
×	CTO	0.0813
		$74,433

The estimate of net operating profit after interest, $74,433, is close to the amount reported on the income statement for the year ending December 31, 2001 ($109,000 − 35,000 = $74,000), which can be found in Table 16-1. Rounding some of the variable cost percent figures calculated in Table 16-1 causes the difference of $433. The important concept here is to understand that as the firm generates sales above the breakeven level, variable costs continue to be paid—these go on forever. Regardless of the level of sales, Brookston pays 91.87 percent of each sales dollar to variable cost.

Assumptions of Volume-Cost Analysis

As the manager begins to utilize the breakeven calculation, it is important to understand the assumptions made for the breakeven procedure. These assumptions are:

1. Fixed costs are constant.
2. Efficiency is unchanged.
3. Input prices are fixed.
4. Product mix is constant.
5. Selling price is unchanged.

The first assumption listed above relates to Figure 16-1. By definition, fixed costs are constant over the operating period. Any decision made, such as the purchase of a new piece of equipment, will change fixed costs and, thus, the breakeven level. The second assumption relates both to fixed and variable costs. The manager assumes the firm is at a specific point on the average fixed cost curve, Figure 16-2, and the firm cannot reduce per-unit variable costs, for example, by taking advantage of volume discounts from suppliers. If efficiency can be increased in some way, a new breakeven level must be calculated.

The third assumption relates to Figure 16-4 where the variable costs of inputs are shown to be a constant percentage of sales. If suppliers or the firm does anything to alter input costs, breakeven changes. The fourth assumption is one that is often overlooked. In a diversified firm there are revenues and costs from several different product categories. Some of these product divisions may have significantly higher percentages of variable costs. Thus, as the firm sells relatively more of this product, the total variable costs for the firm shift. Hence, one makes the assumption of a constant product mix, which means the percentage of sales from each division or department is fixed. The final breakeven assumption is straightforward. Basically, anytime a firm changes selling price, it also changes the contribution margin. Thus, if prices change, breakeven must be recalculated.

Uses of Volume-Cost Analysis

Profit Planning   Volume-cost relationships are useful for much more than just calculating the breakeven point. They also can be used to determine the volume of business necessary to generate certain levels of profit, which is an essential part of profit planning.

Since revenue above variable costs is profit once the overhead has been covered, a similar calculation can be used to determine the additional sales necessary to reach a given profit level. If Brookston has a net operating profit (after interest) goal of $100,000, it will take an additional $1,230,012 sales above their breakeven volume of $4,284,465 to achieve the profit goal. The additional sales needed is calculated as shown below:

$$\text{Additional Sales to Reach Profit Goal} = \frac{\text{Profit Goal}}{\text{CTP}}$$

$$\frac{\$100,000}{0.0813} = \$1,230,012$$

Thus, to reach this profit goal Brookston must generate sales of:

Breakeven Sales	$4,284,465
Additional Sales to Reach Profit Goal	1,230,012
New Total Sales Goal	$5,514,477

Management now must follow through with a marketing and sales program that will allow BF&G to reach its profit goal.

Changes in Costs   Volume-cost analysis also helps answer questions about how changes in the cost structure will affect profit levels. Suppose, for example, that Brookston is considering the purchase of an additional truck that costs $30,000. The truck would be depreciated over a 5-year period. The annual fixed costs for this additional truck, including depreciation, are projected to be $6,500. Variable operating costs for the new truck should be about the same as the current truck. With this information, Brookston management can examine the impact that purchasing this new truck will have on breakeven and profit.

The breakeven calculation with this new cost information shows that Brookston must generate additional sales of $79,951 before it is in any better financial position. The truck purchase, with depreciation over 5 years, would obligate the firm to increase sales by $79,951 each of the next 5 years before profits improve. The calculation to determine the additional sales volume needed is:

$$\text{Additional Sales to Cover New Fixed Costs} = \frac{\text{Additional Fixed Cost}}{\text{CTO}}$$

$$\frac{\$6,500}{0.0813} = \$79,951$$

It is important to note that breakeven does not tell management "what to do." The breakeven relationship simply begins to identify how changes (in this case, the addition of a new truck) have an impact upon the firm. To ensure good management decision making when using this tool, it is imperative that the manager is able to identify the cost changes caused by the new action, in this case, buying a truck. (We will take a closer look at investment decisions in Chapter 17.)

Another use of breakeven analysis is to determine the effect of changes in variable costs. For example, suppose that Brookston's supplier increased prices paid by BF&G for products, so that the cost of goods sold (and variable costs) increased by 1 percentage point, but competition in the area was so keen that Brookston management felt it could not increase the selling price of its products. The result is that the CTO is lowered.

FC = $348,327   VC = 91.87% of Sales or 0.9187

Old CTO   = 1.0 – 0.9187 = 0.0813
Old Breakeven = $4,284,465

New CTO   = 1.0 – (0.9187 + 0.01) = 0.0713
New Breakeven = $4,885,372

This increase in variable costs causes breakeven sales to increase:

New Breakeven = $4,885,372
Old Breakeven = $4,284,465
Additional Sales $ 600,907

This result shows that if Brookston absorbs the cost increase rather than increasing the selling prices of their products, the breakeven point will increase by $600,907. If their volume remains at $5.2 million, profit would fall by $52,000 (1 percent lower CTO times $5,200,000).

Similarly, shrewd purchasing of products or a reduction in manufacturing costs reduces the cost of goods sold and variable costs. In Brookston's case, an improvement in purchasing practices which lowered the cost of goods sold (and variable costs) by 1 percent would increase the CTO to 9.13 cents, and reduce the breakeven point by $469,273.

FC = $348,327   VC = 91.87% of Sales or 0.9187

Old CTO   = 1.0 – 0.9187 = .0813
Old Breakeven = $4,284,465

New CTO   = 1.0 – (0.9187 – 0.01) = 0.0913
New Breakeven = $3,815,192

As variable costs fall, the breakeven sales also falls:

Old Breakeven = $4,284,465
New Breakeven = $3,815,192
Decreased Sales $ 469,273

The impact of any cost change can be anticipated using these techniques. Managers who are skillful in the application of volume-cost analysis find it very useful in projecting probable results of various alternatives before making final decisions. Also, the level of CTO magnifies the cost changes dramatically. With a great deal of grain business, Brookston has a relatively high variable cost and, thus, a small CTO. A little change in costs will make a dramatic difference in Brookston's breakeven level. For example, when Brookston's variable cost increased 1 percent, the breakeven sales level increased over 14 percent.

Changes in Price   Volume-cost analysis is also very useful in analyzing the impact of changing prices. Since CTO is selling price per unit less variable cost per unit, any change in selling price directly impacts the CTO. Let's assume management of BF&G wants to evaluate a 2 percent price reduction in an attempt to increase sales. The new breakeven is:

$$\text{Old CTO} = 1.0 - 0.9187 = 0.0813$$
$$\text{New CTO} = 0.98 - 0.9187 = 0.0613$$

As prices fall, CTO falls and breakeven sales increase:

$$\text{New Breakeven} = \$5,682,333$$
$$\text{Old Breakeven} = \underline{\$4,284,465}$$
$$\text{Additional Sales} = \$1,397,868$$

In this scenario, if Brookston decided to reduce selling price by two percentage points, its breakeven would increase by $1,397,868 or roughly 33 percent. Again, volume-cost analysis does not answer the question of what should be done—this is still the manager's decision. What volume-cost analysis does, in this case, is to show the manager that if BF&G cuts price by two percentage points, BF&G must increase sales by $1,397,868 to be in the same financial position as they were before the price cut. It is the manager's job to decide whether or not this move makes sense in the BF&G market area.

One word of caution about price changes: lowering the price may increase sales only temporarily. For many food and agribusiness firms, total demand may well be reasonably constant in a market area and/or during a season. Competitors may react to a firm's price reduction with a similar price cut. The net effect may be to lower price and CTO for all firms in the market without any appreciable increase in sales. The obvious result is much lower profits (or higher losses).

In only the most unusual circumstances can a firm afford to sell any product at a price lower than its variable cost. The variable cost for a product represents, for all practical purposes, the lowest possible price for that product. To sell when variable cost is not being covered means that some of the variable costs are not covered and none of the overhead can be covered. Losses will be less if the sale is not made at all because in this case only fixed costs are lost at zero sales volume.

So, when would a firm sell product below its variable cost? Two possibilities exist where this decision may make economic sense. First, recall the pricing strategies outlined in Chapter 8. One of these strategies was identified as loss-leader pricing. In this case, the product is priced below its total cost, and sometimes even below its variable cost. For example, in order to gain additional customers, a supermarket may price milk below its supplier cost. The low price would be advertised in the store's weekly flyer for customers to see. The strategy used by the store, in this case, is to get customers into the store by advertising a product at a very attractive price. The store makes money on the other items customers purchase on their trip to the store. This strategy may backfire, however, if the store has a lot of customers who come in only to shop specials.

The second scenario where selling below variable cost makes sense is related to the demand factor of tastes and preferences. Suppose a firm has an inventory of a specific type of farm equipment. The firm is notified that the manufacturer is planning to introduce a new model that has many new features and will sell at the same price. As potential customers become aware of this new model, they have little use for the firm's inventory of the old model. Hence, if the firm is to generate any revenue from this inventory, it will most likely have to cut the selling price—probably significantly. This strategy may make more sense than having to write off obsolete inventory, or sell the equipment later at an even higher level of discount.

Graphical Analysis

Graphical illustration of volume-cost relationships is quite useful for many managers as a means of visualizing how changes in price, various costs, or volume impacts profit. Figure 16-7 illustrates Brookston's breakeven point and graphically shows its profit at the current $5.2 million sales volume. (Note that profit is the difference between total revenue and total costs at a specific volume level.) The impact of almost any cost or price change can be shown by simply drawing in the change and observing the resulting effect on profits.

One of the best features of volume-cost analysis is its simplicity and its applicability to real-world situations. Any manager with an income statement and an understanding of fixed and variable cost concepts can estimate the impact of

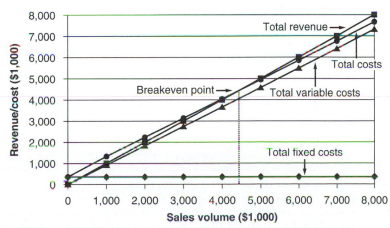

Figure 16-7
Volume-Cost Relationships for Brookston Feed and Grain Company.

various decision alternatives on profit levels. Although the categorizing of fixed and variable costs may not be perfect, experience has shown that a good approximation can be made relatively easily. And the conclusions are fairly accurate so long as the proper definitions of fixed and variable costs are used.

SUMMARY

Decision making is one of the primary responsibilities of agribusiness managers. Professional managers approach this activity in a systematic way—identifying the problem, summarizing facts, identifying and analyzing alternatives, and taking action. Their analysis often uses a wide variety of analytic tools to facilitate this entire process.

Volume-cost analysis is one of the more powerful tools used by agribusiness managers. By separating fixed costs (those not related to volume of business) from variable costs (those directly related to volume), managers can study the impact of a variety of cost and price changes on their profit and determine the amount of business necessary to break even. They can project the amount of business necessary to reach a certain profit level.

Volume-cost analysis helps managers and owners of agribusiness firms determine the impact of a variety of important decisions to the firm. Should a salesperson be added, can we afford to cut selling price, and what happens if we can reduce our input costs? Those interested in applying volume-cost analysis to a particular firm should note that variable costs for agribusinesses can vary dramatically—from less than 50 percent to 90 percent or more. Our example of Brookston Feed and Grain results in a relatively low contribution to overhead (CTO). There is a good reason for this. About 60 percent of Brookston's volume is from grain sales. This is a business that operates with very thin margins and makes money on turnover. Alternatively, if our example firm had been a seed business, CTO would have been higher. It is important to know the nature of the business you evaluate.

Finally, volume-cost analysis is an important planning tool for making operating decisions. This financial management tool is used to help make and/or evaluate decisions for the next operating period. Other financial tools, specifically those related to capital investment analysis, take a longer-term planning horizon. These tools will be discussed in the next chapter.

DISCUSSION QUESTIONS

1. What are the major steps in the decision-making process? What are some of the problems decision makers encounter when trying to use this process?
2. Is volume-cost analysis better used as a short-term or long-term planning tool? Why?
3. Explain the difference between variable costs and controllable costs. Give an example of each of these types of costs.

4. A particular cost can be variable under some circumstances but fixed in others. Given the following expense items, what conditions might make each fixed and what might make each variable? Which do you think each generally is, fixed or variable? Why?
   a. Part-time labor expense
   b. Sales compensation
   c. Truck maintenance expense
   d. Truck rental expense
   e. Cost of goods sold
5. What are the major assumptions made in determining the breakeven point? Which of these assumptions do you feel is likely the most restrictive in practice? Why?
6. The manager of a firm wants his assistant to "take action to lower our breakeven." Describe at least two ways to accomplish this goal. Are these choices good ones for a typical firm? Why or why not?
7. Consider the actions below. How would you examine the impact these changes would have on the firm's breakeven level? Be specific. How would each of these actions affect the firm's breakeven point?
   a. Hire a new salesperson.
   b. Pay additional overtime help.
   c. Increase advertising expense.
   d. Fire one of three assistant managers.

CASE STUDY — Castor Lumber Company

Bill and Judy Castor started a hardwood lumber company in northwestern Ohio about 15 years ago. Sean Jones, the manager for the past 3 years, has been under some pressure recently to make necessary improvements to the firm so it would continue to generate a sufficient return on equity for the owners. Sean is attempting to evaluate several changes in the business to determine the impact on profitability. He has decided to use volume-cost analysis in this evaluation process.

Sean, with the help of the firm's accountant, has determined that Castor's fixed costs or overhead for the coming year will be about $1,800,000. The firm sells a mix of select hardwoods (cherry, oak, walnut, and poplar). Their average selling price is about $7.50 per board foot. The firm pays local timber owners about $3.00 per board foot for standing timber. Sean figures that additional costs that seem to vary with sales amount to about $1.50 per board foot.

Questions

1. Calculate the breakeven point for the Castor Lumber Company in dollars and in board feet of lumber.
2. The owners, Bill and Judy Castor, have about $700,000 invested in this business. Their desired rate of return on this investment is 12 percent. What volume of business (again, in dollars and in board feet of lumber) must be generated by Sean to reach this ROE goal?
3. Sean has entered into preliminary negotiations with a labor union seeking to unionize this firm. He estimates that this would increase costs by about 50 cents per board foot. With no other changes, what impact would this have on the business? (Show your answer in dollars and in board feet of lumber.)

4. The Castors' have been impressed with Sean's enthusiasm for his job. They have decided to give him a raise of $5,000. What does this do to the breakeven point? Be specific about the dollar change in the breakeven point.
5. Sean has investigated the possibility of acquiring new saws that would increase the potential capacity of the business. Sean estimates the total cost, including acquisition and installation, to be about $150,000. He assumes this new equipment can be depreciated over a 5-year period. If the write-off for equipment is equal each year and there is no salvage value, what additional sales will Castor Lumber need for this to be a profitable decision? (Answer this question in both board feet of lumber and in dollars.)
6. Bill and Judy have benefited from Sean's management ability. They have never had an unprofitable year. However, they are interested in ways the firm might increase sales. During their last performance review with Sean, it was noted, however, that a new competitor might be preparing to enter their market. The Castors are concerned about losing market share and have suggested the possibility of reducing prices for their hardwood lines. They hoped this change would increase sales in the near term and help them maintain sales later if the competitor did enter their market. What would the impact of a 10 percent price reduction have on the business? (Answer this question in dollars and in board feet of lumber.)
7. Sean has considered adding a new product line to the business—walnut veneer. This diversification decision would require the firm to purchase new equipment. The new equipment will be depreciated over 5 years with no salvage value. If Sean calculates the new breakeven point with this equipment is 700,000 board feet, what is the cost of the equipment?
8. On a piece of graph paper or using a computer, illustrate the initial breakeven relationship for Castor Lumber Company. You may use dollars or board feet on the horizontal axis. *Note:* A review of Figure 16-7 should be useful.

17

TOOLS FOR EVALUATING CAPITAL INVESTMENT DECISIONS IN AGRIBUSINESS

- Describe the role of capital investment decisions in an agribusiness
- Examine and learn to apply basic tools for analyzing capital investment decisions in agribusiness
- Discuss the payback, simple rate of return, net present value, and internal rate of return methods for evaluating capital investment decisions
- Examine the data needs when using the net present value and internal rate of return methods for evaluating capital investments
- Describe the decision criteria for making and rejecting capital investments when using the net present value and internal rate of return methods

INTRODUCTION

An area of major consequence for agribusiness managers is making capital investment decisions. **Capital investment** refers to the addition of durable assets to an agribusiness, which usually require relatively large financial outlays and will last over a long period of time. Typical capital investments might include trucks, manufacturing equipment, or storage facilities. Even relatively small firms can spend millions each year on capital investments. These investments tie up funds for long periods and release them slowly as the investment produces revenue. The impact of these investment decisions may affect the business for years to come. In addition to deciding whether or not to make a specific investment, the agribusiness manager also has to decide on how to finance the investment. Capital investments can be financed through cash purchase, borrowing, or leasing.

This chapter will explore tools for evaluating capital investment decisions. Both relatively simple tools and more sophisticated tools will be covered. Given

the very large amounts of funds involved, and the long-term nature of the investments, careful analysis of capital investment decisions is another important skill
of the effective manager.

CAPITAL BUDGETING

There are many investment decisions that Brookston Feed and Grain might face
in the normal course of business activity. Some of these include:

1. *Expansion projects.* Would it be profitable to expand the main plant now?
2. *Replacement projects.* Should the custom application floater be replaced
   now, or should it be kept for another year?
3. *Alternative investment projects.* Which would be more profitable, stainless
   steel or mild steel storage tanks?

The procedure for evaluating the effects of an agribusiness manager's investment
choices on the profitability, risk, and liquidity of a business is called **capital
budgeting** or investment analysis. Capital budgeting is an orderly sequence of
steps that produces information relevant to an investment choice. These steps are:

1. Identification of investment alternatives
2. Selection of an appropriate capital budgeting evaluation method
3. Collection of relevant data
4. Analysis of the data
5. Interpretation of the results

These five steps are developed and explained in the following sections.

Capital budgeting focuses on the longer term decisions of an agribusiness.
Photo courtesy of Pioneer Hi-Bred International, Inc.

IDENTIFICATION OF ALTERNATIVES

An important function of management is to identify profitable investment alternatives. This is an essential function because even though an investment may be of high value and be profitable today, capital obsolescence and even price declines may reduce the value of the investment in the not too distant future. Consequently, management must continuously search for investment alternatives.

When money is tied up in a capital investment, management expects to get more back than was invested in the project. The excess of revenue over cost is referred to as the return. The return should be higher than that which could be earned by putting the same money in a "safe" place, such as savings accounts or government bonds. In addition to making a return for the use of the capital that is comparable to other investment alternatives, the return should also compensate owners of capital for any additional risk associated with the investment.

The idea here is relatively straightforward. If a person could earn 4 percent per year on an investment in a bank certificate of deposit (CD), how much more than 4 percent would that person require to put that money into a food or agribusiness firm? The return for the CD is relatively low, but so is the risk—the person will get their principal back plus the 4 percent interest at the end of the year. The investment in the agribusiness is more risky. The general market, the competition, the management, and a host of other factors will influence the investor's return from the agribusiness. This added risk means that investors will want a lot more than 4 percent return to put their money into a food or agribusiness firm.

Some people think that if their business borrows the money to purchase a new piece of equipment, they have transferred the risk to the lender. In reality, this is not true because the bank is sure to be repaid, except in the extreme case of bankruptcy. If the manager makes a bad decision, the total business will suffer the loss because the loan will be repaid from other, more successful ventures. So managers must have a systematic approach for identifying investment alternatives. They must also understand that using financial leverage or debt to acquire investments can change the firm's risk and liquidity positions. (Financing the agribusiness was discussed in Chapter 14.)

The identification of investment opportunities falls into one of the four categories listed below:

- Maintenance and replacement of depreciable capital items
- Cost-reducing investments
- Income-increasing investments
- A combination of the preceding categories

For example, consider a swine finishing operation that is very labor-intensive. The manager is evaluating the changes needed to replace worn-out facilities. One investment alternative to consider is a mechanized feeding system. However, the size of the operation needs to increase to fully use the mechanized system. This investment should cut labor costs, but the manager needs to better understand the cost of operating the mechanized system. In addition, the system

may allow for better control of the nutrition program, which may allow the producer to do a better job of managing carcass quality. This could potentially increase returns if the producer sells the hogs on a carcass-merit (quality) basis.
Given the complexity of the decision, the manager in this situation needs a systematic procedure to evaluate the investment.

CAPITAL BUDGETING EVALUATION METHODS

The next step is to choose an evaluation method that can be used to rank, accept,
or reject the investment alternatives. There are several methods for evaluating
capital investment decisions. Some of them are simple, while others are quite
complex, requiring sophisticated mathematical analysis. Four methods will be
discussed here. Listed in order of complexity, they are: payback period, simple
rate of return, net present value, and internal rate of return.

Concepts of Time Value

The net present value and internal rate of return methods are based on the time
value of money, so an introduction to the concepts of time value is needed prior
to discussing those methods. This section develops the basic ideas underlying
the time value of money. The first of those ideas is that an interest rate serves as
the pricing mechanism for the time value of money. The **interest rate** reflects
investors' time preferences for money and serves as the exchange price between
money received today versus money received at some point in the future. In
essence, a dollar received today can be exchanged for a dollar plus an amount of
interest $(1 + i)$ received one period in the future. Alternatively, one dollar received one period in the future exchanges for $1 \div (1 + i)$ dollars today.

For example, one thousand dollars invested today for 1 year at 10 percent
will be worth $1,100 at the end of the year. So, the present value of $1,100 received in one year, at a 10 percent discount rate is $1,000. The $1,100 investment has been "discounted" to its present value of $1,000. The formula to calcu-
late the present value is:

$$\text{Present Value of Investment} = \frac{\text{Future Value of Investment}}{(1 + i)^n}$$
$$\text{where: } i = \text{Interest Rate}$$
$$n = \text{Number of Periods}$$

So for our example:

$$\text{Present Value} = \frac{\$1,100}{(1 + 0.10)^1} = \frac{\$1,100}{1.10} = \$1,000$$

Suppose that at high school graduation a relative gives a new graduate a note
promising him or her $1,000 four years later upon graduation from college. As-
suming that the time value of money is 7 percent (interest rate), what is the present value of that gift? Although the formula is the same as before, this is more
difficult to calculate.

TABLE **17-1** | **Present Value of $1 (Compounded Annually)**

Year (N)	1%	3%	$(1+i)^{-n}$ 5%	6%	7%
1	0.9901	0.9709	0.9524	0.9434	0.9346
2	0.9803	0.9426	0.9070	0.8900	0.8734
3	0.9706	0.9152	0.8638	0.8396	0.8163
4	0.9610	0.8885	0.8227	0.7921	0.7629
5	0.9515	0.8626	0.7835	0.7473	0.7130
6	0.9420	0.8375	0.7462	0.7050	0.6663
7	0.9327	0.8131	0.7107	0.6651	0.6228
8	0.9235	0.7894	0.6768	0.6274	0.5820
9	0.9143	0.7664	0.6446	0.5919	0.5439
10	0.9053	0.7441	0.6139	0.5584	0.5083
15	0.8613	0.6419	0.4810	0.4173	0.3624
20	0.8195	0.5537	0.3769	0.3118	0.2584
25	0.7798	0.4776	0.2953	0.2330	0.1842

Year (N)	8%	9%	$(1+i)^{-n}$ 10%	12%	14%
1	0.9259	0.9174	0.9091	0.8929	0.8772
2	0.8573	0.8417	0.8264	0.7972	0.7695
3	0.7938	0.7722	0.7513	0.7118	0.6750
4	0.7350	0.7084	0.6830	0.6355	0.5921
5	0.6806	0.6499	0.6209	0.5674	0.5194
6	0.6302	0.5963	0.5645	0.5066	0.4556
7	0.5835	0.5470	0.5132	0.4523	0.3996
8	0.5403	0.5019	0.4665	0.4039	0.3506
9	0.5002	0.4604	0.4241	0.3606	0.3075
10	0.4632	0.4224	0.3855	0.3223	0.2697
15	0.3152	0.2745	0.2394	0.1827	0.1401
20	0.2145	0.1784	0.1486	0.1037	0.0728
25	0.1460	0.1160	0.0923	0.0588	0.0378

$$\text{Present Value} = \frac{\$1,000}{(1+0.07)^4} = \frac{\$1,000}{(1.07 \times 1.07 \times 1.07 \times 1.07)}$$

$$= \frac{\$1,000}{1.3108} = \$762.90$$

To make things simpler, managers often use tables of present values (Table 17-1) or use financial calculators to quickly calculate the present value of any investment. In this example, the present value of $1 four years from now at 7 percent is given in the present value table as 0.7629, so the present value of $1,000 would be:

$$\text{Present Value} = \$1,000 \times 0.7629 = \$762.90$$

In other words, if the relative invests $762.90 today at 7 percent interest, compounded annually, that investment will be worth $1,000 in 4 years.

Compounding is a method of calculating interest earned periodically which is added to the principal and becomes part of the base on which future interest is earned. Compounded annually means that interest earned during the first year ($762.90 × 0.07 = $53.40) will be added to the original investment **(principal)** ($762.90 + $53.40 = $816.30), so that in year two, the entire investment ($816.30) will earn interest at the 7 percent rate. Each year additional interest earned is added to the investment, and it too earns interest. With these ideas of time value of money, present value, and compounding in mind, let's take a look at some of the tools for evaluating capital investments.

Payback Period

The length of time it will take an investment to generate sufficient additional cash flows to pay for itself is called the **payback period.** This simple tool allows the agribusiness manager to compare investment alternatives and determine which will recoup its initial investment in the shortest period of time and so would be the most desirable. The formula to calculate payback period is simple:

Payback Period = (Investment)/(Average Annual Net Cash Flow)

The accept-reject criterion centers on whether the project's payback period is less than or equal to a firm's maximum desired payback period. If two projects have payback periods that are less than the maximum desired period, then the one with the shorter period is accepted.

For example, suppose that BF&G is considering the purchase of a truck-mounted power soil probe to take soil samples, and that this outfit costs $12,000. The firm plans to charge customers for the full cost of these soil samples plus a profit margin. Annual net cash flows to the business before depreciation are estimated to be $3,400. (Note that since depreciation is in itself a method of offsetting initial cost, it must be omitted in calculation of the payback period.) A second investment alternative being considered is a new pickup truck costing

The payback period is the length of time it takes an investment to pay for itself.

$16,000. The truck will be used for immediate deliveries and pick-ups and a fee will be charged for its use in those situations. Estimated annual net cash flow before depreciation is $2,800.

Soil Probe Payback Period $= \text{(Investment)}/\text{(Avg. Annual Net Cash Flow)}$
$= \$12,000/(\$3,400/\text{year}) = 3.5 \text{ years}$

Pickup Truck Payback Period $= \text{(Investment)}/\text{(Avg. Annual Net Cash Flow)}$
$= \$16,000/(\$2,800/\text{year}) = 5.7 \text{ years}$

Generally, the shorter the payback period, the better the investment. So it would appear from this simple method that the soil probe is the better alternative because it will pay for itself in 3.5 years, compared to 5.7 years for the pickup truck.

The payback method indirectly considers risk in the decision process. The payback principle suggests that by accepting the project with the quickest payback, that project has the lowest risk. This is based on the investor desiring the cash flow as soon as possible due to the uncertainty involved in receiving the cash and the opportunity to reinvest that cash sooner with a shorter payback period than with a longer period. Here, the manager is assumed to be comparing similar investment options and time is the primary risk factor under consideration. Thus, an investment that is repaid by cash flows in 3.5 years involves less risk than a project that pays back its original investment in 5.7 years.

The payback period method is widely used, partly because of its simplicity. But it ignores the length of life of the investment. What if the pickup truck lasts for 10 years, while the power soil probe lasts only 5 years with regular cash flows? Now, the decision is more complicated and the payback period doesn't capture the difference in the length of the life of the two investments. We need a more sophisticated investment evaluation method in such a case. Although this limitation is a serious drawback, when cash flow is critical, the payback method is useful.

In a somewhat similar limitation, the payback method ignores cash flows generated after the payback period. Hence, payback gives more weight to those investment alternatives that have the majority of cash flows generated early in the life of the project or investment alternative. Some investments may generate small cash flows early due to various startup costs, but the cash flow may increase over certain periods of their lives. These cash flows would be ignored when compared to an investment that generates the majority of its cash flow in the first few years.

Another weakness of the payback method is a failure to account for the time value of money. Thus, $1 of cash flow received a year from today is given the same "weight" in payback analysis as $1 received 3 years from today. As discussed in the section on time value of money, cash flows received after one period need to be discounted. So if all the cash flows have equal value regardless when received, the cash flows received in 3 years are overvalued relative to those received in 1 year. Other methods for evaluating investments have been developed to help address these limitations of the payback period.

Simple Rate of Return

The **simple rate of return** method refers to the profit generated by the investment as a percentage of the investment. The most common variation of this method uses the average return and the average investment to give a more accurate analysis. The simple rate of return calculation is:

Simple Rate of Return = (Average Annual Return/Average Investment)

Simple rate of return is perhaps the most commonly used method of capital budgeting. This method considers net earnings over the entire expected life of the investment. It is easy to understand and is consistent with ROE goals imposed by management. In fact, many firms have estimated ROE standards that serve as a cutoff point for investment projects. Unless an investment proposal exceeds these minimal standards, it will not be seriously considered. A standard of 15 to 25 percent is common for many food and agribusiness firms.

To illustrate the calculations of the simple rate of return, suppose that Brookston is considering the purchase of a self-propelled fertilizer applicator with flotation tires and an onboard GPS controller to allow variable rate application of fertilizer and/or lime. This new machine would reduce soil compaction, allow for site-specific application of fertilizer and/or lime, and upgrade the services provided to producers. The net returns expected from this $80,000 investment decline over its life because of increasing maintenance and repair costs. BF&G estimates that average net profit after depreciation for the new applicator equals $16,000 and the calculations are shown in Table 17-2. The average rate of return would be:

Simple Rate of Return = (Average Annual Return/Average Investment)
= ($16,000/$40,000) × 100 = 40%

Brookston would rank this investment alternative higher than those with returns under 40 percent.

T A B L E **17-2** | **Fertilizer Applicator Simple Rate of Return Calculations**

Year	Net profit before depreciation	Depreciation
1	$ 36,000	$ 16,000
2	34,000	16,000
3	32,000	16,000
4	30,000	16,000
5	28,000	16,000
Total	$160,000	$ 80,000
Average	$ 32,000	$ 16,000

$$\text{Average Investment} = \$80,000 \div 2 = \$40,000$$

$$
\begin{aligned}
\text{Simple Rate} \atop \text{of Return} &= \frac{\text{Average Net} \atop \text{Income} - \text{Average} \atop \text{Depreciation}}{\text{Average Investment}} \\
&= \frac{\$32,000 - \$16,000}{\$40,000} \times 100 = 40\%
\end{aligned}
$$

The chief limitation of the simple rate of return method is that it fails to consider the timing of cash flows. When the initial net cash flow is high but falls off quickly, the situation is far different than when the initial net cash flow is low but increases over time. A consequence of this problem is that incorrect conclusions may be drawn because the method fails to consider that a dollar returned now is significantly more valuable than a dollar of cash flow received several years later. Also, the simple rate of return is a rate and is influenced by the size of the investment. So, an investment project may be very small, but have a very high simple rate of return. At the same time, a much larger project may have a lower simple rate of return, but generate far more dollars for the agribusiness.

Net Present Value

Net Present Value (NPV) is the current, net value of an investment taking the time value of money into consideration when evaluating costs and returns. As discussed earlier, the time value of money captures the fact that a dollar now is worth more than a dollar received at some future date. In essence, net present value provides a measurement of the net value of a multiyear investment in

Capital equipment, like this fertilizer applicator, generates profit for the agribusiness over time.

Photo courtesy of Ag-Chem Equipment Co., Inc.

today's dollars. The net present value method uses the discounting formulas discussed previously to value the projected cash flows for each investment alternative. The net present value model is set up as presented below:

$$\text{NPV} = -\text{INV} + \frac{P_1}{1+i} + \frac{P_2}{(1+i)^2} + \dots + \frac{P_N}{(1+i)^N} + \frac{V_N}{(1+i)^N}$$

where:

NPV = net present value of the investment alternative
INV = initial investment
P_i = net cash flows attributed to the investment in period i
V_N = terminal or salvage value of the investment
N   = length of the planning horizon
i = interest rate or required rate of return, also called the cost of capital
         or discount rate

Initial Investment   The **initial investment** refers to the initial amount the investor commits to the investment. It is important to ensure that all the costs necessary to make the investment operational are included (i.e., freight, installation, sales taxes, modifications to buildings, etc.) as well as the cost of the asset. All of these costs may not occur in the initial period, so outlays that occur later must be reflected in the periods in which they occur. Also, the trade-in or salvage value of a replaced asset should be subtracted from the purchase price of the new asset.

Net Cash Flows   For the net present value method, net cash flow from the business rather than accounting profits is used as the measure of returns. **Net cash flows** include all the cash inflows and all the cash outflows for operating expenses and any other capital expenditures. In essence, net cash flow is the stream of cash the owner can withdraw from the investment.

Note that net cash flows are not the same as net income as reported on an income statement, because the accrual accounting method differs from the cash basis of accounting. And, a difference between net income reported on the income statement using the accrual accounting method and the net cash flow will very likely occur. This difference occurs because revenues and expenses reported on the income statement are assigned to the fiscal year in which they are earned or are incurred rather than when they are received or paid in cash. Annual depreciation and inventory changes are examples of income statement items that do not directly impact the cash flow. Only future net cash flows are relevant for inclusion in the net present value calculation. (Note that income taxes complicate the analysis a bit. While we won't address this topic here, the normal rule is to use after-tax cash flow figures and an after-tax discount rate.)

Terminal Value   The **terminal value** for the investment is the residual value the investment is expected to have at the end of the planning horizon. For depreciable assets such as machinery, this value is often called the salvage value.

There could also be some appreciation in the value of the investment, which would result in capital gains at the end of the planning period.

Discount Rate   The **discount rate** or **cost of capital** is the interest rate used in capital budgeting, which is the firm's required rate of return on its equity capital. This was referred to earlier in this text as the opportunity cost (Chapter 3). The discount rate (i) used in capital budgeting consists of three components:

- A risk-free interest rate
- A risk premium
- An inflation premium

The risk premium reflects the riskiness associated with the expected net cash flow. The inflation premium reflects the anticipated rate of inflation. This rate should reflect the rate of return that the firm expects to earn on its equity capital.

Planning Horizon   A multiyear planning horizon is used for the investment in capital assets, because the assets generate cash flows for the firm for more than one year. A major factor that influences the length of the planning horizon is the productive life of the asset that is being purchased.

The acceptability of an investment depends on whether the net present value is positive or negative. The criteria used for decision making are:

- If the net present value exceeds zero, accept the investment.
- If the net present value equals zero, be indifferent.
- If the net present value is less than zero, reject the investment.

Keep in mind that when the net present value exceeds zero the method implies that the investment is profitable; however, it is profitable relative to the required rate of return implied by the discount rate. Rejection of an investment, which is based on a negative net present value, implies that an alternative investment with the rate of return (discount) used in the analysis is more profitable than the investment being evaluated.

When several investments are being considered, all are income generating, and all have positive net present values, the size of the net present value is used as part of the decision criteria. In this situation, the investment with the largest net present value is most favored, with the next largest net present value second, etc. If the investment under consideration is cost reducing, then the projected cash flows would reflect cash outlays (i.e., expenses). When cost-reducing investments are compared, the decision criteria are based on the minimum net present value of cash outlays.

An example of how net present value analysis might be used by the agribusiness manager follows. Assume that BF&G is considering the addition of another self-propelled custom fertilizer applicator costing $60,000. BF&G's owners insist that any additional investment return a minimum of 14 percent. Based on internal records from operating costs of current application equipment and the best estimates of customer response to additional service equipment, the net cash inflows shown in Table 17-3 are expected from the new custom applicator.

TABLE **17-3** | Net Cash Inflows for Applicator Investment

Year	Net cash inflow*
1	$22,000
2	15,000
3	14,000
4	12,000
5	16,000**

*The net cash inflow is the excess of cash revenue over cash expenses attributed directly to the investment. Depreciation and other noncash expenses are not included.
**Includes $6,000 from sale of the applicator when scrapped.

The net cash inflow generally declines each year because of increased main-tenance and repair costs as the equipment gets older. Taking this pattern of cash flows and the time value of money into consideration, there is a different present value associated with each year. The net present value of the investment is shown in Table 17-4.

The present value of the cash flows for this investment using a 14 percent dis-count rate is $55,706.50, which is considerably less than the $60,000 cost of the investment, so the NPV is negative. This means that the investment will not meet the owners' 14 percent cost of capital criterion. If a purchase price of $55,000 could be negotiated for the custom applicator, then it would meet the 14 percent discount rate criterion. In this case, the NPV would be a positive $706.50.

The net present value method of investment analysis has several advan-tages. First, it deals with cash flows rather than accounting profits. Second, the method is sensitive to the actual timing of the cash flows resulting from the in-vestment. Third, the time value of money enables the user to compare benefits

TABLE **17-4** | Net Present Value Calculation for Applicator Investment

Year	Net cash inflow	14 Percent discount factor*	Present value
1	$22,000	0.8772 =	$19,298.40
2	15,000	0.7695 =	11,542.50
3	14,000	0.6750 =	9,450.00
4	12,000	0.5921 =	7,105.20
5	16,000	0.5194 =	8,310.40
Total present value			$55,706.50
Initial investment in applicator			−$60,000.00
Net present value			−$4,293.50

*See Table 17-1.

and costs in today's dollars. Finally, because investments are accepted if a positive net present value is calculated, the acceptance of an investment will increase the value of the firm. That is consistent with the goal of managers to maximize shareholders' wealth.

One disadvantage of the net present value method results from the need for detailed, long-term forecasts of cash flows. These forecasts over long time periods are just that, forecasts. Therefore, there is considerable pressure on the manager to forecast cash inflows and outflows as accurately as possible.

Internal Rate of Return

The same equation used to calculate the net present value of an investment can be used to determine the **internal rate of return (IRR)** for the investment. The internal rate of return is called by various names, such as discounted rate of return, marginal efficiency of capital, and yield. However, it is essentially the discount rate that equates the net present value of the projected net cash flows to zero.

To calculate the internal rate of return for an investment, set up the net present value model as outlined above. Determine the projected cash flows and set the net present value to zero, then solve for i as shown below:

$$NPV(\$0) = -INV + \frac{P_1}{1+i} + \frac{P_2}{(1+i)^2} + \ldots + \frac{P_N}{(1+i)^N} + \frac{V_N}{(1+i)^N}$$

The discount rate that satisfies this equation is the internal rate of return.

To illustrate this method, let's find the internal rate of return for an investment that requires a $1,000 initial investment and yields a $1,200 cash payment one year in the future. The IRR model is set up as shown below:

$$0 = -1,000 + \frac{1,200}{1+i}$$

Solving the equation for i results in the following:

$$1 + i = \frac{1,200}{1,000} = 1.2$$
$$i = 0.2$$

So, the IRR for this investment is 20%.

When the planning horizon exceeds one year, the procedure is essentially a trial-and-error approach to determine the discount rate that will yield a zero net present value. Today, spreadsheets are used to determine the IRR given the complexity of the calculation.

Investments can be ranked and accepted or rejected on the basis of their internal rates of return. The ranking is based on the relative sizes of the internal rates of returns, with the largest being the most favored. The acceptance of each investment depends upon the comparison of its internal rate of return with the investor's required rate of return. Acceptability is based on the following decision rules:

- If the internal rate of return exceeds the required rate of return, accept the investment.

- If the internal rate of return equals the required rate of return, be indifferent.
- If the internal rate of return is less than the required rate of return, reject the investment.

Comparing Net Present Value and Internal Rate of Return

The net present value and internal rate of return methods are closely linked because each uses the same discounting procedure. However, the net present value method requires a specified discount rate, while the internal rate of return method solves for the discount rate that yields a zero net present value.

The IRR method gives the same ranking of investments as the NPV method under most circumstances. Although both account for differences in the time pattern of cash flows, occasional differences in ranking can arise because of different assumptions about the rate of return on reinvested net cash flows. The IRR method implicitly assumes that net cash inflows from an investment are reinvested at the same rate as the internal rate of return of the investment under consideration. The NPV method, on the other hand, assumes that these funds can be reinvested to earn a rate of return that is the same as the firm's discount rate.

Which reinvestment assumption is more realistic? The NPV rate has the advantage of being consistently applied to all investment proposals. Also, the NPV rate may be more realistic if the discount rate is determined by the opportunity cost of capital. On the other hand, the internal rate of return from each investment alternative can be compared against a common required rate of return. The internal rate of return also represents profitability in percentage terms, which is often preferred by business managers, even though the increase in wealth measured by net present value more directly reflects the objectives of the firm. In the end, both approaches provide useful insights into the evaluation of capital investment decisions.

SUMMARY

Capital investment decisions are among the most important decisions made by management. Their impact is long range and greatly impacts the flexibility of the business. Professional managers approach these decisions in a systematic way by first identifying the alternatives, selecting the evaluation method, collecting data, analyzing the alternatives, and taking action. The procedure for evaluating investment choices is called capital budgeting.

Four methods for evaluating capital investment decisions are discussed in this chapter. Listed in order of complexity, the four methods are the payback period, simple rate of return, net present value, and internal rate of return.

Simplistic approaches such as the payback period and simple rate of return are widely used methods of making investment decisions. However, they may lead to incomplete or even wrong conclusions. By recognizing the time value of money and using the net present value method or the internal rate of return method, managers can get a much more realistic assessment of investment alternatives.

1. Why are capital investment choices so important to the agribusiness manager? Why are these decisions more complex than most operating decisions?
2. What are the advantages and the disadvantages of the payback period method of evaluating capital investments?
3. What information is needed to calculate the simple rate of return? What are the advantages and disadvantages of this method of evaluating capital investments?
4. Why is net cash flow used in the net present value method rather than net income from a pro forma income statement prepared using accrual accounting?
5. Which type of business would likely have a higher cost of capital—a small biotechnology company or a large food retailer? Why?
6. What are the advantages and disadvantages of NPV and IRR for evaluating capital investments? Why are these approaches superior to the payback period and simple rate of return approaches?
7. What are the similarities and the differences between the net present value method and the internal rate of return method?

CASE STUDY—Columbia Valley Packing Plant

The Columbia Valley Packing Plant packs and ships apples and stores fruit for area growers. Brad Johnson, Columbia Valley's new manager, was being pressured by the owners to evaluate several possible changes in the business, including a major upgrading of the packing line. Perhaps you can help Brad by applying some of the decision tools discussed in this chapter to the decisions he faces.

Columbia Valley has the option of significantly upgrading their plant by adding electronic sizing and sorting equipment. The cost of this equipment is $300,000, with an expected life of 5 years. Equipment of this nature is not expected to have an appreciable salvage or scrap value. There will be significant labor savings involved. Initially, the new equipment will require little maintenance and repair, but as it gets older, the cost of maintenance will go up sharply. Generally, Columbia Valley's owners and management feel that a 12 percent return should be expected from any new investment project before it can be undertaken. Expected net returns before depreciation resulting from the investment and estimated depreciation expenses are shown in Table 17-5.

T A B L E **17-5** | **Estimated Net Profit and Depreciation Expense for New Sizing Equipment**

Year	Net profit before depreciation*	Depreciation
1	$150,000	$60,000
2	120,000	60,000
3	80,000	60,000
4	30,000	60,000
5	20,000	60,000

*Net profit here includes only cash inflows and outflows.

Questions
1. Initially, before any calculations, what is your general reaction? Should Columbia Valley invest in the new equipment?
2. What is the payback period for this investment?
3. What is the simple rate of return for this investment?
4. What is the net present value of this investment, using the 12 percent minimum return required by the owners?
5. What are the limitations of each of these approaches?
6. In the final analysis, do you believe this investment project would be good for Columbia Valley? Why or why not?

PART V

OPERATIONS MANAGEMENT
FOR AGRIBUSINESS

Operations and logistics: this section explores plans and processes for getting products and services developed, produced, and delivered.

18

PRODUCTION PLANNING AND MANAGEMENT

- Develop an understanding of operations management, including the definition, background, and current issues in the area
- Discuss some of the unique characteristics of agricultural commodities and products as they impact operations management and production planning
- Define quality and several of the current, popular quality initiatives
- Summarize the key elements involved in plant and facility location decisions
- Identify some of the factors to consider in determining the capacity of a plant or facility
- Differentiate between process, product, hybrid, and fixed position layouts
- Understand the key elements of job design

INTRODUCTION

Operations management refers to the direction and control of the processes used by food and agribusiness firms to produce goods and services. In the past, operations management was concerned primarily with manufacturing in factories. Today we recognize that service managers in supermarkets, financial institutions, Web-based industries and agribusiness consulting firms have the same questions and considerations of job design, location choice, facility design, purchasing, transportation, and scheduling as does the manufacturing sector. Operations management for food and agribusiness firms can be broken down into two distinct areas: (1) *production planning* and (2) *logistics.* Likewise, Chapter 18 is devoted to production planning as it pertains to both agribusiness service and manufacturing firms. Chapter 19 looks at the key area of logistics in the food and agribusiness industries.

The **production planning** aspect of operations management includes a wide range of decisions and activities including:

- Devising a quality program
- Locating a plant
- Choosing the appropriate level of capacity
- Designing the layout of the operation
- Deciding on the process design
- Specifying job tasks and responsibilities

The **logistics** function of operations management is discussed at length in Chapter 19 but because both planning and logistics are integral parts of operations management, they must be considered in tandem. Logistics in operations management consists of an equally wide range of decisions and activities including:

- Aggregate production planning
- Production scheduling
- Purchasing of materials for production
- Management of the various types of inventories
- Distributing the finished goods or services

Operations management involves a system of interrelated activities and players as shown in Figure 18-1. **Suppliers** provide the inputs to the system. Timely delivery of high quality inputs influences all activities in the system. **Inputs** consist of the human resources (skilled workers and managers), capital (equipment), materials, information, and energy required to produce the desired **outputs.** The transformation or conversion of inputs into outputs involves the system of facilities, processes, and procedures by which goods and services are produced. **Customers** purchase the outputs from the system, and their feedback to the rest of the system is essential in generating and designing new products and services. Finally, managers make decisions and obtain feedback on those decisions to make the total production system flow smoothly.

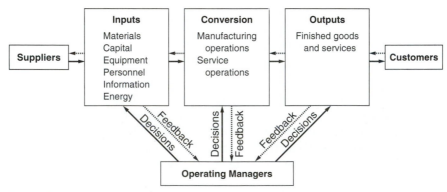

Figure 18-1
Operations Management System.

For example, a dairy begins with raw milk and can produce a variety of milk, cheeses, ice cream, and other products. The technology in a dairy is quite different from the technology used to sell crop protection chemicals and provide agronomic services, yet the problems of scheduling, controlling, storing, shipping, and general management of the two processes are quite similar. Other examples are shown in Table 18-1.

DIFFERENTIATING BETWEEN MANUFACTURING AND SERVICES

Until recently, operations management was also known as production management because of its focus on the manufacturing sector. However, the service sector is now the fastest growing sector of the U.S. economy comprising roughly 40 percent of U.S. GDP. There are, of course, differences between manufacturing and service organizations but the differences are shrinking as techniques for managing manufacturing firms find their way to service firms and vice versa. Some characteristics that distinguish the two sectors are shown in Table 18-2.

T A B L E **18-1** | **Examples of Production Systems and Components**

Manager	Suppliers	Inputs	Conversion	Outputs	Customers
Restaurant manager	Distributors Meat suppliers Equipment mfrs.	Meats Vegetables Food servers Utensils Equipment	Cooking Meal preparation Serving	Prepared foods Happy customers	Hungry diners
Farm producer	Equipment dealer Seed firms Chemical mfrs. Fertilizer mfrs.	Land Labor Equipment Seed Chemicals	Plowing Planting Spraying Harvesting	Fruits Vegetables Grains	Food processors Consumers Exporters

---

T A B L E **18-2** | **Characteristics of Goods Producers and Service Providers**

More Like a Goods Producer	More Like a Service Provider
Physical, durable product	Intangible, perishable product
Output can be inventoried	Output cannot be inventoried
Low customer contact	High customer contact
Long response time	Short response time
Large facilities	Small facilities
Capital intensive	Labor intensive
Quality easily measured	Quality not easily measured

Source: Krajewski and Ritzman.

Goods producers are distinguished from service providers because their products are physical, durable, and can be inventoried. Service firms produce intangibles—ideas, information, performances, etc. Some service organizations like restaurants typically produce a good and service—food and service. Many farm supply retailers not only provide quality products but product information, services such as delivery and product assembly, and credit. Service providers, unlike manufacturers, tend to have high levels of customer contact, and response time to customers is often measured in minutes, rather than days and weeks for goods producers.

Goods producers tend to serve regional or even international markets whereas services tend not to be able to be shipped. Rather, service providers tend to locate near the customer (although this is changing with new communications and information technologies). Quality is more easily assessed with goods producers as it can be measured and tested. Quality in service providers is difficult to assess because it is measured on intangibles such as customer perceptions, communications, and time, each subjective in nature.

UNIQUE ISSUES WITH AGRICULTURAL PRODUCTS

Special problems inherent in the production of agricultural products influence many of the decisions involved in operations management including the choice of plant location, method of transportation, and scheduling. Two such problems are **seasonality** and **perishability.** (A more complete list of these factors was presented in Chapter 1.)

Agribusiness supply firms are engaged in highly seasonal production processes. During the peak season, facilities are often strained to the utmost to produce the services and products producers need to utilize in a very short period of time. While the manufacture and processing of farm supplies is scheduled as evenly as possible throughout the year to maximize efficiency, lack of sufficient storage to accommodate peak season needs places tremendous pressure on production facilities as they struggle to keep up.

Seasonality and perishability of agricultural products present real challenges for operations and logistics managers in the food and agribusiness markets.

Perishability is in some ways, related to the problem of seasonality. Of course, fruits and vegetables are highly perishable and must be processed quickly to prevent spoilage. At the same time, some of these products are so highly seasonal that processors, canners, freezers, and packers are left with virtually idle facilities after a peak season. Some industries—the tomato industry is one—have evolved successful ways of counteracting this dual problem. Storing tomatoes in aseptic, nonrefrigerated, stainless steel vats effectively eliminates problems of deterioration. Operations that produce tomato products are now able to run consistently for 12 months of the year without fear of spoilage, thus reducing costs during the peak season and increasing productivity in the off-season. Fresh tomatoes can be shipped across the country in aseptic railroad cars, which produces significant savings in transportation costs.

Yet another typical problem of agricultural products is **bulk;** the costs for shipping bulky loads of oranges from Florida to the Pacific Northwest can be extremely high. Handlers must also address the physical problems of storing bulky products as they await processing or shipping; as with most such problems, this difficulty is also reflected in costs.

Variability in quality and quantity must still be managed in the agricultural industry although consumer demand for consistency has had significant impacts in this area. Processors, canners, and freezers of lobster and shrimp, for example, must weigh and differentiate among fish of unequal quality. Eggs and milk must be graded, as are many other products. In some cases, product weights must be standardized to fit certain-size packages or selling weights; this can sometimes produce waste.

A final difficulty experienced in the production of agricultural products is variation in value. Price efficiency demands that production managers manufacture outputs with the highest possible value consistent with the costs of production. A dairy processor may find that cheese has a higher market value than milk for table consumption; but if its value is only one-third more than that of milk, and it requires twice the production cost, then price efficiency tends to favor producing the milk. Because there is such a wide variety of value in agricultural products (in different cuts of beef, for example) this is an important area for agribusinesses to explore. This influences decisions related to product mix, which were discussed earlier in Chapter 8.

OPERATIONS MANAGEMENT: A HISTORICAL PERSPECTIVE

Operations management has its roots in the Industrial Revolution, which originated in England in the late eighteenth and early nineteenth centuries. During the Industrial Revolution, many new inventions were created that allowed goods to be manufactured with greater ease and speed. One of the more important inventions was the steam engine, which James Watt patented in 1769. It may be surprising to learn that the steam engine's major impact was on manufacturing, not on transportation.

Methods of planning and managing work were also changing. In 1776, Adam Smith, an economist, proposed that productivity improvements could be achieved by having different workers perform a few of the tasks involved in completing a job rather than having one worker perform all of the tasks. Eli Whitney furthered the Industrial Revolution in 1798 with his introduction of the interchangeable parts concept.

About a century later in the early 1900s, Frederick W. Taylor introduced the concept of **scientific management.** Born in Pennsylvania in 1856, Taylor had been a laborer who advanced to chief engineer. Through his experience, Taylor observed and studied production, focusing on workers, the methods of work, and the wages paid for increased output. He developed a basic set of tenets of scientific management. As a result, Taylor is often viewed as "the father of scientific management." Using his scientific or analytical approach, Taylor believed in, among other things, the following:

1. A standardized set of procedures that were to be used for a job each time it was performed
2. Determination of the most efficient procedures for each job through scientific comparison tests
3. Careful screening, hiring, and training of workers to match them with the most appropriate job
4. A functional division of labor that would allow each person to perform the most suitable task in the production process, with management and workers jointly devising the best system for division of labor

A period of extensive research on operations management followed Taylor's early work. Henry L. Gantt was best known for devising scientific wage-incentive plans and for introducing bar chart analysis as a means of controlling manufacturing and scheduling work. Frank B. and Lillian M. Gilbreth pioneered the field of time and motion study by breaking down activities into smaller, manageable tasks and by analyzing the components of fatigue. In one way or another, all of these people were engaged in the scientific study of work processes, and all made a contribution to what is now known as operations management. As a result, industrialists such as Henry Ford applied Taylor's principles to mass production and were able to produce quality goods that a much greater number of people could afford.

The development of modern computers in the 1940s further advanced operations management. Although the first machines were both expensive and difficult to use, they greatly magnified the scope of numerical computations and quantitative models. George Dantzig introduced linear programming in the 1940s, a mathematical technique that determines the most profitable allocation of resources in situations such as transportation, least cost feed rations, and crop acreage allocations. Production scheduling models and critical path methods were developed in the 1950s. During the 1960s, computer simulations were developed, and the 1970s saw the introduction of a powerful materials management system called **material requirements planning (MRP).** The 1980s and

1990s have seen the incorporation of robotics, flexible-manufacturing systems, and computer-integrated manufacturing systems into the operations management arena.

CURRENT ISSUES IN OPERATIONS MANAGEMENT

Whereas cost and efficiency were the main issues of operations managers in the past, five issues have dominated operations management most recently:

1. A growing service sector
2. Time-based competition
3. Productivity
4. Global competition
5. Quality

The Growing Service Sector

The United States, as well as the rest of the developed world, has seen the growth of the service sector. Between 1975 and 1992 in agriculture and agriculture-related industries, the percentage of jobs in wholesale and retail trade jumped from 46.3 percent to 64.4 percent. This dramatic increase mirrors that of the general economy. Food and agribusiness firms are responding to the increased service demand by adding service-related components to the products that they sell. Pioneer Hi-Bred International, Inc., a leading seed producer, offers annual operating lines of credit to farm producers and precision farming technical specialists to aid producers in utilizing this new technology—both in addition to the traditional seed products the firm produces.

Time-Based Competition

Time-based competition is competing for customers and market share using the dimension of time as a source of competitive advantage. Quicker response to changing customer needs is one example of time-based competition. Technological innovations, especially in telecommunications, have enabled firms to compete on the basis of time. Global competition and higher expectations by consumers are other reasons. Agribusinesses are reducing the time it takes to develop and introduce new products or services. As well, they are responding faster to customer orders, deliveries, service needs, and problems.

Productivity

Productivity can be defined as the dollars (or quantity) of output (goods and services) produced divided by the dollars (or quantity) of resources (materials, wages, equipment cost). Productivity is the prime determinant of a nation's standard of living because employee wages are generally determined by the productivity of human resources. Part of the concern over productivity has arisen because of the shift to a service-oriented economy. Many economists believe that it is more difficult to achieve the same improvements in productivity for the service sector as can be achieved in the manufacturing sector.

Agricultural productivity has historically surpassed the productivity of the nonfarm sector. Between 1948 and 1996, U.S. agricultural output grew at the average annual rate of 1.8 percent compared to 1.2 percent for the nonfarm sector. However, the recent growth rate of productivity in agriculture has varied tremendously (Ball, et al.).

Labor productivity in the retail food industry has increased moderately, but erratically, because of extensive mechanization and technological improvements. Technological improvements in supermarkets include computer-assisted checkouts and data processing. New store concepts (i.e., food warehouses) that provide fewer services (and less labor inputs) have also increased productivity. But, labor productivity has also declined with some food retailers because they have implemented expanded services, such as salad bars, delicatessens, and in-store bakeries that require more labor. Some eating and drinking establishments have also experienced declining labor productivity rates because of enhanced customer service and longer store hours (generating revenues that are marginally higher than input rates).

Global Competition

Globalization is affecting the operations management of agribusiness. As noted in Chapter 6, trade is critical to this nation's agricultural economy. About 20 percent of U.S. agricultural production is exported, and imports account for a varied, but growing share of U.S. consumption of agricultural products (World Agricultural Outlook Board). Exports of U.S. agricultural products (including foods, feeds, and beverages) were roughly $53 billion in fiscal 2001. Operations managers must factor the effects of globalization into their decision making. Decisions such as determining where to place a new production plant are more complex as countries with promising markets and/or low cost inputs must be considered. As well, decisions on the size or capacity of a plant must factor in the potential rapid growth of exports. Exported products may require modification to fit the needs of a different consumer; therefore, plant layouts and equipment must be flexible. Also, domestic markets no longer imply domestic competitors. For example, five of the top thirty food retailers in the United States are affiliates of foreign-owned retailers. Operations managers in retail food operations must determine how to positively respond to these new competitors from abroad.

Quality

High quality products and services at reasonable prices have been the source of success for many food and agribusiness firms, but the importance of quality has escalated with increased competition. The results of poor quality are often increased costs of production and lost market share. Because quality is of such critical importance today, it is the first operations management function discussed in this chapter.

QUALITY MANAGEMENT

For today's operations managers, a key challenge is to provide top-quality products or services efficiently. But, "quality" without reference is a nebulous term. Like "comfort" and "beauty," everyone has his or her own idea of what "quality" is. It becomes the manager's job to understand what type or level of quality their customers expect, want, and need. Quality is a factor that must be present to market any product, but the issues include what kind of quality and how much quality. From a planning perspective, managing that quality is a critical element in any successful food and agribusiness operation.

Today's consumers are unlike the customer of a generation ago or even a decade ago—case in point—the 2000 Summer Olympics. Polls showed a portion of the general public was unhappy with the TV media coverage of the Australian Olympiad due to the 15-hour time delay between the United States and Australia. Events were shown in the United States up to a full 24 hours post occurrence. While the previous Olympic coverage other years had been time delayed, the TV consumer of 2000 was a new customer. They now had access to the Internet to find out details and immediate results of their favorite teams, events, or athletes. Quality in this instance was measured in immediacy. Some viewers wanted immediate results. Those were available via another medium, so they got those results from another "supplier"—the Internet.

In addition to a changing buyer, the sheer number of choices available to buyers of almost every good or service has a major impact on the importance of quality. As mentioned earlier, the world is a global market. If the consumer can't find a product or service at one location, they shop at another—if not in Indiana, then in Kansas, if not in Kansas, then perhaps in Canada or Mexico, or Germany or India. Essentially, there are more products and services on the market than ever before. That means there are more options/choices for consumers and buyers.

These types of issues, including a new definition of convenience, the number of choices available, and an overriding demand for value in purchases has placed unprecedented demands on operations managers. The challenge for the operations manager is to ensure that quality is effectively delivered consistently, at a price buyers are willing to pay.

Quality Definitions

The definition of quality may differ depending upon to whom one is talking. Consumers typically define quality as fitness for use. Currently, many companies have come to interpret fitness for use as "meeting or exceeding customer expectations." Consumers also tend to talk about value, or, how well the product or service meets its intended purpose at the price the consumer is willing to pay. On the other hand, manufacturing managers offer a different quality definition: conformance to specifications.

T A B L E **18-3**	Consumer Requirements and Expectations of Quality Products and Services

Products	Services
Performance	Time and timeliness
Features	Completeness
Reliability	Courtesy
Durability	Consistency
Serviceability	Accessibility and convenience
Aesthetics	Accuracy
	Responsiveness

Source: Evan.

Fitness for use and conformance to specifications provide the fundamental basis for managing operations to produce quality products. Marketing and design's role is to understand customer needs and then to translate those needs into appropriate specifications for goods and services. With these specifications, it is up to the manufacturing or service operation to ensure that the specifications are met or exceeded. Poor communication between marketing, design, or operations can result in inferior quality. Table 18-3 summarizes features in products and services that customers perceive as being characteristic of high quality.

What are the implications of good quality products and services? From a marketing standpoint, higher quality can increase profits and/or market share. One important aspect of market share is attracting and retaining customers. A recent study revealed that of the customers who make complaints, more than half will give the firm a second chance if the complaint is resolved. The figure jumps to 95 percent if the complaint is resolved in a timely manner. Higher profits are realized because high quality products and services can typically be priced higher than comparable, but lower quality, ones.

From a cost standpoint, defective products or services can increase the costs of production. Production costs related to quality include the following:

1. Prevention costs
2. Appraisal costs
3. Internal failure costs
4. External failure costs

Prevention costs are the costs associated with stopping defects before they happen. **Appraisal costs** are incurred in monitoring the quality level of products and services during the course of production. **Internal failure costs** are the costs that are generated during the production and/or rework of defective parts and services. Finally, **external failure costs** are incurred when the product or service fails once it is in the consumer's hands.

Quality Management Systems

Food and agribusiness firms have made significant improvements in their product and service quality, which have enabled many to gain a significant advantage over competitors. In order to do this, these firms have instituted one or more quality management systems which have emerged. These quality management systems include the following:

1. Total quality management
2. Deming
3. Juran
4. Crosby
5. Taguchi
6. ISO 9000, ISO 14000
7. HACCP

A brief discussion of each of these quality management systems follows. It is significant to note how quality is defined by each system.

Total Quality Management

Total quality management (TQM) is an integrated management concept directed toward continuous quality improvement. TQM strives to make quality everyone's concern and responsibility. The benefits of this system include improved customer satisfaction, enhanced product and service quality, reduced waste and inventory, increased flexibility, and lower costs. Campbell Soup Company and H. J. Heinz are two food companies that have been successful in incorporating TQM into their organizations. Other food companies that have major quality efforts under way include Coca-Cola, International Multifoods, Inc., Universal Foods Corporation, and Sky Chefs.

The principles of TQM can be summarized as follows:

- Business success can only be achieved by understanding and satisfying the customer's needs.
- Leadership in quality is the responsibility of top management.
- Statistics and factual data are the basis for problem solving and continuous improvement.
- All functions and levels of an organization must focus on continuous improvement in order to achieve corporate goals.
- Multifunction work teams best perform problem solving and process improvements.
- Continuous learning, training, and education are the responsibility of everyone in the organization.

Excellence in applying TQM principles is the basis for the Malcolm Baldridge National Quality Award. Established in 1987, the award recognizes U.S. companies that have made great strides in improving quality and customer satisfaction with their products and services using TQM. Firms applying for the award are evaluated in seven areas: leadership; information and analysis; strategic quality planning;

Quality has many different meanings and interpretations in agribusiness management.

human resource development and management; management of process quality; quality and operational results; and, customer focus and satisfaction.

Deming

Deming defines quality in terms of how well the product or service meets the consumer's needs. W. Edwards Deming, a statistician and consultant, helped improve quality control for the Japanese and for global industry in general from 1950 on. He is considered to be the father of quality control in Japan and many consider Deming to be a pioneer in this field. His analysis and publications were the template from which many other approaches to quality control were built.

Deming's 14 points to quality summarizes his beliefs on quality improvement (Table 18-4). According to Deming, all of the 14 points must be implemented for the company to be successful; that is, none of the 14 points can be used in isolation. Like TQM, Deming promotes the continuous improvement in quality concept, and he especially emphasizes the reduction of variability and uncertainty in the design and manufacturing process.

Juran

Joseph M. Juran became head of industrial engineering at the Western Electric Company in the 1930s. During World War II, he pioneered inventive ways to manage shipment of goods to friendly countries while working for the Roosevelt administration. Focusing on quality, he began a consulting career, writing books and lecturing.

Juran, like Deming, was not discovered by U.S. businesses until the 1980s and was instrumental in Japan's quality improvements prior to his acceptance in the United States. Juran has worked extensively to determine a cost of quality for all aspects of the production and service cycle. In addition, he believes that over 80 percent of quality defects are caused by factors controllable by management. Juran's writings essentially define quality in terms of how well the product or service is fit for use. Juran proposes that quality management consist of a quality trilogy: quality planning, quality control, and quality improvement.

TABLE **18-4** | **Deming's 14 Points**

1. Create and publish to all employees a statement of the aims and purposes of the company or other organization. The management must demonstrate constantly their commitment to this statement.
2. Learn the new philosophy: everyone in the firm, including top management.
3. Understand the purpose of inspection for improvement of processes and reduction of cost.
4. End the practice of awarding business on the basis of price tag alone.
5. Improve constantly and forever the system of production and service.
6. Institute training.
7. Teach and institute leadership.
8. Drive out fear. Create trust. Create a climate for innovation.
9. Optimize toward the aims and purposes of the company, the efforts of teams, groups, and staff areas.
10. Eliminate exhortations for the work force.
11. Eliminate numerical quotas for production. Instead, learn and institute methods for improvement. Eliminate management by objectives (MBO). Instead, learn the capabilities of processes, and how to improve them.
12. Remove barriers that rob people of pride of workmanship.
13. Encourage education and self-improvement for everyone.
14. Take action to accomplish the transformation.

Source: Deming.

Crosby

Phillip B. **Crosby** advocates a goal of zero defects. Crosby's definition of quality is based on how well the product or service conforms to requirements. Crosby's teachings emphasize quality in terms of manufacturing. Unlike Deming and Juran, Crosby believes that the key to quality improvements is a change in top management's thinking. Crosby supports and suggests a plan for implementing cultural change in the workplace. Crosby's book, *Quality Is Free—The Art of Making Quality Certain,* captured much attention with its publication because it was the first to suggest that there were hidden costs of low quality. Hidden costs associated with poor quality included increased labor and machine hours, customer delays, and machine downtime—the costs of which would more than offset the costs of instituting quality improvements.

Taguchi

Building on the quality control techniques of Deming, Professor Genichi **Taguchi** defines quality in terms of a philosophy. Taguchi defines quality as the absence of product variability and social cost. Ideally, Taguchi's quality is closeness to an ideal state—bringing maximum well-being to the product or service consumer. Taguchi's approach to quality uses engineering and statistics to optimize product designs and manufacturing processes. The result: high quality products at lower costs. Taguchi believed that costs increase exponentially as quality characteristics increasingly deviate from their target values.

ISO 9000, ISO 14000

ISO, the International Organization for Standardization, is a worldwide federation of national standards bodies from more than 130 countries. Its mission is to promote the development of standardization and related activates in the world with a view to facilitating the international exchange of goods and services. Without such harmonized standards for similar technologies in different countries or regions, technical barriers to trade mount significantly.

ISO 9000 is a series of three international standards and guidelines on quality management and quality assurance. The standardized definition of quality in ISO 9000 refers to all those features of a product or service, which are required by the customer. Quality management refers to what the organization does to ensure that its products conform to the customer's requirements. First published in 1987, the guidelines may be used by both service and manufacturing firms.

Implementing ISO 9000 does not guarantee a quality product. Instead, ISO 9000 specifies that a company define appropriate quality standards, document its processes, and then show that it consistently adheres to both. More and more companies are finding that having ISO 9000 certification is a key to doing business globally. In fact, registration has become market driven—customers are requesting that their suppliers be ISO 9000 certified.

With **ISO 14000** the same concepts of worldwide standards for quality are also being applied to the environment. ISO 14000 standards provide the framework and systems for managing legislative and regulatory orders for environmental compliance. The components of ISO 14000 are environmental management systems, environmental auditing, environmental performance evaluation, environmental labeling, and life cycle assessments. It is expected that the new standards will be widely accepted—especially in Europe. Benefits may include a worldwide focus on environmental management; more consistency of rules, procedures, and labels across countries to facilitate trade; and a movement by companies toward more environmental sensitivity beyond just complying with current regulatory standards.

Hazard Analysis and Critical Control Point (HACCP)

In the food industry, **Hazard Analysis Critical Control Point (HACCP)** has become an important, recognized food safety program. It has gained rapid acceptance as the result of the public and legislative demands for increased food safety protection. Also, the increased global sourcing of raw materials and distribution of food products has driven the need for a standard food safety program that is acceptable by the international community.

HACCP's basic premise is focused on the prevention of food safety problems rather than controlling them. Although HACCP is primarily helpful in the processing of safe food, it is also applicable to raw materials suppliers. The benefits of HACCP include reducing the risk of manufacturing and selling unsafe products, an increased awareness of hazards in the workplace, and better product quality. The HACCP system consists of seven principles which describe how to

TABLE **18-5** | **HACCP Seven Principles**

1. Conduct a hazard analysis. Prepare a list of steps in the process where significant hazards occur and describe the preventative measures.
2. Identify the critical control points in the process.
3. Establish critical limits for preventative measures associated with each identified critical control point.
4. Establish critical control point monitoring requirements. Establish procedures from the results of monitoring to adjust the process and maintain control.
5. Establish corrective actions to be taken when monitoring indicates a deviation from an established critical limit.
6. Establish effective recordkeeping procedures that document the HACCP system.
7. Establish procedures for verification that the HACCP system is working correctly.

Source: Mortimore and Wallace.

establish, implement, and maintain the system in an operation. The principles are outlined in Table 18-5.

HACCP principles are expected to replace existing FDA and USDA food safety regulations in areas where it is not already being applied. Therefore, an understanding of these principles is critical to those who wish to manage in the meat and poultry industry and other food processing establishments.

LOCATION

Manufacturing firms in the United States typically build more than 3,000 new plants each year and expand 7,500 others. The decision to locate a plant is a strategic decision that will have a lasting impact on other issues such as operating costs, the price at which goods and services can be offered, and a company's ability to compete in the marketplace. A classic example is the meatpacking industry. Meatpackers of the 1960s were located in major railway cities where live cattle were shipped and high wages were paid. The packing plants were old, multistory buildings that did not operate at full capacity because of inconsistent supply. Meatpackers new to the industry at the time saw the opportunity to reduce costs and increase productivity by locating processing plants near cattle production feedlots and employing low-skilled operators in the highly mechanized plants.

Another example—have you ever wondered why White Castle locates its eating establishments near manufacturing plants? White Castle's strategy is to segment the market and cater specifically to a target population of blue-collar workers and away from other competitors such as McDonald's. These two examples illustrate how important inputs (cattle and labor) and customers are to the success of a firm. Thus, location of the plant is a strategic choice, and many factors that influence choice of location should be considered when making these decisions. The following sections outline current location trends in the United States followed by factors that affect location decisions.

Significant U.S. Trends

Geographic diversity, the move to the Sun Belt, and decline of urban areas are three of the more important trends currently affecting plant location. Industries once concentrated in certain geographic regions. For example, most automobile manufacturing was once done in Michigan. Today, geography and distance are becoming irrelevant for many location decisions. Why? Both transportation and communication technologies have improved dramatically. Telecommunications (voice and data) are enabling business firms to conduct business in larger market regions. Also, wage differentials between regions of the United States are gradually disappearing. Although the southeastern part of the United States still has the lowest per capita income, it has moved up from 78 percent to 90 percent of the national wage average in the past three decades.

A second trend is that more service and manufacturing firms are locating to the Sun Belt. Of course, a milder climate is one factor, but other factors include a population shift toward the South (as baby boomers are retiring), lower relative labor costs, less unionism, and a perceived stronger work ethic. Dairy farming has shifted to the Sun Belt states of Florida, Texas, New Mexico, Arizona, and California. Today, more milk is produced in California than in Wisconsin. Factors contributing to dairy farming's shift to the South and Southwest are population movements, the Sun Belt's lower land and facilities costs, favorable climate, ample supplies of high quality hay and forage, and the availability of labor (*Agricultural Outlook*).

The third U.S. trend is the decline of urban areas and the relocation of firms to suburban or rural areas. Reasons for relocating include the high cost of living, a higher crime rate, and an overall decline in the quality of life in larger cities. Obviously, this trend is not lost on urban areas and many have aggressive urban renewal programs in place.

Factors Affecting Location Decisions

When food and agribusiness managers choose a site for their facilities, they consider many factors. Of course, the relative importance of any factor is dependent on the type of product or service produced, plant size, plant type, environmental restrictions, and geographic location. Six of the more important factors in firm location include:

1. Proximity to raw materials and suppliers
2. Location of markets
3. Labor climate
4. Agglomeration
5. Taxes and incentives
6. Proximity to other company facilities

Proximity to Raw Materials and Suppliers

Agribusinesses may wish to have proximity to raw materials and suppliers— especially if one major raw material input is needed and that input is costly to ship in its raw state. Examples of food and agribusinesses that are very dependent on sources of supplies include canning industries, corn processors, and meatpackers. In the beef meatpacking industry for example, cattle would be bulky and costly to ship from Nebraska to New York, so they are slaughtered near finishing feedlots and shipped long distance in the more manageable form of boxed beef. By locating near cattle feedlots and lower wage markets, meatpackers have seen industry cattle kill costs cut by 50 percent and plant utilization rates increased to about 90 percent compared to the 60 percent industry average of early 1960s.

Agricultural industries that transport large volumes of material, often perishable, must also consider locating where bulk transportation resources are available. Most of the large grain elevators are located on either large water or rail systems. Thus, firm location attempts to minimize both the costs of production and transportation of raw materials and supplies.

Location of Markets

There is an old real estate adage that states the three most important factors in real estate are location, location, and location. Location of markets is especially important to retailers because customers will often not travel long distances to buy from a retailer. Agricultural goods that are bulky, perishable, or heavy can receive higher prices if they are located near markets because of higher quality, better service, and lower transportation costs. Agribusinesses that produce inputs for other firms often locate facilities near these buyers to minimize distribution costs and to provide better service.

Locating near markets is very important for service firms. Equipment manufacturers have found that aftermarket (replacement) parts sales and service requires relatively close locations to markets because producers will only drive limited distances for parts and services. Rather than maintaining full dealerships, manufacturers are locating satellite service facilities near these markets to provide for aftermarket parts sales and service.

As mentioned earlier, the White Castle hamburger chain offers an interesting philosophy on location. White Castle always chooses to locate near factories. Even if the location is in an area with few other fast food retailers, the company opts to locate their restaurants near factories with the notion that they will be the first available eating establishment for workers leaving the factory.

Supermarkets located near warehouses and retail fertilizer suppliers located near growers also illustrate the importance of the market location factor in production planning operations. Each market must be evaluated carefully for the operations manager to make the best decisions regarding location.

Labor Climate

In a survey and interview of firms on factors considered in locating a new manufacturing plant in the United States, 76 percent of the respondents indicated that a favorable labor climate was a dominant factor (Schmenner). Labor factors such as wage rates, worker productivity, training requirements, attitudes toward work, and union strength are also important considerations. One other such factor is the availability of qualified labor. Food and agricultural companies that require a great deal of research may find it advantageous to locate near a large land grant university where highly skilled, technical resources are concentrated. In the meatpacking industry example, meatpackers located in unfavorable labor climates and agreements were at a competitive disadvantage because wages and benefits could be as much as 40 percent higher than the competitor's.

Agglomeration

Agglomeration, another factor in facility location decisions, is the accumulation of business activity around a specific location. Firms often locate in close proximity to one another because they can increase the efficiency of services at the business, household, and social level. For example, a firm may locate near a group of existing firms because they can share in the existing infrastructure such as transportation, water, or sewage systems. The amenities of the agglomeration of restaurants, theaters, and/or professional sports teams can also produce social benefits for the employees and management of agribusinesses.

Taxes and Incentives

Increasingly, state and city governments are enticing firms to locate in their areas by offering generous tax and incentive packages. These packages may include items such as special loans for machinery and equipment, state matching funds, state-sponsored employee hiring and training programs, and tax exemptions on corporate or excise taxes. Governments offer these incentives since bringing business to the town, county, or state may help boost the economy in the long run.

The fast and vast expansion in the North Carolina hog industry provides an example of just how a thriving business sector can help give back to the economy. In 1986, North Carolina hog farmers raised 2.4 million hogs and pigs on about 15,000 hog operations. Of those, only about 800 farms had more than 500 head of hogs. But, by 1997, the number of hog farmers in North Carolina shrunk to about 5,800 while the state's total hog production climbed to 10.2 million head. Ninety-eight percent of those hogs were raised on about 1,600 farms, each handling 500 head or more. Overall, North Carolina pork production, including growing, packing, processing, and their associated industries, was responsible for an estimated 43,250 full-time jobs in 1998—representing a growth of 73 percent between 1993 and 1998. In addition, during that same period, North Carolina pork production generated more than $6 billion in sales and almost $1.92 billion in value-added income.

What did this mean to the state's tax picture? In 1997, hog production generated more than $62 million in income tax revenues. Property tax revenues payable to North Carolina counties from swine production in 1998 came in at more than $6.4 million. However, such growth typically has costs, and this expansion in the North Carolina pork industry has fueled protests over the social and environmental costs of the expansion.

Proximity to Other Company Facilities

Food and agribusinesses may also locate new facilities considering their proximity to other company facilities. For example, agribusinesses may opt to open a new service facility in a region that is not adequately serving its customers currently with existing company facilities. Agribusinesses affected by climatic conditions may pursue a risk management strategy of having facilities spread out across a variety of geographic, temperature, and soil conditions. Seed firms locate production facilities in different areas (even countries) to assure themselves of seed production and to minimize the uncertainty of poor weather. Some agribusinesses may locate new facilities in relative proximity to other company facilities because they supply parts to one another or share management and staffs.

Location Decisions for Service Businesses

Location decisions in service businesses can even be more important than in manufacturing businesses. Factors such as convenience to customers, traffic volumes, income levels, and residential density are important sales indicators and thus influence location decisions. For example, most people shop at the supermarkets located on the way home from work or at markets near their homes.

Location of competitor stores can also be an important consideration for service firms—a new supermarket located near a new housing development and away from competitors may provide a distinct competitive advantage for that firm. In contrast, it is interesting to note that locating near competitors can sometimes be advantageous as well. A greater number of total customers (or a critical mass) may be attracted to a group of stores clustered in one location versus the total number of customers who would stop at the same stores at scattered locations. A cluster of fast food restaurants off an interstate exit typically produces a critical mass.

A restaurant chain with operations in North America and Japan considers six primary factors when determining the location of a quick service (fast food) restaurant.

1. Area employment
2. Retail activity
3. Proximity to successful competitors
4. Traffic flow
5. Residential density
6. Accessibility and visibility

Area employment is important because the chain is targeting 20- to 45-year-old workers on lunch breaks. The chain also confirms that area firms allow workers to take lunch breaks off premises. Retail activity, the second factor, is important because a lot of quick service eating is done on impulse by shoppers. Being near successful competitors is seen as an advantage because they signal that the area is a good market.

Traffic flows are also important because most business is from people in cars. In fact, traffic flow of 16,000 cars every 24 hours is considered good. A residential population of about 20,000 upper-middle-class residents within a 2-mile radius is considered to be adequate to ensure weekend and evening business. Finally, accessibility and visibility are also important factors in considering location. Specifically, traffic direction, intersections, traffic backups during rush hour, and surrounding buildings and signs are important accessibility and visibility factors that are considered.

CAPACITY PLANNING

Capacity planning includes the activities that are undertaken to determine what the appropriate size of a manufacturing plant or service location should be, so that a certain quantity can be produced over a specific time period. A food or agribusiness firm must balance the cost of having excess capacity against the potential of lost sales due to too little capacity. If not anticipated correctly, lost sales and disappointed customers may have serious effects on a firm's short-term and long-term profitability. Accurate forecasts of present and future demand are critical in capacity planning. Thus, an understanding of the product life cycle can assist in forecasting overall demand for a product (Chapter 8). Many factors should be considered in determining the appropriate capacity, five of which are detailed here:

1. Economies of scale
2. Flexibility
3. Seasonality and other patterns of production
4. Fluctuating demand
5. Multiple versus single shifts

Economies of Scale

According to the **economies of scale** principle, larger plants usually result in lower per unit costs because fixed costs may be spread out over a larger quantity of output. Volume-cost analysis can be used to determine how much volume is needed to cover fixed costs (Chapter 16). Economies of scale are often associated with firms utilizing large quantities of expensive processing equipment (such as large breweries) where large, bulk quantities are produced. Construction costs are also reduced when a facility producing 10,000 units is built versus two separate facilities producing 5,000 units each.

Flexibility

An alternative view to economies of scale is flexibility. One concept of bringing flexibility to production processes is the **focused factory,** first proposed by Wickham Skinner in the early 1970s. Shorter product life cycles, the importance of quality, and needed flexibility brought about this alternative view of plant size and capacity. The focused factory concept holds that several smaller factories will improve individual plant performance because the operations manager can concentrate on fewer tasks and can more easily motivate a smaller work-force toward one goal. There are also fewer layers of management, the plants are more flexible in introducing new products, and a team approach can be taken to problem solving. Further, smaller firms may offer flexibility in terms of proximity to sources of raw materials and proximity to market destinations, which in turn could result in lower transportation costs.

Seasonality

Highly seasonal agricultural products can produce special headaches for an operations manager. Manufacturers of farm inputs such as seed, fertilizers, and chemicals see heavy demand for their products in the spring. Grain elevators experience peak demand in the fall. On the food side, holiday items such as whole turkeys and cranberries see their demand peak in the weeks preceding Thanksgiving.

In some cases, peripheral operations can be done or other products can be produced in the off-season to balance the resource needs of the food or agribusiness. Where this is not possible, a plant large enough to handle peak productivity levels becomes a costly operation when output is significantly reduced in the off-season. In such cases, it may actually be more economical to run several smaller plants and to close down plants that are not needed during the off-season. This does not reduce the overhead costs associated with an unused facility, but it does limit the day-to-day expenditures of running an unnecessary part of a larger operation.

Fluctuating Demand

Fluctuating demand in manufacturing can be solved by building up inventory during slow demand periods. However, the perishability and bulkiness associated with agricultural products may preclude this strategy. A distinct difference between manufacturing and service is that services are typically produced and consumed simultaneously. Thus, service firms must build enough capacity into their facilities to meet peak periods of demand. For example, a supermarket must have enough checkout counters to meet not only average demand but also for peak demand periods (such as the week before Thanksgiving). Of course, service firms can shift or smooth out demand by altering consumer behavior. Offering discounts or other incentives during off-peak hours are some ways that are used to shift consumer demand.

Multiple Shifts Versus Single Shifts

Multiple shifts may be an alternative to operating a facility at maximum capacity—if the labor is available. Theoretically, it is possible to produce twice as much in a plant with double shifts, while limiting the need for space by spreading out the working hours. Agricultural-chemical companies require multiple shifts because of the nature of their products and large investments in equipment. Likewise, corn and soybean processing facilities typically run 24 hours a day, 7 days a week—stopping only for emergencies or routine maintenance. In plants where this is not the case, managers must carefully evaluate the costs of a larger facility versus the increased costs of sometimes less productive and possibly more accident-prone labor.

LAYOUT PLANNING

Layout planning refers to the specific design of the physical arrangement within a facility. When planning the physical layout of a plant, consideration must be given to utilize space and labor effectively, minimize material handling, and maintain flexibility for future changes in product or demand. The goal of layout planning is to allow both the workers and equipment to run the operation in the most efficient and effective manner. There are four basic categories of facility layout:

1. Process
2. Product
3. Hybrid
4. Fixed position

Process Layout

A **process layout** arranges activities by function. Thus in a process layout, regardless of the product being created or assembled, all like functions are grouped in the same place—that is, canning equipment with canning equipment, inspectors with inspectors, etc. The process layout is common where different goods or services are produced intermittently to serve different customers (Figure 18-2). As a result, general resources and equipment are utilized to make the process layout more flexible to changes in product mix or marketing strategies. Disadvantages of the process layout include slower overall processing rates (because of the switching between products), higher levels of inventory, increased time lags between specific operations, and higher costs for material handling.

Product Layout

A **product layout** is geared specifically to the continuous production process because it produces one product at a time, step by step, with one function following another in sequence as the product is assembled, and with much fewer variations in product (Figure 18-3). Workers grading and packing peaches on a

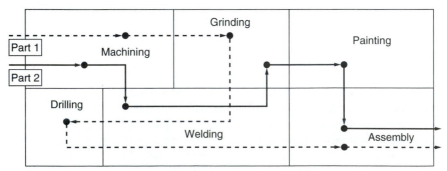

Figure 18-2
Process Layout—Farm Implement Manufacturing Facility.

conveyor belt, for example, are operating in a product layout framework. One person may remove debris, the next may be responsible for sorting, while another oversees packing into crates, and so forth.

Advantages of product layouts include faster processing rates, lower inventories, less skilled labor required, less time lost to product changeovers, and less materials handling. One significant disadvantage is that less flexible and more capital-intensive resources are dedicated to specific products that may have uncertain life cycles. In addition, a line can work only as fast as its slowest operation and downtime can be expensive when one machine on the production line breaks down.

Hybrid Layout

A **hybrid layout** combines the process and product layouts to balance the advantages of each (Figure 18-4). Managers may choose this form of layout when introducing flexible manufacturing systems. In group technology applications, different machines are brought together to produce a group (or family) of similar parts in an assemblylike fashion. Some manufacturing firms have reduced work-in-process inventories by up to 90 percent using such systems. The hybrid layout form is very popular in operations such as plastics molding and even supermarkets. In a supermarket for example, the manager may place similar merchandise in the same location to aid customers in finding the desired items (a process layout). However, the supermarket layout is designed to *lead* customers through the entire store (up and down aisles) maximizing their exposure to the full line of goods (a product layout).

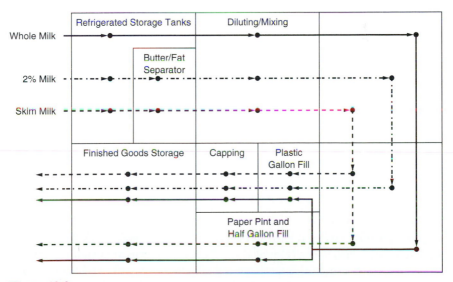

Figure 18-3
Product Layout—Fluid Milk Facility.

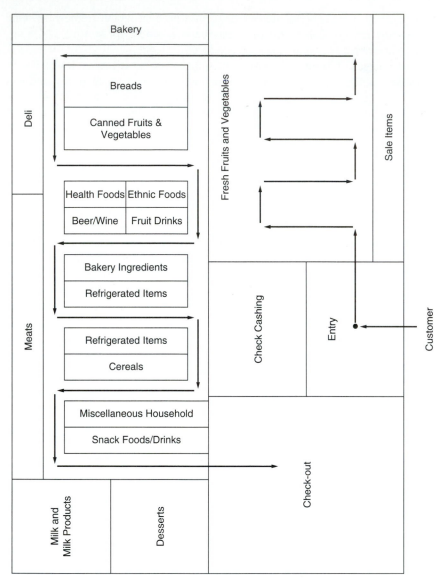

Figure 18-4
Hybrid Layout—Supermarket.

Fixed Position Layout

A **fixed position layout** is used in the construction of large items such as farm buildings, silos, and stores (Figure 18-5). This layout is used in situations where the item being constructed is too large or bulky to move easily or where the item is to be built fixed in place. Construction is accomplished in one place, and tools and materials are brought to the assembly location. A fixed position layout elim-

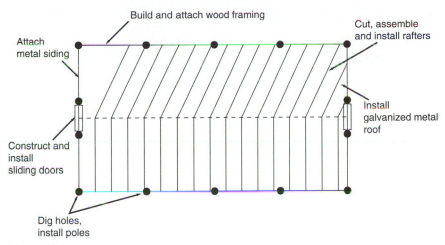

Figure 18-5
Fixed Position Layout—Farm Building.

inates or minimizes the number of times that a product must be moved, and it is often the only feasible layout solution.

PROCESS DESIGN

Process design is selecting the specific inputs, operations, and methods that are to be used to produce the good or service. Decisions concerning process design are very important because of their implications on cost of production, level of personalized customer service, convenience, and flexibility. The first step in process design is outlining a step-by-step description of each operation that is to be completed. Four general aspects are considered in process design:

1. Capital intensity
2. Resource flexibility
3. Vertical integration
4. Customer involvement

Capital Intensity

The determination of **capital intensity** is simply choosing which tasks humans will perform and which tasks machines will perform. Many levels of capital intensity are available today with the advent of new technology and computer-assisted equipment. Hazardous or difficult operations in meatpacking companies have been automated reducing the risk of injury on employees. The level of capital intensity significantly impacts production costs. Customer interaction is also impacted. For example, some Arby's franchises installed computerized ordering devices in their quick service restaurants where customers simply punch in their orders. Although increased computer sophistication and labor shortages and costs drove the change, customer acceptance and ease of use were critical success factors.

Resource Flexibility

Resource flexibility can be defined as the ease with which employees and equipment can accomodate a wide variety of products, functions, and levels of output. Shorter product life cycles require more general equipment and a broader range of employee skills. Thus, more job training may be required for employees. However, the benefits of a flexible workforce are real: improved productivity, higher morale, and a wider product offering to customers.

Vertical Integration

Vertical integration is associated with owning some or all of the various steps that are taken to produce a product from start to finish (i.e., from raw materials to the delivery of the finished product). The more processes a firm in the chain owns, the more vertically integrated the firm is said to be. Vertical integration can be advantageous to a firm if the firm has the relevant skills for each step it controls and if the company views the industry into which it is integrating as being important to its future success. For example, Kroger Company, the largest food retail chain in the United States, has pursued vertical integration aggressively. Kroger's operations include 2,300 supermarkets in 30 states, 800 convenience stores in 15 states, 390 jewelry stores, and 38 manufacturing plants producing branded products, from water to spaghetti sauce, dairy products to peanut butter.

Customer Involvement

Customer involvement is the degree with which the customer is allowed to interact with the process. Particularly in service industries, customer contact is critical to customer satisfaction. Customer involvement can take the form of interaction with self-service options, product selection, product design, time, and location of service. High levels of customer involvement tend to be found in processes that are less capital intensive and more flexible.

Customer involvement in business management decisions greatly increased during the last 5 years. A good example of customer involvement is a farm equipment manufacturer that has designed and implemented a system that allows a farm producer to use their personal computer to dial a local dealership and access parts inventory. When used in the evening, the producer can check stock, order the part, and pick it up the next morning. The system also allows the producer to cross-reference over 400,000 competitive equipment parts that correspond to the manufacturer's parts. This allows the producer to consolidate parts runs into one stop.

The self-service level of customer involvement extends from the fuel pump (pay at the pump), to grocery or discount store self-checkout, which enable the customer to scan items themselves and to pay via credit card, debit card, or cash. Automated teller machines (ATMs) involve the customer interaction through the selection of menu items. Each of these examples is relatively recent. Through technology advancements, the customer has changed—and what they expect and what they are willing to do for the sake of convenience has continued to drive changes in this area.

JOB DESIGN

Detailed job design follows process design and is distinguished by the focus on the individual operator. Job design can be defined as the broad set of activities that determine the tasks and responsibilities of each employee's job, the employee's surrounding work environment, and the detailed order of operations that will be used to complete the tasks required to meet production requirements. The objectives of job design are twofold: (1) meet the production and quality goals of the organization; and (2) make jobs safe, satisfying, and motivating for the employees.

Traditional job design began with Frederick Taylor and his scientific management approach. Operations were divided into their individual components and studied to determine the most efficient manner to perform an operation. Today, job design can be thought of as comprising two major categories: (1) the social or psychological environment, and (2) the physical environment.

The Social Environment

The nonphysical or behavioral issues in the workplace are included in the social or psychological environment. Items such as training, the proper level of supervision, job expectations and responsibilities, and performance feedback are all subsets of an employee's social environment. In particular, motivation of the employee to consistently perform at high levels is the result of a positive social environment. Many theories of work motivation have been developed over the past 75 years.

Psychologists contend that highly repetitive jobs lead to boredom, injuries, and poor job performance. Declining job performance is often associated with absenteeism, high turnover, grievances, and poor quality. New approaches to job design attempt to counter poor job performance. These approaches include job rotation, job enlargement, job enrichment, and work teams.

Job rotation allows operators to perform different jobs during their shifts. Rotation increases the overall skills of the employees and minimizes work-related injuries (as long as the jobs differ in the motions used). Job enlargement is simply increasing the number of tasks that each operator performs. Here, job repetition is reduced because the cycle time of each job has increased. Job enrichment expands the duties and responsibilities of the operator. For example, a technician at a brewery might be responsible for monitoring the filtering process. Job enrichment increases those responsibilities to include taking and analyzing quality measurements, inspecting the filters, and deciding when the filtering equipment needs maintenance.

Work teams are another method of giving employees more responsibility and decision-making authority. Quality circles, used extensively in Japan, are teams that meet to work out solutions to quality issues. Other teams are often formed for special purposes. Issues such as new work policies, the details of an employee bonus plan, or a labor-management problem are examples of issues that these teams have attacked. Self-managing work teams are also common. Teams of 3 to 15 employees meet to organize and run a definable work area. Team members learn all job tasks, rotate between jobs, schedule vacations and overtime, order supplies, and make hiring decisions.

The Physical Environment

The physical environment for employees can have substantial effects on employee morale, productivity, health, and safety. The physical environment in job design consists of such items as safe working conditions, lighting, noise, temperature, and ergonomics. Ergonomics, or human factors, is the study of the interaction between the human and the workstation, tools, and equipment. The objectives of ergonomics are to improve human performance by increasing speed and accuracy; to reduce fatigue; to reduce accidents due to human error; and to improve user comfort and acceptance.

Examples of ergonomics in the food and agribusiness markets abound. Farm producers using tractors see the effects of ergonomics in their tractor cabs: cabs placed in a location where they can best view the operations of the tractor; comfortable but supportive seats; grips on the steering wheel; levers positioned for easy access; and dials that are easy to read.

SUMMARY

Operations management involves a complex network of decisions that affects agribusinesses in the production of goods and services. In considering the production planning aspects of operations management, agribusiness managers must consider the uniqueness of agricultural production: seasonality, perishability, bulkiness, and variations in quantity, quality, or value. Production involves planning and logistics, and must be viewed as a total system.

Production planning decisions entail quality management; location, capacity, layout, and process design of the plant or facility; and job design. Logistics decisions include purchasing, production scheduling, inventory control, and distribution. There are many factors that affect each of these sets of decisions, and food and agribusiness managers must consider all factors and their effect on the total system before making a decision.

DISCUSSION QUESTIONS

1. List and describe some of the activities that an operations manager may be engaged in. How would you describe the agribusiness operations manager's role? How does this role relate to the systems approach outlined in Figure 18-1?

2. In your view, what are the most difficult operations management problems for the following types of firms: (a) a veterinary practice that serves both the farm and companion animal markets; (b) a large California dairy operation; and (c) a retail food store located in Minnesota.
3. Why is Frederick Taylor called "the father of scientific management"?
4. What is your definition of quality? How are the different approaches to quality management presented in this chapter similar? How are they different?
5. Differentiate between the four types of plant or facility layouts and give an example of a specific agribusiness that might employ each layout. Visit a food or agribusiness firm of your choice. Map the layout of the facility and identify the type of facility layout used by the food or agribusiness firm.
6. What factors are involved in the choice of plant or facility location? Name some food and agribusiness companies whose locations reflect consideration of these factors.
7. What are some of the factors that may be considered in determining the optimal capacity for a food or agribusiness firm?
8. For either a job you have held, or for a job held by someone you know, carefully explore the design of the job. Evaluate both the social environment and the physical environment of the job you have selected.

CASE STUDY — **BioAg—Part I**[1]

BioAg's Company History   BioAg is a small midwestern firm manufacturing agricultural chemicals that are environmentally sensitive. Important patents help protect BioAg's 60 percent market share in its market segment. Although still small, the niche market for environmentally sensitive agricultural chemicals has been growing about 50 percent per year since 1991.

BioAg has experienced rapid growth in sales because of an increasing societal concern about environmental issues (see Table 18-6 for a summary of BioAg's sales and marketing activities). In 1991, the company started with two products. By 1998, the BioAg product line included 36 different chemicals for use on wheat, barley, corn, soybeans, and specialty crops. BioAg has no retail outlets, but rather sells their products through farm input supply firms. As the number of products has grown, BioAg has sold its products through more and more supply firms. The company now sells its products through 420 supply firms in 24 states, up from only eight supply firms in one state in 1991. They anticipate selling in international markets within 3 years.

Questions
1. What are some of the production planning and management issues that BioAg has had to face as its product line has expanded and the firm has grown?
2. Why would quality management be an important area for BioAg given the type of product it produces? What are some of the characteristics of the firm's products that would make quality management important?

[1]This case originally published as "Supply Chain Issues: A Case Study of BioAg" in the *International Food and Agribusiness Management Review*, Frank Dooley and Jay Akridge.

T A B L E **18-6** | **Sales and Marketing Statistics for BioAg**

Item	Year							
	1991	1992	1993	1994	1995	1996	1997	1998
Sales (million $)	.50	.75	1.25	2.50	3.75	7.50	12.50	18.00
Number of products	2	4	5	7	8	12	20	36
Number of farm supply dealers	8	12	22	40	70	140	280	420
Number of states where product sold	1	2	3	6	8	12	16	24

Reprinted from *International Food and Agribusiness Management Review,* vol. 1, Issue 3, "Supply Chain Issues: A Case Study of BioAg" by Frank J. Dooley and Jay T. Akridge, pp. 435–441. Copyright 1998 with permission from Elsevier Science.

3. BioAg is considering an addition to its production capacity to support additional growth. The firm is considering adding to the current site, and building a new plant in west Texas. What are some of the issues the firm will need to consider as it explores these two options?

4. BioAg management is also working to decide just how large the addition to capacity should be. What suggestions do you have for the firm's management? What issues will they need to consider as they determine how much to expand?

REFERENCES AND ADDITIONAL READINGS

Ball, Eldon V., Jean-Pierre Butault, and Richard Nehring. U.S. Agriculture, 1960–1996: A Multilateral Comparison of Total Factor Productivity. USDA-ERS Technical Bulletin 1895, May 2001.

Crosby, Philip B. August 1992. *Quality Is Free: The Art of Making Quality Certain.* New York: New American Library, 1979 Mass Market Paperback.

Deming, W. Edwards. 1986. *Out of the Crisis.* Cambridge, MA: MIT Press.

Dooley, Frank, and Jay T. Akridge. 1998. "Supply Chain Management: A Case Study of Issues for BioAg." *International Food and Agribusiness Management Review* 1(3):435–441.

Evan, William M. *Organization Theory: Research and Design.* New York, NY: Macmillan Publishing Company, 1993.

Juran Institute. http://www.juran.com

Krajewski, Lee J., and Larry P. Ritzman. 2000. *Operations Management: Strategy and Analysis.* 5th edition. Reading, MA: Addison-Wesley Publishing Co. p. 435.

Mortimore, Sara, and Carol Wallace. 1998. *HACCP—A Practical Approach.* 2nd edition. London: Chapman and Hall 1994.

Schmenner, Roger W. 1993. *Operations Management from the Inside Out.* 5th edition. New York, NY: Macmillan College Division, USA.

Skinner, Wickham, and David C.D. Rogers. *Manufacturing Policy in the Oil Industry: A Casebook of Major Production Problems.* Homewood, IL: R.D. Irwin, 1970.

USDA Economic Research Service. July 1995. "Agricultural Outlook." ERS-AO-221. pp. 13–14.

World Agricultural Outlook Board. February 2001. "Outlook for U.S. Agricultural Trade— Summary." ERS-AES-29

MANAGING LOGISTICS IN AGRIBUSINESS

OBJECTIVES

- Outline the various objectives and functions of logistics management
- Understand the steps involved in the logistics management process
- Describe the role of, and activities involved in, purchasing/procurement in food and agribusiness firms
- Distinguish between MRP and JIT production operations systems
- Understand the basic ideas behind PERT and CPM scheduling tools
- Identify the role of inventory in the agribusiness and the different types of inventory firms hold
- Describe the different types of inventory tracking systems used in agribusiness
- Explore the role of physical distribution in the agribusiness
- Consider how information technology, especially the Internet, impacts the logistics management process

INTRODUCTION

The previous chapter provided an overview of the production planning aspects of operations management in the agribusiness. Production planning functions include determining the location, layout, capacity, and specific processes to be used by the agribusiness in producing goods and services. With this understanding of planning the physical production process, we now turn to operating the agribusiness. A critical aspect in running the agribusiness is managing the flow of materials into and out of the firm—or **logistics management.**

Logistics management has been described as the combination of materials management and physical distribution management—how inputs to the agribusiness are scheduled and procured, and how the finished products are stored and distributed to customers. More specifically, logistics management can be defined as the process of planning, implementing, and controlling the efficient, cost-effective flow and storage of raw materials, in-process inventory, finished goods, and related information from point of origin to point of consumption for the purpose of conforming to customer requirements (Council of Logistics Management). While this definition is long and complex, you see that the essence of logistics management is focused around moving and storing inputs, finished products, and information.

IMPORTANCE OF LOGISTICS MANAGEMENT

Some companies such as Wal-Mart have made logistics management the cornerstone of their total business strategy. Its increasing importance has also motivated agribusiness firms to restructure their organizations creating logistics departments whose manager reports directly to the company president. But why is logistics management so important?

Three examples show its importance to a company's total profitability and success. Take the critical function of inventory control, for example—just one part of the total logistics management area. Inventory typically represents 35 to 50 percent of the current assets in the average company. With such a significant share of assets, decisions about how to best manage inventory play a critical role in the total profitability of the agribusiness. Other logistics functions involve purchasing inputs for use in production, coordinating the smooth production of goods and services, determining the appropriate levels of inventory, and probably most importantly, serving as the linkage between the agribusiness, its suppliers, and its customers. This last set of functions involves distribution of the agribusiness firm's products and services.

This brings us to another reason why logistics has become very important. More agribusinesses are going outside their own operations and linking their materials and physical distribution systems directly with supplier and customer systems. They are realizing greater efficiencies as a result. The set of activities which make such linkages work is called **supply chain management.** One important food industry concept of supply chain management is a business strategy called **efficient consumer response (ECR).** It has been estimated that the food industry can save over $30 billion annually using ECR. The Kroger Company alone has estimated that over $150 million (on $24 billion sales) was saved in one year using the principles of ECR.

A third reason logistics has become important is a concept called **cycle-time-to-market.** Also called "quick response," cycle-time-to-market has become a central strategic focus for many agribusinesses. The goal here is to reduce the time it takes for a product to go from an idea to a finished product in the customer's hands. To reduce cycle-time-to-market, a team of sales, logistics, marketing, design, and manufacturing employees work closely together to plan, design, produce, and market new products. Logistics plays a critical role in reducing cycle-time-to-market. Logistics works with suppliers to integrate them into new product development, minimizing time and supplier production issues. Logistics provides the equipment and information infrastructure beginning with suppliers and purchasing, through production, and finally to the physical distribution systems. Many firms have developed integrated information systems to reduce cycle-time-to-market.

LOGISTICS MANAGEMENT OBJECTIVES

Logistics, as it is defined today, is a relatively new field of study. While logistics as a business function actually began when the first goods were traded, it has evolved dramatically in the past 40 years. In the past, agribusinesses narrowly focused their efforts on the physical distribution of their finished products, attempting to optimize inventory levels and reduce transportation costs.

In the past 20 years, firms have integrated their materials and distribution functions to operate and control the agribusiness. The result has been a joined system of purchasing, scheduling, production control, inventory, and distribution activities. Today, the importance and resulting rapid growth of logistics management within the firm has expanded these basic objectives to include corporate objectives. Several of the more current, expanded logistics management objectives include:

- Improved communications and information technology support
- Better customer service
- Reduction of risk
- Globalization of operations

Communication and Information Technology

Fundamental to the increased attention on logistics management are the improvements in communications and information technology that have been made. These improvements have enabled the agribusiness to link suppliers, distributors, and end customers. And, new technology developments continue to transform logistics capabilities. Examples include electronic mail systems, efficient consumer response (ECR), and teleconferencing. For example, electronic mail systems allow agribusinesses to communicate, share files, and obtain information from suppliers, distributors, and customers located worldwide. From the consumer viewpoint, Internet access allows customers to access technical product information, determine product availability, and order merchandise from their homes 24 hours per day.

Efficient Consumer Response

As mentioned earlier, many in the retail food industry are focusing efforts on **efficient consumer response (ECR).** ECR is a responsive, accurate, information-based system that links distributors and suppliers together with retail supermarkets. ECR has one goal: to maximize customer satisfaction. The four aspects of ECR include:

- Replenishment of products as they are used
- Providing the right assortment of products
- Introducing new products
- Developing effective promotion strategies

The potential benefits of ECR are increased sales, fresher products, less damage, reduced cycle time, more accurate invoicing, increased cash flow, and fewer instances of out-of-stock products.

Customer Service

Customer service has become another important objective for logistics managers in agribusinesses. Customer service through logistics can be one tactical method of differentiating the goods or services of an agribusiness. For example, timely information about the status of a delivery or a speedy inventory replenishment system for efficient and effective customer service is one way of differentiating a firm's product or service. Logistics options can also enable customers to determine the level and types of services they desire. As mentioned earlier, the primary goal of efficient consumer response (ECR) is to achieve customer satisfaction.

Risk Reduction

Agribusinesses are also collaborating more with other firms to achieve risk reduction. Logistics can provide a linkage between the agribusiness, suppliers, and customers to reduce risk. Logistics managers may establish contracts with suppliers or customers allowing for shared information and shared investments in transportation, new technologies, inventories, and product distribution. Agribusiness can leverage their strengths and the strengths of other firms to reduce risk as well. For example, one agribusiness may have expertise in distribution, while a supplier may have expertise in producing and servicing a certain component for that agribusiness. Sharing expertise allows agribusinesses to emphasize their core competencies and to best utilize the competencies of partners.

Globalization

Globalization is certainly on the minds of most agribusiness firms today as we saw in Chapter 6. Markets continue to expand abroad while competitors with foreign affiliation enter domestic markets. However, a global enterprise can be managed only if an effective logistics system is in place. On the input side, agribusinesses are increasingly going outside the United States to source raw materials. Information and methods of shipment must be developed to guarantee timely arrival of inputs. On the output side, agribusinesses seeking to export their products or to expand their operations globally must develop their own market channels or form partnerships with foreign firms. To do so, agribusinesses must understand the special needs and wants of the foreign consumer. Relationships must be developed with new suppliers, distributors, and retailers that often operate under different regulatory and cultural environments.

LOGISTICS MANAGEMENT FUNCTIONS

This chapter provides the framework for understanding the role of logistics management in the agribusiness and its interrelationship with the other functions of an agribusiness. Figure 19-1 shows the different activities of the logistics man-

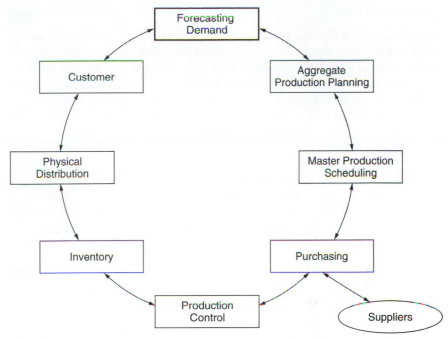

Figure 19-1
Logistics Management Process.

agement process. As you view this figure, keep in mind that the unique features of agricultural products such as perishability, seasonality, and bulkiness can drive the way these functions are performed.

The logistics function begins where the marketing function of forecasting demand ended. **Aggregate production planning** transforms the long-term demand forecasts into general production plans for the next quarter and year. Master production scheduling follows with a final production schedule detailing the exact timing and specific quantities to be produced. Purchasing uses the master production schedule to perform the primary function of procuring the necessary inputs for the production of the finished goods. With the right quantity of inputs available, production control directs the internal logistics function creating detailed internal scheduling of subassemblies, machines, and coordinating the overall production process. Production control is focused on any last-minute changes in the quantities of products produced, and accounts for production lost through scrap or downtime.

Once the products have been produced, they are then shipped for distribution or put into inventory. Many factors determine inventory levels as will be seen later. Distribution can take on many facets in terms of how goods are channeled to the end customer. This brings us full circle to complete the logistics function. Customers purchase the finished goods and they drive forecasted demand for additional products. This quick overview of the activities of logistics management is followed now by more detailed descriptions of the various logistics functions.

FORECASTING DEMAND

Forecasting demand, a marketing function, was described in Chapter 9 and is briefly summarized now to put its role into perspective with the logistics functions that follow. Highly seasonal demand, unpredictable weather, volatile market prices, and an evolving market characterize the agricultural industry. These market characteristics require extensive planning on the part of firm management. As described earlier, the general economic, market, and sales forecasts provide the foundation for many management planning decisions including logistics management decisions.

General economic and market forecasts (i.e., long-range forecasts) drive decisions concerning distribution facilities, warehouse size, transportation equipment, and long-term supplier contracts. Market and sales forecasts (i.e., short-range forecasts) drive the short-term purchasing of raw material for conversion into finished goods, inventory control decisions, and aggregate production planning.

AGGREGATE PRODUCTION PLANNING

Aggregate production planning is the process of developing the specific production quantities and rates, and workforce sizes and rates, while balancing customer requirements and capacity limitations of plant and equipment. Monthly, quarterly, and annual aggregate production plans are developed using the longer-term general economic and market forecasts. Besides maximizing customer service, aggregate production plans strive to minimize inventories, minimize changes in workforce levels and production rates, and maximize utilization of the plant and equipment.

For example, a food company may be forecasting future demand of its packaged cereals and cereal bars. Examining the past quarter's consumption of food products in the United States, the production planner notes the increased consumption of high-energy cereal bars. In addition, upcoming promotions and increasing interest by stores in stocking cereal bars provides valuable customer demand information. In aggregate production planning, these demand forecasts are combined with current production capacity limitations, inventories, and other information to plan how much ingredients, labor, and equipment are required for the next 3 to 12 months. Where aggregate planning shows a bottleneck in an item such as a piece of equipment or a specific labor skill, plans can be put into place to alleviate the shortage temporarily or permanently.

The aggregate production plan does not specify the exact quantities of specific flavors or types of cereal bars that are going to be required. Rather, it estimates the rough quantities of total product types that are going to be needed, so that approximate quantities of raw materials, labor, and equipment can be determined. The exact quantities of individual **stocking keeping units (SKUs)** demanded are determined from sales forecasts (provided by sales representatives) and actual orders. These two items form the basis of the detailed master production schedule, the next step in the logistics function.

MASTER PRODUCTION SCHEDULING

The **master production schedule (MPS)** is created once specific orders for products have been received and/or short-range sales forecasts have been determined. The MPS details the final quantities of SKUs that are to be made in specific blocks of time. The MPS typically provides weekly product requirements over a 6- to 12-month time horizon. In the average production plant, the next 6 weeks of production are not changed to allow purchasing and production control time to finalize their purchases and plans. (Note: This can vary dramatically with the product being produced.) Beyond the approximate 6-week time frame, the quantities of specific products have not been determined but will firm up as the time span to production shortens. The **master scheduler** must consider the total demand on the resources and capacities of the agribusiness as well as the supplier's capacities. Figure 19-2 displays a simplified MPS process.

There are three basic master production schedule strategies. A **make-to-stock strategy** means the firm is producing for inventory, and the production occurs before the actual sale is made. Examples here would be the production of agricultural chemicals, seed, and many food products. Under an **assemble-to-order strategy,** a firm has (or uses) previously manufactured components, and these components are assembled when an order is placed. Farm tractors and machinery would be a good example of this strategy. This is also the "buy direct" strategy made famous by Dell Computer. When employing a **make-to-order**

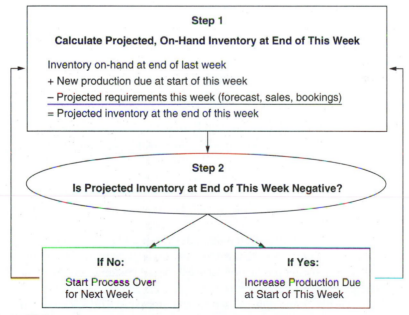

Figure 19-2
Simplified Master Production Schedule Process.

strategy, the firm actually builds to order. A good example here would be firms in the dairy equipment business who build sophisticated milking systems precisely to a producer's specifications. Clearly, each of these strategies would employ the master production schedule idea in different ways.

Purchasing

With a realistic master production schedule completed, **purchasing** begins its tasks of procuring the inputs necessary to meet the requirements of the production schedule. Purchasing is a common function for every organization in the agricultural industry. Although it varies by industry, the typical firm spends about 40 to 60 percent of its total income from sales on purchased materials and services. As a result, a small percentage reduction in such a huge portion of the overall sales for an agribusiness firm can have a significant effect on the firm's bottom line. For example, if net profits in the agribusiness are 5 percent of sales, then a 1 percent reduction in purchasing costs will fall through to the bottom line, thus increasing profits by 20 percent over their previous level. This example illustrates the relationship between costs of production (purchasing) and the firm's profits.

Some agribusinesses have centralized buying where materials are purchased in large quantities, often at substantial discounts. Other firms have decentralized purchasing where each location purchases just the goods that are needed when they are needed. Both have their advantages and most agribusinesses use a combination of both. An agribusiness typically purchases four kinds of products.

- Products used for further processing
- Products that are resold
- Products used directly in the firm's final products
- Products used to make the product but not used in the product itself

The purchasing function performs many activities, which include the following:

- *Receive a purchase requisition.* Purchasing examines these documents to determine if a less costly item can be substituted or eliminated entirely.
- *Select a qualified supplier.* Purchasing selects suppliers based primarily on four factors: price, quality, timeliness, and service.
- *Place the order.* Ordering procedures can be very routine or very time-consuming depending on the nature of the purchase and the buyer-supplier relationship.
- *Track the order.* Purchasing monitors deliveries and production schedules and expedites orders as necessary. Follow-up is particularly important when a delay could disrupt production, cause a loss of customer goodwill, and/or a potential loss of future sales.
- *Receive the order and approve payment.* Purchasing works with receiving and accounting at this point to ensure all goods are received satisfactorily before authorizing payment.

The purchasing activity of selecting suppliers is critical because of the impact of input prices on the agribusiness's bottom line. As mentioned earlier, the four primary factors used in selecting suppliers are quality, price, timeliness, and service. Quality is a very important consideration. An agribusiness that uses a purchased part only to find out later that it is defective incurs numerous costs, including costs internal to the firm as well as potential loss of goodwill and future sales.

The price of purchased inputs is fundamentally important in selecting a supplier because the typical firm spends nearly half of its total income on purchased items. Timely delivery of purchased inputs from suppliers is also very important. Parts priced at just pennies can cost the firm thousands of dollars for every minute that a production line is not running. The opportunity costs of such a problem can be enormous, as an agribusiness does not want to miss sales because they lack finished goods. On the other hand, having too many goods in inventory can be very costly. Carrying costs, physical storage space, recordkeeping expenses, taxes, insurance, and interest lost on capital tied up in inventory can eat up as much as 35 percent of the inventory's purchase value. Consequently, service is another critical but often overlooked item. Shorter lead times and on-time delivery of purchased items help the agribusiness to provide higher levels of customer service with less inventory.

PRODUCTION CONTROL

With the materials for production ordered, the logistics responsibility now shifts to **production control.** Control of production flow is one of the most important daily functions of the operations manager. The responsibilities of production control include the following:

- Controlling raw materials inventory
- Providing detailed production scheduling information
- Controlling work-in-process inventory
- Communicating changes to master production scheduling and purchasing
- Controlling finished goods inventory

The tasks of production control are essential for smooth, continuous flow of work through the agribusiness. Two production operations systems that can facilitate the smooth flow of production are material requirements planning (MRP) and just-in-time (JIT) systems. MRP is often referred to as a "push" system whereas JIT is a "pull" system. We will explore MRP and JIT to better understand why push and pull are very descriptive of these two types of production operations systems. In addition, we will consider two important project management techniques—program evaluation review technique (PERT) and the critical path method CPM).

MRP Production Operations Systems

Material requirements planning (MRP) is a computerized information system for managing production operations. Its purpose is to ensure that the right materials, components, and subassemblies are available in the right quantities at the right time so that finished products can be produced according to the master production schedule.

MRP operates in this manner. First, the finished product requirements from the master production schedule are broken down into the requirements for individual parts or subassemblies. These requirements are compared with on-hand inventory to determine the quantities of units that still need to be produced. Schedules are created, inputs are purchased and received, and production begins. Parts produced in-house are combined with purchased component parts to form the most basic subassemblies. These are then "pushed" to the next level of subassemblies. These subassemblies receive further operations and are "pushed" to the next step of the manufacturing sequence to form larger subassemblies. The process of completing operations and pushing to the next operation continues until the finished product is completed. The entire production system is based on producing components and pushing them to their next stage of production.

JIT Production Operation Systems

Another production operations system is **just-in-time (JIT).** The goal of a JIT system is to produce or deliver goods just as they are needed, in effect, attempting to completely eliminate inventories. JIT is more of a philosophy than a production operations system. The goals here are to eliminate waste, inefficiencies, and unproductive time. In fact, JIT almost always requires changes in process design, scheduling, and inventory. When used effectively, JIT reduces inventory levels, reduces costs, improves quality, reduces set-up time, and smoothes the flow of production. There are four components of a JIT system.

- Kanban
- Layout and production methods
- Total quality control
- Suppliers

Kanban is basically the information system for JIT that helps to pull production through a shop (Kanban is the Japanese word for "cards"). A combine, for example, consists of literally several thousand parts. As the combine is entering final assembly, a cab assembly, the main body section, an engine, a head assembly, tires, and several bolts and nuts are needed. As the final assemblers "pull" these parts for assembly, they replace each part used with a Kanban (card) specifying the number and type of parts and subassemblies pulled. These Kanbans (cards) then serve as notice to replenish the parts that have been taken. This pulling process begins at final assembly and works itself backward through all of the stations in the process, eventually continuing to suppliers. Thus, the entire production process is synchronized to the final assembly schedule that pulls parts as they are needed. This is why JIT is referred to as the "pull" method of production.

JIT is very much related to total quality management (TQM) in that it encourages continuous improvements in quality and processes. A system of **total quality control** must be in place if JIT is to be successful. A high rate of defective parts would make JIT ineffective because bad parts, small lot sizes, and no safety stock would disrupt the smooth flow of production. As a result, much effort is applied toward preventing poor quality, training operators, and self-inspection.

Suppliers also play a critical role in JIT systems. Although the principles of JIT and Kanban begin with the requirements from the final assembly area, they eventually work back through the production system to suppliers of inputs. In JIT systems, suppliers make smaller deliveries on a frequent basis (sometimes several times each day), which has encouraged some suppliers to locate close to the agribusiness.

PERT and CPM Production Planning

The scheduling methods that an agribusiness uses for planning and implementing projects, new designs, or new facilities depend upon the size and complexity of the specific operation. A small firm will have a relatively straightforward method of scheduling. Among the larger operations, however, some of the so-called "network models" are used. Two of the most famous of these are PERT and CPM. Many of the numerous project scheduling software programs available today operate on the principles of PERT and CPM.

The **PERT (program evaluation review technique)** model was developed by the U.S. Navy in the 1950s to schedule operations on the massive *Polaris* missile project. It is estimated that PERT saved the Navy two years of work on that project, and production managers for large companies continue to use this sophisticated technique.

PERT involves a diagrammatic representation of the network of activities in which an operation is engaged (Figure 19-3). The circled letters (nodes) represent activities to be completed. The numbers adjacent to each circle or node represent

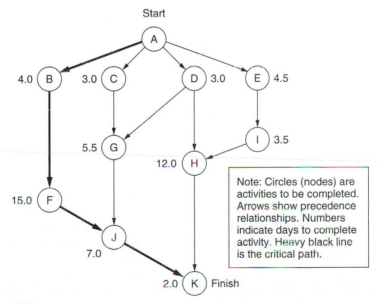

Figure 19-3
PERT Scheduling Diagram.

how long that activity will take. The arrows indicate the required sequence of activities—that is, activity B must be completed before activity F, which must be completed before activity J. The path from the beginning to the end of the sequence that is bolded is the **critical path,** or path that has absolutely no slack time. Any other path of activities has what is known as slack time. If a path has slack time, it simply means that delays along this path (to a point) won't delay the total project.

With all this information in hand, agribusiness managers can determine the least amount of time it will take to complete a project, the greatest amount of time needed (the critical path), and the slack-time options that are available. PERT is a way of understanding the relationships among interdependent scheduling variables, and determining which steps in the process must be given preference in order to complete the project.

The **CPM (critical path method)** of scheduling was developed at about the same time that the U.S. Navy developed PERT. As the name indicates, CPM networks bear similarities to those of PERT. And, while there are some differences in the two techniques, these differences are relatively minor in today's applications.

Although PERT and CPM are useful in planning and analyzing large projects, there are some drawbacks to the systems. Project managers sometimes focus too much on the critical path activities ignoring other, near-critical paths. Also, clear beginning and ending points for activities do not always exist. In addition, activities can have highly uncertain times for completion. Finally, the requirement that certain activities precede others is often restrictive because some work from the two activities may overlap.

INVENTORY

Beginning with suppliers and moving through the production process to the end customer, some form and level of inventories are maintained. Furthermore, the activities of purchasing and inventory control are closely related. Operations managers keep their inventory counts current and accurate, so that they can alert the purchasing department when more product is needed and so that they can track inventory turnover.

Inventory is important to an agribusiness firm from a financial and operations standpoint. Recall that inventory is an asset and in the average company, it can represent 35 to 50 percent or more of the current assets. Although inventories are an asset from an accounting perspective, they are really a cost of doing business. The dollar value of goods held in inventories could be invested in an interest-bearing account or in a productive piece of equipment. Thus, the interest or **opportunity costs** of inventory can be high. Storage and handling costs for holding inventory can also be high. Large physical storage spaces are often required to store inventory preventing such space from being used for more productive purposes.

So why have inventories? One principal reason for maintaining inventories is to provide customer service because products are available immediately for

T A B L E **19-1** | **Reasons for Different Inventory Levels**

Reasons for smaller inventories	Reasons for larger inventories
Interest or opportunity costs	Customer service and varying demand
Storage and handling costs	Ordering or set-up costs
Improved quality	Labor and facility utilization
Property taxes and insurance premiums	Transportation costs
Shrinkage costs—obsolescence, pilferage	Quantity discounts

Source: Krajewski and Ritzman, pages 545–547.

consumption. Another reason is to minimize the effects of varying demand for finished goods—especially in agricultural products. For example, it may be forecasted that 10,000 units of a short-season popcorn variety will be demanded for the spring planting season in Indiana. A late spring that delays planting forces producers to switch to this short-season variety, doubling demand to over 20,000 units. A shortage of the popcorn seed may force customers to wait for shipment from another location or to purchase seed from a competitor. Lost sales opportunities now or in the future may be the result (i.e., customers may switch to a more reliable supplier).

Inventories are maintained sometimes for other reasons (Table 19-1). Ordering costs can be minimized because larger quantities can be purchased less often. Costs to set up equipment and run a second batch of product can sometimes make holding extra inventory advantageous. Transportation costs per unit can be reduced if a truck is filled with product and delivered full versus a partial shipment. Finally, inventories are sometimes held to maximize equipment utilization.

In contrast, many agribusinesses have found that inventory levels can be reduced significantly without reducing customer service because better information systems and equipment technologies have reduced the time between product purchase and order fulfillment and replenishment. Smaller inventories can improve quality levels because the feedback loop between production and use is shorter. Not surprisingly, better quality and the flexibility to produce smaller quantities of products have increased labor and equipment utilization. As a result, the savings combined from all of these improvements can offset higher ordering costs and loss of quantity discounts. Of course, obsolescence costs and pilferage are reduced with smaller levels of inventory.

Inventory can be broken down into three general categories used for accounting purposes. Inventory in the procurement or **raw material** phase comprises about 30 percent of total inventory. For example, various grains, vitamin supplements, by-products, and preventive health ingredients are all raw materials used to make hog feed. A second accounting category of inventory is **work-in-process (WIP)** or inventories that are in the conversion phase. These inventories make up an additional 30 percent of total inventory on average.

Examples of WIP are engine subassemblies that are awaiting final assembly into a farm implement. The third category of inventory is **finished goods,** the remaining 40 percent of inventory. Finished goods are the outputs from the conversion process, which are now ready for usage. Breaded veal cutlets in cold storage would be an example of finished goods inventory.

Types of Inventory

Using an example, let's focus on some different types of inventories. Rita Tyner is vice president in charge of production for a power equipment manufacturing company located in Indiana. Though the firm manufactures several products, one of its most profitable items is a garden tiller. Rita's company sells garden tillers at an average rate of 25 per week. When the warehouse stock of garden tillers reaches a certain, agreed-upon level of depletion, called the **order point,** Mike Torres, the warehouse manager, sends Rita an order for more. The total process takes two weeks (i.e., paperwork, receipt and filling of the order, transportation, and receipt at the warehouse). The amount of stock that Mike has on hand when he reaches that critical order point depends on the kind of inventory being used.

Pipeline inventories are the minimum amount of inventory needed to cover the period of time between the warehouse's reordering and its receipt of the additional stock, or **lead time.** Since the average demand for garden tillers is 25 per week, and it takes Mike two weeks to receive the order, he will need at least 50 garden tillers in stock to avoid depletion and lost sales during the lead time.

Despite the fact that Mike and Rita's lead time is two weeks, it is not necessary for Mike to issue an order every two weeks. For example, suppose that each truck were capable of loading and delivering 100 garden tillers. In the pipeline inventory system, this would mean that a truck delivering every two weeks would go from factory to warehouse half-empty. Rather than use this costly method of operation, and go through twice as much paperwork, Mike would probably order 100 garden tillers every 4 weeks. Since **cycle** or **lot-size inventories** may be considerably larger than pipeline inventories, it goes without saying that use of this method is subject to constraints of available storage space, and dependent on projected future sales.

We have been assuming that Mike and Rita are dealing with rock-steady weekly demands, invariable lead times, and standard truck capacities. In the real world of agribusiness enterprise, this is never the case. Demand varies from week to week, as does the length of time it takes to get an order filled. These differences are taken into account by **safety stock** or **buffer inventories,** which provide additional stock to offset potential variations.

As explained in Chapter 18, agribusiness products are often highly seasonal. Mike and Rita might sell as many as 200 garden tillers a week in peak season, or as few as two during the off-season. Rita has the choice of planning for much higher labor costs at peak season or for costly storage of extra garden tillers in off-season. Both these risks can be avoided by **anticipation** or **seasonal inventories.** Under this system, storage space is provided for products whose demand is not constant so that more are available during peak seasons.

Because of the investment and cost of inventory, effective inventory management is a key factor driving profitability for many agribusiness firms.
Photo courtesy of USDA.

Inventory Management Systems

Logistics management systems typically look at demand in two different ways—independent demand and dependent demand. Retail merchandise, services, and finished goods are types of goods driven by independent demand. The demand for a combine is an example of independent demand—individual customers demand combines. Dependent demand, on the other hand, is derived from the demand for another item or from production decisions. For example, the demand for tires for new combines is derived from production schedule decisions and ultimately from the demand for another item—combines. There are two basic inventory tracking systems for products with independent demand: periodic or continuous. Although most agribusinesses use a combination of both systems, firms are increasingly relying on continuous tracking systems that are automated.

Periodic inventory is an actual physical count of stock on hand, conducted at regular intervals of, say, 1 month (Figure 19-4). This kind of inventory has the disadvantage that, taken by itself, it leaves the day-to-day status of the inventory in doubt. However, recordkeeping is minimized and the periodic inventory is very reliable at that time. A good example of this system is the store-door delivery of soft drinks and salted snacks to supermarkets by suppliers. On a regular basis, weekly or more frequently, the inventory of these products is reviewed and orders placed.

Continuous inventory is exactly what its name indicates, a constant monitoring of stock on hand (Figure 19-5). Whenever sales are made or inventories are replenished, the total amount is subtracted from or added to the previous inventory total. Usually, a count of inventories to be received is also kept, to eliminate the dangers of ordering too much. Several forms of automated continuous inventory tracking systems exist. Bar code scanners at supermarkets, for example, are a technology that enables inventories to be tracked continuously and product to be replenished as it is purchased.

One advantage of continuous systems is that the inventory count is always current. The disadvantage is that even with a sophisticated, computerized inventory

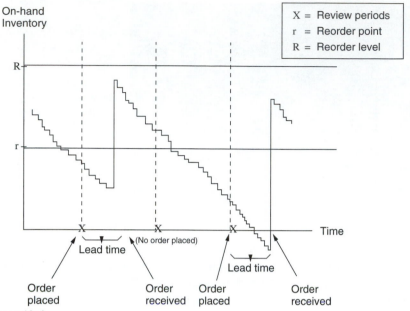

Figure 19-4
Periodic Inventory System.

system, the reliability of the inventory quantity on hand is only as good as the people who update it. The integrity of the system will be lost if employees forget or bypass the system. Theft and shrinkage also will affect the accuracy of the continuous inventory system.

DISTRIBUTION

Customers, markets, and competitors are always changing. Manufacturers have numerous options and resources available to them to distribute their products to the ultimate consumer. Therein lies the challenge of managing the distribution element of logistics management. **Physical distribution systems (PDS)** are the series of channels through which parts, products, and finished inventory are stored and moved from suppliers, between outlets and, ultimately, to consumers. PDS encompasses transportation, storage/warehousing, and delivery of the finished product to the consumer. PDS must be carefully managed and coordinated to meet delivery requirements of business customers successfully. Manufacturers, wholesalers, distributors, retailers as well as the consumer are all affected by the adequate and accurate execution of successful distribution.

Physical distribution systems comprise a costly component of the overall business plan. In many U.S. industries, assets required by distribution can account for more than 30 percent of total corporate assets. Physical distribution costs are often estimated to be as much as 30 to 40 percent of total costs. "Doing

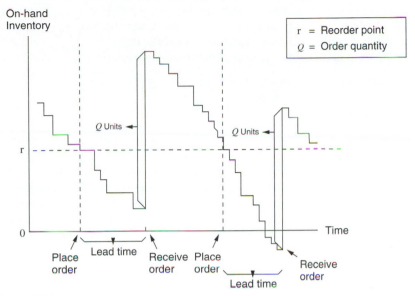

Figure 19-5
Continuous Inventory System.

the math" here shows that PDS can exceed 25 percent of each sales dollar at the manufacturing level.

Today's agribusiness manager will benefit from an understanding of logistics management of physical distribution systems even if they are not directly involved in the management of those resources. Effective PDS can mean opportunities for improving customer services and for reducing costs. For many businesses, making the product available to the customer is the road to customer satisfaction. While those opportunities for improvement and growth are true, so are the opportunities for failure or inefficiency. A repair part delivered a few hours late may cost a customer thousands of dollars in lost production time. The PDS will vary dramatically by product. The role of PDS for Hormel salami will differ from the role of PDS for Monsanto's Roundup® herbicide due to the nature of the product, and the types of warehousing, transportation, and storage that can be used in delivering the product to the customers.

Distribution and Marketing

Place, or **distribution,** was one of the four critical marketing decisions (four P's) examined in Chapter 8. We will briefly summarize that chapter's discussion in the context of the overall logistics function. The choice of a marketing channel is very important to an agribusiness because a trade-off must be made between achieving the dual and competing objectives of customer service (product available wherever and whenever) and minimizing distribution costs. Distribution adds a time and place dimension to marketing that can provide an agribusiness with a competitive advantage. In addition, the special problems of bulkiness, seasonality, and perishability in agribusinesses can make the choices of where to store finished goods and how to transport them difficult.

Chapter 8 detailed the three basic distribution systems:

- Manufacturer-direct distribution system
- Dealer-distribution system
- Distributor system

The choice of distribution system depends on the geographical distribution of customers, the amount of personal and specialized customer service that is required, and the need to control or influence the image or relationship with the customer. Note too that e-business systems have emerged that facilitate distribution in a variety of ways. Such systems may be focused on the end consumer (business to consumer, or B2C) or they may be targeted at facilitating distribution between firms in the distribution channel (business to business, or B2B). E-business is discussed further in the final section of this chapter.

In logistics management, physical distribution systems can be broken down into three basic categories. Each of these categories involves choices and decisions for the logistics manager in making the optimum decisions for the operation. Three basic components of distribution are:

- Warehousing—where finished goods will be stored
- Transportation—how goods will be moved
- Carrier—the scheduling, routing, and selection of the carrier

Warehousing

Manufacturing plants, warehouses, distribution centers, and retail outlets all fall into this category of where goods are "warehoused" or stored before delivery. **Warehousing** can represent a huge cost in terms of the operating costs and capital investment in buildings and facilities required to properly store a company's products. Consideration must be given to the location, size, service levels, proposed new or additional facilities, and so forth.

Efficient and effective handling and storing of inventory can also provide huge savings to a firm. Product movement may be affected by packaging. A change in either product movement or in the packaging can impact the type of warehousing necessary—and again, the bottom line. A new type of package for a meat product may mean easier shipment as well as less costly warehousing. The trucking container which can be moved by rail, highway, or by sea may provide managers with a delivery-ready, cost-effective means for warehousing products.

In considering warehousing, there are two basic choices in determining where finished goods are stored. Forward placement of finished goods locates finished goods near customers at a warehouse, wholesaler, or retailer. The advantages of forward placement are faster delivery times after receipt of order and increased sales (demand is stimulated by product availability). For example, a farm equipment manufacturer may allow inventories of selected parts to be stocked directly on the farm in anticipation of a producer's future needs. Periodically, the inventory will be checked and the producer will be invoiced for parts he/she has used.

Backward placement of finished goods holds goods at the production facility. The primary advantage of holding goods at the manufacturer location is that a centralized warehousing system can reduce the total number of goods held in inventory. However, a primary disadvantage is a lower level of overall customer service.

Transportation

The mode of **transportation** must also be chosen. The five basic modes of transporting goods are highway, rail, water, pipeline, and air. The advantages and disadvantages of each can be evaluated by considering their length of transit time, geographic and product flexibility, cost, and the number of times goods are rehandled. For example, Monsanto initially chose to deliver bovine somatotropin to dairy farmers via an overnight delivery system because it allowed for fast delivery and a centralized warehousing system. On the other hand, a bulk commodity like anhydrous ammonia is usually transported via pipeline. Water transportation and rail are the transport methods of choice for bulk commodities such as grain where cost is a major factor and time of lesser importance.

Developing cost-effective transportation is essential for the physical distribution system to be complete and successful. Managers must carefully weigh the issues surrounding physical distribution of products in terms of getting the product

Whether by highway, rail, pipeline, water, or air, decisions about the transportation of products is an important area for logistics managers.

delivered successfully and in a timely manner to the customer. Take, for example, a food processing equipment company that made changes in the shipping method of some components from air transport to sea transport. The move saved over $15 million dollars in a single year in shipping costs. However, the slower delivery meant product parts were not available when needed, which meant dealers didn't have parts they needed when or where they needed them, which eventually cost the company lost sales. The bottom line on the change in shipping was that it was not cost-effective.

Carrier Scheduling, Routing, and Selection

Finally, the scheduling, routing, and selection of the finished goods carrier can be a rather complex decision because of varying rate differentials, destinations, routing methods, and the mixing of shipments. As a result, sophisticated transportation models have been developed to enable firms to provide finished goods sooner, consolidate freight, take advantage of vendor discounts, and minimize administrative and warehouse labor costs.

Effective scheduling, routing, and selection require a network of information that transcends the physical distribution system. Order processing, transportation, delivery status, inventory turnover, and system performance are each itemized and carefully calculated in the efficiently run distribution system. Today's information systems in the distribution network often involve state-of-the-art bar coding and computerized tracking methods unimagined just 10 to 20 years ago. Compare how years ago your parents and grandparents ordered goods using a Sears catalog mail-in order form. Today, toll free operators stand by to offer more immediate alternatives to customers over the telephone. The order can be processed in a more timely manner and the appropriate distributor can get the product to the customer from the location nearest them for the most cost-effective distribution. Even this level of convenience has been exceeded with a host of Internet-based systems that offer 24/7 ordering and customer service capabilities.

The scheduling, routing, and shipping of products must be coordinated with the inventory supply management of goods to meet demand. Needless to say, information-processing technology available today helps to effectively communicate timely and accurate information (on inventory, location, delivery status, etc.) to the appropriate channels. Knowledge of delivery efficacy can help with overall customer satisfaction. It can also help measure customer service outcomes. Information backflow is an often overlooked, yet critical component of the physical distribution system. Accurate tracking of distribution and inventory information can help the company accomplish distribution objectives, and measure customer service quality, as well as examine, identify, and/or choose management alternatives.

TECHNOLOGY AND OPERATIONS

Technology is simply any manual, automated, or mental process used to transform inputs into products and services. Each operation within a food or agribusiness employs some technology, even if it is manual. Agribusiness managers invest in technologies to improve productivity and quality. They also must consider how technology relates to their operations strategy. For example, if a firm wants to be the low-cost producer, it may possibly want to invest in automation and a product layout that can produce high volumes of parts at low costs.

Technology factors may also influence plant locations by altering the manufacturing process. For example, genetic technologies and new processing technologies that have extended the shelf life of food products without freezing or refrigeration have allowed some industries to move away from end-use markets. Tomatoes imported from Mexico are ESL tomato varieties. ESL (extended shelf life) varieties contain a gene that has been manipulated to allow the tomato to stay on the vine until 90 percent of the fruit is pink or red. ESL tomato varieties last longer—they can stay on the supermarket shelf a week longer than other varieties.

Technology is important for several reasons. New technological innovations and intense competition have necessitated continued improvements in productivity and higher levels of technology, especially information technology (IT). Shorter product life cycles and increased marketing pressure are requiring more flexibility on the part of service and manufacturing operations—again, requiring better technology. Products and materials are also becoming increasingly complex to build, so new technologies to support more challenging production processes are needed.

Information Technology

Information technology advancements including speed of processing, data storage, data transmission, and improved software are changing how business is conducted. The Internet is one such information technology that has exploded onto the food and agribusiness scene. USDA's National Agricultural Statistics Service found that in 1999, 72 percent of U.S. farms with over $250,000 in annual sales owned or leased a computer. Of farmers in that category, 65 percent used the computer for their farm business and 52 percent used the Internet. Those figures may not seem significant until compared with figures from just two years earlier when only 31 percent of all farms owned a computer and only 13 percent had access to the Internet.

Information technology and the Internet have combined to make an entire new way of doing business possible: e-business. The growth in e-business—the use of the Internet and information technology to conduct business—has been staggering. As one example, e-commerce activity (dollars) in the food and agriculture segment was forecast to increase 1,700 percent between 1999 and 2003. More specific than e-business, e-commerce refers to trade that actually takes place over the Internet. For example, in 1999, 2.7 percent of all new car sales in the United States were done through the Internet—an example of e-commerce. However, it is estimated that during the same year, 40 percent of new car sales in the United States involved the Internet at some point in the sales process. This is the concept of e-business: to use the Internet in order to deal with business issues, whether to sell, to inform, or to lower the cost of a transaction.

The Internet is also helping producers who are located in geographically distant areas. Agricultural producers can access markets, product information, organizations, and government and university information. Data files can be exchanged—facilitating a level of communications that was nearly nonexistent among farmers even 10 years ago. And, the bandwidth, or size of the information pipeline, is rapidly improving in rural areas. This makes it possible to send much richer information such as video to isolated areas.

Access and availability dramatically change the way business can and will be done thanks to the round the clock availability of information through the World Wide Web. In many regards, the farmer located in an isolated rural location in central Kansas may no longer be limited to shopping at the single equipment dealer in the county. Agriculture's seasonal nature means farmers (customers) are often not available to make purchases during daylight hours—when the rest of the business world is working. Availability of information on the Web means the issue will no longer prevent farmers from researching, shopping, and purchasing products and services. Classifieds, bulletin boards, and chat rooms posted on the Web are connecting farmers who have never been exposed to one another before. These rapid changes in how business is done will no doubt impact agribusiness managers in the years to come, primarily because issues of access and availability do and will impact logistics strategies. For some companies or suppliers, these changes will create opportunity. For others, such developments will bring intense challenges.

Despite all of the promise and potential of e-business, it is important to remember that physical logistics is still important. UPS may now ship a particular animal health product directly from a centralized warehouse to the farm, or groceries ordered online might be picked up at the supermarket by a consumer trying to save time. In both cases, logistics systems are involved throughout the process of moving the product from the manufacturer to the buyer. E-business may change how these steps are performed, and provide sometimes radical new ways for doing them, but the functions of logistics do not go away.

Many other examples of advanced technologies impacting production and logistics can be found in the food industry. Electronic data interchange (EDI) is a technology that allows business interactions between firms to take place electronically using a standard language format. The standardization enables EDI to be used by any company and has improved accuracy, shortened response time, and has even cut inventory. The food retail chain, SUPERVALU, Inc. estimates that about $5,000 to $6,000 per week is saved by using EDI.

As mentioned earlier, efficient consumer response (ECR) is the more recent information technology to take hold in food retailing. More comprehensive than EDI, ECR is a consumer-driven information system that links distributors, suppliers, and retailers. ECR focuses on the four items of product assortment, new products, product replenishment, and product promotions to achieve customer satisfaction. Such systems will continue to reshape the area of logistics for food and agribusiness firms.

SUMMARY

Understanding the functions of logistics management is imperative to the agribusiness manager's success. Logistics management has become a key business function over the last few years and has grown in both relevance and importance in the last decade alone. More agribusinesses now find greater efficiencies through going outside their own operations and linking their materials and physical distribution systems directly with supplier and customer systems.

Inventory control, supply chain management, and cycle-time-to-market illustrate how a company's profitability can be impacted through effective logistics management. Other logistics functions include purchasing the inputs for production, coordinating the production of goods, determining the levels of inventory, and probably most importantly, serving as the link between the agribusiness, suppliers, and customers.

Agribusiness production managers who wish to maintain control of their operations will have to forecast trends; purchase in-line with these trends; control the amount of inventory on hand; and employ sophisticated scheduling methods and quality controls. Purchasing is an activity that requires skill, knowledge, and honesty in dealing with bids and negotiations from suppliers. Such factors as quantity, quality, price, time, and service are elements of efficient purchasing.

The amount of inventory and kind of inventory systems used depend upon an agribusiness's size, complexity, and management sophistication. Physical distribution systems (PDS) include warehousing, transporting and scheduling, routing, and selecting product delivery. The right technology applied at the right time and in the right place can help logistics managers increase overall efficiency while expediting the accurate and timely delivery of top quality products and services to the customer. The manager who is able to exercise control over these facets of logistics will have a tightly run operation that stands a good chance of maximizing profits.

DISCUSSION QUESTIONS

1. Why is logistics management important in an agribusiness firm? Describe the various functions and responsibilities of logistics management.
2. What are some of the objectives of an efficient logistics management system? How have these objectives changed over time?
3. What are the principal activities of purchasing? Explain some of the factors to consider in selecting a supplier.
4. Why might inventory levels be minimized and why might some level of inventory be acceptable?
5. What are the advantages and disadvantages of a MRP system? A JIT system? In your opinion, which would be better for a fast food restaurant?

6. For the PERT chart shown in Figure 19-3, verify that the bold path is the critical path. Pick a process that you are familiar with—this could be a business process or a school process (e.g., writing a report). Construct a PERT chart for the process. What is the critical path for your process? How long should the process take, beginning to end?

7. Pick a food or agribusiness of your choice. Describe its physical distribution system. Try to describe each of the components of a physical distribution system outlined in the chapter for the business you choose.

8. "The Internet will revolutionize the physical distribution of groceries as online retailers displace traditional retail supermarkets." Do you agree with this statement? Why or why not? In your opinion, how will the Internet impact the retail distribution of food?

CASE STUDY—BioAg—Part II[1]

Supply Chain Management   The top management of BioAg consists of a microbiologist (Abe) and a cereal scientist (Ben). They are excellent research scientists and have shared leadership of the company. Abe and Ben recently attended the American Crop Protection Association (ACPA) annual meetings, which featured several sessions about supply chain management.

During one session, Farmco, a large regional input supply firm and BioAg's major reseller, announced their intentions to move aggressively with supply chain management initiatives. In a pilot project with a seed company, Farmco lowered inventory stocks by 10 percent and increased customer service (as measured by on-time deliveries) by 5 percent. Farmco feels there are three keys to building a successful supply chain relationship. First, the buyer and seller must move from an adversarial to a partnership-type relationship. Second, to manage inventory levels, the two firms must coordinate and streamline their logistics. Finally, information systems must allow them to share key information, thereby improving communication, allowing quicker action, and better decision making.

Upon their return, Abe and Ben convene in a meeting with four key BioAg managers, Cindy, Dee, Ernie, and Fred. Cindy is the sales manager, Dee runs information systems (IT), Ernie heads research and development (R&D), and Fred leads logistics. Abe and Ben show the other four a graphic from the ACPA meeting and ask how BioAg should proceed to adopt this type of change with Farmco and other customers (Figure 19-6).

"I think this would be a huge mistake," starts Cindy, the sales manager. "We have been aggressively expanding our supplier base. Look at how I've expanded the number of farm input supply firms we sell through in the past four years. It is the competition among the dealers that is leading to our growth. Besides, my job is to sell. Others are supposed to service accounts."

Ben replies, "I understand that, but Farmco has $500 million in sales. They account for 35 percent of our sales and 20 percent of our 420 dealers. They are among our most important customers."

[1]Reprinted from *International Food and Agribusiness Management Review,* Vol. 1, Issue 3, "Supply Chain Issues: A Case Study of BioAg" by Frank J. Dooley and Jay T. Akridge, pp. 435–441. Copyright 1998 with permission from Elsevier Science. It may be useful to review BioAg—Part I in Chapter 18 before working on this case.

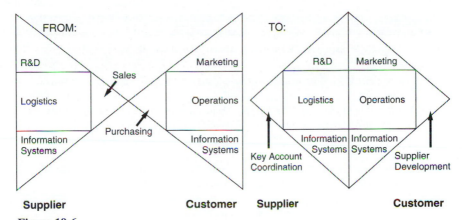

Figure 19-6

Changes in Procurement with Shift to Supply Chain Management.
Source: Peck, 1998.

"I'm also skeptical," adds Ernie. "We are a company living with proprietary technology. Your idea seems to imply that we share our R&D with our customers. How do I know that our ideas won't end up in the hands of our competitors?"

Abe says, "That is an important question, Ernie. But we're looking at our relationship with Farmco as a partner. Don't we have to trust our partners?"

Dee from IT then offers, "I'm sure that we could eventually get there. However, most of our efforts are focusing on upgrading existing computer systems. I'm sure that Farmco and our competition won't move for at least two years."

Abe responds, "I think you may be wrong. According to Farmco people, they have made upgrading their information technology a priority for the next fiscal year."

Abe and Ben then turn to Fred who has remained silent throughout the meeting. Fred frowns and says, "I don't know how I can improve service while cutting inventory levels. Our forecasting ability has already diminished because of the number of products we now offer and markets we serve. Remember that we just built two new distribution centers. And don't forget that we plan to expand to 1,000 dealers in the next two years. I guess I'm with the others. This is not a good fit for BioAg."

Ben answers, "Fred, are you suggesting that our distribution capabilities drive the future direction of the firm? I'd also like your appraisal of Farmco's pilot study. They said they would share the results with us. How did they do it?"

After the meeting Abe and Ben are silent for a few moments. Ben says, "This seemed like such a good idea at the ACPA meeting." Abe replies, "Yeah, and I overheard that Farmco plans to move forward with our competitors if we aren't interested. Can we wait?" They decide to reconvene their team and consider the following questions.

Questions
1. What are the potential risks and benefits to BioAg in pursuing the supply chain management initiative with Farmco?
2. Farmco cites three main factors in moving toward supply chain management. Which factor will be easiest to achieve in this case? Which factor will be the greatest obstacle? Does a successful supply chain relationship require all three factors?
3. Assume that BioAg and Farmco move forward with their supply chain initiative. How might this affect BioAg's relationship with its other dealers?

References and Additional Readings

Council of Logistics Management. "What It's All About." Oak Brook, IL.
http://www.clm1.org

Davis, Tom. Summer 1993. "Effective Supply Chain Management." *Sloan Management Review.*
pp. 35–46.

Dooley, Frank, and Jay T. Akridge. 1998. "Supply Chain Management: A Case Study of Issues for
BioAg. *International Food and Agribusiness Management Review* 1(3):435–441.

Fisher, Marshall. March–April 1997. "What is the Right Supply Chain for Your Product?" *Harvard
Business Review.* pp. 105–116.

Krajewski, Lee J., and Larry P. Ritzman. 2000. *Operations Management: Strategy and Analysis.*
5th edition. Reading MA: Addison-Wesley Publishing Co.

Peck, Helen. 1998. "Partnerships in the Supply Chain: Introducing Co-Managed Inventory at
Guinness GB." Oak Brook IL: Council of Logistics Management.

PART VI

HUMAN RESOURCE MANAGEMENT
FOR AGRIBUSINESS

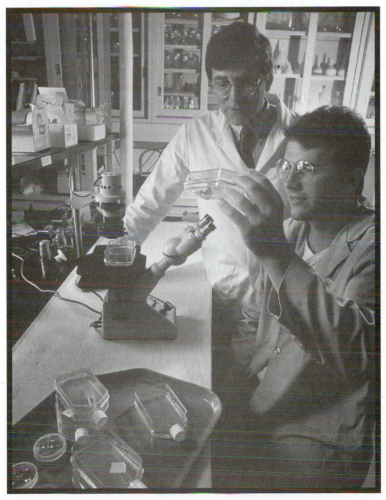

Human resources: this section develops some of the concepts behind the human element of management, and the management of the potential value and contributions individuals can make to an organization.

Photo courtesy of Agricultural Research Service, USDA

20

MANAGING ORGANIZATIONAL STRUCTURE

- Differentiate between responsibility, authority, and accountability relationships in agribusiness
- Describe formal organizational structures used in agribusiness
- Understand organizational principles as they apply to the development of agribusiness organizational structures
- Discuss the impact of informal organizational relationships on the success of the agribusiness
- Identify various leadership styles employed by agribusiness managers
- Review widely used theories of managing and motivating employees
- Understand the use of transactional analysis as a tool for understanding basic human behavior
- Discuss the types of recognition and their role in motivating employees

INTRODUCTION

Food and agribusiness firms are comprised of people working together toward a common purpose. As soon as an agribusiness involves more than one person, a variety of organizational, personnel, leadership, and motivational issues inevitably arise. The larger the organization, the more complex and critical these issues become. One of the fundamental responsibilities of management is to acquire, organize, motivate, lead, and manage its human resources to accomplish its business objectives as effectively as possible. Given this charge, management must address whatever complex issues may challenge or potentially disrupt that responsibility.

Managing the human resources in an agribusiness has many dimensions. First, management must develop an organizational structure where the responsibilities, authority, and accountability of individuals are clearly defined. By **organizational structure,** we mean the formal way that employee responsibilities are assigned in a firm—who reports to whom, who has responsibility for what, etc. Basically, we are talking about the firm's organizational chart. Management must also concentrate on directing and supervising the day-to-day activities of employees. This involves the total personnel function—recruiting, hiring, training, evaluating, promoting, administering compensation and benefits, firing, and, in some agribusinesses, working with organized labor (see Chapter 21).

This chapter will focus on issues of organization and leadership. The organizational structure may well determine whether or not an agribusiness succeeds or fails. There are many important decisions to make when developing the organization's formal reporting relationships. And, creating a structure that is flexible, responsive to the marketplace, affords employees growth opportunities, and manages accountability and responsibility is a major challenge. Leadership is critical to business success as managers seek to motivate and manage human resources to maximize productivity. Managing people successfully requires more than a charismatic personality; it requires an understanding of the basic concepts of supervision and leadership. We will explore these issues in this chapter, then turn our attention to the more focused area of personnel management in Chapter 21.

GREENTHUMB, INC.: MANAGING PEOPLE IN A GROWING BUSINESS

Marie and Bob Jordan founded GreenThumb, Inc. 6 years ago. Capitalizing on Marie's experience in managing a retail lawn and garden center, Bob's experience in a commercial greenhouse, and some funds from their extended family, Marie and Bob set out to make their dream a reality. Beginning modestly with a small greenhouse and an attached lawn and garden shop, they quickly established their business in the community. Marie's creative ads in the local newspaper, her flair for creative in-store displays and landscape ideas, and her outgoing personality really drove the sales and marketing side of the business. Bob's natural talent with growing plants and his previous experience as a field supervisor for a large corporate greenhouse provided a steady supply of quality bedding plants and flowers for sale in the store.

At first, the business was simple. During the first 2 years, Marie and Bob did just about everything themselves, filling in labor needs with the part-time help of a retired hobby gardener and several high school students. Things were going so well that in the third year they decided to purchase six adjoining acres and begin growing some of their own nursery stock. By the end of their third year, the business was growing so rapidly that Marie and Bob had to begin adding full-time help—they just could not take the long hours required to handle all the work that the expanded business required. First, GreenThumb hired three managers: one each for the nursery, the greenhouse, and the retail store. Several workers reported to each of these managers, the exact number depending on the season. Bob acted as general manager, and Marie assumed the role of assistant general manager. Marie handled everything when Bob was away on buying trips and kept the books and records. In addition, she maintained her role as chief salesperson and marketer for the growing company.

Adjusting to this expanded operation was a genuine struggle for both Marie and Bob. It became harder and harder to stay in touch with customers, and it seemed as if Marie and Bob were no longer able to do some of the things that had been so much fun in the business initially. They seemed to spend their time doing more administrative things—hiring people, meeting with the managers, working on the firm's finances. Less and less time was spent in the greenhouses, or with customers. Yet the business was successful and profits were good, so they continued to expand and grow.

In their fifth year, things began to go wrong. By this time GreenThumb employed about 50 people during the peak season. Recently, a number of troubling problems had required a lot of attention from Bob and Marie. Some of these problems included: inventory damage in the cramped warehouse, high turnover among part-time help, and unexplained escalating costs. Most serious of all was an increase in insect and disease problems in the nursery and greenhouse. While none of these problems were huge, together they were most disturbing—especially since they were leading to some dissatisfied customers, and lagging sales.

Both Marie and Bob felt they had lost touch with the business. Bob simply could not be on top of the technical production problems because he was tied up with management problems. In fact, many of the disease problems they were encountering were beyond his experience and expertise. Similarly, Marie had no formal training in accounting, but she did realize that GreenThumb needed better financial information to make good management decisions. She had spent a lot of time learning the computerized records program they used, but this was just not her area. This became extremely clear when financial statements illustrated that higher costs and lower sales had significantly lowered profits—and neither Bob nor Marie knew exactly why. To make matters worse, all the time with the books kept Marie from putting her marketing and sales skills to work.

After a careful deliberation of the problems faced by GreenThumb, Bob and Marie interviewed and hired an aggressive young woman who had a degree in horticulture and had completed an internship with a large grower. They also hired an accountant who had graduated from a business college 5 years before and had worked as an office manager in a local electronics store. Both Marie and Bob felt that these two new staff members, although relatively costly additions to the payroll, would pay off in improved productivity and financial control within GreenThumb.

This plan worked well for GreenThumb for about a year. But then, early in March of her second year, the horticulturist walked into Bob's office and quit. She said she was fed up with the way things were done at GreenThumb. After a rather emotional discussion, it became clear that her frustrations resulted from a series of incidents where she had strongly recommended several cultural practices to the nursery manager, only to be ignored. Finally, after she discovered that some new stock had been planted too deep and was likely to die, she had taken matters into her own hands. The nursery manager had been away for the morning, and she felt it necessary to handle the problem immediately, so she told two part-timers to spend the morning resetting the stock. When the nursery manager returned and found that the horticulturist had pulled the workers off the job he had assigned, he was livid and confronted her. There had been quite a scene in the field, and several workers seemed to enjoy the very, very heated argument.

GreenThumb has reached a point where it is truly riddled with people problems. The business has been profitable and has potential for considerably more growth. But, Bob and Marie are almost overwhelmed—how did the people side of their thriving business get so out of control? How should they structure the business to address these problems?

THE FORMAL ORGANIZATION

Organizations depend on two kinds of structure to operate efficiently. We first look at the **formal structure** that serves as the foundation for all activities. Then, in the next section, we will look at the informal structure, or the undercurrent of informal interpersonal relationships that also affects how work gets done.

In the U.S. business economy, the owners of a firm provide the firm with the financial resources with which to operate. They also set the general direction for the organization either directly or through duly elected representatives (the board of directors). The owners in turn delegate to managers the authority to make decisions, and management is held accountable for the success of the business. Management then develops an organizational structure specifying the various responsibilities, authority, and accountability of employees. Employees help develop and execute plans for accomplishing business objectives. Of course, in practice things are rarely so simple.

Responsibility, Authority, and Accountability

Responsibility is the obligation to see a task through to completion. It may be a contractual obligation or it may be voluntarily assumed, but it cannot be given away. Responsibility can be shared with another person or group, but it can never be passed downward with no further obligation. The obligation remains undiluted with its originator.

Authority is the right to command or force an action by another. Authority allows instructions to be given to another individual with the expectation that they will be carried out explicitly. Authority is a derivative of responsibility, since it must come from the ultimate source of responsibility.

Accountability involves being answerable to another for performance. Associated with accountability is the notion of a reward for acceptable behavior or a penalty for unacceptable results. Accountability, too, is derived from responsibility.

The formal organizational structure of an agribusiness defines areas of responsibility and authority, and delineates who is accountable to whom and for what. Historically, the larger the agribusiness, the more formalized and structured its organization is likely to be. In fact, in very large businesses it is not uncommon for specialists in organizational development to constantly review the organizational structure, with an eye toward changes that may facilitate the total management process. Today's trend, even among large agribusiness firms, is to flatten this organizational chart in an attempt to facilitate communication and responsibility from top to bottom, and to keep the costs of administration as low as possible.

Principles of Organization

There is no shortage of books, articles, magazines, and websites on "how to be successful" in business management. Many factors can be cited, but perhaps the most far-reaching is the human element: culture, value systems, structures, communications, and personalities. Although organizations are always in a fluid state because they are continually changing, there are several key **principles** that are useful in developing an effective organizational structure.

The **span-of-control principle** states that the number of people who can be supervised effectively by one individual is limited. The maximum number depends on many factors, including the frequency of contacts that must be made, the type of work, the level of subordinates, and the skill of the supervisor. In military organizations, the number of individuals directly supervised is seldom more than four to seven, while on assembly lines where work is routine, a supervisor may oversee 30 to 40 people. Information technologies such as e-mail, cell phones, and the Internet have in general expanded span of control, with managers likely supervising more people than they did a few short years ago.

The **minimal layer principle** states that the number of levels of management should be kept as low as possible, which is consistent with the goal of maintaining an effective span of control. As organizations grow, there is a tendency for the levels of management to proliferate, but each additional level increases the complexity of communications and the opportunities for breakdowns. More recent trends, consistent with a growing span of control, is for organizations to "flatten," taking out layers of management, and giving each layer more responsibility and authority.

The **delegation downward principle** states that authority should be delegated downward to the lowest level at which the decision can be made competently. This allows upper management to concentrate on more important decisions. At the same time, delegation of authority never relieves the delegator of the original responsibility. The supervisor always remains responsible for everything that happens under his or her supervision. This principle has become extremely important as firms have grown, and the market has demanded local responsiveness.

The **parity of responsibility and authority principle** states that a person should have enough authority to carry out assigned responsibilities. It would be totally unfair to hold employees accountable for areas in which they have not been granted enough authority to make the decisions that affect the outcome. While this principle makes good common sense, it is not unusual to see it violated in practice.

The **flexibility principle** states that an organization should maintain its structural flexibility so that it can adjust to its changing internal and external environments. Organizations are always changing. Since the organizational structure involves people, changing the power structure can be a very delicate issue as people often feel threatened by change and tend to resist it. Periodic reviews and a mechanism for changing structure are necessary for healthy organizations. At the same time, changing the organizational structure is highly disruptive and takes the people's focus off day-to-day operations—just ask any food or agribusiness employee who has been through a major merger or acquisition!

Types of Organizational Structures

The foregoing organizational principles give rise to three primary types of organizational structure: line, line and staff, and functional. There are actually many, many variations on these basic types, and a pure organizational type seldom occurs. Generally speaking, the larger the business, the more complex its structure. But, most are designed around these three basic types.

Line Organization   A **line organization** is a structure in which there is one simple, clear line of authority extending downward from top management to each person in the organization. Each subordinate reports to only one person, and everyone in the organization is directly involved in performing functions that are primary to the existence of the business. Figure 20-1 shows an early organizational chart for GreenThumb, an example of a line organization.

GreenThumb, Inc.
Line Organizational Chart
2nd year

Figure 20-1
Line Organization.

The line organization is ideal for smaller agribusinesses, such as Green-Thumb, in their first few years of doing business. In such cases, lines of authority and accountability are simple and cannot be easily confused, communication channels are clear and rapid with few people involved. Top management are often the owners of the business, who find it easy to stay informed and make effective business decisions with good information. Marie and Bob Jordan, as the managers of GreenThumb, Inc., could react quickly and personally to a shoplifting problem in the retail store or a disease problem in the nursery, because of their relative proximity to the problem area.

But as a business grows, it becomes more complex. The more people there are and the more levels of management that are added, the less effective the pure line organization becomes. Lengthening the chain of command reduces the speed of important decisions and increases the probability of communication breakdowns. If an assistant nursery manager is hired and nursery workers are specialized into field hands, delivery persons, and mechanics, the owner/manager may not learn of a feud between two delivery people until one of them quits, unless the lines of communication are working exceptionally well.

Furthermore, there is really no place in a line organization for specialists. As a business grows, it requires a more diversified structure. As GreenThumb grew, it became obvious that an office manager/accountant would be needed, as well as a professional horticulturist to advise on matters of disease control, plant propagation, and field practices. In a line organization such specialists should be in a position to offer advice, but typically are not directly involved in the line of authority over activities in the nursery or greenhouses.

Line and Staff Organization   The **line and staff organization** is a variation of the line organization. The difference is that it includes a place for specialists, sometimes known as staff (Figure 20-2). In this type of organization, staff personnel have direct accountability to key line managers and are responsible for offering advice on problems or providing services in their area of specialization. Typically, these specialists or staffs have no authority except over assistants who may be assigned to them. Their advice can be accepted or rejected by line managers, who retain responsibility for all decisions.

GreenThumb management is likely to find the advice of a staff horticulturist and accountant invaluable in making technical decisions and analyzing operating costs. The highly trained horticulturist can work with the nursery manager and the greenhouse manager, advising them on disease problems in the nursery or propagation problems in the greenhouse, or assisting them with customer problems in the retail store. Under this structure, no one is required to accept all the specialists' suggestions, but professional advice and services can be extremely beneficial to the business. Staff specialists can be advisory, they can handle service problems, and/or they can provide control functions, as do quality inspectors in a processing plant. In any case, they typically do not have line authority over others in their organization, and this is one major drawback of the line and staff organization.

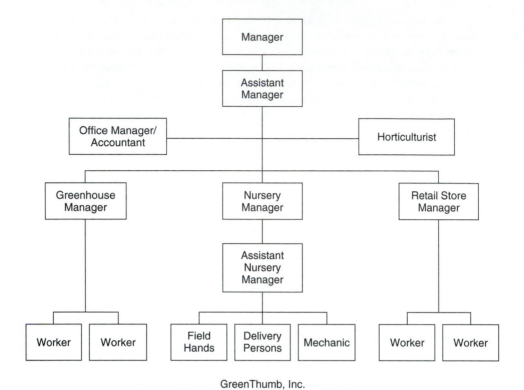

GreenThumb, Inc.
Line and Staff Organizational Chart
5th year

Figure 20-2
Line and Staff Organization.

Staff specialists must be positioned to avoid undermining the authority of line management. This is difficult, since their position and special knowledge often give them considerable status and prestige that can easily be misused. Line managers or workers may not welcome changes or policies that these staff specialists may legitimately suggest. Specialists may feel so strongly about a particular issue that they may apply pressure or go around the normal chain of command. If GreenThumb's horticulturist tells nursery workers to begin treating growing beds with insecticide, such a request may be in direct conflict with the nursery manager's established work schedule, creating much confusion among employees and ill feelings with management.

Functional Organization   A **functional organizational structure** meets the problem of staff specialists' authority head-on by granting them authority in the areas of their specialty (Figure 20-3). The horticulturist who sees the need for an immediate insecticide treatment of bedding plants has the responsibility and the authority to command workers to make the application.

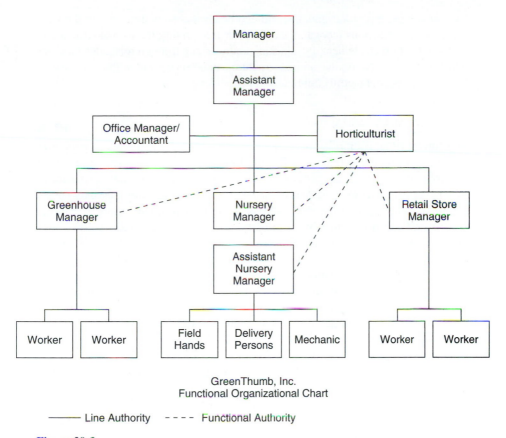

GreenThumb, Inc.
Functional Organizational Chart

——— Line Authority   - - - - Functional Authority

Figure 20-3
Functional Organization.

Of course, a functional organization offers an almost unlimited potential for conflict and confusion. Who has the highest authority? From whom do workers really take orders? The key to making the functional structure work is coordination of staff and line management efforts. A cooperative attitude and good communication are absolutely essential for this organizational structure to work. While complex, this organizational structure, or a variant, is very common in agribusinesses, especially larger ones. Such businesses have found that the advantages of functional organization outweigh the disadvantages.

Communication: The Key to Success   No matter how well thought-out the organizational structure, there will be times when it breaks down. Agribusinesses, large and small, are complex operations that do not always work the way management plans. No organizational structure is better than the people in it. Their understanding of the structure and the employee's willingness to work within it is essential. Yet even when people are trying to make it work, problems arise. People have emotions, misunderstandings, and ego needs that sometimes get in the way. Effective organizations recognize the need for interpreting the formal structure in terms of the human element, by adjusting and working through misunderstandings when they occur. No organizational structure can be successful without a constant focus on honest but tactful communication at all levels.

THE INFORMAL ORGANIZATION

The formal organization is concerned primarily with the activities of people as they perform their job functions. But the formal organizational structure cannot possibly control all of the personal relationships that exist in any agribusiness. Whenever people come together, informal relationships develop that may have a significant impact upon the effectiveness of the formal organization. The **informal organization** is primarily concerned with interpersonal relationships among people, that is, their emotions, feelings, communications, and values. This informal organization is an important part of the organization's culture.

The informal organization is crucial to the success of any firm, since it contributes greatly to the fulfillment of individuals' personal needs. A great many (if not most) professional people and workers spend as many or more hours on the job as they do at home or with their families. And for some, the hours spent on the job may be a primary source of ego fulfillment and social relationships. Their role on the job, their status among peers, and their personal feelings about their job are critical to their well-being; therefore, all these factors directly affect how well the formal organization works. Personal lives of employees can affect an organization's efficiency positively or negatively.

Managers who cultivate a positive informal organization will discover a desirable fringe benefit: the formal organizational structure will be much more productive. Any gaps caused by unexpected situations can be handled far more easily. Communication will be facilitated through informal channels. The span of control can be lengthened because people will work together more effectively. The end result is that more work gets done better as people feel a commitment to the common goal.

On the other hand, where the informal organization is ignored, results are less predictable, people are less flexible, and they may spend considerably more time on activities that are counterproductive to company goals. Managers who ignore the power of the informal organization, who dictate without regard to sensitive interpersonal factors, may even experience "malicious obedience" among employees, that is, a situation where an employee spitefully carries out a superior's command to the nth degree, even when the employee knows that such literalism is not called for in the situation and will produce a negative result.

Informal organization begins with the primary group, or the people with whom an employee works most closely. Within this group, as in the larger organization, relationships develop as the result of status, power, and politics. **Status** is the social rank or position of a person in a group. Status and symbols of status exist in every group. **Symbols** include title, age, experience, physical character-

A positive and vibrant informal organization helps ensure that the formal organization is effective and profitable.

istics, knowledge, physical possessions, authority, location, privileges, acquaintances, and a host of other factors, depending on the situation.

At GreenThumb, Inc., the issues of who gets to drive the newest pickup truck and who has the first locker in the clean-up room have become status symbols among workers. Insensitivity in assigning trucks to work crews could create problems in the organization. Status is not necessarily bad, however, because it helps people fulfill ego needs and can serve as a method by which managers can motivate subordinates.

Power is the ability to control another person's behavior. Power may come from the formal authority issued through the chain of command, or its source may be less formal, arising out of respect, knowledge, status, proximity to a source of power, or control of a resource. Effective agribusiness managers recognize the importance of power—both their own power and that of others in the organization. They can use power productively to accomplish corporate objectives, or they can abuse their power to emphasize their own importance. Managers must recognize that by withholding full cooperation, employees can exert a great deal of unofficial power themselves.

Marie and Bob Jordan must recognize the power plays currently operating between the nursery manager and their staff horticulturist. Both are struggling for power they believe is rightfully theirs. The nursery manager may feel threatened by the horticulturist's higher position and knowledge, or he may undervalue her because of her gender. The Jordans must deal with the informal aspects of the power struggle as well as the structure of the formal organization.

And, they may well find, as many agribusiness managers do, that managing the informal organization is much harder than structuring the formal organization.

Politics is the manner in which power and status are used. Politics involves the manipulation of people and situations to accomplish a particular goal. Everyone is at least superficially familiar with the formal political process of give-and-take, individual and group coalitions, and negotiating. The formal political process in the state or national sense is a complex but systematic application of power and persuasion. Within the agribusiness, however, politics is an intricate part of the informal organization. Although it is likely to be more obvious in larger organizations, politics is necessary for getting things done when one lacks total authority to dictate an action, and can be useful even when authority is present.

The greenhouse manager at GreenThumb understands the role of politics, so he will plan his request for an automatic overhead sprinkler system very carefully. He knows that investment funds are limited and that he will be in direct competition for these funds with the retail store manager and the nursery manager. If the office political situation is such that the accountant can and does influence the Jordans' investment decisions, the greenhouse manager may well cultivate a relationship with the accountant and subtly look for ways of gaining this individual's support.

Politics has positive value: it encourages compromise and offers a way of solving problems without direct, explosive confrontation. In larger organizations, politics offers a method for accomplishing things that would be cumbersome to handle through purely formal channels. While politics may seem manipulative or even unethical to some, it is a fact of life and influential to some degree in all organizations. Effective managers recognize its existence and deal with it realistically.

Leadership

Managers are designated leaders, formally appointed by the chain of command that originates with ownership. But leadership is much more than issuing authoritarian commands. It requires working with personnel and motivating them to accomplish the firm's objectives. **Leadership** is in part the ability to combine personal and organizational goals in correct proportions.

Styles of Leadership

Leadership and leadership styles is a heavily studied area. Obviously, business could profit greatly from a consistent way of identifying good leaders, or of teaching selected individuals how to be effective leaders. Although a great deal is known about leadership, there are no consistent or simple answers to the question, "what makes a good leader?" What makes a good leader in one situation does not necessarily work in another. Some research suggests that a strong will, extroversion, power need, and achievement are key variables in a leader. Other work identifies intelligence, social maturity, breadth of development, inner motivation, and a positive human relations attitude as necessary ingredients to leadership.

All these factors are subjective and difficult to measure. The conclusion is that leadership is not static but dynamic, and must adapt to the specific circumstances of the situation. However, there are a number of leadership styles that have been identified, and understanding these styles can help in assessing what type of leadership may be most appropriate in a given situation. One popular classification scheme describes managers as authoritarian, democratic, free rein, and transformational.

Authoritarian

Authoritarian, also called **autocratic,** is a leader-centered style, in which the thoughts, ideas, and wishes of the leader are expected to be obeyed completely, without question. Authoritarian leaders seldom consult subordinates before making decisions. Also, decisions may be changed suddenly for no apparent reason. The hard-nosed autocrat is coercive in relationships with subordinates, and often threatens them if they do not perform to the expectation level of the autocrat.

Yet there are other styles of authoritarian leadership. **Benevolent autocrats** convince followers to do what they want by being so well liked that no one would consider being disloyal or of "letting the chief down." The benevolent autocrat gives so much praise that employees are shamed into obedience. Another form of autocratic style is that of the **manipulative autocrat,** who creates the illusion that employees are participating in the decision-making process. This leader's motto is "Make them think they thought of it."

Note that in each autocrat type outlined above, all decisions originate with the autocrat, and the autocrat maintains firm control. The only difference among the three is the manner in which the control is exercised. Some argue that the autocratic style of leadership is always bad and leads to poor performance and ill feeling, but this is not so. Although the authoritarian style would seldom be the only method of operation for any given leader, those who are primarily autocratic can show excellent results, particularly when the leader has a good feel for the situation. Over a period of time, autocratic leaders tend to surround themselves with subordinates who enjoy not having responsibility; they just like to "follow orders." This combination of leadership and following is often quite productive, at least temporarily. However, it is limited completely by the individual autocrat's ability and can result in utter chaos if that leader becomes incapacitated, or if the situation changes in a way that does not play to the strengths of the autocrat. The autocratic style is common and is most likely seen in smaller, owner-managed businesses.

Democratic

The **democratic** or participative leadership style favors a shared decision-making process, with the leader maintaining the ultimate responsibility for decisions while actively seeking significant input from followers. Research shows that while this style may require a considerable amount of skill, it stimulates employees' involvement and enhances favorable employee attitudes toward their jobs. The only real disadvantage is that a democratic leadership style requires management skills and time that may not always be available—so this leadership style may not be appropriate for all situations.

Free Rein

Free rein or **laissez-faire** leadership literally relinquishes all decision making to followers. The leader essentially abdicates his or her responsibility to the group and simply joins the group as an equal; thus, the group decides what to do. Although free rein leadership may work with some decisions, it seldom leads to timely or consistently good decisions, and often results in a poor outcome and frustration among employees. This leadership style is highly dependent on the quality of the employee team, and it takes a very capable group of individuals to make this work.

Transformational

As discussed in Chapter 2, managers' jobs include planning, organizing, directing, and controlling. And, as the agribusiness markets have evolved over time, it has become increasingly clear that a new type of leader is required to accomplish these tasks in an environment of rapid change and intense competition. One new style of leadership, **transformational leadership,** has been identified as particularly helpful in guiding organizations and employees through such changes. Transformational leadership, as defined by Bernard Bass in his book *Leadership Performance Beyond Expectation,* is a form of leadership that motivates followers to work for transcendental goals and for aroused higher-level needs of self-actualization rather than for immediate self-interest. Essentially, transformational leaders (TLs) motivate individuals in an organization to proactively work for a "transformation" in how the company does some aspect or aspects of its business.

Transformational leadership involves three steps:

1. Recognizing the changes in the market
2. Creating a vision of the firm in this new market
3. Institutionalizing the change

Transformational leaders are usually found in a chaotic marketplace. At a time of upheaval, it is transformational leaders who in the midst of struggle can identify those issues that are most critical. They create a vision for dealing with the situation, usually by "thinking out of the box," that is, thinking beyond the normal ways in which such situations are handled. But it doesn't stop there. TLs follow through by making the change a reality. They foster a company culture that understands the need for change and embraces the vision.

TLs have the ability to influence major changes in organizations by influencing changes in followers' attitudes and assumptions. They typically exhibit some combination of two or more of the following four characteristics: idealized influence, inspirational motivation, intellectual stimulation, and individualized consideration.

Idealized Influence TLs often act as role models. They have a high degree of self-confidence and strong conviction to their own ideas and beliefs. They consider others' needs, share risk and are consistent and responsible, and demonstrate accountability.

Inspirational Motivation   TLs inspire followers to exert themselves beyond what is expected. They are good at generating enthusiasm, team spirit, and optimism by demonstrating self-determination and commitment.

Intellectual Stimulation   TLs encourage creativity by acting as teachers. They offer a breath of fresh air by analyzing old problems in new ways and by using intelligence and reasoning instead of unsupported opinions. They encourage followers to think about actions before taking action.

Individualized Consideration   TLs give each follower individual attention. They find a follower's individual talent, offer criticism and provide opportunities for learning. This one-on-one coach or mentor notion makes followers feel accepted, valued, secure, and may help them become a more loyal follower.

Agribusinesses continue to face changes and challenges in the ever-more competitive business climate. As a result, today's agribusiness managers may have even more opportunities in the future to apply the principles of transformational leadership—or some of its components—in their organizations.

Beyond these four basic styles of leadership, there are a host of refinements and behavioral patterns that can, when studied in detail, contribute greatly to leadership effectiveness. Some of these theories provide classic insights into human behavior and managerial effectiveness. We will review two here: Theory X and Theory Y, and management by objectives (MBO).

Theory X, Theory Y

Douglas McGregor has suggested a dual theory about human behavior in business that has important implications for management style and methods of motivation. McGregor based his theory on what has come to be known as the "dual nature of humankind." For centuries, philosophers have noted that people have the capacity for love, warmth, kindness, and sympathy, but at the same time can exhibit hate, harshness, and cruelty. This dichotomy led McGregor to an explanation of management style that is based on assumptions managers make about the people they supervise. McGregor referred to his dual theories as **Theory X** and **Theory Y.**

Theory X   McGregor believed that traditional management practices and methods are based on the following set of assumptions about employees:

1. Most employees have an inherent dislike for work and will avoid it if at all possible.
2. People will work only when they are coerced, threatened, or at least controlled in all their activities.
3. Most people actually prefer to be closely controlled because they dislike responsibility and have little ambition.

4. Most people are basically self-centered and selfish.
5. Security is very important to most employees, and they are threatened by change.
6. Most employees are gullible, will believe anything, and are not very intelligent.

McGregor believed that for centuries these assumptions had served as the basis for most management and leadership styles: strong control with little concern for employee needs. Most of the supervisory practices resulting from Theory X assumptions have failed to tap the human resource potential in the employees being supervised.

Theory Y   At the other end of the continuum is another set of assumptions derived from a belief in the natural goodness and creativity of human beings. Theory Y assumptions include the following:

1. Work, both physical and mental, is as natural as play or rest.
2. People can and will exercise self-control and evidence motivation to achieve goals to which they are personally committed.
3. The level of commitment is dependent on the rewards received for reaching the goals.
4. People basically like responsibility and will seek it.
5. People are naturally creative and have far more capability than is generally utilized.

Theory Y assumptions lead to a very different management and leadership style. Theory Y managers focus attention on employees themselves, attempt to draw out their creativity, and appeal to their willingness to accept responsibility. These managers do not abdicate their responsibility and authority to their employees. They are quick to enter into a dialogue that will help the manager and employee arrive at mutually acceptable goals, and to establish a reward system that is fulfilling to the employee. Theory Y management is not an easy style. It requires patience and a great deal of people skill on the part of managers.

Obviously, there is much appeal to the Theory Y management approach, and McGregor was a strong proponent of it, especially early in his career. If one fully accepted these appealing assumptions, there would be little need for time clocks, fixed working schedules, production quotas, and the like. Most organizations have made a strong attempt over the past two decades to move much closer to a Theory Y style of management, though there are still holdovers.

However, it should not be assumed automatically that Theory X management is bad and Theory Y management is good. It is said that McGregor used the Theory X, Theory Y terminology interiorly so as not to imply superiority of one style, even though he preferred the Theory Y approach. In fact, McGregor himself softened his strong support of the Theory Y position later in his career when he became a university administrator and personally experienced some of the frustrations associated with applying strong Theory Y management assumptions. Some employees hold Theory X assumptions about themselves and so perhaps prefer to be treated in a manner consistent with this belief. But clearly there is great merit in the Theory Y approach, and agribusinesses have generally moved in this direction.

Management by Objectives

MBO (management by objectives) concentrates on results rather than on the process by which they are achieved. It is a system of management whereby the supervisor and the subordinate jointly:

1. Set objectives that both agree are reasonable and in line with corporate goals.
2. Determine how performance will be measured in each area of major responsibility.

First introduced by Peter Drucker in 1954, the MBO process usually begins with a very broad set of objectives and sense of direction as established by the board of directors. Top management develops its own objectives, consistent with the board's framework, and submits them to the board for reaction. Top management then asks middle management to develop and submit goals that are consistent with overall corporate goals as interpreted and applied to the middle managers' respective departments. Then top management and middle management coordinate these goals through a negotiating process until an acceptable set of goals and yardsticks to measure progress have been established. Next, middle management works with lower management, and so on down the line until the process has been repeated throughout the organization.

There are obvious advantages to such a system. First, since it accords individuals a certain amount of input into their own goals, employees tend to become emotionally committed to accomplishing whatever they have targeted for themselves. Second, with MBO, the criteria for evaluation become crystal clear and are agreed on by all concerned, which eliminates many problems later. It is important that objectives be clearly established and clearly measurable in the MBO process. As manager and subordinate meet for evaluation, judging whether or not objectives have been met should not be subjective.

The process is also quite time-consuming, however. Often, proposed objectives are submitted, negotiated, and resubmitted repeatedly before agreement is reached. Although the MBO system is intended to enhance communication, its success is built on assumptions of preexisting communication and mutual trust. If subordinates intentionally establish goals that are unrealistically low, so that they can exceed their goals easily and look good, the entire system breaks down. Additionally, there must be commitment to the system from the top down. Whenever top management resorts to dictating all the objectives, others learn quickly what the game is and play accordingly. Yet the appeal of an effective MBO system is strong and if done properly, it can be highly successful.

MOTIVATION

Motivation is a stimulus that produces action, and directed action is a primary function of management. This is why managers are so concerned with the concept of motivation. There have been many arguments about whether it is possible for a manager to actually motivate employees. Some say that a good manager is skillful in stimulating others to accomplish targeted objectives. But others argue that all motivation is really self-motivation, which comes from within each individual. Those holding this view say that managers affect the environment and provide the stimulus, but it is ultimately up to the employee to decide to act. No matter which point of view seems more logical, management is responsible for results and must somehow stimulate, encourage, or coerce the employee behavior necessary to accomplish organizational goals and objectives, or else it must create an environment that will produce the same effect.

Motivation has many dimensions. It can be negative, as when an employee's attitude results in activity that undermines company goals, or it can be so positive that the employee becomes emotionally and personally involved in completing assignments at the highest level of their capabilities. Employees can be stimulated by positive rewards or by fear of undesirable treatment or lost privileges.

Like leadership, motivation has been studied extensively. As a result, there are a number of theories that have been suggested to explain how motivation works within individuals and organizations. Most researchers agree that in some way motivation comes down to a matter of rewards and punishments. By controlling rewards and punishments, management can significantly affect employee performance.

The basic problem comes in trying to determine what rewards and punishments work in which circumstances. The situation is further complicated by the wide variation in people. A simple "thank you" may be motivating to one person, while a new title may be of paramount importance to another. Some employees place high priority on monetary rewards, others on getting challenging assignments, while still others value perks like company cars and expense accounts. This very, very wide range in what employees value demonstrates the scope of the problem in developing a successful motivation system within the agribusiness.

A brief overview of some widely used theories of motivation provides useful ideas for managing and motivating people in agribusinesses. We will explore two here: Maslow's needs hierarchy, and Herzberg's motivators and hygienic factors.

Maslow's Needs Hierarchy

Motivation focuses attention on the personal needs of human beings, so some understanding of human needs is necessary for effective supervision. Abraham H. Maslow developed one widely used model of human needs. **Maslow's needs hierarchy** is based on the idea that different kinds of needs have different levels of importance to individuals, according to the individual's current level of satis-

Figure 20-4
Maslow's Needs Hierarchy.

faction (Figure 20-4). Usually the needs of one level must be met before the next level becomes a motivating force. Needs basic to human survival take priority over other needs, but only until survival has been assured. After that point, other needs form the basis for the individual's behavior.

Survival   Maslow suggested that every human's most basic concern is physical survival: food, water, air, warmth, shelter, and so forth. Obviously, humans cannot live unless these needs are at least minimally fulfilled. These most basic needs are immediate and current. It takes only moments for a person gasping for breath to be more concerned about survival than about anything else in the world. Survival here and now is even more critical than survival over the days or weeks ahead. For example, the call of "fire" prompts immediate action—notions such as financial concerns or etiquette are quickly forgotten.

Safety   Once immediate survival has been assured, humans become concerned about the security of their future physical survival. Today, this concern often takes the form of income guarantees, insurance, retirement programs, and the like. In a down agricultural economy, farmers will likely buy more conservatively than they would in more prosperous times. They're less likely to be innovators and more concerned about getting a "tried-and-true" product at the lowest cost. In such an environment, safety is a much more important motivator than the prospect for larger financial rewards.

Belongingness   After they have achieved a reasonable degree of confidence in their safety, people become concerned with their social acceptance and belonging. Social approval and peer acceptance can cause a great deal of self-applied pressure and can be the source of strong motivation. If others are shifting to a new product or practice, farmers may be motivated to try the new idea to "keep up" with their neighboring producers. Likewise, food consumers may be interested in trying out the latest energy drink as they "keep up" with other members of the health club where they work out.

Ego Status   With a comfortable degree of social acceptance, most individuals become concerned with their self-esteem or status relative to their group. Group respect and the need to feel important depend heavily on the responses of other group members. Recognition from superiors and peers is a compelling drive for many people and may be responsible for the upward movement of a great many successful people. When salespeople help customers excel, they're appealing to their self-esteem or egos. For example, a livestock or poultry producer may be motivated to have the top rate of gain, or lowest disease incidence, or fewest days to market. An incentive/recognition program such as "Salesperson of the Year" may work well for salespersons in a food manufacturing firm who are highly motivated to be the very best in the region.

Self-Actualization   The highest level of need, and one that becomes important only when lower-level needs have been relatively well satisfied, is **self-actualization,** the feeling of self-worth or personal accomplishment. This category of needs is highly abstract and takes a multitude of forms among different individuals. Self-actualization may be achieved through creative activities such as art, music, helping others in community activities, or building a business. The results of self-actualization embrace an attitude of "I feel I've made an important and worthwhile contribution."

In our society, most people's survival and safety needs are relatively well taken care of, so these are seldom strong motivators. Only when these needs are threatened do they become much of an issue. Although employees may feel good about a retirement plan, retirement benefits are seldom a factor in motivating them to higher levels of productivity. Even when job security becomes the focal point of labor negotiations, few would argue that such factors motivate employees to a very large degree, since they are becoming expected as normal and reasonable.

Belongingness or group acceptance can exert somewhat more pressure on people. New line employees may seek out belongingness as they try to fit in with other workers on their shift. The dynamics of the group will determine whether or not this attempt to fit in results in positive or negative outcomes from the firm's standpoint. Supervisors must recognize and encourage peer group acceptance and manager acceptance as important to employee performance.

Ego status or recognition is one of the most common and most productive needs through which agribusiness managers supervise and motivate employees.

Serving as a consultant or advisor to customers can help fulfill the need for self-actualization.

Photo courtesy of USDA.

A great many people seem to be responsive to ego need fulfillment. Nearly everyone feels they want to be important and recognized. Managers who give employees recognition frequently, honestly, and positively are often rewarded with highly motivated personnel. Such recognition can take many forms. The simplest kind may be verbal: compliments are very effective when they are honest. They can be nonverbal: simply a smile or nod of approval. Listening is also very powerful. Caring enough to listen to subordinates and to consider their opinions and feelings is one of the most powerful ways to recognize employees. Every successful manager has his or her own way of recognizing people effectively, but no matter what the technique, it is not accidental. It is an important management activity that relates directly to firm productivity.

Self-actualization is not quite as easy to deal with in a management context because it comes from within the individual. However, this need can be a very important one for individuals who have "done it all." Examples here might be the department manager who has demonstrated exceptional performance year after year, and is now looking for the next challenge; or, the veteran salesperson who has won every major sales award and is looking for what's next. Management can best address self-actualization by:

1. Recognizing that some employees will be motivated by their own sense of self-accomplishment
2. Allowing this to happen
3. Encouraging the proper environment for it, whenever possible

Herzberg's Motivators and Hygienic Factors Theory

A popular but somewhat controversial theory of employee motivation is one developed by Frederick Herzberg. **Herzberg's motivators and hygienic factors theory** argues that two separate factors explain employees' level of motivation. Each of these sets of factors is a separate continuum that affects employees' attitudes and performance. One class of these factors is called hygienic factors; the other is called motivators.

Hygienic Factors   **Hygienic factors,** or pain relievers, are conditions necessary to maintain an employee's social, mental, and physical health. Research suggests that there are several such factors, including company benefits and policies, working conditions, job security, supervision, and pay. Herzberg argues that the presence of these factors does not make a worker happy, but if they are lacking, the employee is likely to become unhappy. That's why hygienic factors are called pain relievers (Table 20-1). On a continuum, hygienic factors range from dissatisfaction to no dissatisfaction.

These factors in and of themselves do not motivate people, but they are part of the environment, and when absent they can cause considerable employee discontent. For this reason, hygienic factors are sometimes referred to as demotivators. Low pay, constant fear of losing one's job, or unappealing working conditions can, over a period of time, lead to disgruntled employees and poor performance, especially when these conditions are worse than those of another peer group.

Motivators   **Motivators** are reward producers, or conditions that encourage employees to apply themselves, mentally and physically, more productively to their jobs. Motivators bring about commitment to the task. Most motivators are predominantly psychological, such as recognition, advancement, responsibility, challenging work, and the opportunity for further growth. The job itself becomes the major source of motivation. On a continuum, as shown in Table 20-2 these factors range from no satisfaction to satisfaction.

T A B L E **20-1** | **Continuum for Hygienic Factors**

Hygienic Factors

Dissatisfaction <————————————————> No Dissatisfaction

T A B L E **20-2** | **Continuum for Motivators**

Motivators

No Satisfaction <————————————————> Satisfaction

One of the controversies surrounding Herzberg's theory is whether compensation is a hygienic factor or a motivator. Many agribusinesses use compensation as an incentive to motivate employees to higher levels of performance. Proponents of such plans feel strongly that both salespeople on commission and field or line workers on wages tied to productivity are highly motivated to maximize their performance.

Yet Herzberg argues that these payment techniques are hygienic because they do not have the effect of increasing commitment to the job. He says that pay incentives may work well temporarily, but once a particular pay and performance level has been reached, and the employee adapts to it, the pay itself is no longer a reason for productivity. In fact, if something happens to cut the pay level, the employee is likely to become frustrated and dissatisfied. Salaried employees seem particularly susceptible to this phenomenon. Although they may work harder to achieve a pay raise, soon after it has been accomplished, their standard of living and expectations move up to form a new base point of normalcy: at the same time, their performance often regresses to a more comfortable level.

Although there is controversy surrounding Herzberg's idea that money is not truly a motivator, there is much to be gained from his theory. The message is that agribusiness managers should work to *enrich* the job. He suggests doing this through providing constant feedback, variety in tasks, and a means of making the job more important. This, he feels, will help the employee become more enthusiastic about the work.

TRANSACTIONAL ANALYSIS

Managing people is centered on managing relationships. And, personal interactions in the workplace often become much like those of the interactions with one's family. One practical approach to understanding and supervising employees is Dr. Eric Berne's concept of **transactional analysis (TA)**. Transactional analysis was developed as a therapeutic tool, based on the concepts of psychotherapy. However, Berne's concepts have been applied widely to management and offer a useful, simple explanation of communication and behavior in business dealings as well as personal situations.

P-A-C Model

Transactional analysis (TA) divides human behavior into three classes of ego states: parent, adult, and child (P-A-C). These ego states have nothing to do with age, but are simply descriptive of various types of feelings, thoughts, and behavioral patterns exhibited by individuals. In any situation, a person assumes one of these three ego states as either an automatic response based on past experiences, or a conscious response of controlled behavior.

Parent Ego State   The **parent ego,** sometimes referred to as the learned concepts of life, represents the values, attitudes, and behavioral patterns established early in life from authority figures, primarily parents (Figure 20-5). The parent

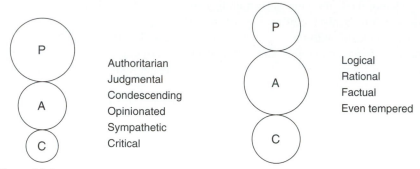

Figure 20-5
Strong Parent Ego Portrait.

Figure 20-6
Strong Adult Ego Portrait.

provides a role model from which the young child learns values and responses that usually serve as a basis of behavior throughout life. Three categories of the parent ego state that are important to management include the prejudiced parent, the nurturing parent, and the critical parent states.

The **prejudiced parent** represents unfounded attitudes and opinions that are held without any logical basis or analysis. Prejudices may be either positive or negative, but in either case, they can drastically affect behavior. The **nurturing parent** represents attitudes and behavior that nurture and take care of others, just as parents offer sympathy, understanding, and warmth to their children. Nurturing behavior is common and appropriate to the supervisory role at times. In a very real way, managers must spend time consoling, encouraging, and listening to employees, just as a parent does with a child. Although nurturing may be difficult for some individuals, it can be a powerful and effective way for a supervisor to relate to subordinates. The **critical parent** refers to a stance involving critical attitudes and behavior. It is also thought to be learned from authority figures early in life. The manner by which one expresses critical feelings—tone of voice, gestures, posture, facial expressions, and words—is often similar to one's parents' behavior under similar conditions. Part of a supervisor's role includes reprimanding subordinates when necessary, which is purely the critical parent role.

Adult Ego State   The **adult ego state** is often referred to as one's personal computer, because it focuses on logical thinking based on facts, like a computer (Figure 20-6). In the adult state, a person analyzes, estimates probabilities, computes dispassionately, and makes decisions very objectively. These decisions are not based on emotions or tradition, although both of these may be considered. Successful managers spend considerable professional time in their adult ego states.

Child Ego State   The **child ego state** embodies people's feelings and emotions, or related childlike behavioral patterns. Nearly everything that happens in one's personal and professional life is filtered through a screen of emotions that significantly affects the resulting behavior (Figure 20-7).

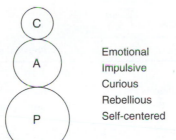

Emotional
Impulsive
Curious
Rebellious
Self-centered

Figure 20-7
Strong Child Ego Portrait.

The **natural child** is the first and most significant subset of child ego state patterns. The natural child represents the full gamut of emotions and behavioral patterns that are natural even to very young children, such as happiness, sadness, anger, and fear. When a grown person is caught up in any of these basic emotions, the resulting behavioral patterns are likely to express these feelings in much the same manner as when that person was young. Managers who recognize the role of the natural child can adjust their management styles to better cope with the circumstance.

The **little professor** is a special set of childlike behavioral patterns that is especially important to managers. First, the little professor represents the natural creativity that seems to dwell in all people. Unfortunately, parents and other authority figures often suppress this creativity in others by forcing them to fit into the same mold, and to think very much alike. But there does exist in all people a tendency for creativity, if it can be released.

The **adapted child** represents that time in children's early years when they did exactly what they were supposed to do. The adapted child suppresses natural tendencies and follows authority figures' wishes without question. Naturally, such people are easy to supervise, but they seldom originate new ideas or handle things on their own initiative.

Recognizing Ego States

Managers who can recognize ego states should be able to exercise appropriate supervisory action. Although the parent, adult, and child ego states cannot always be precisely defined or identified, many behavioral clues will provide good indications of ego state. Managers can learn to watch for these clues and adjust their responses accordingly. Clues include frequently used words and phrases, tone of voice, facial expressions, gestures, and posture (Table 20-3). One clue does not generally give a strong indication of ego state, but a cluster of consistent clues is often meaningful, and a good basis for choosing management action.

When an employee is clearly acting from a child ego state, her or his behavior is somewhat predictable, especially if the manager has the advantage of past experience with the employee. A worker who is angry about a minor incident is not likely to be thinking rationally, but instead is emoting. The manager has several choices: to nurture the worker, listen to the worker's problem with a sympathetic

TABLE **20-3** | Ego State Clues

Words and phrases	Tone of voice	Physical expression
Parent		
You should, ought, must,	Condescending	Hands on hips
You can't . . .	Consoling	Drumming fingers
Don't . . .	Critical	Pointing finger
Never, always	Sympathetic	Looking down nose
Maybe	Judgmental	Arms folded
Because I said so	Worried	Frowning
If you ask me . . .		
Adult		
How many?	Even	Eye contact
When, where, why?	Interested	Level, alert
It's my opinion that . . .	Intent	Leaning forward
It seems to me . . .	Natural	Not tense
	Moderated	
Child		
Wow! Gosh!	Expressive	Nervous
Do I have to?	Excited	Frightened
Why me? I won't!	Happy	Nail biting
No!	Sad	Active
Please, can I?	Angry	Showing off
Swearing	Whining	Spontaneous
Look at me	Defiant	
Well, OK . . .	Curious	
I wish	Compliant	
I hope	Impulsive	

ear, and calm the worker; or to be very firm, and order the employee to "shape up" or be fired. The manager could try to tease the employee out of a bad mood, or might try to "hook" the worker's adult ego state by asking questions that require logical reasoning to answer. No one solution is best—it depends on the situation, the individuals, and the history of the problem. But the key is for the manager to recognize that there are alternatives and to choose the style of supervision that has the best chance of working.

Transactions Analyzed

A **transaction** is anything, verbal or nonverbal, that happens between two people. It involves both people's ego states. Analyzing transactions between people can be a powerful tool for managers who recognize that communications are an important part of successful supervision. Whenever a person speaks, the communication originates from a specific ego state, and it is directed to a specific ego state in the other person. By understanding their ego state, and that of the employee, the manager has a much better chance of handling the situation or the transaction in the appropriate fashion.

EMPLOYEE RECOGNITION AND MOTIVATION

Everyone needs to be noticed. Agribusiness managers will find that effective communication with employees regarding their particular situation, a current project, or their state of being can be a powerful management tool. Effectively addressing individuals as individuals can greatly enhance motivation and performance.

Communication

Effective communication is critical to good supervision. It is the manager's responsibility, not the subordinate's, to ensure good communication. Although one would hope that the subordinate would try to communicate objectively, the supervisor has no assurance of that, and so must assume that responsibility personally. Approaching communication from the adult perspective can help the supervisor avoid getting hooked and losing control of the situation. Professionals need to maintain control.

Strokes

Every person has the need to feel valued and recognized by others. Any recognition of another person's presence is known as a **stroke.** It could be a word, simply "hello." It could be a gesture—a smile or nod. It could be a touch—a handshake or a pat on the back. A stroke is anything that says, "I know you're there."

Strokes are necessary for growth and survival. Some people need many strokes in order to feel socially secure. For a great number of people, a primary source of strokes is their job. They work hard and do a good job to receive strokes—both from their peers and their supervisor. In fact, hunger for strokes probably affects everyone's behavior on the job to some extent.

Positive Strokes   **Positive strokes** are any form of recognition that leaves another person feeling good, alive, and significant. Friendships and close relationships are often built around positive strokes. People like those who stroke them, who give them attention and care enough to listen. Positive strokes give real meaning to life for most people, and there are many kinds of positive strokes.

Positive Unconditional Strokes   **Positive unconditional strokes** convey the spoken or implied message, "I like you . . . you're OK with me." There are no conditions to the acceptance. This is thought to be the most meaningful kind of stroke; it is common among family members and close friends, but is frequently found in the business world as well. It is reflected in all kinds of behavior and conversation. Unconditional strokes may be affectionate or complimentary:

"Good morning!"

"I like that shirt!"

"I just don't know how we'd get along without you."

"You really know how to close a sale."

"Come join me for coffee."

"Your decision was very well thought-out."

Note that unconditional strokes enhance people's feelings about themselves. If the strokes are honest, genuine, and not overdone, they are of great benefit to a person's ego. Unconditional strokes cannot really be considered a management tool, since there cannot be any attempt to manipulate another's behavior with them. These are given freely, without strings. Yet managers who easily give recognition and show genuine approval fulfill important needs of subordinates, and the result is often positive motivation.

Positive Conditional Strokes **Positive conditional strokes** are recognitions given to intentionally modify another's behavior. Since it is management's job to motivate and guide the behavior of employees, positive conditional strokes are widely and appropriately used as a management tool. Conditional strokes essentially say, "You're OK, if . . . ." Acceptance of the other person is conditional on some expected behavior.

"If you keep that up, you'll get that bonus we discussed."

"I'm impressed when you work overtime."

"You look good today, for a change."

"If you make $1 million in sales, you'll receive a trip to Aruba."

Though they qualify acceptance with an "if," positive conditional strokes are not unethical—quite the contrary. They are a very popular and effective management and motivational tool—positive conditional strokes can help people feel good about themselves.

Negative Strokes Negative forms of recognition also give attention and status to people. **Negative strokes** do not make people feel as good as positive strokes, but at least the person is getting some attention, and that is usually preferable to being ignored. Negative strokes most often come in the form of verbal criticism and reprimands. They are meant to squelch and usually to hurt another person:

"What's the matter with you?"

"That's the trouble with this next generation."

"Can't you do anything right?"

"OK, I'll just do it for you!"

Negative strokes are an effective way of getting attention. People who are ignored sometimes learn that inappropriate behavior can get them attention much faster and more predictably than any other type of action.

It is difficult to manipulate others into giving positive strokes, but it is easy to get negative strokes. As strange as it may sound, some workers may be so accustomed to negative strokes, that negative strokes are the only thing they know how to get or how to handle. They may subconsciously create circumstances in which they get much-needed attention. It is possible that a person who is habitually late for work is looking for strokes. Positive strokes are usually considered a more effective management technique than negative strokes. Clearly, however, recognition of employees in general is an extremely powerful form of motivating and supervising.

SUMMARY

Managing human resources is a fundamental responsibility of agribusiness managers. Managers must develop an organizational structure in which the responsibilities, authority, and accountability of individuals are clearly defined. Management must then direct and supervise daily activities, leading and motivating employees to maximize productivity.

Many agribusinesses use a formal organizational chart to clarify the responsibilities, authority, and accountability of employees. Line organizations are structures in which everyone is in the chain of command and has direct responsibility for the primary functions of the business. In the line and staff organizational structure, specialists without authority are added to advise line managers. Finally, in the functional organizational structure, specialists and advisors are given the authority to implement ideas in their area of special responsibilities.

Leadership is a challenging task for most agribusiness managers. There are many different styles of leadership, ranging from authoritarian to democratic to free rein to transformational. Management theorists have developed many models and theories to explain effective leadership. McGregor's Theory X and Theory Y and Drucker's management by objectives are examples of such models that have helped agribusiness managers become more effective.

Motivating means stimulating employees to act in a specific way. Maslow explained people's basic needs as a hierarchy, and suggested that fulfillment of these basic needs is what motivates people. Herzberg pointed out that while some factors motivate people, other factors, such as money, become part of people's expectations. Such factors, if not maintained at expected levels, can lead to dissatisfaction.

Transactional analysis is one of many models for understanding employees and what motivates them. It offers a simple language and framework for discussing various leadership problems. Transactional analysis is a powerful tool that helps managers understand employees, but it should be considered only one additional tool. Managers need to examine many such tools to find concepts that may be helpful to them.

In the end, there are no fixed formulas or precise answers on the best way to manage people. Management is a complex process based on the individual characteristics of the leader, subordinates, and the situation. But there are many approaches and theories that can help the agribusiness manager develop a leadership style that results in improved firm performance, content employees, and a satisfied management team.

DISCUSSION QUESTIONS

1. The principle of parity of responsibility and authority is often violated in business. What are the likely consequences of violating this principle? Have you personally experienced this situation? Explain.
2. In which type of formal organization would you prefer to work? Why?
3. Identify power sources and status symbols in an organization in which you have worked or with which you have some familiarity. What impact, good or bad, did each have on the organization?
4. Suppose you start a new job. Under which leadership style would you wish to work? Why?
5. What is your own feeling about Theory X, Theory Y? Where on this continuum would you likely place yourself? Which type of leader do you believe is more effective, and why?
6. List at least five factors that you feel would motivate you on the job. List five factors that you consider hygienic. Do you think money is a motivator? Why or why not?
7. Briefly explain the purpose and use of transactional analysis.
8. What are some common ways to identify an individual's ego state? Give examples using each ego state.
9. How might an agribusiness manager use strokes to have a positive effect on employees?
10. Sometimes people set others up to give them negative strokes. How can this make sense? What can a manager do about the situation? Why would anyone seek negative strokes? If employees expect negative strokes, what might this say about the firm?

Case Study — **GreenThumb, Inc.**

Review the case of GreenThumb, Inc., the business started by Marie and Bob Jordan discussed in the chapter, then answer the following questions.

Questions
1. How do you recommend GreenThumb be reorganized, and why?
2. Why have the current problems at GreenThumb surfaced? Could they have been avoided? How?
3. What are some of the possible sources of power in the informal organization at GreenThumb?
4. How does GreenThumb demonstrate the increasing problems of internal communication as a business grows?

References and Additional Readings

Bass, Bernard. 1995. *Leadership Performance Beyond Expectation.* New York: Free Press.

Drucker, Peter F. 1954. *The Practice of Management.* New York: Harper and Row.

Drucker, Peter F. November 20, 1990. "Marketing 101 for a Fast-Changing Decade." *The Wall Street Journal.* p. 19A.

Herzberg, Frederick, Bernard Mausner, and Barbara Snyderman. 1959. *The Motivation to Work.* New York: Wiley.

Maslow, Abraham H. 1970. *Motivation and Personality.* New York: Harper and Row.

McGregor, Douglas. 1960. *The Human Side of Enterprise.* New York: McGraw-Hill.

Megginson, Leon C., Donald C. Mosley, and Paul H. Pietri, Jr. 1992. *Management: Concepts and Applications.* New York: HarperCollins. pp. 434–437.

Robbins, Stephen P., and Mary Coulter. 1996. *Management.* 6th edition. New Jersey: Prentice-Hall. p. 534.

Tichy, Noel M. 1986. *The Transformational Leader.* New York: Wiley.

Yukl, Gary A. 1986. *Leadership in Organizations.* New Jersey: Prentice Hall.

21

MANAGING HUMAN RESOURCES IN AGRIBUSINESS

- Explain how human resource management and managing employee motivation in agribusiness are related
- Discuss the functions of human resource management in agribusiness
- Understand the difference between job specifications and a job description
- Describe recruitment methods and the process of employee selection in agribusiness
- Explain the major steps of job orientation in an agribusiness
- Identify the important criteria that determine employee compensation and fringe benefits
- Discuss the performance evaluation process in agribusiness
- Describe the purpose and types of training programs in agribusiness
- Outline valid criteria for promotion and employee advancement in agribusiness
- Explain the purpose, format, and importance of an exit interview
- Describe the role and impact labor unions have on the agribusiness sector

INTRODUCTION

When a group of food and agribusiness managers is asked to name their firms' assets, the most likely response is a listing of the firm's physical assets—land, equipment, trucks, cash, and the like. Too often the forgotten element on this list is any business's most important asset: its people. No matter what the physical or financial assets of a firm, it is the way people use those assets and execute the firm's business strategy that determines the firm's success or failure. Because some managers fail to view employees as an asset, they are unwilling to invest adequate time and money in the **human resource (HR)** function. To such managers, human resource management is focused on filling positions and keeping them filled. However, this lack of attention to personnel can cost the firm dearly in the longer run.

Managing human resources is all about helping people reach their maximum potential.

Firms that embody this attitude often do not reach their full potential. No matter how much management spends to develop the firm's marketing program, or refines the technology and methods for producing a product or service, they fall short. In the final analysis, most agribusinesses offer services and products that do not differ significantly from those offered by other companies. Many customers view these products and services as "commodity" items. That is, they are fairly homogeneous, so it is difficult to make these products and services different from those of competitors. When products and services are similar, it is the execution of the strategy that matters—and the organization's human resources are the critical element in the execution of any strategy.

In many ways, administrative tasks and physical work are easier than "people work." People cannot simply be analyzed by a computer or turned off and on like a processing machine. Rather, human resource management is much like production agriculture, partly art and partly science, requiring the right balance of a variety of resources, and the right conditions, to yield a bumper crop. Every person in an organization is a unique individual with individual needs, feelings, and emotions. And all of these variables change daily—sometimes moment-to-moment in today's intense competitive environment. This makes the HR management function complex and difficult. However, investments here can yield high dividends through increased productivity and financial success for the firm. In this final chapter, we will consider the most important elements in managing this most important of resources.

THE SCOPE OF HUMAN RESOURCE MANAGEMENT

Basically, **human resource management** can be divided into two separate but closely related areas:

1. Managing human resource functions
2. Managing employee motivation

In other words, the mechanics of managing people and the finer points of motivating people are highly interdependent. If the person and the job are not carefully matched, no amount of motivation is likely to make much difference. The emphasis in this chapter will be on the functions of human resource management. In Chapter 20 you explored the important areas of organizational design, leadership, and motivation.

Key Issues

Three major issues affect the management of human resources in most organizations: the size of the firm, knowledge of the human resource functions, and top management's philosophy toward human resources.

Sometimes, particularly in small firms, the human resource (or **personnel,** as it may be called in smaller firms) function is carried out more by accident than by design. Regardless of the size of the agribusiness, jobs must be defined, people hired, legal requirements met, employees trained, wages set, and customers served. Because a small agribusiness usually cannot afford to hire qualified full-time people to handle HR management, this task is often left to the general manager. In these cases, people may only get attention when a crisis occurs. Examples of such crises include a key employee leaving the company, a lawsuit filed by an employee, or customer service slipping dramatically. Unfortunately, in some agribusiness firms, human resource management is almost completely ignored until such problems reach crisis stage.

There is no absolute size of firm, in terms of number of employees or sales volume, which dictates when a full-time person should be designated to handle human resource management. The kind of agribusiness, its complexity, the diversity of its jobs, and its degree of seasonality are all factors. A rough rule of thumb is that once an agribusiness has reached 75 to 100 full-time employees, management should examine its need for a full-time human resource manager. Another helpful guide is that a full-time HR department can be profitable when 1 to 2 percent or less of total wages will support that department's budget.

Examining a growing midsize cooperative, the Culpen County Farmers' Co-Op, illustrates some of the key issues involved in managing human resources. The general manager, Kevin Staton, has been with the firm about 5 years. The firm primarily focuses on crop inputs and fuel. The cooperative sells and services a complete line of agronomic inputs including fertilizer, crop protection chemicals, and seed. In addition, Culpen sells gasoline, diesel fuel, and LP gas to farmers, small businesses, and homeowners in the four-county area. The cooperative has 35 full-time employees, plus another 8 to 10 part-time employees during the busy season. While the business has grown somewhat, the number of employees has remained relatively constant over the last couple of years because new equipment and technology has allowed the firm to increase productivity.

Until recently, the labor force had been stable, and Kevin felt that the organization really had no personnel problems. Yet in the last year, a number of events have occurred that radically changed his perspective: the co-op's work-ers' compensation insurance rate had increased substantially; the firm was cited for six violations of safety laws; a new plant which manufactured auto parts for Toyota had just opened in an adjacent county; and several of his departmental managers had been forced to spend time driving trucks, running the plant, and the like. Because of a work-related accident in which an employee lost an arm, Kevin had to cancel a trip to Europe that he and his family had been planning for two years. While 10 of the 35 full-time employees have been with the organiza-tion for over 10 years, records show that last year 18 new people were hired. The plant employees have no fringe benefits other than those required by law. The firm obviously has an employee turnover problem, and Kevin expects this situation to only get worse given the benefit package offered by the new auto parts plant. He knows that Culpen is not hiring employees that have the experi-ence and work attributes that they need to keep the cooperative competitive.

Culpen County Farmers' Co-Op has no formal or written human resource policies. Just a few short years ago, the firm had only one location and a dozen employees, and such a policy seemed like so much red tape. A rudimentary form is used as a job application. Applicants are interviewed and hired by whatever manager has the time to talk to the individual applying for the job. There are no well-defined job qualifications, and job functions have not been described for-mally. Training and orientation usually consists of, "You go out and help in the plant—whoever is there will show you the ropes." Because total employee costs are high, the wage policy consists of "pay as little as you can." Except for a token raise each year for everybody, raises are generally given only when a per-son wants to resign, as an inducement to stay. At that point it is often too late to change the employee's decision.

Larger firms can also be guilty of having human resource programs almost as haphazard as those of the Culpen County Farmers' Co-Op. Sheer size often forces a much more fully developed, formal human resource program in larger companies, but such a program can be almost as ineffective as Culpen's if poorly organized and managed. A relatively large agribusiness with over 2,000 employees provides a good example. The current personnel director was pro-moted to his job some 15 years ago, primarily because he was available at the time. He had been in the training department (which was not well run) and had no formal HR management training. More importantly, since taking the position, he has never tried to learn more about the job in any organized way. His ap-pointment clearly mirrors the attitude of top management—top management be-lieves the HR function is relatively unimportant. "Find the bodies" and "pay them as little as possible" are the two implicit, but primary, personnel objec-tives. In this firm, the HR director is not regarded as top management, and is not included in important planning and policy-making processes. In fact, the ware-house superintendent and maintenance supervisor have far more influence than the HR manager does. Labor productivity is down, turnover and absenteeism are high, grievances continue to increase, strikes are anticipated, and top manage-ment wonders why.

Clearly, things must change in both firms. Without major changes in their human resource management approaches, both firms will continue to lose business to competitors or fail to survive. People problems affect productivity, costs, and profits. Both of these firms provide examples of the major issues (size, skill, and knowledge of the HR staff, and managerial philosophy and attitude) that must be addressed in successfully managing the human resource area. Let's dig a bit deeper into the human resource management function and explore ways to help the agribusiness manager maximize the returns from the investment in people.

THE FUNCTIONS OF HUMAN RESOURCE MANAGEMENT

There are a number of steps in an effective human resource management program. These steps focus on finding the right people for the right jobs, and then providing for their continued productivity and development. They include:

1. Determine the firm's human resource needs.
2. Find and recruit people.
3. Select and hire employees.
4. Orient new employees to their jobs.
5. Set terms of compensation and fringe benefits.
6. Evaluate performance.
7. Oversee training and development.
8. Provide for promotions and advancement.
9. Manage terminations or transfers.

Literally thousands of people enter the job market each month. Some of them have never worked before; others have many years of experience and training. Each of these prospective employees has the potential for performing a lifetime of service as a productive and satisfied employee. Unfortunately, each employee has the same potential for being dissatisfied, unhappy, and unproductive. The agribusiness manager's challenge is to tap the hidden resources of each new employee and guide these resources to benefit both the firm and the employee. This process begins with an evaluation of the firm's human resource needs.

Defining the Jobs to Be Performed

The first step in human resource management is often the one missed by those new to formal people management. The reason this step is neglected is simple—practically speaking, it does not seem absolutely necessary. One can hire people without knowing specifically what job they are supposed to do, or why or how they are supposed to do it. It is easy to go into a hiring situation with only a vague idea of what the new hire should do. It is also easy to hire a person to fill a vacant position with no real thought about whether or not the job responsibilities should be changed. The challenge of defining the job rests on a sound, well-developed organizational plan (Chapter 20). Every position should have job goals that contribute to the firm's success. The job should be defined two ways: (1) job specifications and (2) a job description.

Job Specifications   **Job specifications** spell out the qualifications needed to perform a job satisfactorily. A set of job specifications should be developed for each job, regardless of the size or kind of agribusiness. In the Culpen County Farmers' Co-Op, one of the department managers, perhaps the office manager, should be put in charge of the HR function. Delegation of this responsibility to one person will help free top management from crises that occur because of inadequate personnel programs. The person to whom the HR responsibility is delegated can develop a set of job specifications that use the following format:

1. *The purpose of the job.* What are the job goals? What activities are necessary to accomplish those goals?
2. *The type of job.* How is the employee supervised? What are the responsibilities and training opportunities? Is this a single level job or one that offers growth and promotion opportunities?
3. *The requirements of the job.* What educational level is required? What experience, special skills, physical strength or condition, emotional or personality factors would be helpful for this position?
4. *Other factors.* What testing would provide useful information? What work experience or other factors might determine the applicant's ability to do the job?

Every manager wants someone who is as creative as Thomas Edison, as intelligent as Albert Einstein, and as motivated as Michael Jordan, but such perfect people just do not exist. Qualifications that are set too high can be just as bad as those that are set too low. For this reason, job specifications should reflect the actual needs of the job. Generally speaking, a series of concise statements will suffice for a job specification. The person who supervises the job in question can often provide the information the HR manager needs. If the agribusiness manager cannot inventory the requirements of a job, it may be wiser to leave it unfilled, since the manager will never know whether or not it is being done well. Table 21-1 provides an example set of job specifications for a specific position with Culpen County Farmers' Co-Op.

Job Description   At the same time that job specifications are being developed, a **job description** should also be formulated. A job description stresses a job's activities and duties. Like the job specification, the immediate supervisor of the job in question is the best source of information about what activities are important to the job. The format of a job description is relatively simple (Table 21-2):

1. A brief summary paragraph of the job and the goals it is intended to accomplish
2. A list of the duties, responsibilities, and attendant authority
3. A statement defining lines of authority
4. An indication of how and when job performance and standards will be evaluated

TABLE **21-1** | Example of Job Specifications

Culpen County Farmers' Co-Op: Job Specifications

Job: Millhand

1. *Purpose of job:* To service customers' needs in unloading and loading grain, mixing and loading feed, and filling customers' orders.

2. *Responsibilities:* To direct and assist farmers in unloading grain rapidly and efficiently; to operate feed-mixing equipment; to load mixed and bagged feed; to maintain clean and safe conditions in the mill and elevator at all times; and to service customers promptly and courteously.

3. *Requirements:* Education—*High school diploma (minimum)*
   Experience—*Farm background or 1 year grain elevator experience*
   Physical—*Good health and free from dust allergies*
   Personality—*Must be friendly and outgoing, and work well with others*

4. *Opportunities:* This position provides the necessary experience to move into management as mill supervisor and eventually into sales or general management.

5. *Special Considerations:* Any applicant must be willing to work in dusty and noisy conditions and to work as many as 4 to 6 hours overtime per day during the harvest season.

TABLE **21-2** | Example of a Job Description

Culpen County Farmers' Co-Op: Job Description

Job: Millhand

Brief Description: The primary responsibility of the millhand is to assist customers in unloading and loading grain, grinding and mixing feed to customer specifications, bagging or loading feed onto trucks, and filling customer orders for premixed feeds and animal health products. The job will require significant overtime hours during harvest season. The millhand has a great deal of customer contact and is responsible for providing friendly, efficient service to farmers.

Duties and Responsibilities

1. Unload and load grain.
2. Grind and mix feed to customer specifications as per work order.
3. Load feed.
4. Fill customer orders as per work order.
5. Maintain clean and orderly conditions in stock room, feed grinding, and elevator areas at all times.
6. Sweep and clean all work areas daily.
7. Maintain and enforce all safety regulations at all times.
8. Take inventory of products weekly.
9. Perform routine maintenance of unloading and mixing equipment as scheduled.
10. Maintain helpful, friendly attitude toward customers at all times.

Supervision: Millhands report to and receive instruction from the mill supervisor. The supervisor will have authority and responsibility for all activities in the elevator and the mill. Suggestions or complaints should be made through the supervisor.

Evaluation: Performance of millhands will be evaluated semiannually. This evaluation will be completed by the immediate supervisor, initialed by the general manager, and discussed with the millhand. This evaluation will establish the millhand's strengths and areas of needed improvement. It will serve as the basis for wage increases and promotions.

While the specifications of the job description should be exact enough to provide guidance for the employee and the supervisor, they should also be flexible enough to allow for special situations, emergencies, or minor changes. It seems obvious that finding the right person for the job requires a careful appraisal of the qualities needed by the person filling the job, and of the duties to be performed and responsibilities to be assumed. These steps are clearly the starting points toward effective HR management.

Finding or Recruiting Employees

Prospective employees can be secured from many sources. The qualifications for the job, its wage or salary, the kind and size of organization, and the location of the agribusiness will play an important role in locating new employees. Just as with customers, there may be no better recommendation for the HR program of a firm than to have present employees tell their friends, "This is a good place to work." Many well-managed firms have long lists of people who want to work for them because of present employees' satisfaction with their jobs.

If some present employees see the available position as a promotion, those employees may be good prospects for the job, assuming that they are qualified. If the job requires special training or education, school or university placement services or counselors can provide help in finding recruits. Private employment agencies and government employment services can often help locate the applicant needed. The wise manager is always alert to the existence of top-notch people in competitors' firms or others who may be interested in a change of jobs. Advertising in newspapers, trade journals, and the Internet, and listing the position with an Internet job-posting site, can all help uncover applicants.

Selecting the Right Person

Each of the available applicants should be screened against both the job specifications and the job description for that particular job (Table 21-1 and Table 21-2). First, the job applicant should be compared to the job specifications. A good job application should elicit information on personal history, education and special skills, experience, personal references, and previous employment—all consistent with the background and skill set required for the position.

Human resource managers also have to be very careful not to seek information that the applicant does not legally have to provide. Care must be taken in designing the application so that it does not violate the Civil Rights Acts, which make it illegal to discriminate against any person on the basis of race, color, religion, gender, age, or national origin. Seeking information like marital status, financial position, number of children, and age can get a firm in a lot of trouble if the firm uses the information in some discriminatory way. And, a firm that is audited for civil rights violations will be glad it has designed a set of job specifications for each job. If these are carefully designed to represent those factors needed to perform the job successfully, and if the person interviewing the applicants has filled them out honestly, the firm will be in a stronger position to defend its hiring practices.

The job application can be simple, but it should provide the applicant with a chance to present qualifications in an organized and fair manner. The task of managing the process for finding and screening applicants should fall to the person(s) responsible for the HR function. There are several reasons for this: (1) the person doing the screening interview develops skill through experience; hence, that person is the most qualified for the task; (2) there is a consistency in interviewing when one person handles all screening interviews; and (3) when a HR specialist handles this function, others are not interrupted in their tasks. Note that for many positions, the HR specialist may well involve a number of others in the full hiring process—the supervisor, coworkers, staff specialists, etc. But, even here, the HR specialist will likely manage or lead the process to insure that the best possible person is hired and that all legal requirements are met.

The ability to learn whatever needs to be known depends on the skills of the interviewer. Matching the job specifications and job description against the application form can provide solid facts, but the interviewer's subjective judgments must be trusted to determine the applicant's attitude, personality, and ability to fulfill the responsibilities of this position. In addition, it is common for larger organizations to use a variety of personality/problem-solving/leadership tests to more objectively evaluate individuals in these more subjective areas.

For best results, the interview should be planned in advance. The interview should be held in a private, quiet place where the applicant recognizes that the manager's interest is real and that the job is important to the organization. When the interview begins, the manager should help the applicant relax. It is critical to establish rapport with the interviewee to give the applicant every opportunity to demonstrate his or her potential. Good icebreaker topics include the last job, or the applicant's aspirations. Then the manager should ask leading questions and be a good listener. A good interviewer does not monopolize the conversation. The applicant should be given an accurate description of the background of the firm and the job in question. The position should not be either undersold or oversold. When interviewing, it is useful for the manager to give applicants a copy of the job specifications and the job description, so that they can refer to this information. Some applicants may eliminate themselves from consideration at this point, thereby saving time and money for both the firm and the applicant.

References can be helpful and should be checked out—after securing permission from the applicant to do so. The manager should check to see whether the information given by the applicant is correct as well as whether the reference is a positive one. Checking by phone has an advantage over checking by letter. People may talk more over the telephone than they will write on paper, and the tone of voice, along with the things that are not said, can be significant. Judgment must be used in interpreting references, and good references from former employers are the manager's best criteria. In today's business environment, firms may be very cautious to do anything but validate employment. It may take some persistent effort to learn more about a job candidate in these situations, but, again, getting some feedback from references is a very important step in the hiring process.

As mentioned, testing can also be helpful in screening applicants for some jobs, but the manager must make sure that the tests really measure the factors important to the job. In any event, tests should not be the only criteria for hiring. Testing is most often done in five general areas: intelligence, aptitude, personality, manual dexterity, and physical condition. Some testing programs are more accurate than others. For example, manual dexterity and physical fitness tests are usually quite accurate in predicting a person's physical performance capabilities, while attitude or personality tests have less accuracy. If the test provides useful insights into areas that are required by the job, the manager's success in matching jobs with people can be improved significantly.

Once the HR manager has reduced the list of candidates to two or three qualified individuals, the immediate supervisor for the vacant position (and perhaps other employees interacting with the candidate) should interview them and make the final selection. This important step builds the relationship between the position supervisor and the new employee in two ways:

1. A new employee tends to have a positive feeling toward the supervisor who personally selected them.
2. The supervisor who has taken part in the selection process tends to have a special feeling of responsibility for helping the new person succeed on the job.

Job Orientation

When the best potential employee has been found, the successful applicant must be introduced to the job. Every time someone is hired, the organization is betting heavily on that person's success. The right start, quick adjustment, and future productivity depend on effective **job orientation.** At the start, the new employee is more receptive than at any other time to developing the attitudes necessary to produce a long-lasting and successful career. Job orientation involves four major steps:

1. Introducing the company to the employee
2. Establishing job relationships and encouraging familiarity with the facilities
3. Helping the employee to begin the job
4. Following up and evaluating the employee's adjustment

The first step in orientation is helping the new employee better understand the company and the job. This is a very important step and the amount of time devoted to it will depend on the position and the company. A part-time employee in a retail farm supply business may get a very brief orientation. A new sales hire in a large, complex agribusiness may well have an orientation program that lasts a full year.

At this point, the history, nature, and scope of the organization, and information relative to hours of work, pay, fringe benefits, rules or restrictions, company policies, regulations, overtime, special programs, and facilities should be reviewed. Of course, the new employee will not be able to remember all this information, so it is helpful for it to be written down for future reference. In a

Guiding new employees through the nuances of a new job can be both productive and rewarding.

small organization, this may involve a couple of copied sheets in a folder; in a large organization, it may take the form of an employee handbook. A tour of the entire facility will help to foster an overview of the organization and an understanding of where the employee fits in. Again, in a larger company, this process may involve an extended period of job shadowing.

The second step is establishing **job relationships.** Here supervisory responsibility should be outlined. The new employee should be introduced to his or her new supervisor, if this person is different than the one who interviewed candidates, fellow workers, and to the union steward if there is one. The physical layout, parking, washrooms, lunchroom, etc., should be pointed out. The immediate supervisor should review work-related matters, such as safety regulations, and job expectations. It is often helpful to assign a new employee a "buddy" or "mentor" who will help that person get acquainted and feel at home in the new job.

The next step is helping the employee to actually begin the job. Supplies, safety devices, equipment operation, work accessories, and the like should be discussed, even if the new employee has previous experience. After these explanations, the supervisor should allow the new employee to try the job. The speed of this phase will vary according to the level and complexity of the job and the employee's skills. The supervisor should not try to push too much information on the new employee at once, but should let learning occur step by step.

The supervisor should be sure to stress the required work habits and the norms and expectations of job performance. Again, a fellow worker assigned as a buddy can be helpful in giving the new person proper orientation. Managers must recognize that, like it or not, much of what the new employee learns about how to do the job and about work habits and attitudes will come from fellow workers. So, if rules are specified on paper but never enforced, the new employee will learn this very quickly. When a new person comes on board, problems often surface in firms that do not abide by the rules and regulations they outline in the employee handbook.

The final step in orientation is follow-up. Good HR managers do not assume that things are going well; they make sure that they are. This routine follow-up is usually best handled in an informal, on-the-job situation. The manager should determine how the new employee is getting along on the job and with fellow workers and supervisors. The manager should learn what problems exist, if any, and help solve them. Follow-up is not only useful in making sure that things are going as they should, and allowing for corrective action, but it also gives management an opportunity to encourage new employees and assure them of the firm's continuing interest in their careers. In many firms, a 30-, 60-, and/or 90-day review is simply specified up front, and is therefore expected by the new employee.

In large companies, several people may handle orientation, while in a smaller firm the manager may perform this important step. Some larger firms have orientation schools or sessions. These may be several days (or weeks) in length, particularly for management or administrative jobs. The more that fellow employees and immediate supervisors are involved, the more they will feel responsible for the new employee's success as a team member. Fellow employees and immediate supervisors should be trained to perform orientation functions. Even the smallest firm should have a planned job orientation program to ensure that new employees get off to a good start. Again, this cannot be overemphasized—this is the time to help the new employee understand the organization and expectations, and to help them to quickly become a part of the team.

Compensation and Fringe Benefits

Attitudes toward **compensation** vary across agribusiness firms. But, the labor market facing a food or agribusiness has a tremendous amount of influence on the level of compensation an agribusiness offers. Firms located near rapidly growing suburban areas may find their wage structure substantially higher than a firm located in a more rural, slow-growing market. The agribusiness firm must consider both its ability to pay and competing compensation rates for similar jobs in the area or industry. Often the local Chambers of Commerce or trade associations can provide guidelines for wage and salary levels. Some agribusiness trade associations may sponsor compensation surveys conducted by a university in the region or by consulting firms.

Please note that the base wage is only the starting point for many agribusiness compensation programs. **Commissions** tied to productivity, bonuses if certain levels of performance are reached, stock options for more senior positions, may all be a part of the compensation package. While exploring all of these areas is beyond the scope of this book, it is important to understand that the compensation of an employee will likely involve more than just the base salary.

Fringe benefits also are important factors in securing motivated employees and keeping them satisfied. And, fringe benefits are growing in importance. They range from those required by law to health benefits, vacations, sick pay, life insurance, and retirement benefits. Management of larger firms will devote considerable time to the development of fringe benefit policies and programs. Their counterparts in smaller agribusiness firms can consult insurance companies and other firms in the area for guidance in determining the kinds and costs of fringe benefits for their employees. Today with the concern about rising health care costs, many prospective employees are as concerned about this area as they are about wages or salary. Leading edge agribusinesses attempt to offer their employees a fringe benefit program that fits their particular needs. Hence, many fringe benefit programs offer employees choices of the benefits that they desire.

Compensation satisfies employees' need for security and provides concrete feedback on the value of their contribution to the organization. Competitive factors, employees' need for recognition and self-esteem, and the skills and knowledge required all play a part in determining the level and the composition of the total compensation. The compensation program of the agribusiness firm must consider the true cost of compensation in terms of the jobs that need doing. The firm's ability to pay must be measured against the person's productivity. Putting together a "package" of wages, incentives, and fringe benefits requires careful study by management.

Evaluating Performance

Performance evaluation is critical to successful human resource management programs. Managers and employees both benefit from a carefully conceived and developed performance evaluation program. The goal of the employee performance evaluation program is to:

1. Improve future performance
2. Identify employees with untapped potential, in order to help them realize that potential
3. Provide employees with a benchmark of their achievements
4. Provide information relating to decisions about promotions, advancement, and pay
5. Give the manager guidelines for helping employees in the future

Performance appraisal should not only evaluate the employee and make them aware of their contributions, but should also focus on results of the workers' efforts and their goals for the next evaluation period. Good evaluation programs concentrate not only on past performance but also on future opportunities.

Both the manager and the subordinate often view evaluation with apprehension. The manager should remember that evaluation is a tool and not a weapon, crutch, or cure-all. The major purpose of evaluation is to improve performance, not to punish employees. The successes as well as the failures of both the manager and those managed are recorded during an effective evaluation.

As with any tool, the effective use of performance appraisal depends on the skill of the operator as well as the quality of the tool. The evaluation instrument itself should be as objective and clear as possible. Letting the subordinate use the document for self-evaluation prior to the session is helpful. A comparison of the management and employee ratings opens the way for good communication flow. Managers must understand that they are not in a position of judgment, but rather are seeking to create an honest two-way discussion aimed at positive reinforcement of morale and improved productivity.

The evaluation should be conducted in a private, quiet place without interruption. The manager should make the process relatively informal and be willing to listen. As in the job interview, the manager should put the employee at ease as quickly as possible. Discussion of personality factors should be avoided, if possible. However, if the subordinate is in the mood for such a discussion and this discussion could improve job performance, then the manager may make the most of the opportunity.

Follow-up discussions after the evaluation give the manager a chance to encourage improvement if and when it occurs. Such discussions or "checkups" also keep the manager current on changes in performance. Follow-up (particularly when a subordinate's evaluation was substandard) allows the manager the opportunity to offer recommendations and suggestions, and, if necessary, to exercise disciplinary practices to improve performance.

New employees should be evaluated formally with great frequency. Established employees should be evaluated at least semiannually to ensure that work performance is progressing satisfactorily. An annual checkup usually is not sufficient to catch problems and change work habits or productivity factors that need improvement. Of course, formal evaluation does not replace the need for day-to-day evaluation and coaching that is a constant part of the manager's job. Employees want and need on-the-spot feedback if they are to enjoy satisfaction from their jobs (Chapter 20).

Evaluation must measure work habits and personal traits, as well as success at reaching job goals. Many agribusiness firms use a management concept called management by objectives (Chapter 20). Table 21-3 provides an example of a simple, traditional employer appraisal form.

Training and Development

The benefits of training are quality work, efficient work, and safe work—all of which are vital to the business and to the employee. Employees also want to feel they are growing personally and professionally in a job, and training plays an important role here. Training needs vary according to the circumstances faced by the firm and the specific job position. The kind of agribusiness, the level of complexity of the job, and the experience and educational level of the employee all determine the kind of training needed. **Training programs** are devised to meet one or more of the following objectives: reduce mistakes and accidents; increase motivation and productivity; and prepare the employee for promotion, growth, and development. Managers can increase each employee's value to the firm through training.

All training programs should be based upon specific objectives and meet the test of the following formula:

What should be known – what is known = what must be learned

The employee's present abilities and future capabilities should be analyzed by the person responsible for this function, with the help of the employee. Today there is a real emphasis on lifelong learning—a longer-term plan for developing the employee through a targeted, ongoing education and training. Regardless of

T A B L E **21-3**	**Culpen County Farmers' Co-Op: Employee Evaluation**

Employee _____ Date _____

Job _____

Work Habits:

Areas of Strength:

1. _____

2. _____

3. _____

Areas for Improvement:

1. _____

2. _____

3. _____

Personal Characteristics:

Areas of Strength:

1. _____

2. _____

3. _____

Areas for Improvement:

1. _____

2. _____

3. _____

Employee Growth Plan for Next Period

1. _____

2. _____

3. _____

Signatures:

Supervisor	Manager	Employee

type, training and development programs generally fall into the following major categories:

1. On-the-job
2. Formal in-house
3. Outside formal

On-the-Job Training   Most training in small agribusiness firms, or at lower job levels in large firms, is handled on-the-job. Here the objective is very simple: to teach the skills or procedures for accomplishing a specific task or job. **On-the-job training** begins by determining what needs to be taught. Next, the manager must decide who is to do the training, and be sure that person understands the job operations, safety rules, etc.

By immersing the individual into the actual job situation, on-the-job training can be very effective for new employees or employees changing jobs.

Photo courtesy of Agricultural Research Service, USDA.

Regardless of the job, the actual process will be one of show, do, judge. The supervisor should show employees step-by-step how to do the job, let them do it, then evaluate and offer suggestions and encouragement until a sufficient level of skill has been developed. The training period will vary with the complexity of the job, the experience level of the employee, the skill of the instructor, and the ability of the employee to learn. The benefits of such training are quality work, efficient work, and safe work, all of which are vital to the business and to the employee.

A specific type of on-the-job training is the apprenticeship. An **apprenticeship** is when a new employee works with and helps a more experienced person, and learns under that person's direction over a period of time. This can be of great help in jobs requiring greater skills. Many agribusiness firms are training salespeople, skilled operators, and managers by means of this system.

Formal In-House Training   In larger firms, there is often a full-time training staff. Managers of smaller agribusiness firms may think they have fewer options for **formal internal training** and development. However, many more training opportunities are available to all sizes of firms today, if managers will just take the time to seek them out. For example, such training programs are provided by suppliers, trade associations, consultants, manufacturers, and schools or universities. Online distance learning courses, films, videotapes, slide programs, audiotapes, correspondence courses, and programmed learning kits are but a few of the options available to all agribusiness managers for enriching their training programs.

Training on every conceivable topic, from financial management to trimming a head of lettuce, is available in some form. Suppliers, trade associations, and extension service personnel at the nearest land grant university can help the trainer locate the latest and best training aides. Committee meetings can provide a convenient and fertile ground for training and development if the agendas are carefully controlled and directed toward this end. Sharing information in this setting can satisfy the need for employee recognition and participation and can help make employees feel a part of the organization.

Formal External Training and Development   Agribusiness abounds with conventions, conferences, lectures, workshops, seminars, and the like, on every known subject. In addition, there are trade schools, extension courses, high school and university courses, and business and trade courses that can provide enhancement and/or enrichment to employees and employers simultaneously. Almost every community has most of these alternatives available. Most agribusinesses will be affiliated with a trade association or group that sponsors educational experiences. The trick is to use sufficient self-discipline to determine what specifically must be learned. Many managers are not discriminating enough in choosing educational experiences to improve their employees' productivity. Employee participation in selecting the best training options is often beneficial.

In today's fiercely competitive marketplace, learning should challenge a firm's people, from top to bottom. This is recommended not only to improve basic skills (the best way of doing anything has not yet been discovered), but also to encourage the positive psychological effect on people of working for a progressive, aggressive firm interested in developing its employees. Training and development involves some cost, but the returns in the form of a more dynamic, effective human resource far offset this expenditure. Training of employees should be regarded as an investment in the future of the firm.

Promotion and Advancement

An agribusiness firm's promotion program and policies are related to the factors previously discussed in this chapter. Finding and hiring promotable people, determining their compensation, and overseeing the training and development programs, all lead to the question of promotion, since promotion may be viewed as another form of compensation. To many employees, a **promotion** says, "I've done a good job—the firm appreciates me!"

At any time a position is open anywhere but the lowest level in the organization, a fundamental question is raised: should the firm promote someone from within, or hire someone from outside the firm? The answer to this question should be whoever has the best qualifications, skills, knowledge, and communication ability for the job. However, a strong case can be made for promoting from within, provided that the right people are available. Such a program of advancement encourages employees to grow in their current jobs—they know they have a potentially higher level position to aspire toward. In other words, the potential for promotion can be a strong motivational factor. At the same time, management must be careful that the firm does not become too inbred—new ideas, approaches, and methods need to be introduced from time to time, and hiring from the outside can help make this happen.

The criteria for promotion, then, become the critical factor. Several criteria must be considered, the most important of which are the qualifications that are needed to fill the job. This is an absolute requirement. Failure to meet this demand will cause the firm to become a victim of something an author described as the Peter Principle, which states that employees tend to be promoted based on past performance, until they hit a job that they simply can't do—their level of incompetence.

Once the needs of the job have been ascertained, seniority or length of service must be considered. In a union operation, this may be the primary consideration. Other factors to evaluate are the merit or contribution of the employee in the past, and the employee's attitude and personality (in particular, the ability to function with the new team). A careful performance evaluation program will prove its value here. The employee's potential for growth must also be considered. Finally, the individual employee must want the promotion and be able to handle the new assignment. Too often managers assume that everyone wants to be promoted, when, in fact, some employees may lack the confidence needed to handle this step, at least at this point in their career. In these situations, perhaps there are other ways of recognizing and rewarding a particular employee besides promotion. Offering added responsibility and pay, without increased stature in the organizational hierarchy, may be a better option in some cases. Other employees are quite comfortable in their current job, and will make their greatest contribution to the company if they are simply left alone.

Agribusiness managers must continually guard against the greatest error made in promotion. Simply because a person excels at one job is no guarantee that the same person will excel at a job involving more responsibility and decision-making skills. For example, a good plant employee may not be a good truck driver, nor will a good salesperson necessarily make a good sales manager. To reiterate: the job and the person must be matched carefully.

Termination and Dismissal

It is inevitable that each employee will eventually leave the firm for some reason. Some will quit, some will be fired, and some will retire. In every case, the agribusiness manager should hold to a minimum the loss to the firm and the possible pain for the employee. The loss of any employee can be seen as a learning experience, if the manager takes the time to conduct an exit interview.

An **exit interview** simply gives the person who is responsible for the HR function the opportunity to do two things: make sure the employee leaves as a friend (if possible), and find out what caused the employee to leave, so that any problem areas influencing this decision could be addressed in the future. Retirees should go through the exit interview, as well. Often they can offer valuable information in terms of job function and firm HR management practices. Retirees may serve as an unbiased judge of firm strengths and weaknesses. The manager should take every opportunity to collect and evaluate their comments. Even if the employee has been fired or has quit in anger, a well-conducted exit interview that allows the person to express dissatisfaction will often allow the employee to vent this frustration. A friend or even a neutral party is better than an enemy.

The following exit interview questions would likely work best with processing line, retail sales floor, production plant, or other lower level job positions within the firm. These questions would need to be altered to fit higher level, management employees. Some possible questions for the exit interview include:

1. Was the selection process adequate? Did the applicant understand the job, the compensation, and working conditions?
2. Were the job description and job specifications useful? Were the person and the job matched carefully during placement?
3. Was orientation to the job adequate? Did the new employee feel at home and a part of the team? Did the employee thoroughly understand the firm's expectations?
4. Were wages, fringe benefits, and working conditions reasonable, fair, and competitive? Were employee evaluations fair? Was there adequate feedback and reinforcement from the supervisor?
5. Was there a personality conflict with the supervisor or fellow workers? Were discipline and work rules fair? Was there any favoritism or erratic or unusual discipline?
6. Was the employee given a chance to grow and develop? Was training given in a meaningful way? Were promotion policies understood and fair?

Turnover is one of the most expensive of all business costs. The cost of filling high-level positions often runs into tens of thousands of dollars (or more, depending on the size of the firm, and the responsibility of the position). There is the direct cost of finding, hiring, orienting, and training a new employee. There is also the indirect cost of the job not being performed at its highest level during the "lame duck period" before the old employee leaves. Finally, there's the downtime, from the time of the vacancy until the new employee develops the skill level to do the job. Often, there is poor morale among employees if the turnover is high. And, customer service and sales efforts can suffer if high levels of turnover make it difficult to establish and support customer relationships.

There is a real loss to all concerned when termination and turnover occur. And, the answers to the questions in an exit interview are quite likely to be biased, especially if the person is quitting or being fired. In today's agribusiness labor market, frequent job changes are a fact of life. Despite the very best efforts of firms to create a supportive work climate, to pay competitive salaries, and to offer opportunities for growth, employees will leave. Still, an exit interview can assist the firm in getting better in all of these areas.

THE LABOR UNION

Many larger agribusiness firms deal with unions, or organized labor. **Labor unions** are organized essentially to protect and promote the interests of their members. As the food and agribusiness industries have consolidated over time, and firms have increased in size, the likelihood of working with a labor union has increased. In the U.S. economy, both managers and workers believe that productivity and profits are essential to firm success. The question on which management and labor are divided is what share of these profits and productivity each should receive. Workers need strong, profitable enterprises to compensate them well, and management needs productive workers to make reasonable profits for firm owners.

Unions have become a potent force only since the 1930s, when the Norris-LaGuardia Act (1932) and the Wagner-Connery Act (1935) gave workers the right to organize and bargain collectively with their employers. Many states also passed legislation called closed shop laws, which, in essence, required everyone working for a company that has voted in a union to belong to that union. Some labor unions are small or local in nature, usually referred to as shop unions. They may encompass as few as one firm or just certain workers within a firm. However, in 2000, most of the more than 16 million union workers in the United States belong to the more than 200 national or international unions, such as the Teamsters (the largest), United Auto Workers, and United Steelworkers.

A union is organized within a firm when employees petition the National Labor Relations Board (a governmental agency) to conduct an election. Laws and regulations require that certain conditions be met, and if everything is in order, the vote is conducted. If over 50 percent of the firm's designated workers vote for a specific union to represent them, the firm must deal with that union and develop a labor contract. These labor contracts can and usually do deal with wages, fringe benefits, other forms of compensation, seniority, promotion, hours of work, layoffs, grievance procedures, and mediation. Most of the areas discussed in HR management are referred to in the contract. It is essential that the agribusiness firm consult competent legal advice if it is confronted by an attempt to organize its workers, or if a union already exists and management is forced into collective bargaining and contractual relations with its workers.

When an agribusiness is organized, it is not a case of management dealing with labor; rather, it is a case of business managers dealing and negotiating with labor union managers. Labor union managers are usually highly skilled and well-paid representatives of the unionized workers. The process called **collective bargaining** is, broadly speaking, one in which employers and representatives of employees attempt to arrive at agreements governing the conditions under which the employees are willing to work for the employer. If the collective bargaining process fails and no contractual agreement is reached, workers then have the right to withhold their services, or **strike.** A strike is used by both parties to apply economic pressures to the other until an agreement is reached.

The labor union and organized labor are a well-established part of the American business economy. Their continued success and growth seems likely, since managers' and workers' ideas of the ideal split of power and of profits from business activities will always be different. Collective bargaining offers a process whereby these divisions are determined while the free enterprise system is maintained.

SUMMARY

Human resources are the most important assets of any agribusiness. The human resource function of management is concerned with administering the mechanics of employment, and creating a working environment that helps employees feel valued, growing both personally and professionally. The larger the agribusiness, the more formal and complex the process, but the HR function must be accomplished in any agribusiness.

HR management begins with determining employment needs. This usually requires defining the job and developing job descriptions so that the right person can be recruited. Recruiting involves searching for qualified candidates, interviewing, and assisting in the selection of the best person. And, once a new person is on the job, it is a HR function to see that he or she gets off to a good start with proper orientation, training, and follow-up.

Another HR function is the development and administration of employee benefit programs. Insurance, retirement, health care, safety, continuing education, and related programs are all designed to increase the welfare and productivity of employees. And, a well-designed benefits program represents a significant investment of both time and money by management. The HR manager is also responsible for ensuring that the firm meets complex government and labor union regulations concerning employees and working conditions.

In most agribusinesses, the human resource function also includes regular evaluation of employee performance and ensuring continued professional growth through ongoing training and development programs. Training may consist of informal, on-the-job training or formal seminars, but regardless of source, the goal is greater productivity, and more satisfied employees. The HR manager must also become involved in the termination of some employees. In this case, the manager must ensure that employees' rights are not violated and that there is just cause for dismissal.

HR managers are often designated to coordinate the agribusiness's relationship with labor unions. Negotiating union contracts and carrying out contract specifications are difficult but important human resource management functions.

DISCUSSION QUESTIONS

1. How can one argue that human resources are the most important resources for the agribusiness enterprise?
2. At what point should an agribusiness hire a full-time HR manager? What are the specific reasons a firm should invest resources in a full-time HR manager?
3. What are the responsibilities of the HR manager?
4. Draw up a set of job specifications and a job description for a food or agribusiness job with which you are familiar. (This could be a job that you have recently held.) Contrast and compare them. What are the main differences? Are there any similarities? How are these two HR tools related? Why are job specifications and job descriptions important to good HR management?
5. What are some key elements to a good job application?
6. What are some common sources of new employees? Develop a list of the best sources for a new, entry-level sales position. Develop a list of the best sources for a senior level marketing position. Compare and contrast your two lists.
7. Explain the steps involved in job orientation. Think about the most recent job you have held. Were these steps followed when you were introduced to the job? Which ones were followed and which ones were not? How could the organization you worked for have improved their job orientation process? (If you have not had the opportunity to have a job that had some type of orientation process, you can do this assignment as an interview with a friend who has.)
8. What role do compensation and fringe benefits play in turnover rate?
9. What is the purpose of a formal performance evaluation program? What are some important elements of an effective performance evaluation program? If you have been through a formal performance evaluation process, describe the process. Was the process useful? How could it have been improved?
10. Evaluate the effectiveness of each of the three major types of training and development programs for agribusinesses of varying sizes. What are the strengths and weaknesses of each?
11. What factors should HR managers consider in determining promotions? What are some of the easiest mistakes to make when making promotion decisions?
12. What is the purpose and format of an exit interview?

CASE STUDY — **Culpen County Farmers' Co-Op**

Review the description of the situation at Culpen County Farmers' Co-Op presented at the beginning of the chapter. Imagine that Kevin Staton called you in as a consultant to help clear up the organization's people problems.

Questions
1. Outline the human resource problems that Culpen County Farmers' Co-Op is currently facing.
2. What would be your first recommendation?
3. How would you rearrange Kevin's current job priorities?
4. Should Culpen County Farmers' Co-Op have a training program? If yes, outline a viable training program for the cooperative. If no, why should no program be offered?
5. What can be done to reduce employee turnover? Outline some ideas for lowering the firm's turnover rate.
6. What can be done to attract qualified personnel? Outline some ideas for attracting qualified personnel to the cooperative.
7. How would you go about modifying their current wage and benefits policy?
8. How do the changes proposed in your answers to questions 1 to 7 relate to Culpen's market competitiveness? How will making the changes you outlined improve the organization's chances for success?

GLOSSARY

account A specific category of financial information such as sales, labor expense, or cash; also can refer to a specific customer who owes the business money for credit purchases.

accountability In organizational structure, being answerable to another for performance.

accounting period A predetermined regular interval for which financial transactions are summarized.

accounts payable The amount of money owed by a firm to outside businesses for any items purchased on credit and for which payment is expected to be made in less than one year.

accounts receivable The total amount customers owe to a business for purchases made on credit.

accounts receivable loan A loan where a firm's accounts receivable are used as collateral.

accrual accounting An accounting method of reporting revenue when it is earned (regardless of when the cash is collected) and reporting expenses when they are incurred (regardless of when the cash disbursement is made).

accruals Those financial obligations that have been incurred, but not invoiced or paid, such as taxes payable and wages payable.

accrued expense An account summarizing the firm's accruals.

acid test ratio See *quick ratio.*

acquisition The process of buying another firm or business unit; a method of direct investment that allows a firm to gain country and market-specific knowledge without incurring a long and costly learning process.

adapted child In transactional analysis, a subset of child ego state patterns that represents the time in one's early years when one did just what one was supposed to do.

adult ego state In transactional analysis, the ego state that is often referred to as one's personal computer, because it reflects logical thinking based on facts.

advertising Mass communication with customers or prospects, usually through public media.

agglomeration The accumulation of business activity around a specific location; firms often locate in close proximity to one another because they can increase the efficiency of services at the business, household, and social level.

aggregate production planning The process of developing time-phased specific production quantities and rates, and workforce sizes and rates, while balancing customer requirements and the capacity limitations of plant and equipment.

agribusiness Any business organization that supplies farm inputs or services, or that processes, distributes, or wholesales agricultural products, or retails them to consumers.

agribusiness management   The management of any firm involved in the food and fiber production and marketing system.

agricultural services and financing   A segment of the input supply sector that focuses on providing services, information, and financing to production agriculture; includes farm management services, veterinary clinics, consulting businesses, and farm lenders.

analyzing facts   A step in the planning process requiring answers to such questions as "Where are we?" and "How did we get here?"; helps pinpoint existing problems and opportunities, and provides insights upon which to base successful decisions.

annual percentage rate of interest (APR)   The effective annual rate of interest which is charged on a loan:

$$APR = \left(\frac{2 \times P \times F}{B \times (T + 1)} \right) \times 100$$

where P = payments per year, F = dollars paid in interest, B = amount of capital borrowed, and T = total number of payments.

anticipation inventories   Inventories held for products where demand is not constant so that adequate supplies are available during peak seasons; also called seasonal *inventories.*

appraisal costs   Costs incurred in monitoring the quality level of products and services during the course of production.

apprenticeship   A form of training where a new employee works with a more experienced person, and learns under that person's direction.

articles of incorporation   Legal documents filed with the state government where the organization is incorporated and which set forth the basic purpose of the corporation and the means of financing it.

assemble-to-order   A process for producing products that have many options after the customer's order has been placed.

asset turnover ratio   A financial ratio measuring efficiency; net sales / total assets.

assets   Items of value, including physical and financial property, that are owned by the business.

authoritarian leadership   A leader-centered style of leadership (also called autocratic), in which the thoughts, ideas, and wishes of the leader are expected to be obeyed completely, without question.

authority   In organizational structure, the right to command or force an action by another.

autocratic leadership   See *authoritarian leadership.*

balance sheet   A financial statement that shows the financial makeup and condition of a business at a specific point in time by listing what the business owns, what it owes, and what the owners have invested in the business.

behavioral segmention   In marketing, a means of classifying or categorizing customers or potential customers by their behavioral tendencies.

belongingness   A component of Maslow's needs hierarchy; suggests that after people have achieved a reasonable degree of confidence in their safety, they become concerned with their social acceptance and belonging.

benchmarking   A technique for internal assessment that involves identifying a noncompetitive firm known for excellence in a particular area, carefully studying that firm to see how they deliver this excellence, and then comparing the findings of this study against the firm's current practices.

benefit   In sales, an interpretation of what a product's advantage is likely to mean to a particular customer.

benevolent autocrat   A leadership style in which leaders convince followers to do what they want by being so well liked that no one would consider being disloyal and giving so much praise that employees are shamed into obedience.

blue-sky laws   State laws which regulate the way in which corporation stock may be sold and which protect the rights of investors.

board of directors   The governing body of a corporation; elected by the stockholders to supervise the affairs of the corporation; the number on the board may vary according to the bylaws of the organization.

bond   A form of loan in which an organization issues a promise to pay back a certain sum of money to the bondholder at some specified future date.

borrowing capacity   The total amount of loan repayment ability possessed by a firm or customer.

brand recognition   The ability of customers and prospects to readily identify a firm's brands in the market.

breakeven analysis   See *volume-cost analysis.*

breakeven point (BE)   The point at which income generated from sales just equals the total costs (both fixed and variable) incurred from those sales.

budget   A forecast of sales, expenses, cash and/or other financial information for a future operating period; primarily used for control purposes.

buffer inventory   Inventory (usually relatively small amounts) held in reserve to accommodate uncertainty in production and/or sales; also known as *safety stock.*

build-up forecast   A sales forecast developed from data collected by the sales force or other employees with considerable customer interaction.

bulk   The size and weight conditions or circumstances associated with products; a challenge associated with packing and shipping agricultural products.

bureaucracy   A highly specialized organizational structure in which work is divided into specific categories, is carried out by special departments, and a strict set of guidelines determines the course of activities to ensure predictability and eliminate risk.

buying group   An informal group (not a cooperative) whose purpose is to buy products for its members.

buying signal   In selling, verbal or visual clues that the customer has decided to buy the product.

bylaws   In a corporation, the legal documents which set forth such rules of operation as election of directors, duties of officers and directors, voting procedures, and dissolution.

capacity   In credit analysis, refers to the earnings and cash flow available to meet financial obligations; in production management, refers to the maximum amount of activity possible for a facility, piece of equipment, or process.

capacity planning   All activities that are undertaken to determine what the appropriate size of a manufacturing plant or service location should be, so that a certain quantity can be produced over a specific time period.

capital   The financial resources of a business, comprising in the broadest sense all the assets of the business and representing both owned and borrowed funds.

capital budgeting   A procedure for evaluating long-term investment decisions which takes into account the timing of the cash flows associated with the investment, and the time value of money.

capital intensity   The proportion of tasks in a production or service process performed by machines, relative to the tasks performed by humans.

capital investment   The addition of long-lived assets, typically used in the production of products or services, to an agribusiness.

capitalism   An economic system based on private ownership of the means of production and distribution of goods; characterized by a free, competitive market and an incentive structure motivated by profit.

capital lease   A long-term contractual arrangement in which the lessee acquires control of an asset in return for rental payments to the lessor.

Capper-Volstead Act of 1922   Important cooperative legislation ensuring the right of farmers to organize and market their products collectively so long as the association conducts at least half its business with members of the association and no member of the association has more than one vote, or the association limits dividends on stock to 8 percent.

cash   Funds that are immediately available for use without restriction, such as checking account deposits in banks and petty cash.

cash accounting   An accounting system of recognizing income when it is actually received and expenses when they are actually paid.

cash budget   A projection of a firm's cash receipts and cash requirements made on a periodic basis to enable management to plan for and control cash.

cash flow statement   An accounting statement reconciling the firm's cash balance between two points in time; typically shows cash receipts and cash disbursements for a prior accounting period.

centralized cooperative   A type of regional cooperative where the local cooperative outlet is controlled by the regional cooperative rather than by a local board of directors.

chain of command   In organizational structure, the authority-responsibility relationships or links between managers and those they supervise.

chain store   A set of seven or more retail store locations operating under one management.

change in demand   A situation where the demand curve shifts to the right or left signaling a change in the quantity demanded at every specific price.

change in supply   A situation where the supply curve shifts to the right or left signaling a change in the quantity supplied at every specific price.

change in the quantity demanded   A movement along a given demand curve.

change in the quantity supplied   A movement along a given supply curve.

channel management   A set of decisions concerned with who owns and controls the product on its journey from manufacturer or producer to the customer, and the roles the different entities will play on this journey.

character   In evaluating creditworthiness, the personal habits, conduct, honesty, and attitude of the credit customer.

child ego state   In transactional analysis, a state that embodies people's feelings, emotions, or related childlike behavioral patterns.

close   In selling, the salesperson's act of securing a commitment from the customer.

cold calls   A method of prospecting that involves simply making sales calls on farms or agribusinesses without the use of screening or prioritization criteria.

collateral   Assets pledged to guarantee payment of a loan.

collective bargaining   A process in which an employer and representatives of employees attempt to arrive at an agreement governing the conditions under which the employees are willing to work for the employer.

commercial banks   An investor-owned institution that accepts deposits and then allows checks to be written against those deposits as a means of making payments; the primary source of borrowed funds for most agribusinesses.

commercial finance companies   Those finance companies or specialized lenders that focus on business and commercial loans.

commission   The fee paid to a salesperson or other individual for providing a service; typically a percentage of the total amount of business transacted.

common-size analysis   A financial analysis technique where the firm's financial statements are expressed as percentages of some key base figure such as total sales or total assets, and then the resulting common-size statements are used to make comparisons over time, or against peer firm data.

common stock   A specific form of equity investment in a corporation that carries with it the right (and obligation) to share in any profits or losses incurred by the business.

communication   The successful exchange of information or ideas between two or more individuals.

compensating balance   A specified amount of money that must be retained in a bank account while a loan is outstanding.

compensation   The total amount of salary, commissions, bonuses, benefits, and perquisites that are given to an employee in exchange for work done.

competitive advantage   The set of firm competencies that are important to customers where the firm has a clear and distinct advantage over the competition.

competitive pricing   A simple pricing method, based primarily on competitors' prices where prices are set at the "going" rate.

compounding   A method of calculating interest in which interest earned periodically is added to the principal and becomes part of the base on which future interest is earned.

consensus forecast   A forecast (usually generated by the Delphi method) where agreement is reached by a group on a set of future outcomes, scenarios, or indicators.

contingency planning   The development of alternative plans in case the assumptions made in the planning process do not materialize.

continuous inventory   In inventory management, the constant monitoring of stock on hand.

continuous production   A total production process that involves the flow of inputs in an uninterrupted way through a standardized system to produce outputs that are basically identical.

contribution margin   Gross margin less variable costs.

contribution to overhead (CTO)   The portion of each unit (dollar) of sales that remains after variable costs are covered and that can be applied toward covering fixed costs.

contribution to profit (CTP)   The portion of each unit (dollar) of sales that remains after variable costs and all fixed costs are covered; CTO at sales volumes greater than the firm's breakeven point.

control   The process of monitoring performance, as compared with pre-established standards, for the purpose of making adjustments to ensure that goals will be accomplished.

controlling   One of four tasks of management; represents the monitoring and evaluation of activities, and assessment of whether the goals and performance objectives developed within the planning function are achieved.

conviction   The customer's belief that a product will be of benefit to them.

cool calls   A method of prospecting that involves calling on those one believes to be likely prospects.

cool leads   Leads with potential customers whose need for a product or service may or may not be current.

cooperative   A distinctive form of corporation that is organized to serve the needs of its member-patrons rather than to make a profit for investors.

cooperative advertising   A common advertising practice in agribusiness in which local advertising is sponsored, and usually funded, jointly by the manufacturer and the local dealer or distributor.

cooperative borrowing   A type of loan made to agribusiness cooperatives from the Banks for Cooperatives, which is part of the Farm Credit System.

corporation   A form of legal organization with the powers, rights, and obligations of an individual; a corporation's assets and liabilities are controlled by the legal entity, not by the owners (stockholders).

cost basis of valuation   A valuation practice by which assets are recorded on accounting records at their acquisition price, not their current market value.

cost-based pricing   A pricing method based on adding a constant margin to the basic cost of the individual product or service.

cost leadership   One of two generic business strategies; involves meeting competitors' product offerings with an offering of comparable quality and features, but beating the competitor on price.

cost of capital   The rate of return a firm requires of its investments.

cost of goods sold   The total cost to the business of goods that were actually sold during a specified time period; includes such items as raw material costs, incoming freight costs, and the cost of any damaged or lost goods that must be absorbed.

cost-plus pricing   See *cost-based pricing.*

critical parent   In transactional analysis, a state involving critical attitudes and behavior; so named because the manner by which one expresses critical feelings—tone of voice, gestures, posture, facial expressions, and words—is often similar to one's parents' behavior under similar conditions.

critical path   In PERT and CPM, a path that has absolutely no slack time.

critical path method (CPM)   A method of scheduling that involves a diagrammatic representation of the network of activities which comprise a production operation.

Crosby   Phillip B. Crosby, one of the pioneers of total quality management (TQM); work advocates a goal of zero defects and emphasizes quality in terms of conformance to requirements.

CTO pricing   A pricing method that encourages extra sales by reducing the price (while still covering variable costs) to generate additional contribution to profit (CTP).

culture   The set of socially transmitted behaviors, arts, beliefs, speech, and all other products of human work and thought that characterize a particular population.

current assets   An accounting term designating actual cash or assets that are expected to be converted to cash during one normal operating cycle of the business, typically one year.

current liabilities   An accounting term designating those outsiders' claims to a business that must be paid within one normal operating cycle, usually one year.

current ratio   A financial ratio measuring liquidity; total current assets divided by total current liabilities.

customers   Those who purchase outputs (products, services, and/or information) from a firm.

custom hiring   A form of leasing that combines labor services with the use of the asset being leased.

cycle inventory   An inventory quantity that is tied to truck capacity, pallet size, manufacturer promotion, or some other factor that increases inventory levels relative to a normal pipeline inventory; also called *lot-size inventory.*

cycle-time-to-market   The time it takes for a product to go from an idea to a finished product in the customer's hands; reducing cycle-time-to-market (also called quick response) has become a central strategic goal for many agribusinesses.

days sales in receivables ratio   A financial ratio measuring efficiency; the average collection period of accounts receivable; (accounts receivable / net sales) × 360.

dealer-distribution system   A marketing channel in which manufacturers sell their products to dealers or retailers, who, in turn, sell products in their own local market.

debenture note   A promise to pay (typically unsecured) issued in exchange for a loan of financial capital.

debt-to-asset ratio   A financial ratio measuring solvency; shows the proportion that lenders are contributing to the total capital of the firm; (total liabilities/total assets).

debt-to-equity ratio   A financial ratio measuring solvency; total liabilities divided by owner's equity.

decision-making process   A logical procedure for identifying a problem, discovering alternative solutions, analyzing them, and choosing a course of action.

decline stage   A period in the product life cycle in which sales decline rapidly.

delegation downward principle   In organizational structure, the idea that authority should be delegated downward to the lowest level at which the decision can be made competently.

Delphi approach   A tool for developing forecasts in which a panel of experts is asked to develop a forecast for an area of interest; the estimates are pooled, reviewed, and any differences noted; and the process is repeated until a consensus forecast is achieved.

demand   The quantity of a product or service that consumers are willing to buy at different prices; the relationship between price and the quantity demanded.

demand curve   A graphic or algebraic representation of a product's demand.

demand elasticity   A measure showing the relative change in quantity demanded as price changes; reflects the responsiveness of quantity demanded to a given change in price.

Deming   W. Edwards Deming, a statistician and consultant who was one of the pioneers of total quality management (TQM); defines quality in terms of how well the product or service meets the consumer's needs.

democratic leadership   A leadership style, also called participative leadership, which favors a shared decision-making process, with the leader maintaining the ultimate responsibility for decisions while actively seeking significant input from followers.

demographic segmentation   A market segmentation approach which groups customers based on demographics such as age, income, size of household, education, number of children, type of employment, etc.

derived demand   A situation where demand for one product (usually an input) is based on the demand for another product (usually the final product).

developing alternatives   A step in the planning process; after performance objectives have been set, the process of generating different ways of achieving the objectives.

development stage   The stage early in the product life cycle in which the market is analyzed, and both the product and the broader marketing strategy are developed.

differential advantage   One of two generic business strategies; involves offering buyers unique products and services which add value for the buyer but are unavailable from other firms.

differentiation value   In pricing, the perceived value of the product's unique attributes.

diminishing marginal utility   A law of economics that states that as more and more of a product is consumed, the extra satisfaction of consuming an additional unit actually declines; translates directly into the inverse relationship between price and the quantity demanded.

direct exporting   Mode of entry in international business in which the agribusiness itself handles the details of exporting their products.

directing   One of the four tasks of management; involves guiding the efforts of others toward achieving a common goal and leading, supervising, motivating, delegating, and evaluating human resources.

directors   See *board of directors.*

direct selling   A process that involves prospecting for new customers, pre-call planning, getting the customer's attention and interest, making presentations, handling objections, and closing the sale in direct, one-on-one contact with customers.

discounted loan   A loan where the amount of interest to be paid is deducted from the amount the lender provides the borrower; amount of loan – amount of interest paid = amount of available capital.

discount pricing   A method of pricing that offers customers a reduction from the published or list price for specified reasons, such as cash payments, volume purchases, or preseason ordering.

discount rate   The interest rate used to determine the present value of a stream of cash flows to be received in the future.

distribution   In marketing, one of the four elements of the marketing mix (also known as place); involves all decisions related to how the product moves from the manufacturer to consumer.

distributor system   A system of channel management that uses both middleman-distributors, or wholesalers, and dealers to move products to customers.

diversification   The process of entering alternative lines of business that pose different risks and opportunities so that the likelihood of a loss in one area will be offset by the possibility of gain in another.

division of labor   The process by which jobs are broken into components and then are assigned to members or groups.

dormant partner   A partner not active in managing the partnership and not known by the general public; retains limited liability.

early adopters   The second group of buyers (next 13½% after innovators) to try a new idea, product, or service; tend to be progressive and well-respected individuals.

early majority   The third group of buyers (next 34% after the early adopters) to try a new idea, product, or service; tend to be deliberate people who see themselves as fairly progressive, but not generally as leaders.

earnings on sales ratio   A financial ratio measuring profitability; (net operating profit / net sales).

economic profit   A measure of profit that takes into consideration alternative uses for the resources; accounting profit less opportunity cost.

economics   The study of how scarce resources (land, labor, capital, and management) are allocated to meet the needs of a society with unlimited wants.

economic value analysis   A method of pricing that is based on the economic value of a product.

economies of scale   A production situation where the larger the scale of the operation, the lower the per-unit cost.

efficiency   A ratio measuring the output or production of a system or process per unit of input.

efficient consumer response (ECR)   An information-based system that links distributors and suppliers together with retail supermarkets; the food industry's concept of supply chain management.

ego status   A component of Maslow's needs hierarchy that suggests that once a comfortable degree of social acceptance has been attained, most individuals become concerned with their self-esteem or status relative to their group.

elastic   In supply and demand analysis, a condition where a change in price brings a proportionately larger change in quantity supplied (demanded).

elasticity   In economics, a measure of the responsiveness of one variable to changes in another.

electronic data interchange (EDI) systems   An information technology that allows business interactions between firms to take place electronically using a standard format.

equilibrium   In economics, a state or situation in which opposing forces balance each other and stability is attained.

equity capital   Funds provided by the owners of a business either directly or by reinvesting profits back into the business; the risk capital of a firm.

evaluating results   A step in the planning process; involves a careful review and assessment that helps determine whether or not the plan is on course.

exchange function   One of three major marketing functions; involves those activities that are concerned with the transfer of ownership in the marketing system.

exchange rates   The relative value of one country's currency compared to that of another country.

exit interview   An interview with a departing employee that is designed to better understand the employee's attitudes and reasons for leaving and to provide suggestions for improving the workplace.

expected product   A part of the total product concept; surrounds the generic product with the minimum set of features/services that the customer expects when they make a purchase.

expenditure   An outlay of cash for operating or investment purposes.

expenses   The value of resources consumed by the business in acquiring, producing, and selling goods and services; includes only those resources consumed by actual sales during the accounting period.

external failure costs   Costs incurred when the product or service fails once it is in consumer hands.

facilitating functions   One of three major marketing functions; involves such activities as managing risk, developing standards or grades, and financing.

factor   An organization that buys accounts receivable from a business at a discounted price and then collects directly from the individuals responsible for the receivable.

factors of production   In economics, land, labor, capital, and management; scarce resources used in the production process.

farm   As defined by USDA, any rural unit grossing $1,000 or more per year in agricultural sales.

farm-food marketing bill   A breakdown of the consumer's food dollar into the proportion the farmer gets for producing the raw material and the proportion the food industry gets for marketing (in a broad sense) the final product.

fast food/quick service restaurants   Limited menu and limited service restaurants focused on providing convenience and speed of service.

feature   In selling, a descriptive, measurable fact about the product.

federated cooperative   A cooperative made up of two or more member associations organized to market farm products, purchase production supplies, or perform bargaining functions; cooperatives whose members are other cooperatives.

financial accounting   Financial recordkeeping conducted in a manner that meets the reporting requirements of governmental units, lending institutions, investors, and suppliers.

financial management   One of the four functions of management; managing the assets, liabilities, and owner's investment in the firm; generating the financial data required to make decisions, and using the tools of finance to make effective decisions.

financial statement   A summary statement of the financial status of a business such as a balance sheet or an income statement; usually prepared by summarizing accounts from the general ledger.

financing   The process of raising or acquiring the funds required to operate a business.

finished goods   Products that are ready for sale.

fixed assets   Those items the business owns that have a relatively long productive life, such as land, buildings, and equipment.

fixed costs   Costs that do not fluctuate with the volume of business; considered the costs of being in business.

fixed position layout   A production process layout used in situations where the item being constructed is too large or bulky to move easily or where the item is to be built fixed in place; construction is accomplished in one place, and tools and materials are brought to the assembly location.

flexibility principle   In organizational structure, the concept that an organization should maintain its structural flexibility so that it can adjust to its changing internal and external environments.

focus group interview   In market research, a guided discussion among users or potential users of a product or service that is designed to discover attitudes and opinions toward those products or services.

focused factory   A production concept intended to operationalize the notion of flexibility; holds that several smaller factories will improve individual plant performance because the operations manager can concentrate on fewer tasks and can more easily motivate a smaller workforce toward one goal.

follow-up   In selling, includes all of the customer's experiences with the company or product after the sale has been made.

food and fiber production and marketing system   The system of firms and organizations that encompass all economic activity supporting production agriculture, producing raw farm commodities, converting raw farm products to consumable food and fiber products, and distributing these final products to end consumers.

food processing and manufacturing   Part of the food sector; includes a broad range of firms that take raw agricultural commodities and turn those commodities into ingredients for further processing, or into final consumer products; examples include meatpackers, bakers, wet corn processors, and brewers.

food retailers   Firms involved in the retail distribution of food; includes supermarkets, warehouse stores, and combo stores.

food sector   The sector of the food production and marketing system in which food processing, marketing, and distribution occur.

food security   The state of affairs in a society where all people at all times have access to safe and nutritious food to maintain a healthy and active life.

food service   Firms providing prepared food, typically ready-to-eat, including traditional restaurants, fast food/quick service restaurants, and institutional food services.

forecast   An estimate of some future business activity or indicator.

form utility   One of four types of utility provided by a marketing system; adding value to a product by transforming the physical form of the product.

formal internal training   A type of employee training; typical in larger firms, this is job- or task-specific instruction conducted by a full-time training staff.

formal structure   In organizational structure, the formal reporting relationships between employees in an organization; one of two kinds of organizational structure required to operate effectively.

forward contracting   A formal, binding agreement to buy or sell products at some established terms for delivery at some specified future date.

free market system   A type of economic system in which consumer wants are expressed directly in the marketplace using a market-determined pricing mechanism as the basis for allocation of scarce resources.

free rein leadership   A leadership style (also called *laissez-faire leadership*) in which the leader literally relinquishes all decision making to followers.

fringe benefits   A nonsalary benefit offered to all employees; ranges from those required by law to health benefits, vacations, sick pay, life insurance, and retirement benefits.

functional organization   An organizational structure that grants staff specialists authority in the areas of their specialty.

futures market   A specialized type of market where promises to deliver or accept delivery of a standardized unit of product at some specified future date are traded; reflects anticipated future supply and demand situations.

gap analysis   An evaluation of the difference between a firm's desired performance at some point in the future and what performance will be if the firm does not change strategies or approaches.

gathering facts   A step in the planning process; involves gathering sufficient information to identify the need for a plan, and systematic gathering of the information needed to make the plan work once it has been developed.

General Agreement on Tariffs and Trade (GATT)   An international organization established in 1947 to reduce global trade barriers through multilateral trade negotiations; evolved into the World Trade Organization (WTO) in 1995.

general economic forecast   A forecast of conditions in the general economy for a specific future time period.

general partner   A partner who is active in the management of the partnership, typically has an investment in the business, and is subject to unlimited liability; unless specified differently in writing, all partners are assumed to be general partners.

general partnership   The association of two or more people as owners of a business in which each individual partner, regardless of the percentage of capital contributed, has equal rights and liabilities, unless stated differently in the partnership agreement.

generic advertising   A form of advertising aimed at promoting a class of products or an industry, not a specific company or brand; also called *institutional advertising*.

generic product   One component of the total product concept; the standard product as offered to the customer with no special services or features attached.

genetically modified organism (GMO)   An organism that has been modified by the application of recombinant DNA technology (a set of techniques enabling the manipulation of DNA).

geographic segmentation   In marketing, a means of classifying or categorizing customers or potential customers by their geographic location.

globalization   The process of incorporating a worldwide perspective into the processes of a firm and into the attitudes and practices of a firm's employees.

goals   The specific quantitative or qualitative aims of the organization that provide direction and standards one can use to measure performance.

grading   Classification of agricultural products into specific standardized classes or categories to facilitate the buying and selling process and improve the efficiency of the marketing system.

greenfield investment   A method of direct investment by a firm into a particular country or new business where the firm handles all elements of the entry, including building plants, developing distribution, locating markets, etc.

gross margin   The difference between total sales and the cost of goods sold; income remaining after paying for raw materials or merchandise to cover operating expenses and generate a profit; gross margin plus service income is sometimes called gross profit.

gross margin ratio   A financial ratio measuring profitability; gross margin expressed as a percentage of sales; (sales – cost of good sold) / sales.

growth stage   A period in the product life cycle of rapid expansion, during which sales gain momentum and firms tend to hold prices steady or increase them slightly as they try to develop customer loyalty.

Hazard Analysis Critical Control Point (HACCP)   An important, widely recognized food safety program; basic premise is the prevention of food safety problems rather than attempting to control them.

hedgers   Those traders in a market who have or desire possession of the physical commodity, but do not want to accept price or market risk.

hedging   Process by which price risk is transferred from one party to another; establishing equal and opposite positions in the cash and futures markets for a specific commodity.

Herzberg's motivators and hygienic factors theory   A popular but somewhat controversial theory of employee motivation developed by Frederick Herzberg which argues that two separate factors explain employees' level of motivation: hygienic factors and motivators.

high value products (HVPs)   Agricultural products that have received additional processing or require special handling or shipping; also known as *value-added products.*

hot buttons   In selling, areas of particular customer interest.

hot leads   Leads with potential customers whose need is believed to be current.

human resource management   One of the four functions of management; acquiring, organizing, motivating, leading, and managing the firm's people resources to accomplish its business objectives as effectively as possible.

human resources (HR)   The talents and assets offered a company by the people it employs.

hybrid layout   A production process layout that combines the process and product layouts to balance the advantages of each; used extensively when introducing flexible manufacturing systems.

hygienic factors   One component of Herzberg's motivators and hygienic factors theory; also called pain relievers, these are conditions necessary to maintain an employee's social, mental, and physical health.

income statement   Financial statement summarizing the firm's revenue, expenses, and profit during a specific period of time; also called *profit and loss statement.*

indirect exporting   Mode of product movement in international business in which a firm uses a trading company or an export management company to handle the logistics of exporting.

indirect selling   Activities outside the sales call interview process; involves providing service and follow-up to customers.

individual proprietorship   See sole *role proprietorship.*

inelastic   In supply and demand analysis, a condition where a change in price brings a proportionately smaller change in quantity supplied (demanded).

informal organization   The firm's organization outside the organizational chart; primarily concerned with interpersonal relationships among people, their emotions, feelings, communications, and values; an important part of the organization's culture.

information technology   Any electronic technology focused on processing, storing, transmitting, and reporting data and information.

initial investment   The initial equity an investor commits to an investment.

innovators   The first group of buyers (initial 2½ %) to try a new idea, product, or service; tend to be venturesome people who like to try new ideas.

input distributors   Organizations engaged in moving products from the input manufacturer to the farm and providing services that ensure productive use of the inputs.

input manufacturers   Organizations engaged in the manufacture and production of inputs for use in production agriculture.

inputs   Resources used to create outputs in a production process; includes human resources (skilled workers and managers), capital (equipment), materials, information, and energy, among other such resources.

input supply sector   The sector of the food and fiber production and marketing system that manufactures and distributes supplies and services to farmers for use in production of agricultural products.

institutional advertising   See *generic advertising.*

institutional food services   Firms and organizations providing reliable, convenient meals and/or the components necessary to make the meals to outlets such as schools, hospitals, and government facilities.

insurance companies   A source of financing for agribusiness; typically provide longer term financing as these firms look for opportunities to invest funds they have collected from policyholders.

interest   The charges made by a lender for the use of money.

interest rate   Interest expressed in percentage form; reflects investors' time and risk preferences and serves as the exchange price between borrower and lender.

intermediate-term loan   A temporary grant of money to be paid back to the lender, usually in one to five years.

internal failure costs   In total quality management, costs that are generated during the production and/or rework of defective parts and services.

internal rate of return (IRR)   A method of ranking and accepting or rejecting capital investments; the discount rate that yields a zero net present value.

introductory stage   The stage in the product life cycle in which the new product first appears on the market.

inventory   Items that are held for sale in the ordinary course of business, that have been purchased for use in producing goods and services to be sold, and work in progress.

inventory turnover ratio   A financial ratio measuring efficiency; shows the relationship between inventory and sales volume; cost of goods sold/inventory.

inverse price ratio   The ratio of two prices; used extensively in economic analysis as it reflects the relative prices of two items.

ISO 9000   A series of international standards and guidelines on quality management and quality assurance; standardized definition of quality focuses on conformance to customer requirements.

ISO 14000   A worldwide standard for quality applied to the environment; provides the framework and systems for managing legislative and regulatory orders for environmental compliance.

job description   A list and description of a job's activities and duties.

job orientation   Process used to introduce a new employee to a job; involves four major steps: introducing the company to the employee, establishing job relationships and encouraging familiarity with the facilities, helping the employee to begin the job, and following up and evaluating the employee's adjustment.

job relationships   The network of individuals an employee interacts with on the job; includes the supervisor, fellow workers, the union steward, payroll and benefits personnel, among others.

job rotation   The practice of allowing operators to perform different jobs during their shifts.

job specifications   A written set of qualifications of an employee that are needed to perform a specific job satisfactorily.

joint venture   A type of collaboration between two or more firms; involves sharing resources in research, production, marketing, and/or financing, as well as costs and risks.

junior partner   A partner who is typically younger and has not been involved in the business as long as a senior partner; receives a smaller portion of the business profits; rarely takes an active role in managing the business affairs of the partnership.

Juran   Joseph M. Juran, one of the pioneers of total quality management; worked extensively to determine a cost of quality for all aspects of the production and service cycle.

just-in-time (JIT)   An operating philosophy and production system with the goal of producing and/or delivering goods just as they are needed, in effect, eliminating inventories.

Kanban   An element of JIT; the information system for JIT that helps to pull production through a shop; the Japanese word for "cards."

key indicators (KI)   See *key performance areas.*

key influencers   Individuals or organizations who have some influence on the purchase decision, but don't actually make the decision.

key performance areas (KPA)   Important points of focus in management control systems; also called *key indicators.*

labor unions   Groups organized to protect and promote the interests of their worker/members.

laggards   The final group of buyers (last 16%) to try a new idea, product, or service; tend to be tradition-bound people who take so long to adopt new ideas that by the time the ideas are adopted, they no longer are new.

laissez-faire leadership See *free rein leadership.*

land lease A contract that conveys control over the use rights of real property from the lessor to the lessee without transferring title.

late majority The fourth group of buyers (next 34% after the early majority) to try a new idea, product, or service; tend to be skeptical people who adopt new ideas only after considerable evidence has been accumulated.

layout planning The specific design of the physical arrangement within a facility.

leadership The process of helping individuals or groups to accomplish organizational goals by unleashing each person's individual potential as a contribution to organizational success.

leads In selling, prospects that have some support from credible sources.

lead time In logistics management, the amount of time necessary to cover the period between the warehouse's reordering and its receipt of the additional stock.

lease A contract by which the control over the right to use an asset is transferred from the lessor (owner) to the lessee (person acquiring control) for a specified period of time in return for a lease payment.

leasing A form of renting, usually involving a contractual agreement that specifies the terms of the arrangement; normally for a longer period of time than a rental agreement.

lessee A person or organization that pays for the privilege of using the property of another for a specified period of time.

lessor A person or organization that leases property to another person or organization but maintains ownership of the property for the period of the lease.

leverage In finance, the concept of using borrowed funds to expand the total capital of the firm beyond the amount of owner equity.

liabilities Anything a business owes to those outside the business, such as accounts payable, loans, bonds, and mortgages; the total debt of the business.

licensing A business arrangement where a firm (licensee) in the target market produces and distributes another firm's (licensor) product; in return the licensor receives a fee or royalty for the right to license the product.

limited liability company (LLC) A form of business organization that closely resembles a partnership, but provides its members limited liability; creditors or others who have a claim against an LLC can pursue the assets of the LLC to satisfy debt and other obligations, but they cannot pursue personal or business assets owned by the individual members of the LLC.

limited partner A partner who has taken steps to limit his/her liability in a partnership; the agreement must be in writing and limited partners may not take an active role in managing the partnership.

limited partnership An association of two or more people as owners of a business in which some partners contribute money or ownership capital to the partnership without incurring the full legal liability of a general partner.

limited returns In cooperatives, the requirement that returns on member equity be limited to a nominal amount, not greater than the going interest rate, or 8 percent (whichever is higher); done to ensure that members of the cooperative do not view the cooperative as an investment in and of itself.

line and staff organization An organizational structure that is a variation of the line organization in that a place for specialists is included.

linear projection A quantitative procedure for projecting past trends into a future time setting for developing forecasts.

line of credit   A commitment by the lender to make available to the firm a certain sum of money, usually for a one-year period, at a specified rate of interest at whatever time the firm needs the loan; also known as a revolving loan.

line organization   An organizational structure in which there is one simple, clear line of authority extending downward from top management to each person in the organization.

liquidity ratios   A class of financial ratios that measures the firm's ability to pay short-term obligations as they come due.

little professor   In transactional analysis, a subset of child ego state patterns that includes childlike curiosity and creativity.

local cooperative   A cooperative that is organized and controlled by local member-patrons to serve their own needs.

logic   A qualitative technique involving the combination of fact, induction, and deduction to solve a problem.

logistics management   One of the four functions of management (along with production management); involves the set of activities around storing and transporting goods and services from manufacturer to buyer.

long-range budget   A budget that is generally 2 or more years in scope.

long-term liabilities   Liabilities that are scheduled to be repaid over one year or longer, such as bonds, mortgages, and long-term loans.

long-term loans   A temporary grant of money from a lender to be repaid over a period of more than 5 years.

loss-leader pricing   A pricing method that involves offering one or more products at a significantly reduced price for the purpose of attracting additional customers who will purchase other regularly priced items.

lot-size inventory   See *cycle inventory.*

macroeconomics   The study of how consumers, businesses, and governments in the aggregate interact to allocate scarce resources; includes topics such as inflation, interest rates, gross domestic product, and fiscal and monetary policy.

make-to-order strategy   A production approach where the firm actually builds to order.

make-to-stock strategy   A production approach where the firm is producing for inventory, and the production occurs before the actual sale is made.

management   The art and science of successfully pursuing the organization's desired results with the resources available to the organization.

management by exception   A management philosophy with the basic premise that management should not spend time on areas that are progressing according to plan, but should identify and concentrate on areas that are not generating acceptable performance.

management by objectives (MBO)   A management philosophy introduced by Peter Drucker in 1954 whereby the supervisor and the subordinate jointly set objectives that both agree are reasonable and in line with corporate goals, and then determine how performance will be measured in each area of major responsibility.

manager   The person charged with the responsibility of planning, organizing, directing, and controlling the activities of a business organization to accomplish the objectives established by the owners.

managerial accounting   Financial recordkeeping for the purpose of helping managers guide the operation of the business and make more effective management decisions.

manipulative autocrat   A leadership style in which the manager creates the illusion that employees are participating in the decision-making process; another form of autocratic leadership.

manufacturer-direct distribution system   A marketing channel in which the manufacturer sells directly to the farmer or food customer.

manufacturers' sales offices and sales branches   Wholesale divisions or subsidiaries typically run by large manufacturing or processing firms to market their own products.

marginal cost   In economics, the additional cost incurred in producing one more unit; the change in total cost divided by the change in output.

marginality   In economics, a concept important in maximizing (profit) or minimizing (cost) a particular objective; focuses attention on the incremental change in the objective caused by an incremental change in a variable or variables.

marginal rate of substitution   In economics, the relationship between two inputs in the production process.

marginal revenue   In economics, the additional revenue generated by selling one more unit of output.

market-driven philosophy   A marketing approach based on a true understanding of customer needs; the focus on customer needs drives all decisions and activities in the organization.

market efficiency   A measure of productivity of the marketing process in terms of the resources used and output generated during the process; involves operating efficiency and pricing efficiency.

market forecast   A forecast for a specific industry or type of products.

market information   Any pertinent details and data such as news and industry-specific research which is used by buyers, sellers, inventory holders, wholesalers, and others in the industry to make decisions.

market mapping   A tool for visually presenting information about a firm's customers and markets by plotting relevant data on a map either manually or using computer-based geographic information systems (GIS).

market penetration   The proportion of a specific market that a firm controls.

market research   The study of customers, competition, and trends in the marketplace, with the goal of providing objective information on which to base marketing decisions.

market risks   Any number of uncertain outcomes involving the market for goods and services; includes the possibility of price declines or increases, changes in consumer preferences, and/or changes in the nature of competition.

market segmentation   In marketing, the process of classifying customers into categories that have members who will react in a common way to a firm's marketing offer.

marketable securities   Stocks, bonds, and other investments that can be easily turned into cash during an operating period (usually one year).

marketing   In the macro sense, the process by which products flow through the U.S. food production and marketing system from producer to final consumer; involves the physical and economic activities performed in moving products from the initial producer through intermediaries to the final consumer; from a managerial perspective, the process of identifying customer needs, and developing, pricing, promoting, and distributing products and services to meet those needs.

marketing audit   An objective examination and review of a company's entire marketing strategy.

marketing channel   The network of firms that manage the flow of products, services, information, and money between producers and consumers.

marketing cooperative   A cooperative that derives most of its total revenue from the sale of members' products.

marketing management   One of the four functions of management; involves managing the total process of identifying customer needs, developing products and services to meet those needs, establishing promotional programs and pricing policies, and implementing a system of distribution to customers.

marketing margin   The difference between prices at two different levels in the marketing channel; for example, the difference between the price paid by consumers for a food item and that received by producers for the raw ingredients going into the item.

marketing mix   In marketing management, the combination of price, product, promotion, and place strategies developed and implemented by a firm to support a specific position in the market.

marketing system   The interrelated set of market and interfirm relationships such as contracts and alliances that add utility or value to a product as it moves from producer to consumer.

Maslow's needs hierarchy   A theory of human behavior based on the idea that different kinds of needs have different levels of importance to individuals, and depend on the needs that are currently being satisfied.

master production schedule (MPS)   In operations management, details the final quantities of stock keeping units (SKUs) that are to be made in specific blocks of time; created once specific orders for products have been received and/or short-range sales forecasts have been determined.

master scheduler   In operations management, the individual(s) that must consider the total demand on the resources and capacities of the agribusiness as well as the supplier's capacities in developing production plans.

material requirements planning (MRP)   A computerized information system for managing production operations; purpose is to ensure that the right materials, components, and subassemblies are available in the right quantities at the right time so that finished products can be produced according to the master production schedule (MPS).

maturity stage   A period in the product life cycle characterized by slow growth or even some decline of sales as the market becomes saturated.

merchant wholesalers   Firms in the market channel that buy products from processors or manufacturers, and then resell those products to retailers, institutions, or other businesses.

microeconomics   The study of how scarce resources are allocated, primarily focused on consumer choice and firm production decisions, and how these are related to markets and prices.

minimal layer principle   In organizational structure, the idea that the number of levels of management should be kept at a minimum, while at the same time maintaining an effective span of control.

mission statement   A firm's declaration of the business they are in, the customers they serve, and the basic concepts or beliefs they hold.

mixed cooperative   A cooperative that has individual farmer members *and* autonomous cooperative members; represents a combination of centralized and federated cooperative structures.

model   An abstract representation of a system or situation, usually in mathematical terms, that exhibits the factors believed to be most pertinent to the system or situation.

motivation   A stimulus that produces action.

motivators   An element of Herzberg's theory of motivation; reward producers, or conditions that encourage employees to apply themselves, mentally and physically, more productively to their jobs.

natural child   In transactional analysis, the first and most significant subset of child ego state patterns; represents the full gamut of emotions and behavioral patterns that are natural even to very young children, such as happiness, sadness, anger, and fear.

negative strokes   In interpersonal communications, an interaction which often takes the form of verbal criticism and reprimands; do not make individuals feel as good as positive strokes, but give the person attention, which may be preferred to being ignored.

net cash flow   The difference between total cash inflows and total cash outflows for an operating period.

net farm income   At the aggregate level, the portion of the net value added by agriculture to the national economy that is earned by farm operators.

net operating profit   The amount remaining after operating expenses are subtracted from gross margin; represents profit from the primary operations of the firm.

net present value (NPV)   A method for evaluating capital investment decisions; involves calculating the current value of a stream of net cash flows to be received over some specified future period of time using the firm's cost of capital as the discount rate.

net profit after taxes   Net profit after adjusting for the federal business profits tax; sometimes called net income.

net profit before taxes   The amount remaining after net operating profit is adjusted for any non-operating revenue or expenses.

net working capital   Total current assets less total current liabilities.

net worth   See *owner's equity.*

new generation cooperatives (NGC)   A form of cooperative requiring an up-front equity investment in equity shares possessing both tradable and appreciable properties; the investment is based on a member's anticipated level of patronage and all members adhere to a legally binding uniform marketing agreement.

nominal partner   A partner in name only; is not active in the partnership and has no investment.

nonprofit corporation   A corporation that is exempt from certain forms of taxation, and that generally cannot directly enrich its owners financially.

notes payable   Short-term loans or liabilities from individuals, banks, or other lending institutions that must be repaid within a year.

nurturing parent   In transactional analysis, represents attitudes and behavior that nurture and take care of others, just as parents offer sympathy, understanding, and warmth to their children.

objection   In selling, a negative reaction or concern expressed by the customer about the product or any selling point that has been made; constitutes a reason for not buying.

objectives   See *performance objectives.*

on-the-job training Training conducted while the employees are actually performing the job they were hired to do.

operating characteristics segmentation A market segmentation approach based on characteristics such as type of operation (e.g., crop versus livestock), size of operation, production technology used (e.g., no-till versus conventional till), and form of ownership (e.g., owner/operator versus cash rent versus crop share).

operating expenses Those expenses associated with the actual operation of the business; does not include expenses not directly associated with operations, such as interest expense, and nonrelated legal expenses.

operating lease A leasing method that is usually a short-term rental arrangement (i.e., hourly, daily, weekly, monthly, etc.) in which the lease charge is calculated on a time-of-use basis.

operational efficiency A measure of marketing system productivity; the raw ratio of marketing output to marketing input.

operations management One of the four functions of management (production logistics management); involves direction and control of the processes used by food and agribusiness firms to produce goods and services.

opinion leaders Well-respected individuals or firms who tend to adopt products and services more quickly than others, and shape the attitudes of others in a market.

opportunity cost The income given up by not choosing the next best alternative for the use of resources; represents the amount that the business forfeits by not choosing a specific alternative course of action.

order point An agreed-upon level of depletion of a specific product in inventory that warrants an order for more.

organizational chart A graphic representation of a firm's formal organizational structure which shows reporting relationships and areas of responsibility.

organizational structure The formal assignment of employee responsibilities in a firm—who reports to whom, who has responsibility for what, etc.

organizing One of the four tasks of management; the systematic classification and grouping of human and other resources in a manner that effectively accomplishes the firm's goals.

other current assets Any of a variety of assets that are likely to be converted to cash, or can be converted into cash, within a year; normally does not include cash, accounts receivable, inventory, or marketable securities.

output-input ratio An efficiency ratio that relates the output of a process to the inputs employed in the process; also called *productivity*.

outputs Any product that is the result of a production process.

owner's equity The value of the owner's investment in the business; the total assets of the business less all obligations to nonowners; also called *net worth*.

parent ego In transactional analysis, a state of human behavior representing the values, attitudes, and behavioral patterns established early in life from authority figures, primarily parents.

parity of responsibility and authority principle In organizational structure, the idea that a person should have enough authority to carry out their assigned responsibilities.

participative leadership See *democratic leadership*.

partnership A form of business organization in which two or more people jointly own the assets and/or manage the business.

patronage refunds A return of net margin to members or patrons of a cooperative.

payback period   The length of time it will take an investment to generate sufficient additional cash flows to pay for itself.

penetration pricing   A pricing strategy that consists of offering a product at a low price, perhaps even at a loss, in order to gain wide market acceptance quickly.

performance evaluation   An appraisal of an employee's performance; makes the employee aware of their contributions and their strengths and weaknesses; normally includes goals for the next evaluation period.

performance objectives   Measurable, definable goals that are set for specific units and/or individuals; provide shorter term performance targets at the unit and/or individual level that are consistent with broader, longer range strategic goals.

periodic inventory   An actual physical count of inventory on hand conducted at regular intervals.

perishability   The characteristic of rapid deterioration in the quality of some agricultural or food products after harvest or processing.

perpetual inventory   See *continuous inventory.*

personal interview   A market research technique that involves direct dialogue with a customer, prospect, influencer, or other individual of interest.

personnel   The people employed in an organization; also can be a department of a firm that deals with employee issues such as hiring, benefits, and records.

physical distribution systems (PDS)   The series of channels through which parts, products, and finished inventory are stored and moved from suppliers, between outlets, and, ultimately, to consumers.

physical functions   One of three major marketing functions; includes adding time, place, and form utility to the product as it is transported, stored, and processed to meet customer needs and wants.

physical risks   The uncertainty and potential damage or losses that can result from such phenomena as wind, fire, hail, flood, theft, and spoilage.

pipeline inventory   The minimum amount of inventory needed to cover the period of time between the warehouse's reordering and its receipt of the additional stock.

place utility   One of four types of utility or ways in which value is added to a product; involves the value a buyer receives from a product that is available where the buyer wants it.

planning   Forward thinking directed at specific objectives about courses of action based upon a full understanding of all factors involved.

points   Service charges, based on the face value of the loan, that are paid to the lending institution to secure the loan.

policy   A rule or guideline that sets the boundaries for handling specific situations that occur frequently.

politics   The manner in which power and status are used in an organization; involves the manipulation of people and situations to accomplish a particular goal.

position   The specific market space, image, set of activities, and/or products and services that a firm wants to be known for among its target customers.

positioning   The process of creating the desired image or position in the customer's or prospect's mind.

positive conditional strokes   In interpersonal communication, any form of interaction that conveys a positive message, yet acceptance of the other person is conditional on some expected behavior.

positive strokes   In interpersonal communication, any form of interaction that leaves another person feeling good, alive, and significant.

positive unconditional strokes   In interpersonal communication, any form of interaction that conveys the spoken or implied message, "I like you . . . you're OK with me," with no conditions on the acceptance.

possession utility   One of four types of utility or ways in which value is added to a product; involves the value a buyer receives from the transfer of ownership of a product from seller to buyer.

potential product   One component of the total product concept; the product as it can become; what the next benefit is that customers will seek from the firm, and how the product/service/information bundle will be managed to add even more value for customers.

power   The ability to control another person's behavior.

practices   Activities and processes as they are actually performed in the agribusiness; may conflict with policies and procedures.

preferred stock   A type of ownership position in a corporation that does not carry the privileges of voting for directors, but has a preferred position in receiving dividends and in redemption in the case of liquidation.

prejudiced parent   In transactional analysis, a specific state of behavior that represents unfounded attitudes and opinions that are held without any logical basis or analysis.

prepaid expenses   Current assets that represent prepayment of an item or service that will be consumed or used during a future accounting period at which point it will become an expense.

preparing the sales call   In selling, the process of developing specific plans for the call, including preparation of any materials that may be needed.

present value   The current value of a stream of payments to be made or received over some specified period of time in the future.

presentation   In selling, the heart of the sales call in which the primary objective is to present the product or service so effectively that the customer will see it as satisfying a particular need.

prestige pricing   A pricing strategy that appeals to an elite image and plays to the attitude that price and quality go together: "You get what you pay for."

pretesting   In market research, evaluating research questions before actually using them with a group of respondents to insure that the questions are effective at gathering the desired information.

prevention costs   In total quality management, the costs associated with stopping defects and errors before they happen.

price   The value of an item or service in monetary terms.

price discovery   The process in which producers and consumers meet in the marketplace, and the equilibrium quantity and price is determined.

pricing efficiency   One component of marketing efficiency; a measure of how adequately market prices reflect production and marketing costs throughout the total marketing system.

principal   The initial sum of money invested or borrowed; the portion of the original loan that has yet to be repaid.

principles   Ideas, values, philosophies, and beliefs regarding an issue or issues.

pro forma financial statements   Projected financial statements for some specified time period in the future.

probing   In selling, the art of asking questions.

procedure   A step-by-step guide to performing a specific activity or function.

process design   Selecting the specific inputs, operations, and methods that are to be used to produce the good or service.

process layout   In production planning, an arrangement in which all like functions are grouped in the same place.

product advertising   Advertising designed to promote a specific product, service, or idea.

product-driven philosophy   A marketing approach based on offering a unique product that satisfies a specific customer need; creating a product that is so good customers will seek it out.

product layout   In production planning, an arrangement that involves a step-by-step sequence of functions as the product is assembled.

product life cycles   The predictable way in which sales and profits of a product unfold as a product is introduced, sales grow rapidly, the market matures, and the product ultimately declines in the marketplace.

product mix   The breadth, depth, and combination of products, services, and information offered to the market.

production agriculture sector   The sector of the food production and marketing system in which purchased inputs, natural resources, and managerial talent come together to produce crop and livestock products.

production control   All activities related to controlling raw materials inventory, providing detailed production scheduling information, controlling work-in-process inventory, communicating changes to master production scheduling and purchasing, and controlling finished goods inventory.

production planning   An important aspect of operations management; includes a wide range of decisions and activities including devising a quality program; locating a plant; choosing the appropriate level of capacity; designing the layout of the operation; deciding on the process design; and specifying job tasks and responsibilities.

productivity   See *output-input ratio.*

profit   A general term for the difference between total revenue and total cost.

profit and loss statement   See *income statement.*

profitability ratio   A class of financial ratios that measure a firm's profitability.

program evaluation review technique (PERT)   An approach to project management developed by the U.S. Navy; involves a diagrammatic representation of a network of activities and a search for the most sensitive or restrictive steps in the network.

promissory note   A promise by a borrower to repay a lender a specific sum of money, loaned at a specific rate of interest for a specified period of time.

promotion   In marketing, an element of the marketing mix; all activities related to communicating the firm's offering to the market; in human resource management, advancement in job responsibilities or position.

proprietorship   See *sole proprietorship.*

prospecting   In selling, the process of identifying and locating potential customers.

psychographic segmentation   A market segmentation approach based on classifying or categorizing customers or potential customers by a combination of their psychological profiles and demographic data.

psychological pricing   A pricing method that involves establishing prices that are emotionally satisfying because they carry the perception of good value—for example, two for 99 cents instead of 50 cents each.

public relations   An element of a firm's promotion (market communications) strategy; somewhat unique as a form of market communications as the target audience is influenced in an indirect way.

pull strategy   Promotion (market communications) strategy to encourage the ultimate consumer to "pull" the product through the marketing channel; focus is on consumers rather than channel members as the target audience.

purchasing   All tasks involved in procuring the raw materials and inputs necessary to meet the requirements of the production schedule.

push strategy   Promotion (market communications) strategy designed to "push" products through the marketing channel; focus is on channel members rather than consumers as the target audience.

quick ratio   A financial ratio measuring liquidity; reflects a firm's ability to meet cash needs with funds that are quickly available; (cash + marketable securities + accounts receivable) / total current liabilities.

ratio analysis   A method of financial statement analysis that uses relationships between key accounting data to better understand the relative position of an organization.

raw materials   Basic goods from which other products are made.

reference value   In pricing, the price of a competing product or the closest substitute; forms the starting point for an economic value pricing strategy.

regional cooperatives   A "cooperative of cooperatives" whose primary purpose is to provide manufacturing, processing, and wholesaling services to local cooperatives.

responsibility   The obligation to see a task through to its completion.

resumé   A written summary of the personal, educational, and professional qualifications of a person seeking employment.

return on assets ratio (ROA)   A financial ratio measuring profitability; measures the return to the total investment in the business; (net profit + interest expense) / total assets.

return on equity ratio (ROE)   A financial ratio measuring profitability; measures the return to the owner's investment in the business; (net profit / owner's equity).

return on investment ratio (ROI)   In the profitability analysis model, a measure of return to assets where interest expense has been subtracted from net profit (net profit − interest expense) / total assets; more broadly, a general term referring to any of a class of financial ratios measuring profitability or the return on some type of investment: (profit / investment).

return on investment (ROI) pricing   A pricing method where a markup sufficient to earn a specified return on investment is added to the basic cost of an individual product or service.

return on net worth (RONW)   See *return on equity*.

return on sales ratio (ROS)   A financial ratio measuring profitability; reflects the return on each sales dollar the organization generates; (net profit / net sales).

revolving fund financing   A unique feature of cooperatives that gives them the option of issuing patronage refunds in the form of stock; periodically, the cooperative revolves the stock by allowing older stock to be cashed in.

risk   A situation where outcomes are unknown, but where probabilities of occurrence can be assigned to each possible outcome.

Rochdale principles   Formal business and organizational principles adopted by the Rochdale Society in 1844 which have served as a model for the development of modern cooperatives.

S corporation   A closely held corporation where it is possible for the owners of the corporation to elect to be taxed as individuals rather than as a corporation.

safety   A component of Maslow's needs hierarchy that suggests that once immediate survival has been assured, humans become concerned about the security of their future physical survival.

safety stock   See *buffer inventory.*

sales   The total dollar value of all the products and services that have been sold for cash or on credit during the period specified on the profit and loss statement.

sales call interview   In selling, the process of establishing customer contact, developing or renewing a positive relationship, discovering the customer's needs, convincing the customer that one can meet those needs, and securing the customer's commitment.

sales call objective   In selling, a tangible, measurable objective of what the salesperson expects to accomplish during the sales call.

sales call strategy   In selling, a plan of action designed to maximize the potential for success with the customer and make the most of the sales call.

sales-driven philosophy   A marketing approach based on intensifying the sales effort and/or reducing prices in order to improve sales.

sales forecast   An estimate of sales in dollars and/or physical units for a specific future period of time.

sales promotions   A form of promotion (market communications) strategy; programs and special offers designed to motivate interested customers to purchase a product or service.

sanitary and phytosanitary (SPS) regulations   Rules and guidelines that are intended to protect human, animal, or plant life or health; instituted in many countries as consumers in developed countries increasingly demand a higher level of food safety.

scenario analysis   A qualitative forecasting technique in which a series of future business possibilities or environments are developed based on a careful understanding of what is known, and systematically varying the possibilities for what is not known.

scientific management   A management concept introduced by Frederick W. Taylor in the early 1900s, focused on standardized procedures, determination of the most efficient procedures through scientific testing, matching worker skills with job requirements, and management and workers jointly devising the best system for division of labor.

seasonal inventories   See *anticipation inventories.*

seasonality   The characteristic of many agricultural goods which can only be produced one time per year, during that product's growing season; also a characteristic of consumption of some goods where use ebbs and flows in a regular pattern across a year.

secret partner   A partner that takes an active role in managing the partnership but is not known to be a partner by the general public.

self-actualization   A component of Maslow's needs hierarchy; the highest level of need, and one that becomes important only when all lower level needs have been relatively well satisfied; the feeling of self-worth or personal accomplishment.

selling   The act of transferring ownership of goods and services.

semivariable cost   A cost that is partly fixed and partly variable.

senior partner   A partner that has an investment in the firm, has major responsibilities for management of the organization, and receives the major portion of partnership profits; typically an individual that helped form the partnership or one who has seniority in the business.

service cooperatives   Cooperatives that provide specialized services related to the business of a certain group of individuals.

short-term budget   A budget that generally covers a time period of one year or less.

short-term loan   A temporary grant of money to be paid back in one year or less; may be a note with regular terms, or a revolving or line-of-credit loan.

silent partner   A partner who has restricted management rights and responsibilities and limited liability for the organization's actions.

simple interest rate   The type of interest charge used on many personal loans; involves a rate of interest applied to an amount available for the entire period of the loan; the amount of interest paid divided by the amount of available capital.

simple rate of return   A commonly used ratio for capital investment analysis; the profit generated by an investment expressed as a percentage of the investment.

simulation   A systematic approach to problem solving that usually involves evaluation of several (sometimes thousands) possibilities derived from a model based on past operating experience and records; used to project the probability of different outcomes.

skimming the market   A pricing method where a product is introduced at a high price that affluent or highly interested customers can afford, then the price is gradually lowered over time to bring less affluent and less interested customers into the market.

sole proprietorship   A form of business owned and controlled by one person.

solvency   The firm's ability to meet long-term financial obligations.

solvency ratios   A class of financial ratios that measures a firm's ability to meet long-term financial obligations.

span-of-control principle   In organizational structure, the concept that there is a limit to the number of people who can be supervised effectively by one individual.

speculators   Traders that do not have nor do they want possession of the physical commodity; typically enter the market solely to profit from changes in the price level.

statement of owner's equity   A financial statement that details the changes in the owner's equity accounts from one operating period to the next.

status   The social rank or position of a person in a group.

stock   Units of ownership in a corporation.

stock certificate   A legal document detailing the amount of an owners' investment in a corporation.

stockholders   Individuals who own the stock of a corporation.

stock keeping unit (SKU)   A specific item of goods for sale for which individual records are kept and that is tracked throughout a firm's logistics system.

strategic alliances   Agreements to collaborate between firms that go beyond normal firm-to-firm dealings, but fall short of merger or full partnership and ownership.

strategic marketing plan   A set of activities intended to help a firm anticipate the needs of targeted customers and find ways to meet those needs profitably.

strategic planning   The process of developing a long-range plan for an organization; tackles the broadest elements of an agribusiness firm's strategy: what countries will we operate in, what businesses will we be in, what plants will we build, etc.?

strength, weakness, opportunity, and threat (SWOT) analysis   An assessment tool used to evaluate the competitive environment facing a business, and the firm's relative position in that environment; requires careful study of general trends in the market, strengths and weaknesses of key competitors, current and anticipated customer needs, and the firm's strengths and weaknesses.

strike   An organized work stoppage by employees of an organization.

stroke   Any recognition of another's presence by word, gesture, or touch.

supermarket   Any retail food store generating more than $2 million in sales per year.

suppliers   Firms providing raw materials, supplies, or other inputs to a business.

supply   The quantities that producers are willing and able to put on the market at various prices; the relationship between price and quantity supplied.

supply chain management   The process of managing the complete input acquisition and output distribution channels for a firm; involves linking the materials and physical distribution systems of a firm directly with supplier and customer systems, resulting in greater efficiencies.

supply cooperatives   Cooperatives that generate most of their business volume from the sale of specialized production inputs such as farm supplies, building supplies, etc.

supply elasticity   A measure showing the relative change in quantity supplied as price changes; reflects the responsiveness of quantity supplied to a given change in price.

survival   A component of Maslow's needs hierarchy; suggests that every human's most basic concern is physical survival: food, water, air, warmth, shelter, etc.

symbols   A variety of factors that help establish the relative position of individuals in the informal organization; may include title, age, experience, physical characteristics, knowledge, physical possessions, authority, location, privileges, acquaintances, and a host of other factors, depending on the situation.

tactical planning   Smaller scale, more immediate plans developed to implement the strategic plan; typically address a time period of one year or less, and may be developed by virtually every employee in an organization.

Taguchi   Genichi Taguchi, a pioneer in quality management; defined quality in terms of a philosophy focused on the absence of product variability and social cost, bringing maximum well-being to the product or service consumer.

target market   A market segment prioritized by a firm and the focus of a marketing effort involving a set of decisions tailored to the unique needs of the segment.

telephone interviews   A market research technique that involves asking questions of individuals over the telephone.

terminal value   The residual value an investment is expected to have at the end of the planning horizon.

test market   A market research technique that involves an experiment designed to test consumer behavior under actual buying conditions: a test city or area with characteristics similar to the target market is selected, a product (typically a new product) is introduced into the test city or area, sales results are measured and evaluated, and these results are then generalized to the target market of interest.

theory x, theory y   Douglas McGregor's theory about human behavior in business; based on the notion of the "dual nature of humankind" or people's capacity for love, warmth, kindness, and sympathy, and at the same time the capacity for hate, harshness, and cruelty.

time-based competition   Competing for customer and market share using the dimension of time as a source of competitive advantage; includes ideas such as cycle-time-to-market, customer responsiveness, and flexibility in operations.

time utility   One of four types of utility or ways in which value is added to a product; value is added through storage of the product to ensure that sufficient quantities of product are available when needed.

total asset turnover ratio   A financial ratio measuring efficiency; shows the sales dollars the firm is able to generate for every dollar invested in assets; net sales / total assets.

total cost   The sum of total variable cost and total fixed cost.

total fixed cost   The total of all fixed costs incurred during an operating period; the total of all costs that are constant during the operating period, regardless of the level of production.

total product concept   A means to relate the customer's definition of value to the agribusiness firm's product/service/information strategy by visualizing the four parts of the firm's product/service offering: the generic product, the expected product, the value-added product, and the potential product.

total quality management (TQM)   An integrated management concept directed toward continuous quality improvement; involves ideas such as: business success can only be achieved by understanding and satisfying the customer's needs, statistics and factual data are the basis for problem solving and continuous improvement, and multifunction work teams best perform problem solving and process improvements.

total revenue   The total income from operations received during an accounting or operating period.

total variable cost   The total of all variable costs incurred during an operating period; the total of all costs that vary directly with the level of production during the operating period.

trade credit   Credit advanced by suppliers and vendors to an agribusiness firm in the process of providing goods and services to the firm.

traditional restaurants   Restaurants that offer menu, sit down, and dine service.

training programs   Learning opportunities for employees devised to meet one or more of the following objectives: reduce mistakes and accidents; increase motivation and productivity; and prepare the employee for promotion, growth, and development.

transaction   In transactional analysis, anything verbal or nonverbal that happens between two people.

transactional analysis (TA)   An approach for understanding and supervising employees developed by Dr. Eric Berne based on the concepts of psychotherapy; divides human behavior into three classes of ego states: parent, adult, and child (P-A-C).

transactions data   Internal firm data derived from business relationships with customers and suppliers.

transportation   The means of getting products from one point to another; five basic modes include highway, rail, water, pipeline, and air.

transportation and storage firms   Firms that acquire or assemble commodities from agricultural producers, and store and transport these products for food manufacturing and processing firms.

trend forecast   A forecasting method that involves projecting sales objectively based on past trends, and then adjusting these projections subjectively to take into account the expected economic, market, and competitive pressures.

trial close   In selling, a technique designed to find out whether the customer is ready to buy.

turnover   A measure of a firm's ability to retain employees once hired; measures the proportion of employees who leave an organization in a period (usually a year) relative to the total employee base.

uncertainty   A situation where outcomes are unknown and where probabilities of occurrence cannot be assigned to each possible outcome.

unit train   A special 100-car train in which grain or other products are moved.

utility   The value added in a marketing system; the four types are form, time, place, and possession.

value   The ratio of what the customer receives (perceived benefits) relative to what they give up (perceived costs).

value-added product   One component of the total product concept; the first opportunity for the agribusiness firm to truly exceed the customer's expectations by going beyond the tangible, physical properties of the product and the minimum services that are typically provided with the product.

value bundle   The set of tangible and intangible benefits customers receive from the products, services, and information an agribusiness provides.

value-based pricing   A pricing method based on pricing a product or service at a level just lower than the estimated perceived value of the value bundle.

variable cost   Those costs that increase directly with the volume of sales; the costs of doing business.

venture capital   Money used for investment in projects (normally new) that involve a high risk but offer the possibility of large profits.

vertical integration   The process of extending a firm's presence in the marketing channel either forward toward the consumer, or backward toward suppliers, usually through ownership of new stages in the channel.

volume-cost analysis   An analytic tool for examining the relationship between costs and the volume of business; also called *breakeven analysis.*

wage efficiency ratio   A financial ratio measuring efficiency; shows the relationship between labor cost and sales volume.

warehouse receipts   A means of using inventory as security for a loan.

warehousing   The process of storing goods for resale.

wholesale agents and brokers   Firms who buy and/or sell as representatives of others for a commission; typically do not physically handle products, nor do they actually take over title to the products.

wholesalers   A firm that buys from one firm and sells to another, in many cases buying from a processor or manufacturer and selling to a retailer.

work-in-process (WIP)   An accounting category of inventory or inventories that are in the conversion phase from raw materials to final products.

World Trade Organization (WTO)   Created in 1995, a body to regulate, monitor, and encourage world trade; with more than 125 member countries, its rules apply to over 90 percent of international trade.

written surveys   A market research technique in which individuals respond to written questionnaires.

INDEX